Biology of Turtles

Biology of Turtles

Edited by

Jeanette Wyneken
Florida Atlantic University
Boca Raton, FL, U.S.A.

Matthew H. Godfrey
North Carolina Wildlife Resources Commission
Beaufort, NC, U.S.A.

Vincent Bels
Muséum of National D' Histoire Naturelle
Paris, France

CRC Press
Taylor & Francis Group
Boca Raton London New York

CRC Press is an imprint of the
Taylor & Francis Group, an **informa** business

Cover photos by Stephen L. Barten, D.V.M., Ann C. Burke, and Jeanette Wyneken.

CRC Press
Taylor & Francis Group
6000 Broken Sound Parkway NW, Suite 300
Boca Raton, FL 33487-2742

Printed in the United States of America on acid-free paper
10 9 8 7 6 5 4 3 2 1

International Standard Book Number-13: 978-0-8493-3339-2 (Hardcover)

Library of Congress Cataloging-in-Publication Data

Wyneken, Jeanette, 1956-
Biology of turtles / Jeanette Wyneken, Matthew H. Godfrey, Vincent Bels.
p. cm.
Includes bibliographical references and index.
ISBN 978-0-8493-3339-2 (alk. paper)
1. Turtles. I. Godfrey, Matthew H. II. Bels, V. L. (Vincent L.) III. Title.

QL666.C5W96 2007
597.92--dc22 2007024320

Visit the Taylor & Francis Web site at
http://www.taylorandfrancis.com

and the CRC Press Web site at
http://www.crcpress.com

Contents

Preface

The first beasts that we identify as turtles emerged abruptly in the Triassic about 220 million years ago. Since then, countless diverse lineages of turtles—terrestrial, marine, freshwater, and in-between—have come and gone, yet many lineages persist for tens of millions to more than 100 million years. All bring with them the suite of traits that are uniquely turtle: a bony shell, usually a retractable neck, lack of trunk musculature, and limb girdles located inside the rib cage (instead of lying outside of it). The reptiles we identify as turtles, or testudines, or chelonians, have captivated many naturalists and other scientists because of these and other unique traits and their consequences.

Some 180 years ago while in Jena, Germany, Ludwig Heinrich Bojanus developed an interest in the anatomy of the European turtle *Emys obicularis.* By dissecting specimens and illustrating their anatomy over the next decade, Bojanus authored one of the most detailed studies on any single vertebrate species, *Anatome Testudinis Europaeae* (1819–1821). This unparalleled book on turtle anatomy represents the intersection of his interests as a naturalist, comparative anatomist, veterinary surgeon, and teacher of anatomical art. It was because of this historical backdrop that the biology, particularly the anatomy, of turtles was visited again fittingly at the Sixth International Congress of Vertebrate Morphology, which was held in Jena in 2001. The symposium and its many contributed papers served as the foundation upon which this volume is based. However, this book goes beyond a morphologically based symposium in recognizing that the structures of turtles are particularly interesting and best understood within the context of their formation, their diversity of functions, their novelty, and their evolution. Whereas several areas of interest (e.g., turtle genetics, sensory systems and behavior, and life history evolution) are not included—as these fields continue to progress rapidly without a clear stable point—we have brought together a wide range of discussions on other novel features of turtles.

For the first time anywhere, we have a robust discussion about the origins, development, and diversity of the shell in chapters by Gilbert et al. (Chapter 1) and Pritchard (Chapter 3). Snover and Rhodin (Chapter 2) synthesize the important work on bone growth and aging, while Llorente et al. (Chapter 4) bring forward an understanding of limb bone strength in this unique group. Turtle locomotion is unique in many ways because of the presence of the shell. Renous et al. (Chapter 5) provide a novel synthesis of the field while Blob et al. (Chapter 6) present an experimental and functional perspective on the motor patterns used by turtles during locomotion. The retractable neck is expertly described by Herrel et al. (Chapter 7) and given functional context beyond the simple view of "protecting the head." Bels et al. (Chapter 8) provide a functional, anatomical, and behavioral overview of feeding in herbivorous versus carnivorous types and aquatic versus terrestrial turtles that has never previously been summarized. Wyneken (Chapter 9) discusses cardiopulmonary anatomy and function from functional perspectives. Reproductive strategies receive a thorough overview by Miller and Dinkelacker (Chapter 10). Hulin et al. (Chapter 11) challenge the reader to think rigorously in their analysis of the consequences of environmental sex determination and turtle sex ratios, and Milton (Chapter 12) reminds us that at least some turtles (particularly the aquatic species that hibernate underwater) have a number of very unique structural and functional adaptations for tolerating low oxygen levels. Rieppel (Chapter 13) examines why the origin of turtles and their relationships to other amniotes is such a great challenge. Despite more than a century of intensive study, our understanding of the phylogenetic origins of turtles remains indefinite. This chapter particularly challenges us to look both inside and outside the shell as we build an understanding of the evolution of turtles. This challenge extends across all fields represented in this volume and those yet to come.

The editors thank the many external referees who responded to our request for a rigorous review of each chapter. We are grateful for the collegial environment in Jena, Germany, and at the Sixth International Congress for Vertebrate Morphology, which stimulated the ideas and topics for this synthesis. Particularly, we acknowledge Matthias Starck for his encouragement and support of the symposium "Turtles: From Structures to Strategies of Life," and to the late Peter Lutz for a perspective on the history of the science and the many novel integrative approaches that have led to understanding the biology of turtles. John Sulzycki and David Fausel at CRC Press/Taylor & Francis provided essential guidance, superb advice, and good humor as this book progressed at a turtle's pace.

Jeanette Wyneken
Boca Raton, Florida, USA

Matthew H. Godfrey
Beaufort, North Carolina, USA

Vincent Bels
Paris, France

About the Editors

Jeanette Wyneken, Ph.D., is an associate professor of biological sciences at Florida Atlantic University in Boca Raton. She earned her B.A. from Illinois Wesleyan University and later, her Ph.D. in biology from the University of Illinois in 1988. She was a research associate from 1988 to 1989 at the University of Illinois before taking a research position at Florida Atlantic University in 1990. She later assumed a position as assistant professor, then associate professor of biological sciences at Florida Atlantic University. Dr. Wyneken is a comparative and functional anatomist and a conservation biologist. Her studies are diverse and deal with growth, energetics, migratory behavior, development, including environmental sex determination, and medical imaging of reptiles. In addition to teaching vertebrate anatomy and development, she developed and taught the Biology of Sea Turtles course offered at Harbor Branch Oceanographic Institute in Fort Pierce, Florida, and worked with Selina Heppell and Larry Crowder to develop a similar conservation-based sea turtle biology course at the Duke University Marine Laboratory in Beaufort, North Carolina. She is the former president of the Annual Sea Turtle Symposium (now The International Sea Turtle Society), the convener of the 7th International Congress for Vertebrate Morphology, and has organized several symposia on various aspects of the biology of turtles.

Dr. Wyneken serves on the editorial boards of two professional journals in her field and is active in the peer review process for a number of other journals. She is a member of several professional organizations including the AAAS, Association of Ichthyologists and Herpetologists, Society for the Study of Amphibians and Reptiles, Herpetologists League, Association of Reptilian and Amphibian Veterinarians, the International Sea Turtle Society, Sigma Xi, the Society of Integrative and Comparative Biology, and the IUCN Marine Turtle Specialist Group. Dr. Wyneken has authored more than 30 peer-reviewed papers, four book chapters, one book (*The Anatomy of Sea Turtles*), and co-edited *The Biology of Sea Turtles, Volume 2.*

Matthew H. Godfrey, Ph.D., is a biologist with the North Carolina Wildlife Resources Commission and an adjunct assistant professor at the Nicholas School for the Environment and Earth Studies at Duke University, North Carolina. Dr. Godfrey received a B.A. (1991) in history and philosophy from the University of Toronto. He went on to receive a M.Sc. (1994) and Ph.D. (1997) in zoology from the University of Toronto in 1997. Between 1997 and 2002, Dr. Godfrey worked as a research fellow at Projeto TAMAR-IBAMA in Brazil and as a researcher/lecturer at Université Paris, France. His research topics include behavioral and evolutionary ecology of reptiles, with a focus on temperature-dependent sexual differentiation. Dr. Godfrey is also interested in the conservation of protected species and its links to social justice. He has authored or co-authored more than 40 research papers and five chapters in edited books. Dr. Godfrey is currently co-editor of the *Marine Turtle Newsletter* and serves on the editorial board of two other scientific journals. He is also a member of the IUCN Marine Turtle Specialist Group and serves on three national and two state scientific advisory committees concerned with protected species management and conservation.

Vincent Bels, Ph.D., is professor at the Muséum National d'Histoire Naturelle (Paris, France). He obtained his B.A. and his doctorat d'Etat from the University of Liège (Belgium) in 1989. Dr. Bels began his work in ethology at the University of Liège and cultivated his interest in comparative and functional morphology of vertebrates. He is now associate director of the research team UMR7179 "*Mécanismes adaptatifs: des organismes aux communautés*" (CNRS – MNHN – Univ. Paris 6 – College de France) at the Department *Ecologie et Gestion de la Biodiversité* at the Muséum National d'Histoire Naturelle. In addition to teaching vertebrate functional morphology at the University of Mons (Belgium) and the Muséum (master's degree program and doctoral school), Dr. Bels's studies concern a large variety of lower vertebrates from a comparative, functional, and evolutionary point of view. His main work focuses on feeding behavior in squamates and turtles. He has studied locomotor behavior in a number of lower vertebrates including fishes, crocodiles, and marine turtles. He has also investigated behavioral and functional mechanisms of behaviors involved in communication in squamates. He is active in the peer-review process for a number of journals. Dr. Bels has authored more than 50 peer-reviewed papers, five book chapters, and edited or co-edited three books on the functional and evolutionary biology of vertebrates.

Contributors

Peter Aerts
Department of Biology
University of Antwerp
Antwerp, Belgium

Ignacio Barandalla
Department of Animal Biology
University of Barcelona
Barcelona, Spain

Sabine Baussart
Département Ecologie et Gestion de la Biodiversité
Muséum National d'Histoire Naturelle
Paris, France

Vincent Bels
Département Ecologie et Gestion de la Biodiversité
Muséum National d'Histoire Naturelle
Paris, France

Richard W. Blob
Department of Biological Sciences
Clemson University
Clemson, South Carolina, USA

Ann C. Burke
Biology Department
Wesleyan University
Middletown, Connecticut, USA

Adrià Casinos
Department of Animal Biology
University of Barcelona
Barcelona, Spain

Judith A. Cebra-Thomas
Biology Department
Millersville University
Millersville, Pennsylvania, USA

John Davenport
Department of Zoology, Ecology and Plant Science
Environmental Research Institute
University College Cork
Cork, Ireland

Julia L. Davenport
Department of Zoology, Ecology and Plant Science
Environmental Research Institute
University College Cork
Cork, Ireland

Marion Depecker
Département Ecologie et Gestion de la Biodiversité
Muséum National d'Histoire Naturelle
Paris, France

Stephen A. Dinkelacker
Department of Biology
University of Central Arkansas
Conway, Arkansas, USA

Scott F. Gilbert
Biology Department
Swarthmore College
Swarthmore, Pennsylvania, USA

Marc Girondot
Laboratoire des Reptiles et Amphibiens
Muséum National d'Histoire Naturelle
Paris, France

Matthew H. Godfrey
North Carolina Wildlife Resources Commission
Beaufort, North Carolina, USA

Jean-Michel Guillon
Laboratoire Ecologie, Systématique et Evolution
Faculté des Sciences d'Orsay
Université Paris Sud
Orsay, France

Anthony Herrel
Department of Biology
University of Antwerp
Antwerp, Belgium

Vincent Hulin
Laboratoire Ecologie, Systématique et Evolution
Faculté des Sciences d'Orsay
Université Paris Sud
Orsay, France

France de Lapparent de Broin
Département Histoire de la Terre, Paléobiodiversité
Muséum National d'Histoire Naturelle
Paris, France

Gustavo A. Llorente
Department of Animal Biology
University of Barcelona
Barcelona, Spain

Jeffrey D. Miller
Department of Biology
University of Central Arkansas
Conway, Arkansas, USA

Sarah L. Milton
Department of Biological Sciences
Florida Atlantic University
Boca Raton, Florida, USA

Ruth M. O'Riordan
Department of Zoology, Ecology and Plant Science
Environmental Research Institute
University College Cork
Cork, Ireland

Peter C.H. Pritchard
Chelonian Research Institute
Oviedo, Florida, USA

Sabine Renous
Département Ecologie et Gestion de la Biodiversité
Muséum National d'Histoire Naturelle
Paris, France

Anders G.J. Rhodin
Chelonian Research Foundation
Lunenburg, Massachusetts, USA

Olivier Rieppel
Department of Geology
Field Museum of Natural History
Chicago, Illinois, USA

Angela R.V. Rivera
Department of Biological Sciences
Clemson University
Clemson, South Carolina, USA

Xavier Ruiz
Department of Animal Biology
University of Barcelona
Barcelona, Spain

Marc Shorten
Department of Zoology, Ecology and Plant Science
Environmental Research Institute
University College Cork
Cork, Ireland

Melissa L. Snover
National Oceanic and Atmospheric Administration (NOAA)
National Marine Fisheries Service
Pacific Islands Fisheries Science Center
Honolulu, Hawaii, USA

Johan Van Damme
Department of Biology
University of Antwerp
Antwerp, Belgium

Carles Viladiu
Department of Animal Biology
University of Barcelona
Barcelona, Spain

Mark W. Westneat
Department of Zoology
Field Museum of Natural History
Chicago, Illinois, USA

Jeanette Wyneken
Department of Biological Sciences
Florida Atlantic University
Boca Raton, Florida, USA

1 How the Turtle Gets Its Shell

Scott F. Gilbert, Judith A. Cebra-Thomas, and Ann C. Burke

If it weren't for the 250 species of turtles living today ... these animals encased in mobile homes could easily be viewed as bizarre evolutionary experiments that were ordained to failure.

Richard Ellis (2003)

CONTENTS

1.1 THE NATURE OF THE TURTLE SHELL

1.1.1 INTRODUCTION TO THE TURTLE SHELL

The turtle shell is a remarkable evolutionary novelty that defines the order Chelonia. The turtle shell is found in three general forms based on the nature and degree of ossification: hardshells, softshells, and leatherbacks. This section will concentrate almost totally on the bony component of those shells of the hardback turtles of the Emys and Chelydae families. This shell is composed of two main parts, the dorsal carapace and the ventral plastron, connected along the midflanks by lateral bridges. Altogether, the shell contains over 50 dermal bones that are homologous to no other bone in any other vertebrate order. Moreover, the presence of this bony casing has necessitated extensive modifications of the tetrapod body plan (Zangerl, 1969). Whereas dermal ossification itself is a primitive character for vertebrates (Smith & Hall, 1993), the turtle shell represents an extreme development of the dermal skeleton among tetrapods.

The shell clearly has adaptive value for turtles as physical protection, but it also serves physiological functions in different species as a site of hematopoiesis, a reservoir for water, fat, or wastes, and a buffer for pH. The embryonic development of the shell involves a dramatic hypertrophy of the dermis in the dorsal body wall and a resultant rearrangement of the typical relationship between the pectoral girdle and the axial skeleton. Thus, turtles are the only vertebrates whose limbs are found deep to the ribs. The paraxial and limb-girdle musculature—the neck and skull—are also greatly modified. As we detail here, the key innovation for the chelonians appears to be the carapacial ridge, a bulge of ectoderm and mesoderm that influences the growth of the ribs (Burke, 1989a). The ribs are enveloped within the dorsal dermis, resulting in their lateral displacement as the dermis rapidly expands. Thus instead of extending ventrally and enclosing the thoracic cavity, the turtle ribs become integrated into the carapacial dermis. The neural arches of the vertebrae also fuse with the midline of the carapace. As the anonymous author (1676) of the letter to the Royal Society of London wrote in 1676:

"The Anatomie of a Tortoise, showing that what were the Ribs in other Animals one upper Shell is in the Tortoise, and that to that upper Shell are firmly fastened the spinal Vertebrae, so that the Animal cannot go out of its Home, as Snails do."

1.1.2 Anatomy of the Turtle Shell

The character and homology of the bony elements of the turtle shell have a long history of controversy. The shell is comprised of the endochondral axial elements of the trunk overlaid by a mosaic of dermal bones and an outer epidermal layer made of keratinous scales (also called *scutes* or shields). All turtles possess 10 trunk vertebrae associated with the carapace. Each vertebra possesses a single-headed rib that often shares an articulation with the next anterior vertebra. The first and tenth ribs are diminutive and normally extend a short distance before making contact with the second and ninth ribs, respectively. The tenth rib is often indistinguishable in both embryos and adults, but the presence of a large tenth rib in embryos is a normal variation. The thoracic ribs enter the dermis of the shell a short distance from their articulation with the vertebrae, and they extend laterally within the carapacial dermis, terminating at the periphery (reviewed by Zangerl, 1969).

In the dermal layer of the shell, there are generally 59 bones: the carapace has 38 paired and 12 or 13 unpaired bones (sometimes the suprepygeal bone is divided and sometimes it is not). The plastron contains one unpaired and eight paired bones. With the exception of a few key taxa, the only real variations in this general scheme occur as individual variations around the neck and tail where the axial skeleton is not closely joined to the carapace. The shapes and relative sizes of the bones determine the general form of the shell in different genera.

The shell's epidermal layer generally consists of 38 scutes in the carapace and 16 in the plastron. However, this can vary depending on the shape of the shell (domed, hinged, flapped, and so on; see Chapter 3). The shield and bone patterns are not in register; each shield covers a particular area of the bony mosaic. The pattern of the sulci that form between neighboring scutes and the sutures that form between neighboring bones form two minimally overlapping patterns. The epidermal shield pattern develops long before the shell bones begin to ossify, and the underlying dermis may play a major role in the formation of the epidermal scutes, similar to the influence of somitic dermis of feather patterns in the chick (Yntema, 1970; Cherepanov, 1989; Alibardi & Thompson, 1999a,b).

1.2 THE FORMATION OF THE CARAPACIAL BONES: HETEROTOPY AND PARACRINE FACTORS

1.2.1 The Dermal Bones of the Carapace

The unpaired midline dermal bones of the carapace, called neurals, are fused with the neural spines of the 10 thoracic vertebrae (Figure 1.1). The costal bones extend from the neurals toward the periphery. There are eight pairs and each is intimately associated with a rib (Figure 1.1E). Generally, there is a one-to-one correspondence between the vertebral spines and the neural bones, and

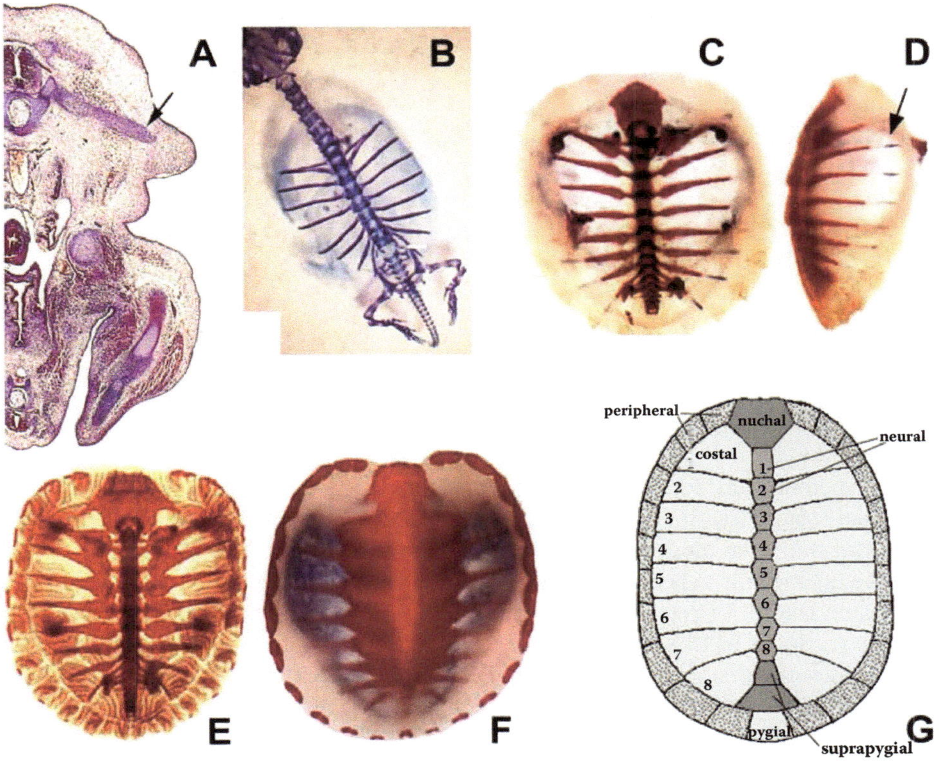

FIGURE 1.1 Development of the carapace. (A). Entry of cartilaginous rib precursor (arrow) into carapacial ridge of *Trachemys* embryo around stage 16. The following show bone formation in *Trachemys scripta*, stained with Alcian blue (cartilage) and alizarin red (bone). (B) 1.2-cm embryo showing cartilaginous ribs forming the outline of the shell. (C). Ventral view of 3.1-cm carapace, showing intramembranous ossification of the nuchal bone and around and in the anterior ribs. (D) Lateral view of the same carapace, showing region of rib chondrogenic growth (blue, arrow) and transition zone (white) between cartilage and bone (red). (E) Dorsal view of 118-day (CL = 3.1 cm) hatchling carapace showing expanded nuchal bone region, the fusion of the anterior costal ossification centers, and the peripheral bone ossification centers that start anteriorly. The pigmentation of the epidermal scutes can be seen. (F) Dorsal view of 185-day (CL = 4.5 cm) hatchling carapace showing fusion of marginal ossification regions anteriorly, as well as the pygal ossification center posteriorly. The costal ossification centers have created bony armor dorsally (the blue staining is beneath the carapace). (G) Predominant pattern of the adult carapacial bones. (Modified from Gilbert et al., 2001; G modified from Zangerl, 1969.)

between the ribs and the costal bones of the carapace. This relationship does not hold in the anterior and posterior ends of the shell, where the vertebral centra are shortened and have little or no contact with the shell. The first costal bone overlies ribs one and two, and the eighth overlies ribs nine and ten (variants have nine pairs of costal bones). The pygal and suprapygal bones form the rear of the carapace. These bones have no contact with vertebra and ribs but project over the sacrum and pelvis. The peripheral bones form the edge of the carapace. There are generally 11 pairs of peripheral bones; before making contact with the costals, they form a socket around the distal tip of ribs two through nine. The nuchal bone forms the anterior margin of the carapace, which overhangs but is not attached to the posterior cervical vertebra. This bone extends laterally around the margins of the carapace to the level of the second rib. It is overlaid by the first three peripheral bones laterally and contacts the first costals and neural bone posteriorly. Each of the carapacial bones is connected by sutures to its neighbors. The distal edge of each costal is attached by suture to the peripheral bones. This contact often does not occur until later stages of post-hatching growth, leaving open a peripheral ring of fontanels that surround the distal tips of the ribs.

Sections across the carapaces of adult turtles show a three-layered arrangement of the bone. The central portion of the bone is a spongy layer containing spherical cavities. On either side of the spongy layer are layers of more compact lamellar bone. This compact bone is thought to form beneath the inner and outer periosteal membranes. The shapes and relative sizes of these bony regions determine the general form of the shell in different genera (Yntema, 1970; Ewert, 1985; Cherepanov, 1997).

1.2.2 Formation of the Carapace

1.2.2.1 The Carapacial Ridge and the Entry of the Ribs into the Dermis

The formation of the carapace involves several steps. The first concerns the entry of the rib precursor cells into the dermis. The turtle egg is laid at the mid-gastrula stage. Turtle gastrulation has not been studied in detail for almost eight decades and presents an interesting contrast to the well-studied avian system (see review; Gilland & Burke, 2004). Later stages of nerulation and somite formation are similar to those processes in the chick (Ewert, 1985; Pasteels, 1937, 1957). The first sign that the organism is to become a turtle rather than some other tetrapod occurs at Yntema stage 14/Greenbaum stage 15 (Yntema, 1968—stages are for *Chelydra*; Greenbaum, 2002—stages are for *Trachemys*. Stage 14/15 is approximately equivalent to Hamburger–Hamilton chick stage 24). At this stage are the first signs of ridges on the lateral surfaces of the embryo, dorsal to the limb buds (Ruckes, 1929). At first, these ridges are seen between the two limb buds, and only later do the ridges extend anteriorly and posteriorly. This structure has been named the carapacial ridge (CR) (Burke, 1989b, 1989c, 1991), and the paired carapacial ridges will eventually form the outer edge of the carapace. The CR is formed by a thickening of the ectoderm and is underlaid by a condensed somite-derived mesenchyme (Yntema, 1970; Burke, 1989b, 1989c; Nagashima et al., 2005).

Ruckes' (1929) observations of turtle embryos described two important features of turtle shell development. First, there is an accelerated lateral growth of the dorsal dermis of the trunk compared to growth in the dorso-ventral plane. Second, there is an apparent "ensnarement" of the growing ribs by the dermis. The involvement of the ribs with the carapacial dermis results in their growth in a predominantly lateral direction (Figure 1.1A). The limb girdles develop in typical tetrapod fashion but because of the growth trajectory of the ribs, the pectoral girdle becomes ventral and deep to the axial elements. Yntema (1970) performed a series of somite extirpation experiments on snapping turtles, confirming a somitic origin for the ribs and dermis of the carapace. Post-otic somite pairs 12 through 21 are involved in forming the carapace in *Chelydra*.

In 1989, Burke proposed that the thickened ectoderm and condensed mesenchyme of the CR is typical of sites of epithelial-mesenchymal interactions. The distributions of the cell adhesion proteins fibronectin and N-CAM in the CR are similar to their locations in other inductive sites such as the early limb bud or feather primordia. Burke (1991) tested the causal relationship between the CR and the growth trajectory of the ribs. In the first set of experiments, she removed the CR by tungsten needles from one side of stage 1 through stage 16 embryos. These extirpations included both ectodermal and mesenchymal components. In those cases where the CR did not regenerate, the growth trajectory of the rib was deflected toward a neighboring region that did have a CR. In a second set of experiments, she placed tantalum barriers between the somite and the presumptive CR. The surviving embryos showed disruptions such that where the CR was interrupted, entire regions of the dermal carapace were missing. The ribs associated with these missing regions interdigitated with those bones of the plastron. Burke concluded that the normal development of the ribs appears to be directed by the CR. In the absence of the CR, these ribs project ventrally into the lateral plate mesoderm like the ribs of non-Chelonian vertebrates.

Loredo and colleagues (2001) were the first to analyze the CR with molecular probes and found fibroblast growth factor-10 (FGF-10) expression in the mesenchyme condensed beneath the *Trachemys* CR. Fibroblast growth factors are paracrine factors that are critical in the patterning, migration, and differentiation of numerous cell types, and they are especially important in determining the fates of cells in the face and in the limbs. Vincent and coworkers (2003) found the turtle homologue of transcription factor *msx1* is expressed in the mesenchyme of the *Emys* CR. This result furthered the notion that the CR was made through mesenchymal/epithelial interactions similar to those that generate the limb bud. The Wnt signaling pathway is used in several embryonic inductions and can mediate the effects of fibroblast growth factors (in the limb bud). By using RT-PCR, Kuraku and colleagues (2005) found turtle orthologs of *Sp5* and Wnt targets *APDCC-1* and *LEF-1* in the CR mesenchyme and ectoderm of the Chinese softshell turtle *Pelodiscus*. They also found CRABP-1 expressed in the CR ectoderm. However, they did not detect the expression of either of the previously reported genes, *msx1*, or FGF-10 in the CR mesenchyme of this species. Species differences might be important in these patterns because the costal bones of *Pelodiscus* might form by different methods from that of the hardshell turtles (Zangerl, 1969), and the pattern of FGF-10 distribution in the limbs of *Pelodiscus* differed from the expression pattern seen in the limbs of *Trachemys*.

The FGF family of paracrine factors is often involved in chemotaxis, and in the chick limb, FGF-10 appears to be critical in directing the endodermal chemotaxis in the lung (Park et al., 1998; Weaver et al., 2000). Cebra-Thomas and colleagues (2005) demonstrated that FGF-induced chemotaxis plays an important role in causing the rib precursors to enter the CR. They cultured eviscerated trunk explants of stage 15 *Trachemys* embryos ventral-side down on nucleopore membranes. At this stage, the CR is visible and the sclerotome has been specified. After three days in culture, the ribs have migrated into the CR, and the ridges are visibly raised. However, if SU5402 (an inhibitor of FGF signaling) is added to the culture media when the explants are established, the CR degenerates and the ribs travel ventrally, like the ribs of non-Chelonians. Cebra-Thomas and colleagues also show that chick rib precursor cells are responsive to FGF-10, and beads containing FGF-10 will redirect chick rib growth in culture. Thus, the CR appears to be critical for directing the migration of rib precursor cells into it. FGF signaling in the CR appears to be crucial in the maintenance of the CR and is either directly or indirectly responsible for guiding the rib precursor cells into the CR.

Another finding of Cebra-Thomas and colleagues (2005) was that the distal tip of each rib expressed FGF-8. High levels of FGF-8 expression have not been reported in the distal ribs of other organisms. Cebra-Thomas and colleagues speculate that FGF-8 (in the ribs) and FGF-10 (in the CR mesenchyme) may establish a positive feedback loop such that the growth of the rib becomes coordinated with the growth of the carapace. Such a positive feedback loop has been shown to be responsible for the coordinated outgrowth of the chick and mouse limb buds (Ohuchi et al., 1997; Kawakami et al., 2001).

1.2.2.2 Costal Bones: The Ossification of the Carapace

The rib precursor cells that enter into the CR are prechondrocytes (Figure 1.1A,B), and the ribs undergo normal endochondral ossification, replacing the cartilage with bone cells (Figure 1.1C,D). Cebra-Thomas and colleagues (2005) have proposed that bone morphogenetic proteins (BMP), which are secreted by hypertrophic chondrocytes during endochondral ossification, are capable of inducing the dermis to ossify as well. Thus, they claim that costal bone formation is caused by the BMP-dependent ossification of the dermis by the ribs. The rib precursor cells enter the dermis of the shell a short distance from their origin in the vertebrae and grow laterally within the carapacial dermis (Ruckes, 1929; Burke 1989b, 1989c; Gilbert et al., 2001). When endochondral ossification takes place, the rib is converted to bone, beginning at the proximal end (Figure 1.1E). However, the distal portion of the rib remains cartilaginous beyond the boundary between pleural and marginal scutes, and they do not make contact with the peripheral bones until later in life. There is an anterior-posterior polarity, in that the anterior ribs begin ossification earlier.

As endochondral ossification ensues, the ribs appear to become the organizing centers for the costal bones that make the plate of the carapace (Gilbert et al., 2001). These costal bones form around the ribs by intramembranous ossification (Burke, 1991; Gilbert et al., 2001; Kälin, 1945). Thus, the carapace is a composite of endochondral axial skeleton (from the ribs) plus intramembranous dermal bone. The costal bones begin to form as the ribs become encased in a thin tube of bone, and trabeculae extend both caudally and cranially from this bony casing. Later, spicules form between the rib and the epidermis, forming a pattern reminiscent of the formation of the mandible around Meckel's cartilage (Suzuki, 1963). The most intense area of costal bone formation is initially located at the sites where the ribs had first entered the dermis.

Bone-forming paracrine factors are secreted by the cartilaginous rib cells during endochondral ossification. In those vertebrates studied thus far (and the turtle is not one of them), Indian hedgehog homolog (Ihh) secreted by the ribs' prehypertrophic cartilage induces BMPs in the perichondrium (Vortkamp et al., 1996). Pathi and colleagues (1999) demonstrated that in chick limbs, perichondrial BMP-2, BMP-4, BMP-5, and BMP-7 are induced by endogenous and ectopic Ihh. Similarly, Wu and colleagues (2001) demonstrated the induction of BMP-2/BMP-4 by Ihh in chick jaw tissue. Both Ihh and BMPs are known to induce bone formation in surrounding competent cells (Barlow & Francis-West, 1997; Ekanayake & Hall, 1997), the competence of dermal cells to respond to BMPs by producing intramembranous bone has been demonstrated in adult dermal and periosteal tissues (Shafritz et al., 1996; Shore et al., 2006).

In turtle embryos and hatchlings, the dermal cells around the rib appear to be responding to BMPs. This was shown (Cebra-Thomas et al., 2005) by using an antibody against phosphorylated (activated) Smad1. (The Smad1 protein is a transcription factor subunit that becomes phosphorylated in response to a BMP's binding to its cell membrane receptor.) Whereas the rib and its perichondrium remain unstained, there was intense staining in the periosteum and in the cells adjacent to it (Figure 1.2). Moreover, when compared to alcian and alizarin-stained adjacent sections (which stain cartilage matrix and bone matrix, respectively), a high level of staining was observed in the cells that were in the area destined to become bone. Thus, it appears that BMP signaling from the rib during endochondral ossification is able to induce intramembranous ossification in the dermal cells surrounding them. Moreover, as the cells ossify they appear to transmit the BMP signal to the cells surrounding them, thereby continuing a cascade through which BMP would be produced by the dermal cells as they ossify.

Although the ribs begin to ossify *in ovo*, the dermal bones of the carapace develop primarily after hatching. The rates of osteogenesis, and perhaps to some degree the pattern, is influenced by environmental conditions (Ewert, 1985). Size and age are both important parameters for bone pattern. Turtles of the same age can be at developmentally different stages, and there is significant variation even among turtles of the same size. Hatching time is also variable, and embryos and juvenile specimens are described by their *carapace length* (CL) as well as their age since the egg was laid. It is also probable that BMP inhibitors in the dermis regulate the progression of ossification because the ossification front slows down and endochondral ossification in the rib is finished long before the fusion of the dermal bones into a carapacial plate (Figure 1.1F).

In the formation of the carapace, one sees heterotopy (change in placement between ancestor and descendent) at several levels. Heterotopy of bone formation is obvious in that these bones are developing in the dorsal dermis, which represents a new site of bone formation. This heterotopy of bone formation is predicated on the heterotopy of the ribs, which have migrated into a part of the body where they do not usually go. This rib heterotopy is further predicated on the heterotopy of FGF-10 expression, which is activated in a tissue that does not usually express this gene.

1.2.2.3 The Nuchal and Peripheral Bones of the Carapace

In *Chelydra* and *Trachemys*, the nuchal bone shows two distinct phases of ossification. We refer to these phases as primary and secondary, referring to both the modes of ossification and the elements

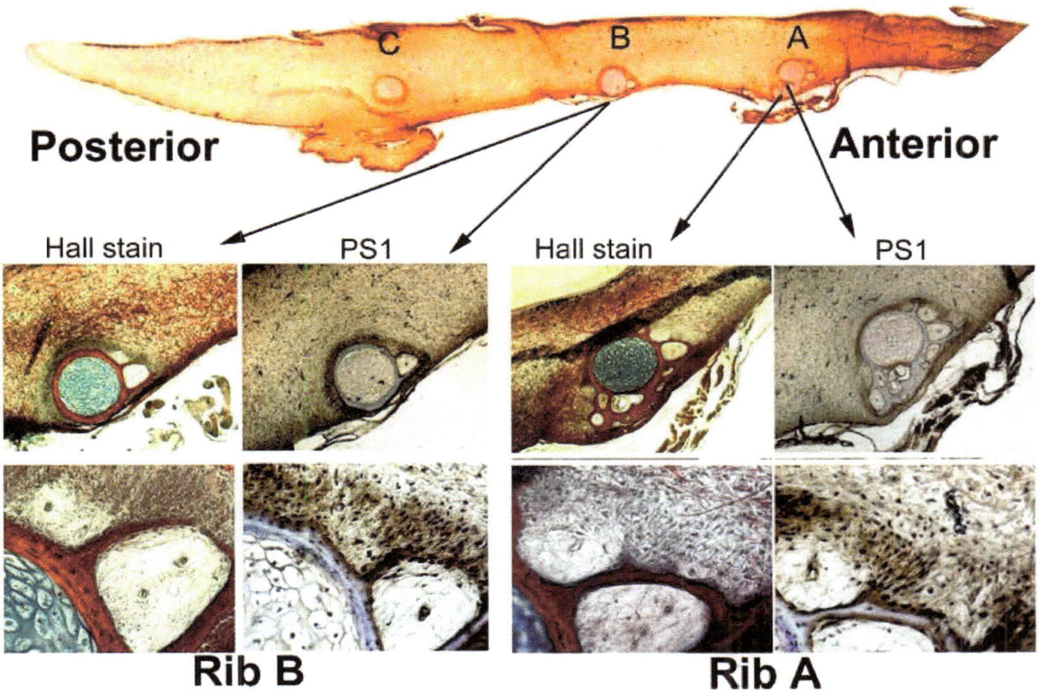

FIGURE 1.2 Formation of the costal bones of the carapace. Sagittal section through the posterior three ribs of a 156-day hatchling *Trachemys* (about a month after hatching). The ribs are at different levels of maturity, the anterior ("A") being the most mature. The sections stained with Hall stain (Alcian and alizarin) are near to the slides stained with antibodies to phosphorylated SMAD1 (PS1). Nuclear expression of phosphorylated Smad1 (brown) is seen in the periosteum of the bone and in the immediately adjacent dermal cells. Below each low-power (200×) is a photograph taken at 400× magnification.

themselves (Burke, 1989a; Gilbert et al., 2001). This pattern of primary and secondary ossification is also seen in the plastron and may have phylogenetic significance.

The primary portion of the *Chelydra* nuchal forms early (CL = 1.4 cm, Yntema stage 20–21), appearing as a thin band of condensed cells within the dermis, continuous across the midline and extending laterally around the margin to the level of the third marginal. The band is visible deep in the dermis before the tissue stains with alizarin, indicating that the well-defined condensation of cells forms well before the deposition of calcium. It underlies the marginal/vertebral sulci, which is clearly visible at this stage. As evidenced by positive staining with alizarin, calcium deposition starts bilaterally at the level of the first marginal scute and spreads along the bars medially and laterally.

The second phase of nuchal ossification involves the nuchal plate, which begins to form in *Chelydra* embryos of CL = 1.8 cm. The nuchal plate forms as a loose lattice work of bone within the carapacial dermis that extends forward over the base of the neck. The pattern of ossification is very similar to that seen in the initial stages of ossification in the skull roofing bones. It begins in contact with the anterior-medial nuchal bar and extends laterally along the bar and posteriorly into the dermis above the neural spines of the last two cervical vertebrae. This posterior extension of secondary dermal bone forms the main body of the nuchal and lies under the first vertebral scute. It will eventually form a suture posteriorly with the first neural bone, which develops around the neural spine of the first thoracic vertebra.

In specimens of CL = 2.6 cm, the nuchal is fully developed and ossified. The lateral bars of the primary ossification extend to the midpoint of the fourth marginal scute, to the level of contact with the cartilaginous distal tip of the second rib. It underlies the sulci separating the marginals from the first vertebral and costal scutes. The lateral extensions of the primary nuchal bone are never in association with the secondary nuchal bone, but rather come to be overlain by the first and second peripheral bones.

The peripheral bones are formed in an anterior-to-posterior manner. Here, small crescents of bone—concave outward—appear in the dermis on the extreme edge of the carapace immediately subjacent to the intermarginal sulci. The first peripheral appears under the sulci of the first two marginal scutes. The ossifications that produce the peripheral bones are also seen to begin in the largest of the new hatchlings. The peripheral ossification centers are first seen in the anterior of the carapace on day 78 *Trachemys* (CL = 3.1 cm) and as the turtle grows, more peripheral ossification centers can be seen caudally on the shell. These ossification centers form on the outer edge of the carapace and expand both laterally and internally as they grow. The pygal bone forms in sequence as the last peripheral and is therefore the last bone to ossify. It is not known what induces these centers to form where they do. It is possible that their positioning is coordinated by the marginal scutes, and that sonic hedgehog, whose gene is expressed in the marginal scute forming region (Lewis et al., 2005) also induces the bone to form there.

Evidence from Gilbert and Cebra-Thomas (Gilbert et al., 2007) suggests that the nuchal bone may form from neural crest cells. This is also a mechanism being proposed for plastron bones and will be discussed later.

1.3 THE FORMATION OF THE PLASTRON BONES: HETEROCHRONY AND NEURAL CREST CELLS

1.3.1 Dermal Bones of the Plastron

The plastron generally is composed of nine bones, formed by intramembranous ossification (Figure 1.3) (Rathke, 1848; Clark et al., 2001). The paired epiplastra and the central (unpaired) entoplastron form the three anterior bones of the plastron. The hyoplastra form the axillary buttresses and the anterior bridge region. The bridge extensions of these bones approach the carapace at the level of peripheral five and rib four. The bilateral hyoplastra meet each other at the ventral midline and form the anterior rim of the central umbilical fontanel. During embryonic development, this fontanel surrounds the yolk stalk that connects to the gut. The paired hypoplastra form the inguinal buttresses, the posterior bridge region, and the posterior rim of the central fontanel. They approach the carapace at the level of peripherals six and seven and ribs five and six. The paired xiphiplastra form the posterior lobe of the plastron.

1.3.2 Ossification of the Plastron

1.3.2.1 Development of the Plastron Bones

The plastron begins to ossify before hatching. In the embryonic turtle (CL = 1.0 cm in *Trachemys*, CL = 2.0 cm in *Chelydra*), the future plastron can be identified by nine ossification centers in the ventral dermis. No Alcian blue staining is seen presaging these sites. In *Trachemys*, the three ossification centers corresponding to the three anterior plastron bones appear to fuse around day 78 (CL = 2.2 cm). The two epiplastral bones form a suture with one another, whereas the entoplastron bone forms more medially and projects caudally. As the hatchling turtle gets larger, the six paired ossification centers of the plastron grow toward one another and form sutures. Condensed mesenchyme is seen in advance of the calcified tissue (Burke, 1989a; Gilbert et al., 2001). These sites contain both alizarin red-stained bony spicules and a region of condensed mesenchyme that has coalesced into the stellate arrays that will later show staining for bone matrix. This is another example of primary ossification, as in the nuchal.

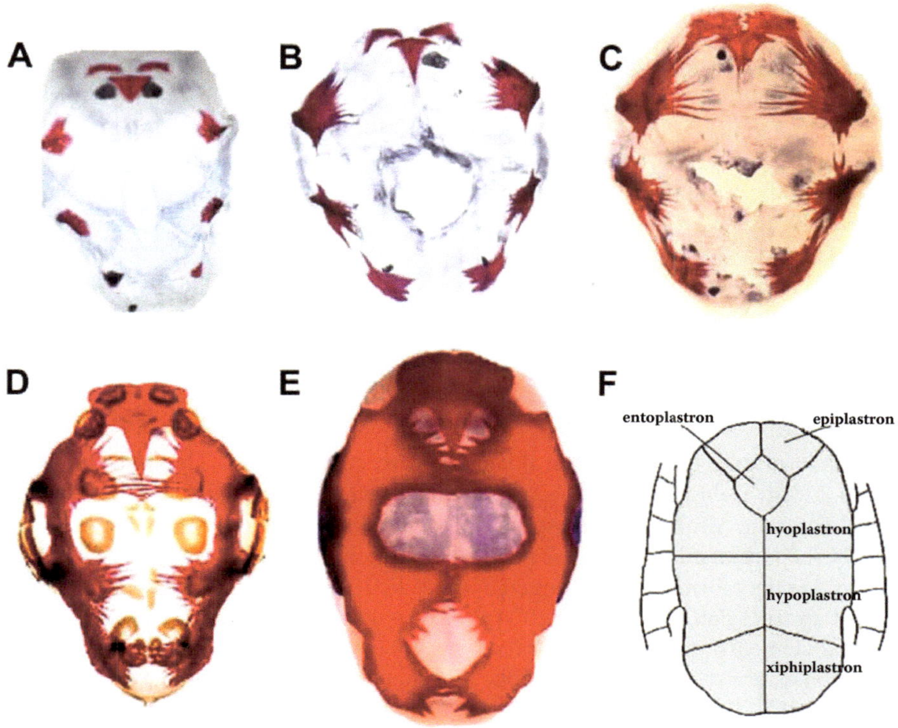

FIGURE 1.3 Dermal ossification of the plastron. (A) 55-day (CL 1.0 cm) *Trachemys* plastron showing the three anterior ossification centers and the three laterally paired ossification centers. The dark blue represents girdle cartilage. (B) 78-day (CL = 2.2 cm) plastron showing spicules radiating from the ossification centers. (C) 78-day (CL = 2.4 cm) plastron showing fusion of the anterior ossification centers. (D) 118-day (CL = 3.1 cm) plastron showing epidermal pigmentation and the crossing of the midline by the spicules. The spicules do not touch but get out of each other's way. (E) 185-day (CL = 4.5 cm) plastron showing fusion of ossification centers and the formation of plastron. No cartilage precursors are seen. Note that (B) and (C) are both 78-day incubations. The hole in the center of the plastron is the umbilical fontanel through which the gut attaches to the yolk stalk. (F) Predominant pattern of plastron bones. (Modified from Gilbert et al., 2001.)

One of the interesting things observed about plastron ossification is that the bony spicules cross the midline. The midline does not appear to be respected by the developing spicules. Moreover, as they crossed the midline the spicules did not immediately fuse. Rather, it appears as if the ossifying spicules on either side avoided one another, altering their course of ossification such that they interdigitate rather than run into each other (Figure 1.3E). This is very likely a prerequisite for continued growth through suture formation.

A similar situation is seen in *Chelydra*. The plastral bones appear with a slight anterior-posterior bias, the epiplastra and entoplastron first and the xiphiplastron last. They are all present in specimens of CL = 1.5 cm, preceded only by the appearance of the primary nuchal bar. Like the nuchal bone, the plastral bones show two phases of development. They first appear as slender bars of condensed cells that then calcify from their centers outward.

The character and homology of the bony elements of the plastron has been extremely controversial (Hall, 2001; Vickaryous & Hall, 2006). In 1834, Carus was perhaps the first to suggest that the carapace and plastron involved both the endo- (endochondral) and the exoskeletal (dermal) bones. He proposed that the plastron formed by overlying the endoskeletal sternum with dermal ossifications. Rathke (1848) argued that the plastron belonged exclusively to the exoskeleton and was in no way homologous to the sternum. However, Owen (1849), adhering to his ideal vertebral archetype, proposed that the plastral bones were homologues of the thoracic vertebral hemapophyses, and as such were part of the endoskeleton. More recent histological studies confirmed Rathke's assessment that the bones of the plastron all ossify intramembranously without any cartilaginous precursors and belong to the dermal exoskeleton (Zangerl, 1939, 1969; Gilbert et al., 2001). Currently, the consensus is that the epiplastra and entoplastron are homologous, respectively, to the clavicles and interclavicle bones of other reptilian lineages (Zangerl, 1969; Cherepanov, 1997; Vickaryous & Hall, 2006; Parker, 1868; Rieppel, 1996), whereas the more posterior plastral bones are homologous to the gastralia ("floating ribs" or "abdominal ribs") of other tetrapods (Zangerl, 1939; Claessens, 2004).

1.3.3 Roles of Neural Crest Cells in Plastron and Nuchal Bone Development

The embryonic origins of the plastral bones are also controversial. The Swarthmore laboratory (Clark et al., 2001; Cebra-Thomas et al., 2007) has put forth the proposal that the plastron bones are derived from the trunk neural crest and form much the same way that vertebrate facial bones form. In 2001, Clark and her colleagues published evidence that the turtle plastron bones are exoskeletal and that they form by the intramembranous ossification of neural crest cells. This assertion has aroused spirited debate (Pennisi, 2004) because trunk neural crest cells are not supposed to form skeletal elements, and cranial neural crest cells (which are skeletogenic) are not supposed to migrate more posteriorly than the collarbone and shoulder based on amniote models like the chick and mouse (Hall, 2005; Matsuoka et al., 2005). Clark and colleagues (2001) showed that the nine developing plastron bones of the 50-day *Trachemys* embryo are formed by cells that stained positively for the cell surface carbohydrate determinant recognized by the monoclonal antibody HNK-1 (Figure 1.4C) and for the membrane receptor protein PDGFRα.

HNK-1 immunoreactivity is the "standard" marker for neural crest cells, and turtle neural crest cells stained positively and strongly for HNK-1 (Hou, 1999; Hou & Takeuchi, 1994). However, in those studies, only early (Yntema stage 12) embryos were examined and the possible migration of neural crest cells to the plastron was not addressed. PDGFRα is a marker for skeletogenic and odontogenic neural crest cells. PDGFRα has been detected on the bone-forming neural crest cells of mice and frogs as well as in teeth and other first branchial arch derivatives. Antibody staining against PDGFRα in the turtle embryo showed its localization in the mandibular mesenchyme, as expected, as well as in each of the developing plastron bones (Clark et al., 2001).

However, neither HNK-1 nor PDGFRα staining are completely specific for neural crest cells and their derivatives. The HNK-1 antibody detects not only cells of the neural crest lineage but also stains the neural tube, cerebellar neurons, motor neurons, and certain leukocytes. In mice, PDGFRα is detected not only on skeletogenic neural crest cells but also on rib precursors and in the embryonic mesenchyme cells contributing to bone, hair, mammary gland, gut, and lung. The definitive identification of neural crest cells can only be confirmed by lineage mapping, Thus, whereas the Clark study strongly suggested neural crest involvement in plastron formation, it did not conclusively demonstrate that these were neural crest cells and, if so, whether they were from the trunk or cranial neural crest.

Cebra-Thomas and colleagues (2007) attempted to find the origin of these plastron-forming HNK-1$^+$ cells and use more markers to identify neural crest cells. They found that stage 17 and stage 18 *Trachemys* embryos (three weeks incubation) had a "staging area" in the trunk carapacial dermis where the HNK-1$^+$ cells resided (Figure 1.4A). The cells in this region were positive not only for HNK-1 immunoreactivity but also for two additional markers for neural crest: the neural crest-specifying transcription factor FoxD3 and the low-affinity neurotrophin receptor, p75. FoxD3 staining of nuclei was seen in the dorsal-most portion of the early stage 17 neural tube as well as in

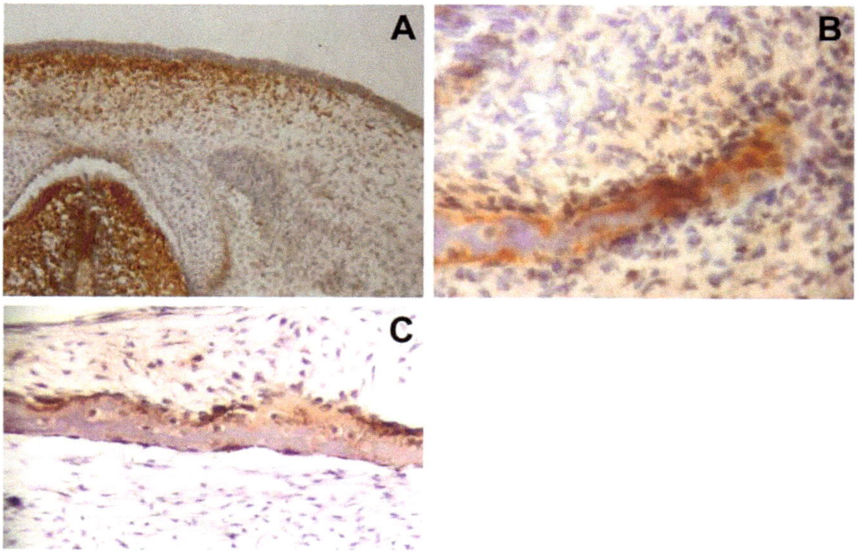

FIGURE 1.4 Late-emigrating HNK-1⁺ cells forming the plastron of *Trachemys*. (A) Dorsal region of stage 17 (three-week) embryo showing the carapacial staging area wherein HNK-1⁺ cells (brown-red stain) reside. (B) Plastron bone being formed by HNK-1⁺ cells in a stage 18 embryo. (C) Hyoplastron of a 50-day embryo. The bone stains with hemotoxylin, whereas the HNK-1⁺ cells are red-brown. (A,B after Cebra-Thomas et al., 2007; (C) adapted from Clark et al., 2001.)

cells in the dermis between the neural tube and surface ectoderm. The fact that these are dorsal cells staining with HNK-1, FoxD3, and p75 makes them excellent candidates to be neural crest cells.

These neural crest cells would represent a very late emigrating population, and they appear to come directly from the neural tube (and not from the neural plate/epidermal boundary) after the first wave of neural crest emigration has already formed the dorsal root ganglia, pharyngeal derivatives, melanoblasts, and enteric neurons. After leaving the dorsal neural tube region, these cells reside within the forming carapacial dermis and by stage 18, these cells form a broad band in the dorsal portion of the carapace. These cells constitute a migratory population, and DiI staining shows them moving laterally and ventrally. In addition, stage 18 embryos also exhibit HNK-1⁺ cells migrating near the vertebrae and migrating down the lateral walls of the embryo within the dermis. These HNK-1⁺ and p75⁺ cells can be seen condensing in the plastral mesenchyme and forming bone (Figure 1.4B). Unlike chick or mouse embryos, the bone-forming neural crest cells (such as those in the head) retain the HNK-1 and p75 markers even as they are forming bone (Clark et al., 2001; Cebra-Thomas et al., 2007).

This pattern of HNK-1 expression is unique to the turtle and suggests that the late emigrating turtle trunk neural crest cells have taken on the characteristics of cranial neural crest cells. In addition to expressing PDGFRα, a marker usually associated with cranial neural crest cells, these late-emerging neural crest cells appear to contribute to the sclerotome-derived vertebral and rib cartilages. Thus, the turtle vertebrae and ribs may have a dual origin—the somite and the neural crest. A bipartite pattern in the cartilage would be expected if the trunk crest cells had the properties of cranial neural crest cells because Le Douarin and Teillet (1974) showed that avian cranial neural crest cells contributed to trunk cartilage when transplanted into the trunk region.

Gilbert and Cebra-Thomas suggest that the nuchal bone and the plastron bones may form totally or predominantly from trunk neural crest cells. The developing plastron and nuchal bones (but not the peripheral carapacial bones of the same turtle) stain positively for neural crest markers. Although HNK-1 reactivity is not specific for neural crest cells (it is also seen in some neurons, leukocytes, and cartilage cells), the observation that the plastron and nuchal bones develop intramembranously (without cartilaginous intermediates), express additional neural crest markers, are near no neurons, and are obviously not made of white blood cells suggests a neural crest origin for them.

How might trunk neural crest cells form bone? In most vertebrates studied, cell labeling studies demonstrated that the dermal cranial and facial bones of the vertebrate exoskeleton (as well as the dentine of the teeth) come from the cranial region of the neural crest, whereas the trunk neural crest is unable to form bone (Smith & Hall, 1993; Matsuoka et al., 2005; Hall, 2005). One distinction between cranial and trunk neural crest cells lies in the expression of *Hox* genes. The neural crest cells that arise from the fore- and midbrain produce Meckel's cartilage and the bones of the skull, face, and jaw do not express *Hox* genes. When *Hox* genes were experimentally expressed in cranial neural crest cells that would normally give rise to the craniofacial skeleton, the resulting chick embryos showed severe skeletal deformities (Creuzet et al., 2002). Smith and Hall (1993) postulated that the ability to form bones was a primitive property that characterized early vertebrates, and Trainor and colleagues (2003) saw the evolution of jaws as resulting largely from the loss of mandibular *Hox* gene expression between the lamprey-like agnathans and the gnathostomes.

Recent evidence has shown that trunk neural crest cells can gain skeletogenic potential if their *Hox* gene expression pattern is downregulated. McGonnell and Graham (2003) found that chick trunk neural crest cells in long-term cell culture can produce osteoblasts and chondrocytes. Moreover, Abzhanov and colleagues (2003) confirmed this observation and demonstrated that the cultured trunk crest cells that had gained skeletogenic potential had also lost their *Hox* gene expression. It is possible that the late emigrating neural crest cells in turtle embryos have lost their *Hox* expression patterns (either by emigrating from the neural tube at a late date or by remaining in the staging area for a prolonged period of time) and have thereby acquired the ability to form bone-like cranial neural crest cells.

The current evidence supports the contention that the trunk neural crest cells of the turtle have gained (or regained) the ability to form a skeleton. Therefore, it is possible that the nuchal bone and the bones of the plastron are formed by neural crest cells using methods similar to forming the calvareum and face. These conclusions can be confirmed by detailed lineage mapping of trunk neural crest cells in turtle embryos.

1.4 EVOLUTIONARY IMPLICATIONS

> "Were there no turtles living, we would look upon the fossil turtles as the strangest of all vertebrates—animals which had developed the strange habit of concealing themselves inside their ribs, for that is literally what turtles do."
>
> **Samuel Williston (1914)**

The order Chelonia emerges abruptly in the Triassic about 210 million years ago with the fossil species *Proganochelys* (Gaffney, 1990). This reptile had the characteristic derived trunk morphology now associated with turtles, including both a carapace and plastron. Based on cranial characters, turtles have traditionally been classified as anapsids, with roots in one of several Triassic forms of "parareptiles." Many of these forms sport extensive dermal armor in the form of bony ossicles that were embedded in the skin.

An evolutionary model where the chelonian costals and other bones were derived from osteoderms that secondarily fused with the ribs and vertebrae was the predominant view among paleontologists for many years (Kälin, 1945; Romer, 1956; Sukhanov, 1964; Carroll, 1988; Laurin & Reisz, 1995; Lee, 1996, 1997a, 1997b). However, among the candidate ancestors—including captorhinomorphs, pareiasaurs, and procolophonids—the fossil record provides no clues to the origin of the unique chelonian rearrangement of the axial and appendicular skeletons. Carroll (1988) comments that their bizarre anatomy might be sufficient to place turtles in their own subclass of the Reptilia.

The anapsid status of turtles has been challenged in recent years. In a recent review, Zardoya and Meyer (2001) analyze six alternative cladograms currently being used to represent the relationships of turtles to other reptiles and birds. In contrast to the traditional paleontologic view that turtles are anapsids, a different view—relying on the physiological and morphometric evidence from extant turtles, as well as from their pancreatic polypeptide sequences, nuclear DNA, and mitochondrial DNA—has caused several groups to argue that turtles are modified diapsids within the reptilian clade. Platz and Conlon (1997) and Hedges and Poling (1999) use sequence data to propose that turtles group with crocodilians among the archosaurs. Further protein sequence data from Iwabe and colleagues (2005) indicate that turtles are a sister group to the archosaur clade. Rieppel (2001) and Rieppel and Reisz (1999) also assign turtles to the diapsida. They propose an aquatic origin of the turtles wherein the ancestor would have already had a plastron-like gastralia to which the newly made carapace could attach. Gastralia are present in numerous orders of reptiles and would probably have already been present in the ancestors of turtles. Claessens (2004) summarizes, "Gastralia may be plesiomorphic for tetrapods, but are only retained in extant Crocodylia and *Sphenodon*, and possibly as part of the chelonian plastron."

Whether one views turtles as anapsids or diapsids, there is a dramatic absence of transitional forms. This raises the possibility that turtles arose saltationally, without intermediate morphologies that would link them to non-Chelonian reptiles. The model proposed by Burke (1989c) sets the timing and position of the CR as the pivotal event in the evolution of the new body plan. It is a safe assumption that epithelial/mesenchymal interactions were the inductive mechanisms for the formation of dermal armor in early amniotes. The precocious initiation of an epithelial/mesenchymal interaction in the dorsal body wall of the early chelonian embryo may have been the initial novelty in the evolution of the dermal carapace. The model proposed by Cebra-Thomas (2005) provides a mechanism for the rapid morphogenesis of the bony shell once the ribs are repositioned into the dermis.

The development of the turtle is full of surprises. Indeed, what we have here is a tentative outline of how the turtle gets its shell, but there are many more questions to ask. If the trunk neural crest cells form the plastron, how are they directed there and what causes them to become bone? What causes some turtles to have a dome-shaped carapace whereas other turtles have a flattened carapace? What causes the sexually dimorphic concavities of the plastron, and how do some turtles develop a hinge in this ventral shell? Developmental biology is just beginning to join paleontology and structural morphology in exploring this fascinating structure, and this union may enable us to see how evolutionary innovations can rapidly emerge and to finally determine the place of the turtle in the history of life.

ACKNOWLEDGMENTS

We wish to thank Ms. Diane Fritz for her assistance in helping prepare this manuscript. Also, we wish to thank the National Science Foundation and the Howard Hughes Medical Institute for supporting much of the recent work reported here.

REFERENCES

Abzhanov, A., Tzahor, E., Lassar, A.B., and Tabin, C.J., Dissimilar regulation of cell differentiation of mesencephalic (cranial) and sacral (trunk) neural crest cells in vitro, *Development*, 130, 4567, 2003.

Alibardi, L., and Thompson, M.B., Epidermal differentiation during carapace and plastron formation in the embryonic turtle *Emydura macquarii*, *J. Anat.*, 194, 531, 1999a.

Alibardi, L., and Thompson, M.B., Morphogenesis of shell and scutes in the turtle *Emydura macquarii*, *Aust. J. Zool.*, 47, 245, 1999b.

Anonymous, letter. *Philos. T. Roy. Soc.*, 11, 743, 1676.

Barlow, A.J., and Francis-West, P.H., Ectopic application of recombinant BMP-2 and BMP-4 can change patterning of developing chick facial primordial, *Development*, 124, 391, 1997.

Burke, A.C., Critical features in Chelonian development: The ontogeny and phylogeny of a unique tetrapod Bauplan, unpublished Ph.D. thesis, Harvard University, Cambridge, MA, 1989a.

Burke, A.C., Development of the turtle carapace: Implications for the evolution of a novel Bauplan, *J. Morphol*, 199, 363, 1989b.

Burke, A.C., Epithelial-mesenchymal interactions in the development of the Chelonian Bauplan, in *Trends in Vertebrate Morphology*, 35, H. Splechtana and H. Hilgers (eds.), Stuttgart: Fischer Verlag, 1989c, 206–209.

Burke, A.C., The development and evolution of the turtle body plan: Inferring intrinsic aspects of the evolutionary process from experimental embryology, *Amer. Zool*, 31, 616, 1991.

Carroll, R.L., *Vertebrate Paleontology and Evolution*, New York: W.H. Freeman, 1988.

Carus, K.G., *Lehrbuch der vergeichenden Zootomie* (2nd ed.), Leipzig: Ernst Fleischer, 1834.

Cebra-Thomas, J., Betters, E., Yin, M., Plafkin, C., Bender, G., McDow, K., and Gilbert, S.F. Further evidence that a late-emerging population of trunk neural crest cells forms the plastron of the turtle *Trachemys scripta*, *Evol. Dev.*, 9, 267–277, 2007.

Cebra-Thomas, J., Tan, F., Sistla, S., Estes, E., Bender, G., Kim, C., Riccio, P., and Gilbert, S.F., How the turtle forms its shell: A paracrine hypothesis of carapace formation, *J. Exp. Zool.* Part B, 304, 558–569, 2005.

Cherepanov, G.O., New morphogenetic data on the turtle shell: Discussion on the origin of the horny and bony parts, *Stud. Geolog. Salmant.*, 3, 9–24, 1989.

Cherepanov, G.O., The origin of the bony shell of turtles as a unique evolutionary model in reptiles, *Russ J. Herpetol.*, 4, 155–162, 1997.

Claessens, L.P.A.M., Dinosaur gastralia: Origin, morphology, and function, *J. Vert. Paleontol*, 24, 89–106, 2004.

Clark, K., Bender, G., Murray, B.P., Panfilio, K., Cook, S., Davis, R., Murnen, K., Tuan, R.S., and Gilbert, S.F., Evidence for the neural crest origin of turtle plastron bones, *Genesis*, 31, 111–117, 2001.

Creuzet, S., Couly, G., Vincent, C., and Le Douarin, N.M., Negative effect of Hox gene expression on the development of the neural crest-derived facial skeleton, *Development*, 129, 4301–4313, 2002.

Ekanayake, S., and Hall, B.K., The in vivo and in vitro effects of bone morphogenentic protein-2 on the development of the chick mandible, *Int. J. Dev. Biol.*, 41, 67–81, 1997.

Ellis, R., *Sea Dragons: Predators of the Prehistoric Oceans*, Lawrence, KS: University Press of Kansas, 2003.

Ewert, M.A., Embryology of turtles, in *Biology of the Reptilia*, C. Gans and F. Billett (eds.), New York: John Wiley, 1985, 74–255.

Gaffney, E.S., The comparative osteology of the Triassic turtle *Proganochelys*, *Bull. Amer. Mus. Nat. Hist.*, 194, 1–263, 1990.

Gilbert, S.F., Bender, G., Yin, M., and Cebra-Thomas, J.A., The contribution of neural crest cells to the nuchal bone and plastron of the turtle shell, *Integr. Comp. Biol.*, 47, 401–408, 2007.

Gilbert, S.F., Loredo, G.A., Brukman, A., and Burke, A.C., Morphogenesis of the turtle shell: The development of a novel structure in tetrapod evolution, *Evol. Dev.*, 3, 47–58, 2001.

Gilland, E., and Burke, A.C., Gastrulation in reptiles, in *Gastrulation*, C. Stern (ed.), Cold Spring Harbor, NY: Cold Spring Harbor Laboratory Press, 205–218, 2004.

Greenbaum, E., A standardized series of embryonic stages for the emydid turtle *Trachemys scripta*, *Can. J. Zool.*, 80, 1350–1370, 2002.

Hall, B.K., Development of the clavicles in birds and animals, *J. Exp. Zool.*, 289, 153–161, 2001.

Hall, B.K., *Bones and Cartilage: Developmental and Evolutionary Skeletal Biology*, San Diego, CA: Academic Press, 2005.

Hedges, S.B., and Poling, L.L., A molecular phylogeny of reptiles, *Science*, 283, 998–1001, 1999.

Hou, L., Effects of local tissue environment on the differentiation of neural crest cells in turtle, with special reference to understanding the spatial distribution of pigment cells, *Pigm. Cell Res.*, 12, 81–88, 1999

Hou, L., and Takeuchi, T., Neural crest development in reptilian embryos, studied with monoclonal antibody, HNK-1, *Zool. Sci.*, 11, 423–431, 1994.

Iwabe, N., Hara, Y., Kumazawa, Y., Shibamoto, K., Saito,Y., Miyata, T., and Katoh, K., Sister group relationship of turtles to the bird-crocodilian clade revealed by nuclear DNA-coded proteins, *Mol. Biol. Evol.*, 22, 810–813, 2005.

Kälin, J., Zur morphogenese des panzers bei den Schildkröten, *Acta Anta.*, 1, 144, 1945.
Kawakami, Y., Capdevilla, J., Buscher, D., Itoh, T., Rodríguez Esteban, C., and Izpisúa Belmonte, J.C., WNT signals control FGF-dependent limb initiation and AER induction in the chick embryo, *Cell*, 104, 891–900, 2001.
Kuraku, S., Usuda, R., and Kuratani, S., Comprehensive survey of carpacial ridge-specific genes in turtle implies co-option of some regulatory genes in carapace evolution, *Evol. Dev.*, 7, 3–17, 2005.
Laurin, M., and Reisz, R.R., A reevaluation of early amniote phylogeny, *Zool. J. Linn. Soc.*, 113, 165–223, 1995.
Le Douarin, N.M., and Teillet, M-A.M., Experimental analysis of the migration and differentiation of neuroblasts of the autonomic nervous system and of neuroectodermal mesenchymal derivatives, using a biological cell marking technique, *Dev. Biol.*, 41, 162–184, 1974.
Lee, M.S.Y., Correlated progression and the origin of turtles, *Nature*, 379, 812–815, 1996.
Lee, M.S.Y., Pareiasaur phylogeny and the origin of turtles, *Zool. J. Linn. Soc.*, 120, 197–280, 1997a.
Lee, M.S.Y., Reptile relationships turn turtle, *Nature*, 389, 245–246, 1997b.
Lewis, K.K., Cebra-Thomas, J., and Gilbert, S.F., The expression of signaling molecules in the developing scute of the turtle *Trachemys scripta*, *Integ. Comp. Biol.*, 45, 1159, 2005.
Loredo, G.A., Brukman, A., Harris, M.P., Kagle, D., Leclair, E.E., Gutman, R., Denney, E., Henkelman, E., Murray, B.P., Fallon, J.F., Tuan, R.S., and Gilbert, S.F., Development of an evolutionarily novel structure: Fibroblast growth factor expression in the carapacial ridge of turtle embryos, *J. Exp. Zool.*, 291, 274–281, 2001.
Matsuoka, T., Ahlberg, P.E., Kessaris, N., Iannarelli, P., Dennehy, U., Richardson, W.D., McMahon, A.P., and Koentges, G., Neural crest origins of the neck and shoulder, *Nature*, 436, 347–355, 2005.
McGonnell, I.M., and Graham, A., Trunk neural crest has skeletogenic potential, *Curr. Biol.*, 12, 767–771, 2003.
Nagashima, H., Uchida, K., Yamamoto, K., Kurahu, S., Usuda, R., and Kuratani, S., Turtle-chicken chimera: An experimental approach to understanding evolutionary innovation in the turtle, *Dev. Dyn.*, 232, 149–161, 2005.
Ohuchi, H., Nakagawa, T., Yamamoto, A., Araga, A., Ohata, T., Ishimaru, Y., Yoshioka, H., Kuwana, T., Nohno, T., Yamasaki, M., Itoh, N., and Noji, S., The mesenchymal factor, FGF10, initiates and maintains the outgrowth of the chick limb bud through interaction with FGF8, an apical ectodermal factor, *Development*, 124: 2235–2244, 1997.
Owen, R., On the development and homologies of the carapace and plastron of the chelonian reptiles, *Phil. Trans. Roy. Soc. Lond.*, 1849, 151–171, 1849.
Park, W.Y., Miranda, B., Lebeche, D., Hashimoto, G., and Cardoso, W.V., FGF-10 is a chemotactic factor for distal epithelial buds during lung development, *Dev. Biol.*, 201, 125–134, 1998.
Parker, W.K., A monograph on the structure and development of the shoulder-girdle and sternum in the vertebrata, *The Ray Society*, 1868.
Pasteels, J.J., Etudes sur la gastrulation des vertébrás méroblastiques, *Archs. Biol. Paris*, 48,105–148, 1937.
Pasteels, J.J., Une table analytique du développement des Reptiles. 1. Stades de gastrulation chez les Cheloniens et les Lacertiliens, *Ann. Soc. Roy. Zool. Bel.*, 87, 217–241, 1957.
Pathi, S., Rutenberg, J.B., Johnson, R.L., and Vortkamp, A., Interaction of Ihh and BMP/Noggin signaling during cartilage differentiation, *Dev. Biol.*, 209, 239–253, 1999.
Pennisi, E., Neural beginnings for the turtle's shell, *Science*, 303, 951, 2004.
Platz, J.E., and Conlon, J.M., And turn back again, *Nature*, 389, 246, 1997.
Rathke, H., *Über die Entwicklung der Schildkröten*, Braunschweig: Friederich Vieweg und Sohn, 1848.
Rieppel, O., Testing the homology by congruence: The pectoral girdle of turtles, *P. Roy. Soc. Lond. B Bio.*, 263, 1395–1398, 1996.
Rieppel, O., Turtles as hopeful monsters, *BioEssays*, 23, 987–991, 2001.
Rieppel, O., and Reisz, R.R., The origin and early evolution of turtles, *Ann. Rev. Ecol. Syst.*, 30, 1–22, 1999.
Romer, A.S., *Osteology of the Reptiles*, Chicago: University of Chicago Press, 1956.
Ruckes, H., The morphological relationships between the girdles, ribs, and carapace, *Ann. N.Y. Acad. Sci.*, 13, 81–120, 1929.
Shafritz, A.B., Shore, E.M., Gannon, F.H., Zasloff, M.A., Taub, R., Muenke, M., and Kaplan, F.S., Overexpression of an osteogenic morphogen in fibrodysplasia ossificans progressiva, *New Engl. J. Med.*, 335, 555–561, 1996.
Shore, E.M., Xu, M., Feldman, G.J., Fenstermacher, D.A., Brown, M.A., Kaplan, F.S., and FOP Res. Consort., A recurrent mutation of the BMP type I receptor ACVR1 causes inherited and sporadic fibrodysplasia ossificans progressiva, *Nat. Genet.*, 38, 525–527, 2006.

Smith, M.M., and Hall, B.K., A developmental model for evolution of the vertebrate exoskeleton and teeth: The role of cranial and trunk neural crest, *Evol. Biol.*, 27, 387–448, 1993.
Sukhanov, V.B., Podklass Testudinata (subclass Testudinata), Osnovy Paleontologii, Zemnovodnye, Presmykayushchiesya I Ptitsy, Moscow: Nauka, 1964.
Suzuki, H.K., Studies on the osseous system of the slider turtle, *Ann. N.Y. Acad. Sci.*, 109, 351–410, 1963.
Trainor, P.A., Melton, K.R., and Manzanares, M., Origins and plasticity of neural crest cells and their roles in jaw and craniofacial evolution, *Int. J. Dev. Biol.*, 47, 541–553, 2003.
Vickaryous, M.K., and Hall, B.K., Homology of the reptilian coracoid and a reappraisal of the evolution and development of the amniote pectoral apparatus, *J. Anat.*, 208, 263–285, 2006.
Vincent, C., Bontoux, M., Le Douarin, N.M., Pieau, C., and Monsoro-Burq, A.H., Msx genes are expressed in the carapacial ridge of turtle shell: A study of the European pond turtle, *Emys orbicularis. Dev. Genes Evol.* 213, 464–469, 2003.
Vortkamp, A., Lee K., Lanske, B., Segre, G.V., Kronenber, H.M., and Tabin, C.J., Regulation of rate of cartilage differentiation by Indian hedgehog and PTH-related protein, *Science*, 273, 613–622, 1996.
Weaver, M., Dunn, N.R., and Hogan, B.L.M., BMP 4 and Fgf10 play opposing roles during lung morphogenesis, *Development*, 127, 2695–2704, 2000.
Williston, S., *Water Reptiles of the Past and Present*, Chicago: Chicago University Press, 1914.
Wu, Q., Zhang, Y., and Chen, Q., Indian hedgehog is an essential component of mechanotransduction to stimulate chondrocyte proliferation, *J. Biol. Chem.*, 276, 35290–35296, 2001.
Yntema, C.L., A series of stages in the embryonic development of *Chelydra serpentina*, *J. Morphol.*, 125, 219–252, 1968.
Yntema, C.L., Extirpation experiments on the embryonic rudiments of the carapace of *Chelydra serpentina*, *J. Morphol.*, 132, 235–244, 1970.
Zangerl, R., The homology of the shell elements in turtles, *J. Morphol.*, 65, 383–410, 1939.
Zangerl, R., The turtle shell, *The Biology of the Reptilia*, *Vol. 1*, C. Gans and A. d'A. Bellairs (eds.), New York: Academic Press, 1969, 311–319.
Zardoya, R., and Meyer, A., The evolutionary position of turtles revised, *Naturwissenschaften*, 88, 193–200, 2001.

2 Comparative Ontogenetic and Phylogenetic Aspects of Chelonian Chondro-Osseous Growth and Skeletochronology

Melissa L. Snover and Anders G.J. Rhodin

CONTENTS

2.1 INTRODUCTION

Form and function are fundamental interdependent strategies of all life. From studies of skeletal and chondro-osseous structure and development, we can gain insights into phylogenetic differences and taxonomic classifications, and we can also better understand how different species—and individuals within species—grow to maturity and respond to the physiological demands of their particular life strategies. Cortical banding patterns within bones correlate to activity patterns of the individual as well as endogenous rhythms, allowing for inferences not only about age and cyclical growth patterns but also previous growth and circumstances that have influenced growth (Suzuki, 1963; Enlow, 1969; Castanet, 2006). Studies of these banding patterns within cortical bone (skeletochronology) have been applied to numerous species of turtles and have allowed us to understand patterns and rates of growth.

In addition to skeletochronology, detailed studies of the chondro-osseous development of appendicular bones have revealed strong similarities among most living chelonians, but with striking differences for certain large, fast-growing sea turtles (e.g., the leatherback, *Dermochelys*) that separates them from all other turtles (Rhodin et al., 1980, 1981, 1996; Rhodin, 1985).

In this chapter, we summarize the application of skeletochronology for estimates of age and growth rates in turtles, review the two basic patterns of bone growth that occur in turtles, and correlate these patterns of chondro-osseous development with phylogeny. Finally, we discuss how these factors influence rates of growth to sexual maturity, highlighting how the leatherback stands apart from other turtles.

2.2 SKELETOCHRONOLOGY IN TURTLES

2.2.1 Background

Skeletochronology has been used to estimate age and growth in numerous species of reptiles and amphibians (Castanet, 1994; Smirina, 1994). Bones are good recording structures, as they contain layers that form with a predictable periodicity and the layers are different in morphology and optical density, making them easily discernable (Klevezal, 1996). In histologic cross-sections of bone are concentric thin layers that stain dark with hematoxylin. Alternating with these concentric thin layers are broad homogeneous light-staining layers (Castanet et al., 1993; Klevezal, 1996). Castanet et al. (1977) introduced the term *line of arrested growth* (LAG) to identify the thin dark lines characteristic of skeletal growth marks (Figure 2.1).

In bone morphology, LAGs are in the general class of cement or cementing lines and are common throughout all vertebrate bones. Resorption cement lines are found around Haversian canal systems (secondarily remodeled bone with vascular ingrowth), differentiating them from cortical bone, and in the lamellar periosteal deposition of secondary endosteal bone. Resting cement lines (the class to which LAGs belong) are found in the layering pattern of periosteal deposition of new cortical bone (Enlow, 1969; Francillon-Vieillot et al., 1990).

Many skeletochronological studies of herpetological species indicate that LAGs are formed as a result of low metabolism and slowed or no growth associated with seasonal climatic changes. This is likely true but serves only as a partial explanation, considering that LAGs also occur in the hard structures of nonhibernating mammalian species (Klevezal, 1996; Castanet 2006). Castanet et al. (1993) extended the terminology of LAGs to both poikilotherms and endotherms as a general description of a resting cement line marking periodicity in growth. Castanet et al. (1993) also proposed that the formation of LAGs is likely to be endogenous while still potentially synchronized to environmental conditions.

Cyclical formation of LAGs appears to be a universal phenomenon in vertebrates (Castanet et al., 1993; Klevezal, 1996; Simmons, 1992), and there is evidence for endogenous control (Schauble, 1972; Castanet et al., 1993; Simmons, 1992; Esteban et al., 1999). Bone formation and remodeling rates are hormonally controlled and synchronized to circadian patterns (Simmons, 1992). Parathyroid hormone (PTH), calcitonin, and vitamins A, C, D, and K have been found to influence rates of bone formation and remodeling (Buchanan & Preece, 1991; Narbaitz et al., 1991). Specifically, PTH—which stimulates bone resorption—is secreted in response to serum calcium levels.

Studies have demonstrated seasonal variability in skeletal growth rates, not just in poikilotherms (Schauble, 1972; Snover & Hohn, 2004) but also in endothermic mammals (Klevezal, 1996; Castanet, 2006). These patterns may potentially be evolutionarily related to an increased availability of vitamins A, C, and D, with the onset of spring in temperate climates or the wet season in tropical climates (Buchanan & Preece, 1991; Simmons, 1992). However, there is substantial evidence that the spring surge in growth rates is also under endogenous control, as animals that are maintained in captivity also demonstrate this pattern. Schauble (1972) amputated limbs from the newt, *Notophthalmus viridescens*, at different times of the year and observed the regeneration rates. She found that regeneration rates were significantly higher in the spring or early summer months, followed by summer, late summer, early fall, and winter, respectively. As temperature, light levels,

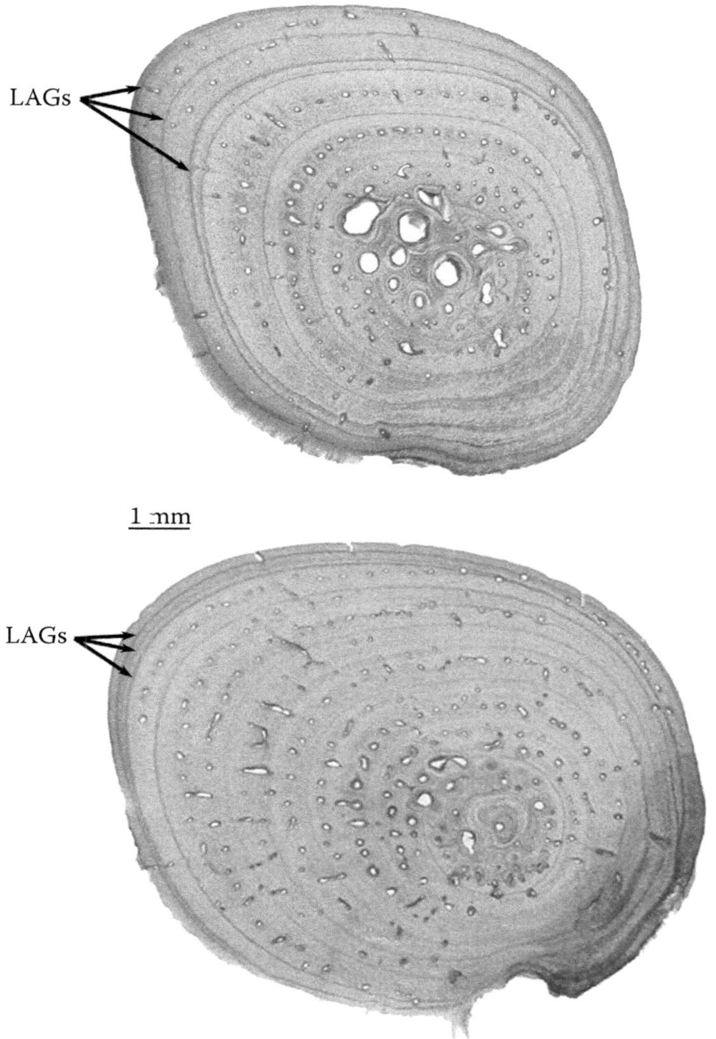

FIGURE 2.1 Cross-sections from humeri of two terrapins (*Malaclemys terrapin*) that have been decalcified and stained with Ehrlich's hematoxylin. Arrows highlight the thin, darkly stained lines of arrested growth (LAGs), and the lightly stained region between LAGs is termed the growth zone and together one LAG and one zone comprise a growth mark. Note how the LAGs are beginning to compress at the outer edge of the lower image. The upper image is from a 15.1-cm straight carapace length (SCL) female, and the lower is from a 16.5-cm SCL female.

and food availability were controlled, these factors could not have played a role in the regeneration rates, suggesting that the results imply the influence of an internal biological rhythm, either endocrine or nonendocrine in nature.

Another line of evidence for seasonal variability in skeletal growth rates is Snover and Hohn's (2004) analysis of bone-growth increments past the last complete LAG in Kemp's ridley humeri relative to stranding date. They found a significant and positive relationship between the amount of new bone deposited after the last LAG and the June–November timeframe. From November to June, the relationship was not significantly different from zero, suggesting that very little new bone deposition occurs during the winter and that LAGs are deposited in the spring for Kemp's ridleys along the U.S. Atlantic coast.

2.2.1.1 Validating Annual Deposition of LAGs

Three common methods can be employed to directly validate the annual deposition of skeletal growth marks: the study of known-age animals, mark-recapture studies, and mark-recapture studies that incorporate fluorescent marking (Castanet, 1994). All three of these methods have been applied to turtles (Castanet & Cheylan, 1979; Klinger & Musick, 1992; Coles et al., 2001; Snover & Hohn, 2004; Curtin, 2006; Snover et al., 2007b). Snover and Hohn (2004) looked at humeri from known-age Kemp's ridley sea turtles (*Lepidochelys kempii*) that had been tagged as hatchlings and released into the wild. The turtles from their study were subsequently recovered as dead strandings and allowed for validation of annual LAG formation and the recognition of an annulus, or diffuse mark rather than a distinct LAG, that represented an annual growth mark. Curtin (2006) used bones from known-age desert tortoises (*Gopherus agassizii*) from mark-recapture studies to test and validate back-calculation methods to account for LAGs lost to resorption in older animals. Snover (2007a) used humeri from dead stranded loggerhead turtles (*Caretta caretta*) that had been previously captured and tagged to validate that carapace length can be back-calculated from the dimensions of earlier LAGs. Castanet and Cheylan (1979) used fluorescent marking to validate that growth marks were annual in Hermann's tortoises (*Testudo hermanni*) and Greek tortoises (*Testudo graeca*). Klinger and Musick (1992) injected wild loggerheads with oxytetracycline and released them. Bone biopsies were taken from turtles recaptured 1 to 2 years later to validate annual LAG formation. A turtle from that same study was found stranded dead 8 years after injection and presented additional validation (Coles et al., 2001).

2.2.1.2 Resorption of LAGs

As bone increases in size during growth, it is constantly remodeled and reshaped (Enlow, 1969). Hard bone tissues cannot grow through internal expansion, but rather they grow by appositional processes (on periosteally derived cortical bone) with the deposition of new tissue on the surface together with endosteal resorption (Enlow, 1969). This process of resorption results in the loss of the innermost (earliest) growth marks and is a serious limitation in estimating age using skeletochronology. While not a serious issue for shorter-lived amphibians and reptiles, it is especially problematic in long-lived turtles, and the problem is noted to be extreme in age-estimate studies of marine turtles (Klinger & Musick, 1995; Zug et al., 1995, 1997, 2002; Parham & Zug, 1997; Zug & Glor, 1998; Snover & Hohn, 2004; Snover et al., 2007b), resulting in the development of several methods of back-calculation to estimate the number of growth marks lost.

Back-calculation techniques in sea turtles rely on the concept that the spatial pattern of the LAGs is representative of the growth of the animal, and to confirm this assumption a correlation must be established between bone dimensions and body size (Hutton, 1986; Klinger & Musick, 1992; Leclair & Laurin, 1996; Snover, 2002; Snover & Hohn, 2004). Using loggerhead turtles, Snover (2007a) demonstrated that the relationship between carapace length and humerus diameter can be used to accurately estimate carapace length at the time of earlier LAG deposition.

Most back-calculation procedures applied to turtles have not been validated and make assumptions about early growth rates (Klinger & Musick, 1995; Zug et al., 1995, 1997, 2002; Parham & Zug, 1997; Zug & Glor, 1998). Curtin (2006) was able to test and validate back-calculation procedures for the desert tortoise using humeri from known-age animals. She tested two methods presented by Parham and Zug (1997), the ranking protocol, and the correction factor methods and found that the correction factor method provided the most accurate age estimates for juveniles and subadults; however, it underestimated adult ages. For adult tortoises, the ranking protocol provided the most accurate estimates.

2.2.1.3 Skeletochronology and Growth Lines on Scutes

For most species of freshwater and terrestrial turtles, age is most commonly estimated from counts of growth lines on the scutes of either the carapace or the plastron (Germano & Bury, 1998; Wilson et al., 2003). This is a powerful technique as, unlike skeletochronology in turtles, it can be applied to living animals and used to understand the age structure of populations. However, many studies that apply this technique do not provide any validation (Castanet & Cheylan, 1979; Wilson et al., 2003) and in a literature review, Wilson et al. (2003) found that of the studies that did attempt validation, 37% were unable to do so. Similarly, Berry (2002) found that even in juvenile desert tortoises, age could not be accurately determined through scute counts alone. Hence, it appears that whereas counting scute growth lines may be a viable method of age estimation in some turtles (i.e., Stone & Babb, 2005), it is not accurate for all turtles and assumptions should not be made that the method is applicable to a given species without validation. While not strictly valid when used in conjunction with each other, skeletochronology and scute growth line counts from dead turtles can serve as supporting evidence of the annual nature of the two methods (Castanet & Cheylan, 1979; Hart & Snover, unpublished data).

Even when scute growth line counts accurately estimate age, an advantage of skeletochronology over scute growth line counts appears with older adult animals. As growth slows to nearly immeasurable rates in older animals, growth lines can no longer be differentiated on scutes (see Wilson et al., 2003, for review), hence only minimum ages can be estimated. However, in histological preparations of bones LAGs can be generally differentiated even in older animals with near cessation of growth (Snover & Hohn, 2004), allowing for estimates of adult growth rates and longevity (Figure 2.1) (Snover, 2002; Snover & Hohn, 2004; Snover et al., 2007b).

2.2.2 Application of Skeletochronology to Turtles

2.2.2.1 Freshwater Turtles

Freshwater turtles were the first turtles to have skeletal growth marks recognized in their long bones. Mattox (1936) noted skeletal growth marks in the long bones of painted turtles, *Chrysemys picta marginata*, and found a correlation between counts of the marks and turtle size. Peabody (1961) and Hammer (1969) documented periosteal cyclical rings in snapping turtles, *Chelydra serpentina*. Suzuki (1963) and Enlow (1969) found them in the slider, *Trachemys scripta*. Hart and Snover (unpublished data) compared skeletochronology preparations of humeri with plastron scute growth line counts to demonstrate the strong comparison of the two techniques in the brackish-water diamondback terrapin (*Malaclemys terrapin*). Counting of growth lines on plastron or carapace scutes remains the primary means of estimating age for freshwater turtles.

2.2.2.2 Terrestrial Turtles

The first study to validate the annual nature of skeletal growth marks was conducted with two species of tortoises. Castanet and Cheylan (1979) used fluorescent marking to validate annual growth marks in Hermann's (*Testudo hermanni*) and Greek (*Testudo graeca*) tortoises. Recently, skeletochronology has been applied to desert tortoises (*Gopherus agassizii*): Curtin (2006) validated the annual nature of the LAGs in humeri from known-age animals and developed correction techniques to estimate the number of LAGs lost to resorption. Similar to the freshwater turtles, growth lines on scutes continue to be a primary means of estimating age in this group of turtles.

2.2.2.3 Marine Turtles

Of all of the turtle groups, skeletochronology has been applied most frequently to marine turtles. The scutes of the plastron and carapace do not retain growth lines like the freshwater and terrestrial turtles (however, see Tucker et al., 2001). Hence, skeletochronology has been the primary means of estimating age and inferring growth rates in these turtles.

To date, skeletochronology has been applied to five of the seven species of marine turtles, the loggerhead (*Caretta caretta*: Zug et al., 1986, 1995; Klinger & Musick, 1992, 1995; Parham & Zug, 1997; Coles et al., 2001; Snover, 2002; Bjorndal et al., 2003; Snover & Hohn, 2004), the leatherback (*Dermochelys coriacea*: Zug & Parham, 1996), the Kemp's ridley (*Lepidochelys kempii*: Zug et al., 1997; Snover & Hohn, 2004; Snover et al., 2007b), the green (*Chelonia mydas*: Bjorndal et al., 1998; Zug & Glor, 1998; Zug et al., 2002), and the olive ridley (*Lepidochelys olivacea*: Zug et al., 2006). The annual deposition of LAGs has been validated for loggerheads (Klinger & Musick, 1992; Coles et al., 2001; Snover & Hohn, 2004) and Kemp's ridleys (Snover & Hohn, 2004).

With the exception of leatherbacks, all of these studies used the humerus bone. Generally, LAGs are most clearly visible in the long bones, and the humerus is ideal as it is easily removed from dead animals and it has muscle insertion scars that create landmarks that allow for the identification of sectioning sites that are consistent (Snover & Hohn, 2004). Humeri of leatherbacks are morphologically different from the hard-shelled turtles, and a high level of vascularization and bone remodeling is characteristic of the leatherback skeleton (Rhodin, 1985). This high level of vascularization may limit the usefulness of long bones to skeletochronology studies. However, Rhodin (1985) documented two wide cyclical growth zones in the periosteal bone of the humerus of an adult female leatherback turtle that suggested the possibility of growth cycles related to migration or nesting patterns (Figure 10 in Rhodin, 1985). Zug and Parham (1996) predicted age at sexual maturity of leatherbacks by skeletochronology based on LAGs found in scleral ossicles; skeletochronology of leatherbacks has also been conducted by Avens and Goshe (unpublished data). However, the possible annual nature of these marks has not been validated, and they may instead simply represent the cyclical result of varying rates of bone deposition and growth related to feeding or migration cycles in this high-metabolism species.

2.3 COMPARATIVE CHONDRO-OSSEOUS DEVELOPMENT IN TURTLES

Form and function are indeed fundamental interdependent strategies of all life. This is especially apparent in the patterns of skeletal growth in turtles as seen in the chondro-osseous development of their appendicular bones, particularly in the patterns of endochondral bone growth. In this section, we review and summarize the two basic patterns of bone growth that occur in turtles and correlate these patterns of skeletal morphology with phylogeny as well as the rate of growth to sexual maturity.

We intend to concentrate this review primarily on the leatherback (*Dermochelys coriacea*), focusing on the morphology and growth of its bones and cartilage. We provide additional detail on its unique vascular cartilage canals that apparently help the leatherback to grow its skeleton rapidly to a large body size. Though related to the hard-shelled chelonioid sea turtles in a number of primitive plesiomorphic features, the leatherback has developed an array of unique derived features that doubtlessly render it the most remarkably specialized turtle in the world.

Unique among living sea turtles in its nearly exclusively pelagic habitat, the leatherback regularly migrates into frigid oceanic waters where it feeds almost exclusively on jellies, diving to incredible depths unequalled by other sea turtles or marine mammals (Eckert & Luginbuhl, 1988; Eckert, 1992; James & Herman, 2001; James et al., 2006). It is well adapted for deep dives, with its hemoglobin, myoglobin, and blood oxygen carrying capacity all greater than in other sea turtles—and more similar to marine mammals (Ascenzi et al., 1984; Lutcavage et al., 1990, 1992). It has a higher metabolic activity than other sea turtles and maintains its body temperature well above surrounding water temperatures, a result of gigantothermy, the ability to use large body size, heightened metabolism, and physiological adaptations to avoid heat loss (Frair et al., 1972; Lutcavage & Lutz, 1986; Paladino et al., 1990; Lutcavage et al., 1992; Penick et al., 1998; James & Mrosovsky, 2004).

Like marine mammals, the leatherback has developed heat retention mechanisms of thickened subcutaneous fibro-adipose tissue, combined with countercurrent heat exchangers in intertwined multiple arterial and venous vascular bundles in its flippers, so as to avoid heat loss in cold waters (Greer et al., 1973). Its body is covered with a corselet of dramatically and uniquely reduced carapacial and plastral shell bones that are reinforced instead with a layer of small irregular intercalated dermal bones (Gervais, 1872). The leatherback skeleton is also unique in having an unusually high degree of neotenic retention of thick cartilages, which are further uniquely specialized through the ingrowth of vascular cartilage canals, a condition totally unlike all other living turtles studied to date (Rhodin et al., 1980, 1981, 1996; Rhodin 1985).

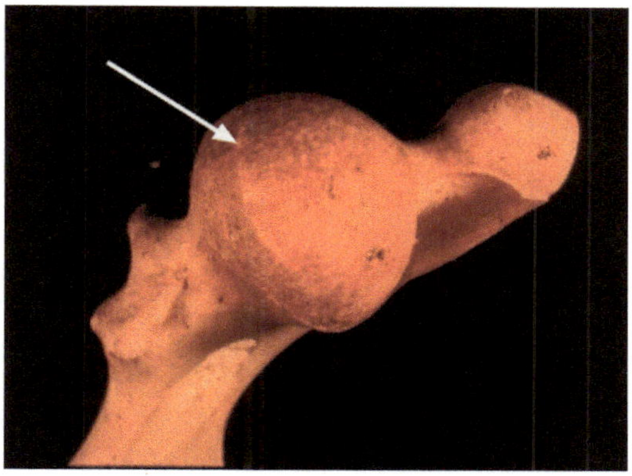

FIGURE 2.2 Proximal humerus, adult *Chelonia mydas*, dry bone preparation, showing smooth articular subchondral surface at the arrow.

Starting out as tiny hatchlings weighing only 30 g and measuring 6 cm in carapace length (CL) (Van Buskirk & Crowder, 1994), leatherbacks grow into the world's largest turtles, with some enormous animals having been recorded at more than 900 kg in weight (Eckert & Luginbuhl, 1988). Leatherbacks reach sexual maturity at about 250 kg with a minimum CL of 120 to 140 cm, about an 8000-fold increase in mass to reach maturity (Márquez, 1990; Van Buskirk & Crowder, 1994). The rate at which that growth is achieved is extremely rapid—much faster than any other reptile (Andrews, 1982)—and similar to the growth rates of some marine mammals.

Based on captive growth studies and patterns of bone growth, Rhodin (1985) previously hypothesized that leatherbacks might reach sexual maturity in as little as 3 to 6 years. More recent skeletochronology work by Zug and Parham (1996) has partially validated that hypothesis and demonstrated that the minimum size at maturity can possibly be obtained as early as 5 to 6 years, with 9 years interpreted as an average minimum age of maturity, and 13 to 14 years considered the average age at maturity. For a turtle of this size, that is phenomenally rapid growth.

How does the leatherback achieve such rapid growth? To understand its function and life strategy, we must look at the underlying form and uniquely specialized structure of its skeletal growth patterns. The work we present here is a review of previous work by Rhodin and colleagues (Rhodin et al., 1980, 1981, 1996; Rhodin, 1985) with new material presented on phylogeny and growth comparisons.

All living turtles studied to date*, except for the leatherback, have bones with articular surfaces that have smooth subchondral joint surfaces, covered by thin avascular cartilage (Figure 2.2). The surface of the subchondral bone is smooth in adult turtles but in growing subadults (and in adults or fossils where the superficial smooth subchondral bone has been worn off), multiple uniformly small holes represent the small metaphyseal vascular channels associated with endochondral bone formation (Rhodin, 1985). None of these very small uniform holes represent vascular channels penetrating into the overlying cartilage.

The leatherback has bones with subchondral articular surfaces that have roughened joint surfaces, with several large holes representing blood vessels penetrating into the thick overlying cartilage from the underlying bone (Figure 2.3) and small holes representing the metaphyseal vascular channels associated with endochondral bone formation. In the longitudinal cross-section of the

* Living turtles studied to date include *Dermochelys coriacea*, *Chelonia mydas*, *Caretta caretta*, *Eretmochelys imbricata*, *Lepidochelys kempii*, *L. olivacea*, *Carettochelys insculpta*, *Podocnemis unifilis*, *Geochelone nigra*, *Macrochelys temminckii*, *Dermatemys mawii*, *Platysternon megacephalum*, *Apalone spinifera*, *Sternotherus odoratus*, and *Chelodina parkeri* (Rhodin, 1985), as well as the genera *Trachemys*, *Homopus*, *Testudo*, *Graptemys*, *Pelusios*, *Chrysemys*, *Emys*, and *Terrapene* (Suzuki, 1963; Haines, 1969).

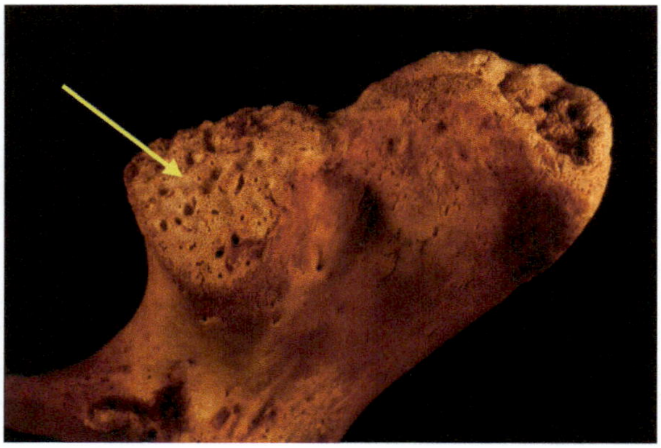

FIGURE 2.3 Proximal humerus, adult *Dermochelys coriacea*, dry bone preparation, showing rough articular subchondral surface at arrow with large holes indicating transphyseal vascular channels penetrating into the cartilage above and small holes indicating small metaphyseal vascular channels associated with subphyseal endochondral bone formation.

proximal humerus joint surface in a fresh bone, regular turtles have thin avascular cartilages (Figure 2.4), whereas leatherbacks have thick and vascularized cartilages with multiple blood vessels coursing through the cartilage (Figure 2.5). Sectioned, preserved adult leatherback bones also show light-colored endochondral bone cones alternating with dark-colored periosteal bone cones in a pattern of minimally remodeled amedullary bone, similar to the pattern seen in marine mammals and unlike other hard-shelled sea turtles (Figure 2.6). The leatherback has many bone growth features that are remarkably similar to marine mammals. Both marine mammals and leatherbacks have epiphyseal cartilaginous vascularization, endochondral and periosteal bone cones, minimally remodeled amedullary bone, and well-vascularized compact bone (Felts & Spurrell, 1965, 1966).

What do we know about actual bone growth patterns in turtles? Prior to the work reviewed here on leatherbacks and other large turtles, studies had only been carried out on small freshwater slider turtles, and their skeletal growth patterns had been assumed to be the pattern typical for all turtles. Work by Suzuki (1963) and Haines (1969) characterized bone growth in small turtles and served as the foundation for work on larger turtles. Rhodin and colleagues (Rhodin, 1985; Rhodin et al., 1996) then investigated the histology of chondro-osseous development in a variety of large turtles, including leatherbacks and giant tortoises and large hard-shelled freshwater and marine turtles, for which details of their bone growth follows.

The appendicular bones of most turtles, including hard-shelled sea turtles, are laid down as cartilaginous anlagen with a diaphyseal periosteal cuff of lamellar cortical bone (Figure 2.7), which is

FIGURE 2.4 Proximal humerus, adult *Caretta caretta*, fresh bone preparation, showing thin avascular joint cartilage at the arrow.

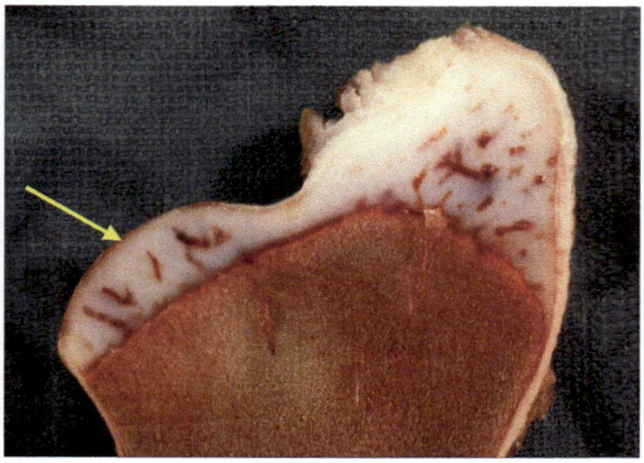

FIGURE 2.5 Proximal humerus, adult *Dermochelys coriacea*, fresh bone preparation, showing thick vascularized joint cartilage at the arrow.

followed rapidly in young post-hatchlings by central diaphyseal cartilaginous cell hypertrophy and calcification. This is followed by vascular ingrowth from the central nutrient artery perforating the mid-diaphyseal periosteal bone cuff (Figure 2.8). This leads to the formation of a central primary ossification center in the mid-diaphysis. In juvenile turtles, this expands toward each metaphysis, while simultaneously cartilage cells in the physeal zone between the epiphysis and metaphysis undergo hypertrophy, calcification, and vascular ingrowth, forming a subphyseal plate that gradually closes, isolating a cone of metaphyseal cartilage that is gradually replaced by bone (Figure 2.9). In subadults, the epiphyseal joint cartilage is relatively thin and avascular with a reasonably smooth underlying subphyseal bone plate (Figure 2.10). Adults have very thin cartilage and very smooth bony subphyseal surfaces.

This pattern of bone growth is typical for all species of living turtles—except the leatherback—such that bone growth proceeds in a unique and specialized pattern. Hatchling leatherback bones are laid down in similar fashion to other hard-shelled turtles (Figure 2.11), and post-hatchlings also undergo initial central diaphyseal vascular ingrowth, leading to a primary diaphyseal ossification center (Figure 2.12). However, at this point further development in the leatherback diverges markedly from other turtles, with rapid ingrowth of vascular canals into the cartilage of the metaphysis, coursing rapidly toward the epiphysis (Figure 2.13). Each of these cartilage canals is associated with a cuff of rapid chondroblastic proliferation and hypertrophied cartilage cells that undergo calcification and rapid ossification.

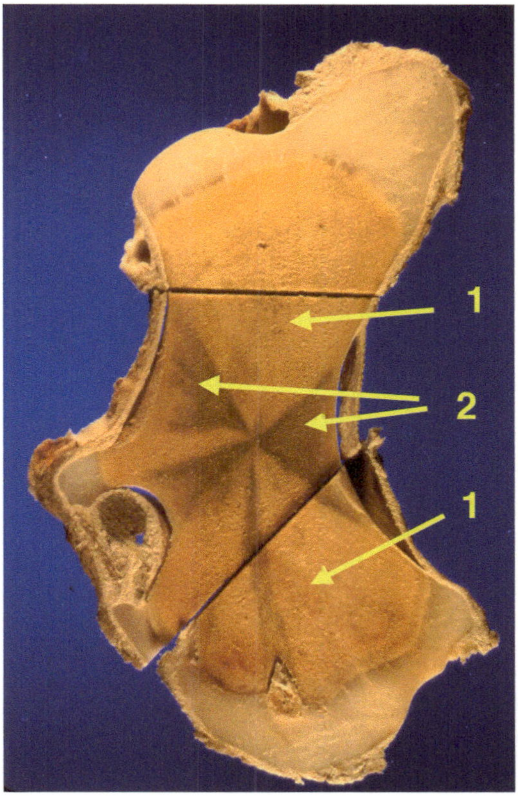

FIGURE 2.6 Humerus, adult *Dermochelys coriacea*, fresh bone preparation, showing (1) light-colored endochondral bone cones and (2) dark-colored periosteal bone cones.

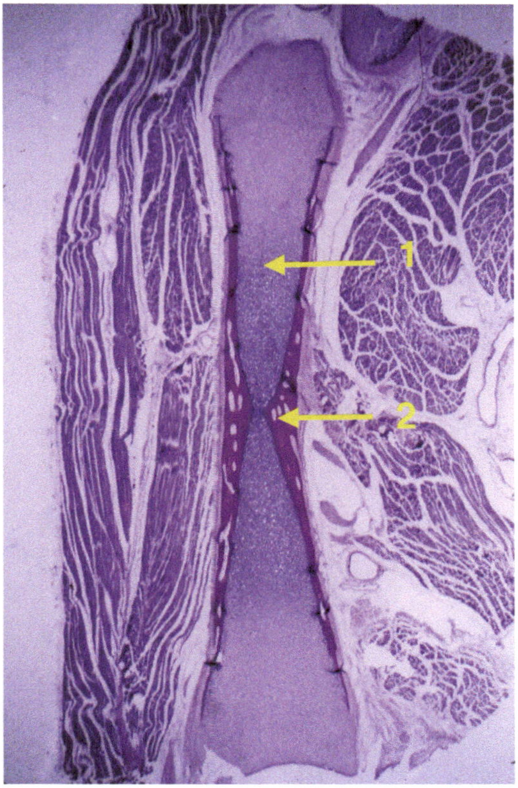

FIGURE 2.7 Ulna of hatchling *Caretta caretta* (SCL = 4.6 cm, H & E stain), showing cartilaginous anlagen and early periosteal cuff of cortical bone at the arrows.

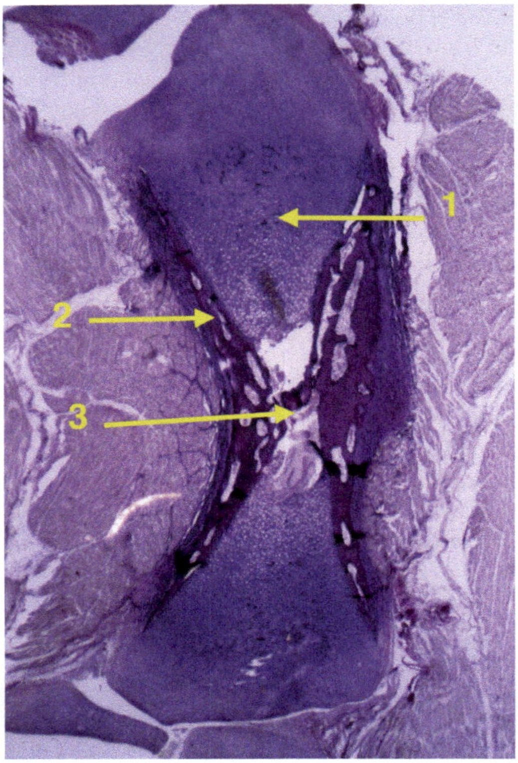

FIGURE 2.8 Radius of hatchling *Caretta caretta* (SCL = 4.4 cm, H & E stain), showing early primary ossification center at the arrow.

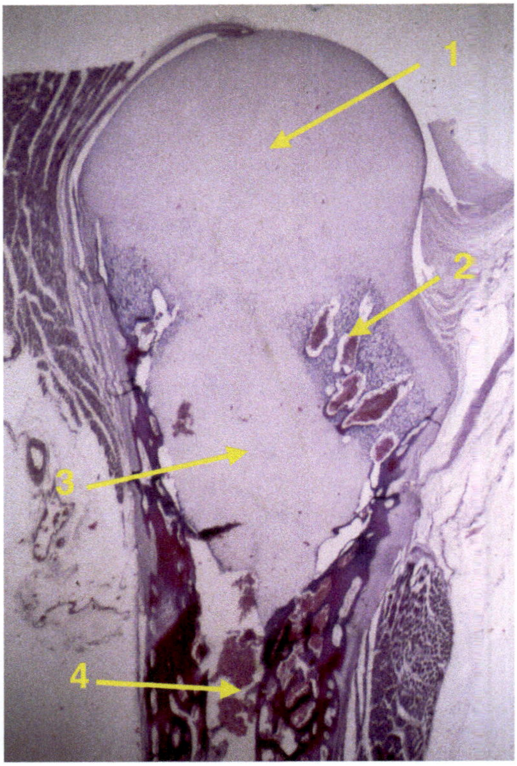

FIGURE 2.9 Proximal femur of juvenile *Caretta caretta* (SCL = 7.0 cm, H & E stain), showing (1) avascular epiphyseal cartilage, (2) vascularized subphyseal ossification plate arising from advancing periosteal ring between metaphysis and epiphysis, (3) avascular metaphyseal cartilage becoming isolated by advancing subphyseal ossification plate, and (4) primary diaphyseal ossification center.

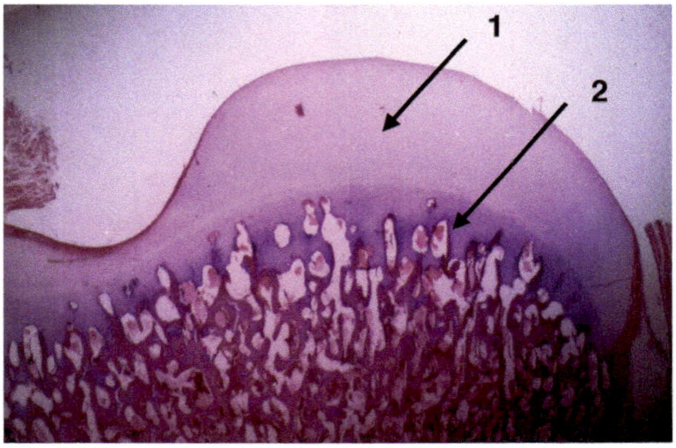

FIGURE 2.10 Proximal humerus of subadult *Lepidochelys kempii* (SCL = 27.5 cm, H & E stain), showing (1) avascular joint cartilage and (2) subphyseal bone plate with advancing calcification and ossification, including small vascular channels associated with subphyseal endochondral ossification.

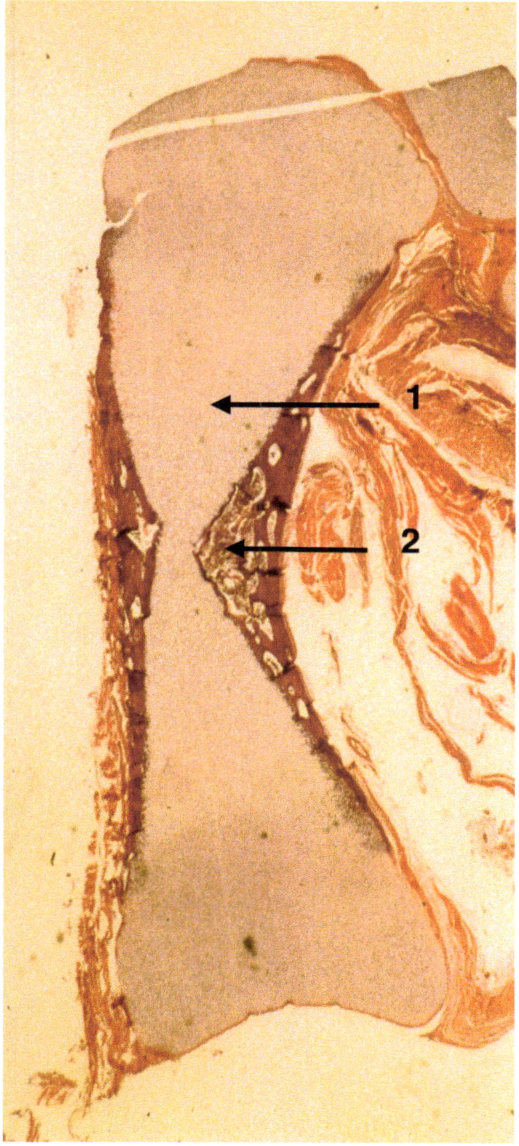

FIGURE 2.11 Metacarpal of hatchling *Dermochelys coriacea* (SCL = 6.5 cm, H & E stain), showing (1) cartilaginous anlagen and (2) early periosteal cuff of cortical bone.

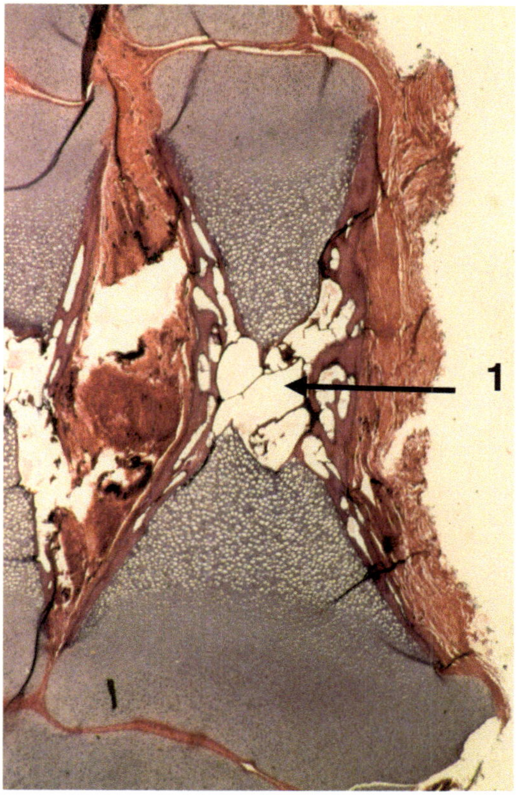

FIGURE 2.12 Metacarpal of juvenile *Dermochelys coriacea* (SCL = 7.0 cm, H & E stain), showing early primary ossification center at the arrow (1).

As ossification proceeds in the metaphysis, vascular cartilage canals penetrate through the subphyseal plate and grow into the thick epiphyseal cartilage under the joint surface (Figure 2.14), leading to the presence of large blood vessels in the joint cartilage (Figure 2.15), which creates the dramatic appearance of blood-filled red vascular canals traversing the bony subphyseal plate and entering deep into the thick white cartilage (Figure 2.5 and Figure 2.16).

The ultrastructural detail of the tips of these vascular canals growing into hatchling cartilage was investigated by Rhodin et al. (1996). Light microscopy of one of those leading vascular buds demonstrates a concentrated active growth cone of specialized chondroclastic tissue boring its way into the cartilage matrix much like a drill with a burr at its tip (Figure 2.17). Below the growth cone tip, the canal is less specialized, filled with vascular channels and hematopoietic cells. A close-up of the growth cone itself located at the tip of the canal shows feeder arterioles that give rise to a cap-like glomerulus of anastomosing large sinusoidal capillaries, drained by venules (Figure 2.18). Between the capillaries and the surrounding cartilage matrix—which shows marginal decreased metachromatic staining indicative of

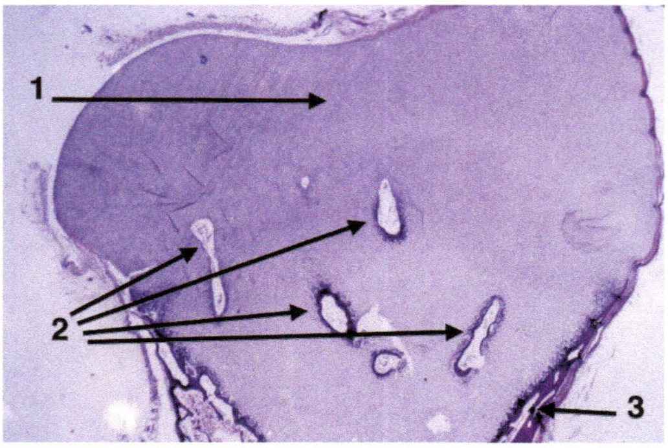

FIGURE 2.13 Proximal humerus of juvenile *Dermochelys coriacea* (SCL = 7.0 cm, H & E stain), showing (1) epiphyseal cartilage that is still avascular, (2) proliferation of vascular canals penetrating the cartilaginous metaphysis, and (3) advancing ring of periosteal bone.

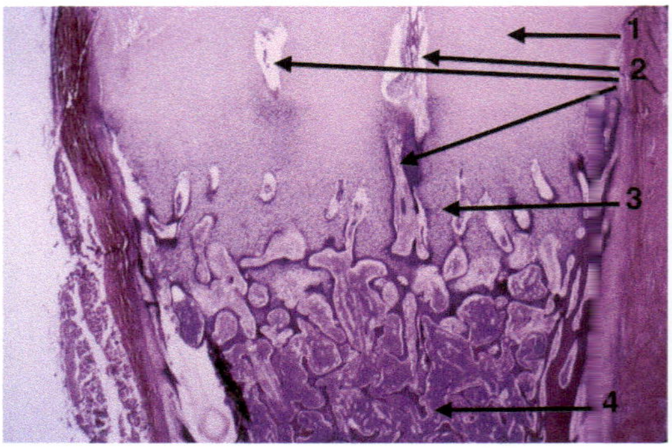

FIGURE 2.14 Metacarpal of larger juvenile *Dermochelys coriacea* (SCL = 40.5 cm, H & E stain), showing (1) epiphyseal joint cartilage, (2) vascular canals penetrating through the (3) subphyseal bone plate into the epipyseal cartilage from the (4) underlying metaphyseal primary ossification center.

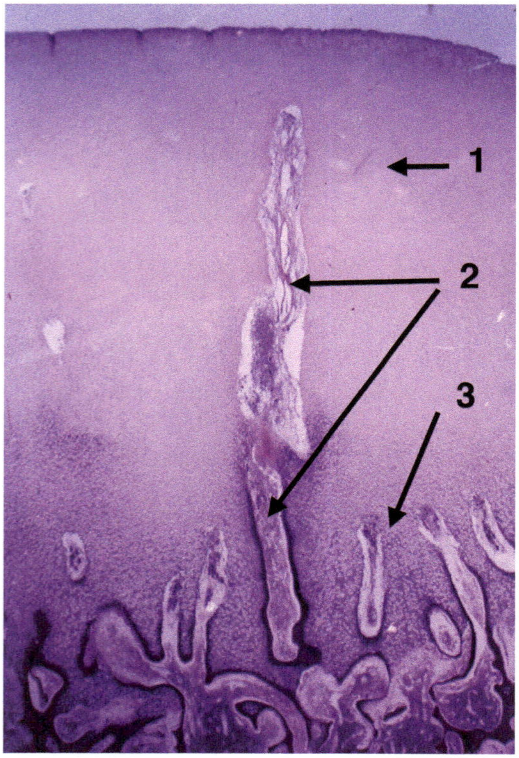

FIGURE 2.15 Proximal humerus of adult *Dermochelys coriacea* (CCL = 135.0 cm, H & E stain), showing (1) epiphyseal cartilage close to the joint surface, a (2) large vascular canal penetrating the (3) subphyseal bone plate into the joint cartilage, surrounded by smaller holes representing metaphyseal vascular channels associated with endochondral bone formation.

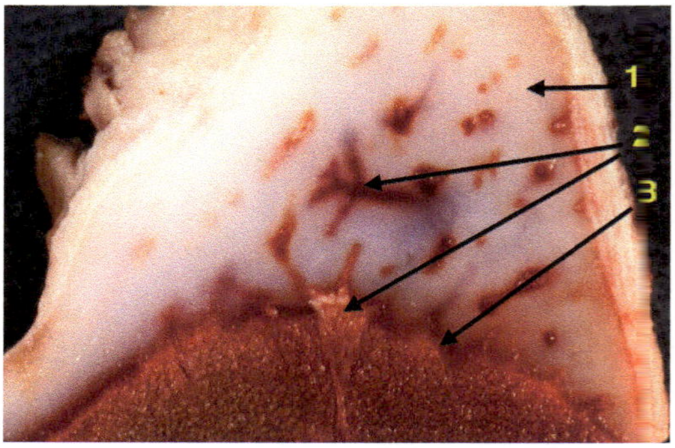

FIGURE 2.16 Humerus of stranded adult *Dermochelys coriacea* (fresh), showing (1) cartilage in the trochanter and (2) large vascular canals penetrating the (3) subphyseal bone plate.

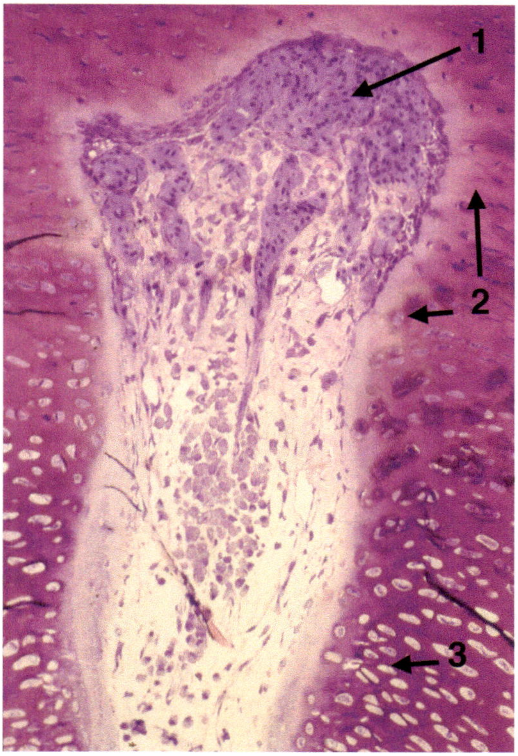

FIGURE 2.17 Proximal humerus of hatchling *Dermochelys coriacea* (SCL = 6.3 cm, H & E stain), showing (1) advancing growth tip of vascular canal bud penetrating into (2) undifferentiated metaphyseal cartilage, causing (3) cartilaginous hypertrophy (see text for detail).

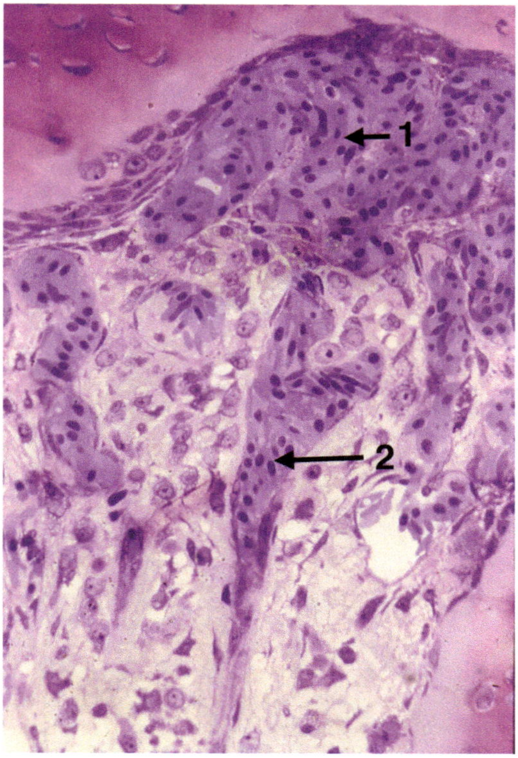

FIGURE 2.18 Humerus of hatchling *Dermochelys coriacea* (H & E stain), detail from Figure 2.17, showing (1) sinusoidal capillaries at the growth tip of the vascular canal and a (2) feeder arteriole (see text for detail).

proteoglycan removal—are layers of small cells that function in the active removal of the cartilage matrix to allow the canal to penetrate into the cartilage.

Using electron microscopy at low magnification, one can further elucidate the ultrastructure of the growth cone tip of the canal (Figure 2.19), which shows the anastomosing sinusoidal capillaries at the tip of the canal penetrating the surrounding cartilage. The cells marked with asterisks along the vascular canal margin are fibroblasts, macrophages, and chondroclasts. Several chondrocytes in the surrounding matrix (marked with dots) demonstrate pyknotic nuclei and signs of cell death.

A close-up view of this area (Figure 2.20) shows a fibroblastic cell within the canal at the very edge of the cartilage, showing a large bulbous cytoplasmic process (at the arrow) penetrating the cartilage matrix. This cell probably actively synthesizes chondrolytic enzymes (Rhodin et al., 1996). The cell next to it is a macrophage with phagolysosomes, active in the removal of cartilage matrix. Also active in the process of cartilage removal are multinucleate chondroclasts (Figure 2.21), where the cell is in such intimate contact with the cartilage matrix that no cell membrane can be discerned at the contact zone marked by red arrows.

Many chondrocytes near the growth cone tip of the canal demonstrate nuclear pyknosis and signs of cell death (Figure 2.22). The cell on the right is a healthy chondrocyte, the one on the left—closer to the vascular canal tip—has died. This cellular death in advance of the leading tip of the canal is probably caused by the release of chondrolytic enzymes by marginal growth cone

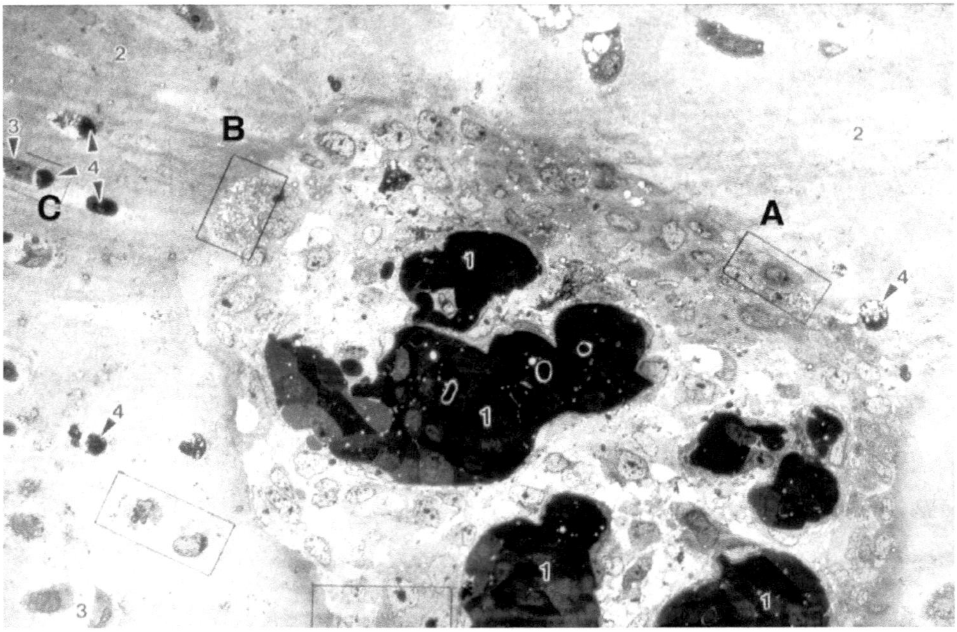

FIGURE 2.19 Humerus of hatchling *Dermochelys coriacea* (electron microscopy), cross-sectional detail of tip of vascular canal bud from Figure 2.17, showing (1) sinusoidal capillaries in tip of canal bud, (2) surrounding cartilaginous matrix, (A, B) rectangles marking chondroclastic cells along the canal edge (see Figure 2.20 and Figure 2.21 for close-ups), (3) hypertrophying chondrocytes, (4) dying chondrocytes, and (C) rectangle marking dying chondrocytes (see Figure 2.22 for close-up) (see text for detail).

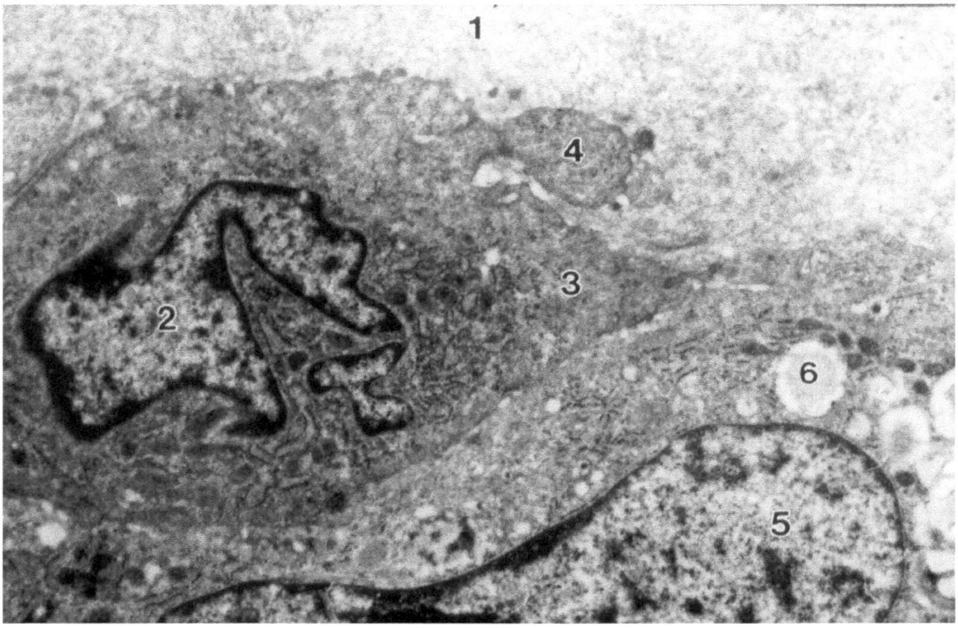

FIGURE 2.20 Humerus of hatchling *Dermochelys coriacea* (electron microscopy), cross-sectional detail of tip of vascular canal bud from Figure 2.17, close-up of rectangle A in Figure 2.19, showing (1) cartilaginous matrix and chondroclastic cells along the canal-cartilage contact zone, including a (2) fibroblast with (3) endoplasmic reticulum and a (4) bulbous cytoplasmic process invading the cartilage, and a (5) macrophage with (6) phagolysosomes.

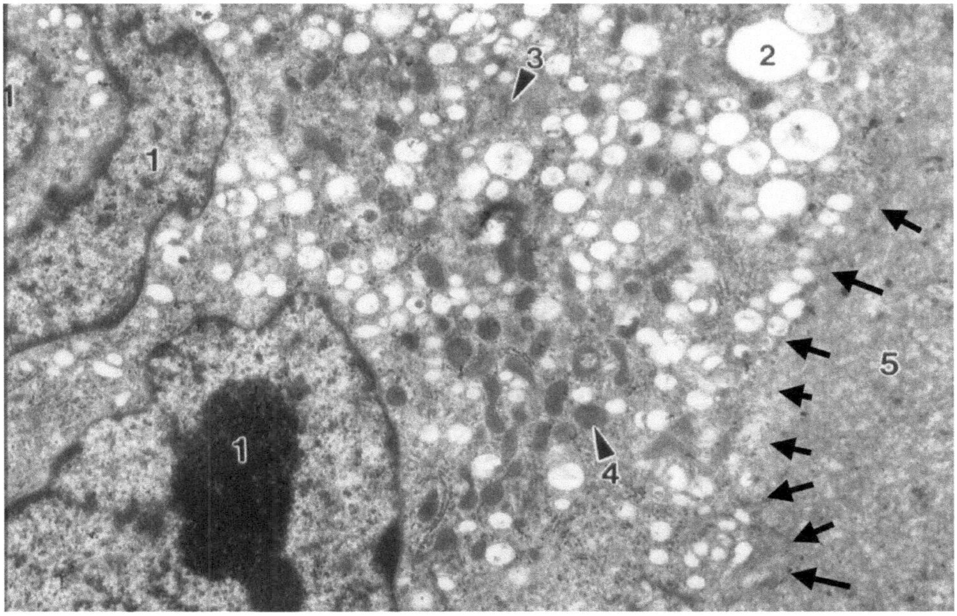

FIGURE 2.21 Humerus of hatchling *Dermochelys coriacea* (electron microscopy), cross-sectional detail of tip of vascular canal bud from Figure 2.17, close-up of rectangle B in Figure 2.19, showing a (1) multinucleate chondroclast with (2) phagolysosomes, (3) primary lysosomes, and (4) mitochondria, in such close contact with the (5) cartilaginous matrix that no cell membrane is visible along the contact zone at the arrows.

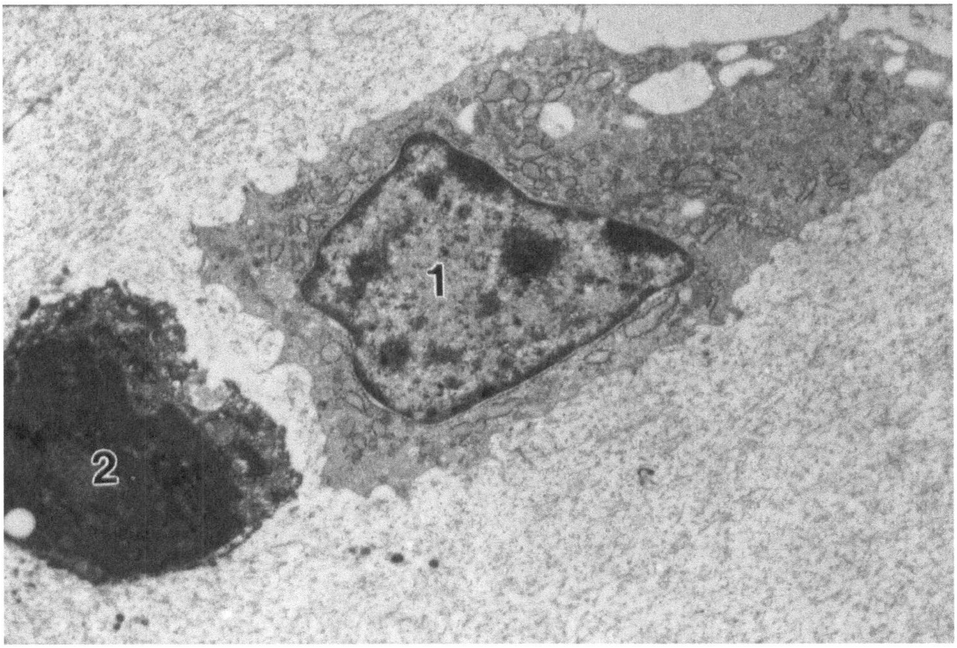

FIGURE 2.22 Humerus of hatchling *Dermochelys coriacea* (electron microscopy), cross-sectional detail of tip of vascular canal bud from Figure 2.17, reversed close-up of rectangle C in Figure 2.19, showing a (1) hypertrophying chondrocyte and a (2) dead chondrocyte closer to the vascular canal (left).

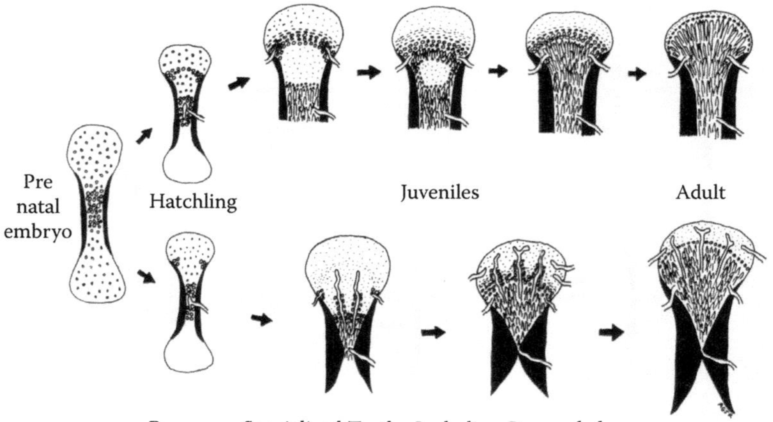

FIGURE 2.23 Schematic representation of the two patterns of skeletal growth that occur in turtles. (From Rhodin, 1985. With permission.) (a) The common pattern in typical turtles is typified by *Caretta*, and the (b) specialized pattern by *Dermochelys*.

fibroblasts (Rhodin et al., 1996). This finding of cellular death near the canal tip had not previously been reported in other vertebrate species with vascular cartilage canals and may be unique to the leatherback (Rhodin et al., 1996). The mechanism probably facilitates more rapid matrix resorption as the vascular canal advances into the cartilage because the dying chondrocytes cannot maintain the proper biochemical environment of the cartilage matrix in advance of the canal tip, which allows for more rapid penetration of the cartilage canal as it grows into the cartilage of the rapidly growing leatherback hatchling. This finding has provided additional support for the hypothesis that leatherbacks have developed cartilaginous vascularization as a specialization related to their rapid growth to a large body size (Rhodin, 1985; Rhodin et al., 1996).

In schematic representation, two patterns of skeletal growth occur in turtles (Figure 2.23, from Rhodin, 1985). The upper pattern, which occurs in hard-shelled sea turtles and all living turtles studied to date (except the leatherback), is characterized by thin avascular cartilage and slow skeletal growth to either small or large body size. The bottom pattern, which occurs only in the leatherback and a few other large extinct marine turtles, is characterized by rapid vascular ingrowth into thick cartilage accompanied by rapid skeletal growth to a large body size.

2.3.1 Implications for Phylogeny

No other living reptile shares the leatherback's pattern of skeletal growth. Though some lizards (notably, large monitor lizards of the family Varanidae) vascularize their cartilages as well, their osteochondral growth mechanisms are different, characterized by perichondral rather than transphyseal ingrowth and the development of secondary ossification centers as opposed to the retention of chondroepiphyses (Haines, 1969). Mammalian and avian patterns of cartilage vascularization are also different from the leatherback and are usually characterized by perichondral rather than transphyseal ingrowth and by the development of secondary ossification centers (Kugler et al., 1979; Moss & Moss-Salentijn, 1983; Kuettner & Pauli, 1983).

Although no living reptiles share the leatherback's specialized chondro-osseous development, what about extinct turtles such as *Stupendemys*, the largest turtle that ever lived? Did its huge body size require vascular cartilages as in the leatherback? Examination of its bones (Figure 2.24) indicates that its joint surfaces had slow-growing avascular smooth cartilaginous epiphyses just like all

FIGURE 2.24 Distal humerus joint surface of adult fossil *Stupendemys geographicus* (SCL midline = 218 cm, straight parasagittal length = 235 cm; Wood, 1976), the largest turtle that ever lived, showing a smooth subchondral surface indicative of overlying thin avascular joint cartilage at the arrow.

other living turtles, and therefore probably reached its huge size slowly.

However, certain extinct sea turtles had skeletal bone structure apparently identical to the leatherback. The giant Cretaceous protostegid turtle *Archelon* had vascular channels penetrating the subphyseal plate from bone into the cartilage above (Figure 2.25), as did the somewhat smaller dermochelyid sea turtle *Corsochelys* (Figure 2.26) and certain other extinct dermochelyid genera, notably *Eosphargis* and *Psephopherus*. Based on these bone structures, these extinct dermochelyid marine turtles would have reached their large size quickly through fast skeletal growth as in the leatherback. However, some dermochelyid turtles did not have vascular cartilages, as in the extinct sea turtle *Desmatochelys*, with typical smooth joint surfaces indicative of thin avascular cartilage (Figure 2.27).

If we look at the phylogenetic distribution of this specialized derived character state of vascularized cartilages on a cladogram of the superfamily Dermochelyidae (Figure 2.28) as hypothesized by Weems (1988), we see that vascularized cartilage would have had to either evolve four separate times or undergo several reversals. Either scenario does not appear likely. Another phylogenetic view of the same genera as presented by Hirayama (1992) would improve the apparent distribution of this character state (Figure 2.29) but would still have it evolving twice, which may also be unlikely.

Instead, perhaps the unique feature of vascularized cartilage serves as a shared derived character uniting these genera into a monophyletic clade, as hypothesized here (Figure 2.30). Clearly, further work on elucidating overall relationships of multiple character states in these genera will be necessary before any definitive conclusions about phylogeny can be reached, but the unique character state of vascularized cartilages offers a potential key to understanding some of their relationships.

2.3.2 Implications for Growth

The physiologic role of these vascularized cartilage canals in leatherbacks is clearly to facilitate rapid skeletal growth to a large body size. How fast do leatherbacks grow and, most importantly, how soon do they reach sexual maturity? Bone growth studies have helped us answer these questions. Based on captive growth studies and these patterns of bone growth, Rhodin (1985) previously hypothesized that leatherbacks might be able to reach sexual maturity in as little as 3 to 6 years.

No leatherback has ever been followed from hatchling to adulthood, so we do not yet know exactly how long that growth actually occurs. However,

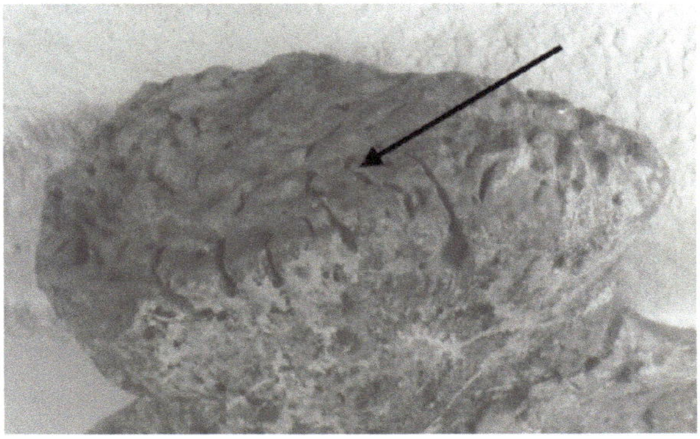

FIGURE 2.25 Proximal carpal bone joint surface of adult fossil *Archelon ischyros*, the largest sea turtle that ever lived (SCL ≈ 190 cm), showing a rough subchondral surface (at the arrow) with large holes indicative of vascular channels extending into overlying thick vascularized joint cartilage.

FIGURE 2.26 Proximal humerus joint surface of fossil *Corsochelys haliniches*, a large sea turtle, showing a rough subchondral surface (at the arrow) with large holes indicative of vascular channels extending into overlying thick vascularized joint cartilage.

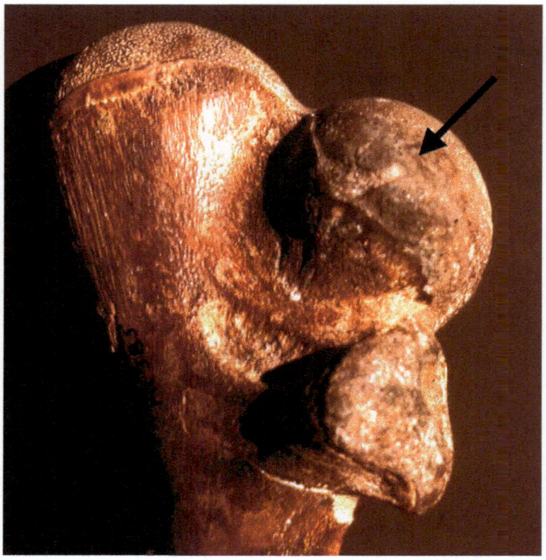

FIGURE 2.27 Proximal humerus joint surface of fossil *Desmatochelys lowi*, a large sea turtle, showing a smooth subchondral surface (at the arrow) indicative of overlying thin avascular joint cartilage.

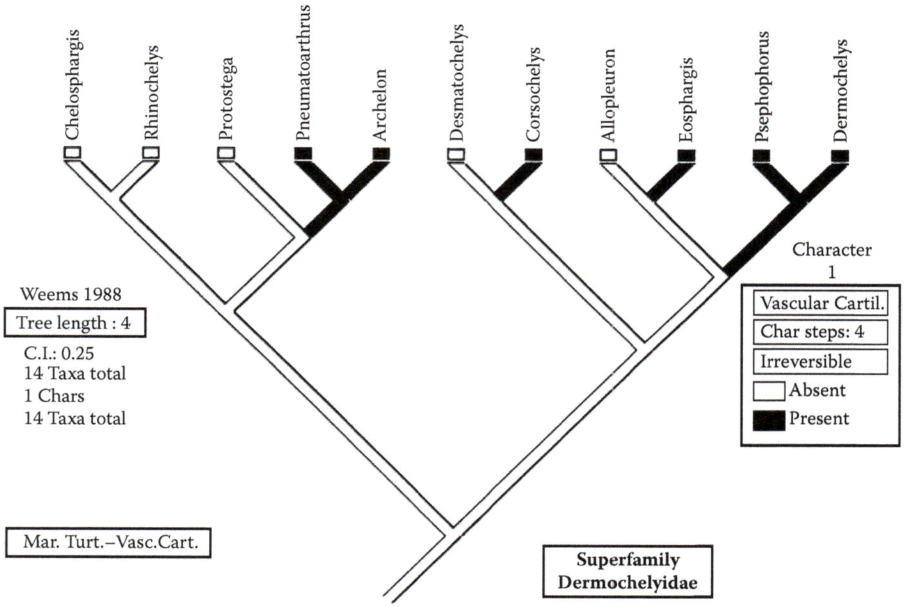

FIGURE 2.28 Cladogram of the superfamily Dermochelyidae as hypothesized by Weems (1988), showing the distribution of the character state of the presence (black branches and squares) or absence (white branches and squares) of vascularized cartilage.

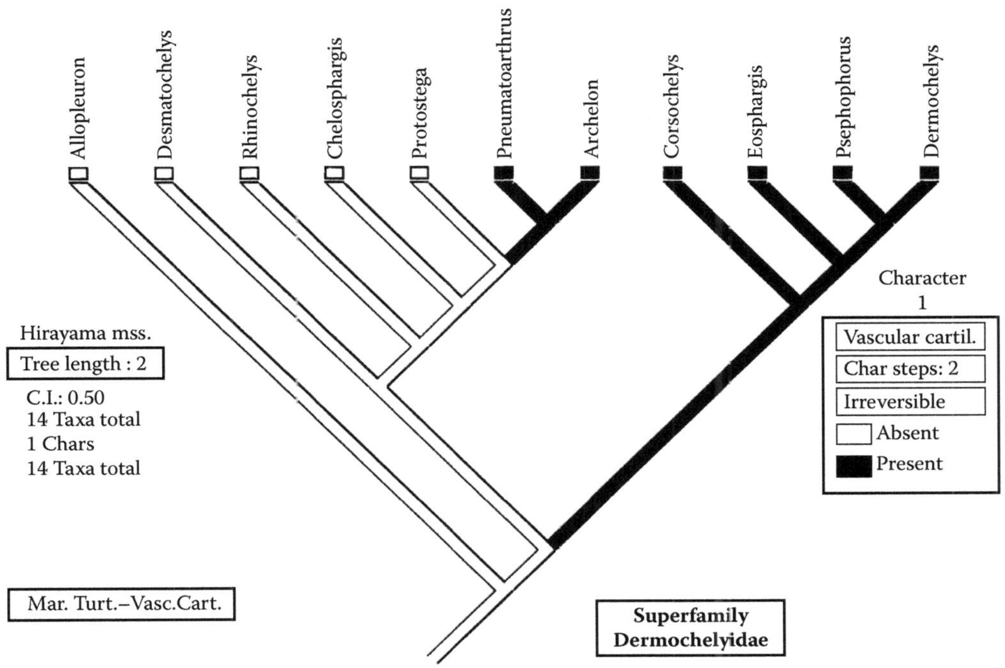

FIGURE 2.29 Cladogram of the superfamily Dermochelyidae as hypothesized by Hirayama (1992), showing the distribution of the character state of the presence (black branches and squares) or absence (white branches and squares) of vascularized cartilage.

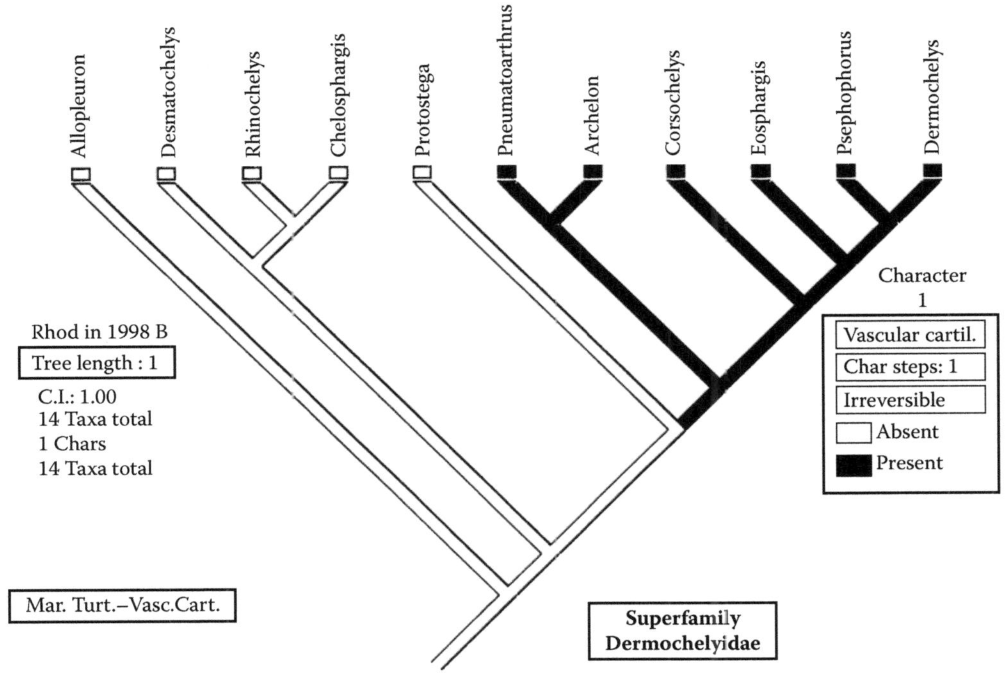

FIGURE 2.30 Cladogram of the superfamily Dermochelyidae as hypothetically developed by Rhodin for the single character state of the presence (black branches and squares) or absence (white branches and squares) of vascularized cartilage. This is an illustrative example only and not a fully predictive phylogeny.

skeletochronology work on scleral ossicles by Zug and Parham (1996) partially validated the earlier hypothesis and demonstrated that the minimum size at maturity could possibly be reached in as little as 5 to 6 years, and averaging about 9 years to the minimum age at maturity, with 13 to 14 years being an average age of maturity of their studied population of Pacific leatherbacks. New recent studies on captive-reared leatherbacks (Jones, unpublished data) and new skeletochronology work using scleral ossicles from Atlantic leatherbacks (Avens & Goshe, unpublished data) combine to suggest that the age at maturity could be as low as 5 to 10 years but also possibly as late as 25 to 30 years. Based on this uncertainty, for the purposes of our analysis we consider possible ages at maturity for leatherbacks of 5, 10, 15, 20, and 25 years. We follow Zug and Parham (1996) and consider 1445 mm curved carapace length (CCL) as the mean size at sexual maturity. Even with this uncertainty in age to maturity, we can still address the question of how fast leatherbacks grow in comparison to other large sea turtles and marine mammals. Are they more similar to other sea turtles or to marine mammals?

To assess how differently leatherbacks might grow in comparison to other chelonians and marine mammals, we reviewed the literature for information on age and size (length) at sexual maturity for numerous species within these groups, focusing on small cetaceans within the marine mammal group. For turtles, the age at sexual maturity is considered the age when the first clutch is laid. For marine mammals, sexual maturity is when the animal is first fertile, as opposed to age at first reproduction. Table 2.1 details the results of the literature review. With regard to turtles, lengths are typically given as the length of the carapace, either straight length or curved. Lengths of cetaceans are measured from the tip of the jaw to the notch of the tail fluke. Hence, the actual lengths are not necessarily comparable between the two lines, but the general trend in growth rates and where leatherbacks fit are demonstrated.

We ran ordinary least-squares regression on the age at sexual maturity and carapace length for all turtles in Table 2.1, using the values of 5, 10, 15, 20, and 25 years as the age at sexual maturity for leatherbacks (five separate regressions were run). For each regression, we determined Cook's distance values, D_i, for each observation $i = 1 \ldots n$, where n is the number of observations (Cook & Weisberg, 1982). These values measure the effect of deleting the ith observation, and observations with larger D values than the rest of the data have unusual leverage and are likely outliers. Fox (1991) suggests that Cook's distance values greater than $4/(n-p-1)$, where p is the number of parameters, should be considered outliers. In our regressions, this critical value is 0.21, and D_i values for leatherbacks at all the ages to sexual maturity we considered were greater than this value and at least three times higher than the rest of the D_i values. Without leatherbacks, a linear regression through all of the data points for turtles was significant and explained 51% of the variability in length (Figure 2.31). Based on this regression, to have growth rates consistent with the rest of the turtles leatherbacks would have to mature on average at 56 years (95% C.I., 36 to 102 years). The histology of their bone growth, captive growth rates (see references in Rhodin, 1985), and the high rates of recovery recorded on nesting beaches in St. Croix (Boulon et al., 1996; Dutton et al., 2005) do not support such delayed maturity.

In comparison with the relationship between the age at sexual maturity and length for small cetaceans, leatherbacks that mature at 5 to 10 years of age would have growth rates similar to these marine mammals, especially considering that length in leatherbacks is only measuring carapace length and not including the length of the head, as in marine mammals. At ages to sexual maturity of 15 to 25 years, growth rates of leatherbacks would not be similar to marine mammals; however, they would still remain well above those of hard-shelled turtles and fall somewhere between growth rates of chelonids and small cetaceans.

The only living reptiles that approach the leatherback in growth rate on a gram-per-day basis are large crocodilians, with *Alligator mississippiensis* growing at an average rate of about 36 g/d to maturity (Andrews, 1982), and giant Galapagos tortoises, *Geochelone nigra*, growing at rates of as much as 47 g/d in captivity (Case, 1978). The hard-shelled marine turtles grow at average rates of

TABLE 2.1
Estimates of Female Age and Size at Sexual Maturity for Various Turtle Species and Small Marine Mammals*

Species	Age at Maturity (years)	Size (mm)	Source
Marine Turtles			
Kemp's ridley turtle (*Lepidochelys kempii*)	10[a]	600 S	Shaver & Wibbels (2007)
Olive ridley turtle (*Lepidochelys olivacea*)	13[b]	600 S	Zug et al. (2006)
Green turtle (*Chelonia mydas*)	35–40[c]	900 S	Balazs & Chaloupka (2004)
Loggerhead turtle (*Caretta caretta*)	30[b]	900 S	Snover (2002)
Pacific leatherback turtle (*Dermochelys coriacea*)	13–14[b]	1445 C	Zug & Parham (1996)
Freshwater Turtles			
Painted turtle (*Chrysemys picta*)	5–6[d]	160–165 S	Iverson & Smith (1993)
Spotted turtle (*Clemmys guttata*)	12–15[c]	103 S	Litzgus & Brooks (1998)
Snapping turtle (*Chelydra serpentina*)	10–12	280–290 S	Iverson et al. (1997)
Snake-necked turtle (*Chelodina rugosa*)	6.5[e]	210 S	Kennett (1996)
Australian snapping turtle (*Elseya dentata*)	13.5[e]	220 S	Kennett (1996)
Blanding's turtle (*Emydoidea blandingii*)	14–20[f]	192–225 S	Congdon et al. (1993)
Wood turtle (*Glyptemys insculpta*)	17–18[c]	185 S	Brooks et al. (1992)
Mud turtle (*Kinosternon subrubrum*)	5.3–7.3[d]	80-85	Iverson (1979)
Musk turtle (*Sternotherus minor*)	8[c]	80 S	Cox et al. (1991)
Mud turtle (*Kinosternon hirtipes*)	6–8[d]	95–100 S	Iverson et al. (1991)
Terrestrial Turtles			
Texas tortoise (*Gopherus berlandieri*)	4–8 (5)[d]	131 C	Hellgren et al. (2000)
Steppe tortoise (*Testudo horsfieldi*)	9–17 (12.6)[d]	124–177 (148) S	Lagarde et al. (2001)
Box turtle (*Terrapene carolina*)	8[d]	150 C	St. Clair (1998)
Ornate box turtle (*Terrapene ornata*)	8[d]	128 C	St. Clair (1998)
Desert tortoise (*Gopherus agassizii*)	26[b]	190 S	Curtin (2006)
Porpoises			
Harbor porpoise (*Phocoena phocoena*)	3.6	1420	Lockyer et al. (2001)
Dall's porpoise (*Phocoenoides dalli*)	3.8–4.4	1720	Ferrero & Walker (1999)
Finless porpoise (*Neophocaena phocaenoides*)	6–9	1350–1450	Shirakihara et al. (1993)
Dolphins			
Spinner dolphin (*Stenella longirostris*)	3.7–5.0	1650	Perrin et al. (1977)
Common dolphin (*Delphinus delphis*)	8	1707–1728	Ferrero & Walker (1995)
Pacific white-sided dolphin (*Lagenorhynchus obliquidens*)	8.3	1775	Ferrero & Walker (1996)
Northern right whale dolphin (*Lissodelphis borealis*)	9.7–10.4	1998–2011	Ferrero & Walker (1993)

(continued)

TABLE 2.1
Estimates of Female Age and Size at Sexual Maturity for Various Turtle Species and Small Marine Mammals (continued)

* For turtles, size is measured as carapace length; S indicates straight and C indicates curved. Age in turtles was estimated by either skeletochronology, growth line marks on scutes, mark-recapture direct (animals tracked throughout their life), mark-recapture indirect (growth curves estimated from growth measurements). For cetaceans, size is measured as length from the tip of the jaw to the notch in the rear fluke. All cetacean ages were determined from counts of dentinal growth layer groups. If the authors reported a range of data with a mean, the range is reported here followed by the mean in parentheses.

[a] Age determined by the observation of turtles nesting that had been raised in captivity for the first year of life, marked, and released.

[b] Age determined by skeletochronology.

[c] Age determined indirectly by analysis of growth records from mark-recapture study.

[d] Age determined by counts of growth lines on scutes.

[e] Age determined by a combination of growth line on plastral scutes and a growth model based on recapture data.

[f] Age determined by a combination of scute growth lines and mark-recapture.

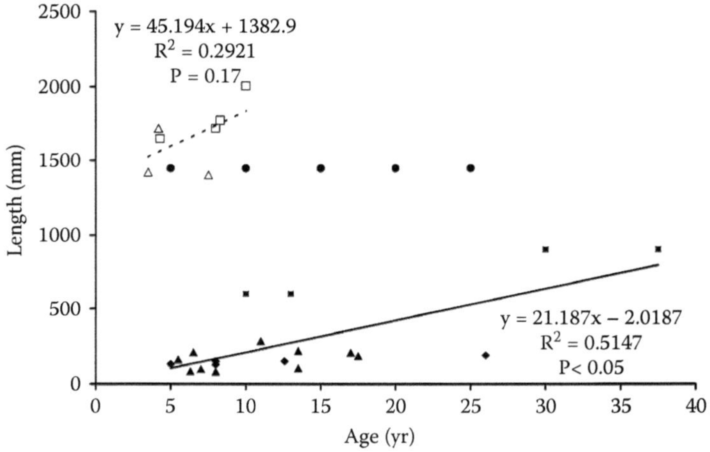

FIGURE 2.31 Plot of age versus size at sexual maturity for turtles and small marine mammals from Table 2.1. Where a range of values was reported, if the authors reported a mean value that value was used in the plot; otherwise, the median value was used. Filled triangles represent freshwater turtles (n = 10), filled diamonds represent terrestrial turtles (n = 5), filled squares represent marine turtles except for leatherbacks (n = 4), and the filled circles represent leatherbacks with the uncertainty in their age at sexual maturity accounted for by considering 5, 10, 15, 20, and 25 years. Based on Cook's distance values, all of these potential ages to sexual maturity for leatherbacks were outliers in the regression of length on age. The solid line represents a linear regression of length on age for the turtles excluding leatherbacks. For marine mammals, the open triangles represent porpoises (n = 3), the open squares represent dolphins (n = 4), and the dotted line is a linear regression between all of the cetacean data.

about 16 to 22 g/d to maturity (Case, 1978; Andrews, 1982), but leatherbacks grow at much higher rates, ranging from a possible 137 g/d if they reach maturity (250 kg) at 5 years of age, 76 g/d if at 9 years, 68 g/d if at 10 years, 46 g/d if at 15 years, 34 g/d if at 20 years, and 27 g/d if at 25 years. Of the smaller marine mammals, fur seals (*Callorhinus ursinus*) grow at about 80 g/d and porpoises (*Phocoena phocoena*) at about 164 g/d, very similar but slightly faster than leatherbacks (Case, 1978).

By comparison, humans grow at the relatively slow rate of about 8 g/dy (Case, 1978). What we have in the leatherback is not only the world's fastest-growing turtle but also its fastest growing reptile. Its bone and cartilage morphology allow that fast growth through its specialized vascularization of rapidly growing cartilage, stimulated no doubt by its heightened metabolism, gigantothermy, and the energetic needs of its pelagic long-distance migratory life.

In summary, the leatherback represents a unique and remarkable life form characterized by specialized and unique morphology. Reptilian in ancestry, testudine in derivation, and chelonioid in affinity, the leatherback has reached a degree of biological specialization unparalleled by other living turtles or reptiles. Its unique specializations make it appear to be converging on the biological regulatory mechanisms evolved by marine mammals; however, given the leatherback's longer evolutionary history, one might more reasonably infer that marine mammals appear to be converging on the leatherback. The leatherback's specialized biology and marvelous adaptations are fertile ground for further studies to increase our understanding of its unique life strategies—providing we can successfully save it from the global threats that are affecting its survival and threatening it with possible extinction, especially in the Pacific (Spotila et al., 1996, 2000; Seminoff et al., 2007). The leatherback represents a rich and unique biological resource whose loss would be both profound and irreplaceable.

REFERENCES

Andrews, R.M., Patterns of growth in reptiles, in *Biology of the Reptilia*, Vol. 13, C. Gans and F.H. Pough (eds.), New York: Academic Press, 1982, 273–320.

Ascenzi, P., Condo, S.G., Bellelli, A., Barra, D., Bannister, W.H., Giardina, B., and Brunori, M., Molecular and functional properties of myoglobin from a marine turtle (*Dermochelys coriacea*), *Bio. Biophys. Acta,* 788, 281–289, 1984.

Balazs, G.H., and Chaloupka, M., Spatial and temporal variability in somatic growth of green sea turtles (*Chelonia mydas*) resident in the Hawaiian Archipelago, *Marine Biology*, 145, 1043–1059, 2004.

Berry, K.H., Using growth ring counts to age juvenile desert tortoises (*Gorpherus agassizii*) in the wild. *Chel. Cons. Biol.*, 4, 416–424, 2002.

Bjorndal, K.A., Bolten, A.B., Bennet, R.A., Jacobson, E.R., Wronski, T.J., Valeski, J.J., and Eliazar, P.J., Age and growth in sea turtles: Limitations of skeletochronology for demographic studies, *Copeia*, 1998, 23–30, 1998.

Bjorndal, K.A., Bolten, A.B., Dellinger, T., Delgado, C., and Martins, H.R., Compensatory growth in oceanic loggerhead sea turtles: Response to a stochastic environment, *Ecology*, 84, 1237–1249, 2003.

Boulon, R.H., Jr., Dutton, P.H., and McDonald, D.L., Leatherback turtles (*Dermochelys coriacea*) on St. Croix, U.S. Virgin Islands: Fifteen years of conservation, *Chel. Cons. Biol.*, 2(2), 141–147, 1996.

Brooks, R.J., Shilton, C.M., Brown, G.P., and Quinn, N.W.S., Body size, age distribution, and reproduction in a northern population of wood turtles (*Clemmys insculpta*), *Can. J. Zool.*, 70, 462–469, 1992.

Buchanan, C.R., and Preece, M.A., Hormonal control of bone growth, in *Bone, Vol 4: Bone Growth-A*, B.K. Hall (ed.), Boca Raton, FL CRC Press, 1991, 53–90.

Case, T.J., On the evolution and adaptive significance of postnatal growth rates in the terrestrial vertebrates, *Quar. Rev. Biol.*, 53, 243–282, 1978.

Castanet, J., Age estimation and longevity in reptiles, *Gerontology*, 40, 174–192, 1994.

Castanet, J., Time recording in bone microstructures of endothermic animals; functional relationships, *C.R. Palevol. (General Palaeontology)*, 5, 629–636, 2006.

Castanet, J., and Cheylan, M., Les marques de croissance des os et des ecailles comme indicateur de l'age chez *Testudo hermanni* et *Testudo graeca* (Reptilia, Chelonia, Testudinidae), *Can. J. Zool.*, 57, 1649–1665, 1979.

Castanet, J., Francillon-Viellot, H., Meunier, F.J., and De Ricqles, A., Bone and individual aging, in *Bone, Vol. 7: Bone Growth-B*, B.K. Hall (ed.), Boca Raton, FL: CRC Press, 1993, 245–283.

Castanet, J.H., Meumier, F.J., and de Ricqles, A., L'enregistrement de la croissance cyclique par le tissu osseux chez les vertebres poikilothermes: Donnees comparatives et essai de synthese, *Bulletin Biologique de la France et de la Belgique*, 111, 183–202, 1977.

Coles, W.C., Musick, J.A., and Williamson, L.A., Skeletochronology validation from an adult loggerhead (*Caretta caretta*), *Copeia*, 2001, 240–242, 2001.

Congdon, J.D., Dunham, A.E., and Van Loben Sels, R.C., Delayed sexual maturity and demographics of Blanding's turtles (*Emydoidea blandingii*): Implications for conservation and management of long-lived organisms, *Cons. Biol.*, 7, 826–833, 1993.
Cook, R.D., and Weisberg, S., Residuals and influence in regression, *Monographs on Statistics and Applied Probability*, New York: Chapman & Hall, 1982.
Cox, W.A., Hazelrig, J.B., Turner, M.E., Angus, R.A., and Marion, K.R., A model for growth in the musk turtle, *Sternotherus minor*, in a north Florida spring, *Copeia*, 1991, 954–968, 1991.
Curtin, A.J., Bone growth strategies and skeletochronological age estimation in Desert tortoise (*Gopherus aagssizii*) populations, Ph.D. dissertation, Drexel University, Philadelphia, 2006.
Dutton, D.L., Dutton, P.H., Chaloupka, M., and Boulon, R.H., Increase of a Caribbean leatherback *Dermochelys coriacea* nesting population linked to long-term nest protection, *Biol. Conserv.*, 126, 186–194, 2005.
Eckert, K.L., and Luginbuhl, C., Death of a giant, *Marine Turtle News.*, 43, 2–3, 1988.
Eckert, S.A., Bound for deep water: Leatherback turtles can pursue their prey half a mile straight down, *Nat. Hist.*, 92, 28–35, 1992.
Eckert, S,A., Eckert, K.L., Ponganis, P., and Kooyman, G.L., Diving and foraging behavior of leatherback sea turtles (*Dermochelys coriacea*), *Can. J. Zool.*, 67, 2834–2840, 1989.
Enlow, D.H., The bone of reptiles, in *Biology of the Reptilia*, C. Gans, d'A. Bellairs, and T.S. Parsons (eds.), London: Academic Press, 45–80, 1969.
Esteban, M., Garcia-Paris, M., Buckley, D., and Castenet, J., Bone growth and age in *Rana saharica*, a water frog living in a desert environment, *Ann. Zool. Fen.*, 36, 53–62, 1999.
Felts, W.J.L., and Spurrell, F.A., Structural orientation and density in cetacean humeri, *Am. J. Anat.*, 116, 171–203, 1965.
Felts, W.J.L., and Spurrell, F.A., Some structural and developmental characteristics of cetacean (Odontocete) radii. A study of adaptive osteogenesis, *Am. J. Anat.*, 118, 103–134, 1966.
Ferrero, R.C., and Walker, W.A., Growth and reproduction of the northern right whale dolphin, *Lissodelphis borealis*, in the offshore waters of the North Pacific Ocean, *Can. J. Zool.*, 71, 2335–2344, 1993.
Ferrero, R.C., and Walker, W.A., Growth and reproduction of the common dolphin, *Delphinus delphis* Linnaeus, in the offshore waters of the North Pacific Ocean, *Fish. Bull.*, 93, 483–494, 1995.
Ferrero, R.C., and Walker, W.A., Age, growth, and reproduction patterns of the Pacific white-sided dolphin (*Lagenorhynchus obliquidens*) taken in the high seas drift nets in the central North Pacific, *Can. J. Zool.*, 74, 1673–1687, 1996.
Ferrero, R.C., and Walker, W.A., Age, growth and reproductive patterns of Dall's porpoise (*Phocoenoides dalli*) in the central North Pacific Ocean, *Mar. Mam. Sci.*, 15, 273-313, 1999.
Fox, J., Regression diagnostics: An introduction, in *Quantitative Applications in the Social Sciences Series No. 79*, Thousand Oaks, CA: Sage Publications, 1991.
Frair, W., Ackman, R.G., and Mrosovsky, N., Body temperature of *Dermochelys coriacea*: Warm turtle from cold water, *Science*, 177, 791–793, 1972.
Francillon-Vieillot, H., de Buffrenil, V., Castenet, J., Geraudie, J., Meunier, F.J., Sire, J.Y., Zylberberg, L., and de Ricqles, A., Microstructure and mineralization of vertebrate skeletal tissues, in *Skeletal Biomineralization: Patterns, Processes and Evolutionary Trends, Vol. 1*, J.G. Carter (ed.), New York: Van Nostrand Reinhold, 1990, 471–530.
Germano, D.J., and Bury, R.B., Age determination in turtles: Evidence of annual deposition of scute rings, *Chel. Cons. Biol.*, 3, 123–132, 1998.
Gervais, P., Ostéologie du Sphargis luth (*Sphargis coriacea*), *Memoires des Nouvelles Archives du Museum d'Histoire Naturelle, Paris*, 8, 199–228, 1872.
Greer, A.E., Lazell, J.D., Jr., and Wright, R.M., Anatomical evidence for a counter-current heat exchanger in the leatherback turtle (*Dermochelys coriacea*), *Nature*, 244, 181, 1973.
Haines, R.W., Epiphyses and sesamoids, in *Biology of the Reptilia, Vol. 1*, C. Gans, d'A. Bellairs, and T.S. Parsons (eds.), London: Academic Press, 1969, 81–115.
Hammer, D.A., Parameters of a marsh snapping turtle population, Lacreek Refuge, South Dakota, *J. Wildlife Mgmt.*, 33, 995–1005, 1969.
Hellgren, E.C., Kazmaier, R.T., Ruthven, D.C., and Synatzske, D.R., Variation in tortoise life history: Demography of *Gopherus berlandieri*, *Ecology*, 81, 1297–1310, 2000.
Hirayama, R., Humeral morphology of chelonioid sea-turtles; its functional analysis and phylogenetic implications, *Bull. Hobetsu Museum*, 8, 17–57, 1992.
Hutton, J.M., Age determination of living Nile crocodiles from the cortical stratification of bone, *Copeia*, 1986, 332–341, 1986.

Iverson, J.B., Reproduction and growth of the mud turtle, *Kinosteron subrubrum* (Reptilia, Testudines, Kinosternidae), in Arkansas, *J. Herpetol.*, 13, 105–111, 1979.

Iverson, J.B., Barthelmess, E.L., Smith, G.R., and de Rivera, C.E., Growth and reproduction in the mud turtle *Kinosternon hirtipes* in Chihuahua, Mexico, *J. Herpetol.*, 25, 64–72, 1991.

Iverson, J.B., Higgins, H., Sirulnik, A., and Griffiths, C., Local and geographic variation in the reproductive biology of the snapping turtle (*Chelydra serpentine*), *Herpetologica*, 53, 96–117, 1997.

Iverson, J.B., and Smith, G.R., Reproductive ecology of the painted turtle (*Chrysemys picta*) in the Nebraska sandhills and across its range, *Copeia*, 1993, 1–21, 1993.

James, M.C., and Herman, T.B., Feeding of *Dermochelys coriacea* on medusae in the northwest Atlantic, *Chel. Cons. Biol.*, 4, 202–205, 2001.

James, M.C., and Mrosovsky, N., Body temperatures of leatherback turtles (*Dermochelys coriacea*) in temperate waters off Nova Scotia, Canada, *Can. J. Zool.*, 82, 1302–1306, 2004.

James, M.C., Davenport, J., and Hays, G.C., Expanded thermal niche for a diving vertebrate: A leatherback turtle diving into near-freezing water, *J. Exp. Mar. Biol. Ecol.*, 335, 221–226, 2006.

Kennett, R., Growth models for two species of freshwater turtle, *Chelodina rugosa* and *Elseya dentate*, from the wet-dry tropics of northern Australia, *Herpetologica*, 52, 383–395, 1996.

Klevezal, G.A., *Recording Structures of Mammals: Determination of Age and Reconstruction of Life History*, Rotterdam: A.A. Balkema, 1996.

Klinger, R.C., and Musick, J.A., Annular growth layers in juvenile loggerhead turtles (*Caretta caretta*), *Bull. Mar. Sci.*, 51, 224–230, 1992.

Klinger, R.C., and Musick, J.A., Age and growth of loggerhead turtles (*Caretta caretta*) from Chesapeake Bay, *Copeia*, 1995, 204–209, 1995.

Kuettner, K.E., and Pauli, B.U., Vascularity of cartilage, in *Cartilage. Vol. I. Structure, Function, and Biochemistry*, B.K. Hall, (ed.), New York: Academic Press, 281–312, 1983.

Kugler, J.H., Tomlinson, A., Wagstaff, A., and Ward, S.M., The role of cartilage canals in the formation of secondary centres of ossification, *J. Anat.*, 129, 493–506, 1979.

Lagarde, F., Bonnet, X., Henen, B.T., Corbin, J., Nagy, K.A., and Naulleau, G., Sexual size dimorphism in steppe tortoises (*Testudo horsfieldi*): Growth, maturity, and individual variation, *Can. J. Zool.*, 79, 1433–1441, 2001.

Leclair, R., Jr., and Laurin, G., Growth and body size in populations of mink frogs *Rana septentrionalis* from two latitudes, *Ecography*, 19, 296–304., 1996.

Litzgus, J.D., and Brooks, R.J., Growth in a cold environment: body size and sexual maturity in a northern population of spotted turtles, *Clemmys guttata*, *Can. J. Zool.*, 76, 773–782, 1998.

Lockyer, C., Heide-Jørgensen, M.P., Jensen, J., Kinze, C.C., and Buus Sørensen, T., Age, length and reproductive parameters of harbour porpoises *Phocoena phocoena* (L.) from West Greenland, *ICES J. Mar. Sci.*, 58, 154–162, 2001.

Lutcavage, M., and Lutz, P.L., Metabolic rate and food energy requirements of the leatherback sea turtle, *Dermochelys coriacea*, *Copeia*, 1986, 796–798, 1986.

Lutcavage, M.E., Bushnell, P.G., and Jones, D.R., Oxygen transport in the leatherback sea turtle *Dermochelys coriacea*, *Physiol. Zool.*, 63, 1012–1024, 1990.

Lutcavage, M.E., Bushnell, P.G., and Jones, D.R., Oxygen stores and aerobic metabolism in the leatherback sea turtle, *Can. J. Zool.*, 70, 348–351, 1992.

Márquez, M.R., Sea turtles of the world. An annotated and illustrated catalogue of sea turtle species known to date, *FAO Fisheries Syn.*, 11(125), 1–81, 1990.

Mattox, N.T., Annular rings in the long bones of turtles and their correlation with size, *Trans. Il. State Acad. Sci.*, 28, 255–256, 1936.

Moss, M.L., and Moss-Salentijn, L., Vertebrate cartilages, in *Cartilage. Vol. I. Structure, Function, and Biochemistry*, B.K. Hall (ed.), New York: Academic Press, 1983, 1–30.

Narbaitz, R., Effects of vitamins A, C, D, and K on bone growth, mineralization, and resorption, in Bone, *Vol 4: Bone Metabolism and Mineralization*, B.K. Hall (ed.), Boca Raton, FL: CRC Press, 141–170, 1991.

Paladino, F.V., O'Connor, M.P., and Spotila, J.R., Metabolism of leatherback turtles, gigantothermy, and thermoregulation of dinosaurs. *Nature*, 344, 858–860, 1990.

Parham, J.F., and Zug, G.R., Age and growth of loggerhead sea turtles (*Caretta caretta*) of coastal Georgia: An assessment of skeletochronological age-estimates, *Bull. Mar. Sci.*, 61, 287–304, 1997.

Peabody, F.E., Annual growth zones in living and fossil vertebrates, *J. Morphol.*, 108, 11–62, 1961.

Penick, D.N., Spotila, J.R., O'Connor, M.P., Steyermark, A.C., George, R.H., Salice, C.J., and Paladino, F.V., Thermal independence of muscle tissue metabolism in the leatherback turtle, *Dermochelys coriacea*, *Comp. Biochem. Physiol.*, 120A, 399–403, 1998.

Perrin, W.F., Holts, D.B., and Miller, R.B., Growth and reproduction of the eastern spinner dolphin, a geographical form of *Stenella longirostris* in the eastern tropical Pacific, *Fish. Bull.*, 725–750, 1977.

Rhodin, A.G.J., Comparative chondro-osseous development and growth of marine turtles, *Copeia*, 1985, 752–771, 1985.

Rhodin, A.G.J., Ogden, J.A., and Conlogue, G.J., Preliminary studies on skeletal morphology of the leatherback turtle, *Mar. Turtle News.*, 16, 7–9, 1980.

Rhodin, A.G.J., Ogden, J.A., and Conlogue, G.J., Chondro-osseous morphology of *Dermochelys coriacea*, a marine reptile with mammalian skeletal features, *Nature*, 290, 244–246, 1981.

Rhodin, J.A.G., Rhodin, A.G.J., and Spotila, J.R., Electron microscopic analysis of vascular cartilage canals in the humeral epiphysis of hatchling leatherback turtles, *Dermochelys coriacea*, *Chel. Cons. Biol.*, 2, 250–260, 1996.

St. Clair, R.C., Patterns of growth and sexual size dimorphism in two species of box turtles with environmental sex determination, *Oecologia*, 115, 501–507, 1998.

Schauble, M.K., Seasonal variation of newt forelimb regeneration under controlled environmental conditions, *J. Exp. Zool.*, 181, 281 –286, 1972.

Seminoff, J.A., Paladino, F.V., and Rhodin, A.G.J., Refocusing on leatherbacks: Conservation challenges and signs of success, *Chel. Cons. Biol.*, 6, 1–6, 2007.

Shaver, D.J., and Wibbels, T., Head-starting ridley sea turtles, in *Biology and Conservation of Ridley Sea Turtles*, P.T. Plotkin (ed.), Baltimore, MD: Johns Hopkins University Press, 2007, 297–323.

Shirakihara, M., Takemura, A., and Shirakihara, K., Age, growth and reproduction of the finless porpoise, *Neophocaena phocaenoides*, in the coastal waters of western Kyushu, Japan, *Mar. Mam. Sci.*, 9, 392–406, 1993.

Simmons, D.J., Circadian aspects of bone biology, in *Bone, Vol 6: Bone Growth-A*, B.K. Hall (ed.), Boca Raton, FL: CRC Press, 1992, 91–128.

Smirina, E.M., Age determination and longevity in amphibians, *Gerontology*, 40, 133–146, 1994.

Snover, M.L., Growth and ontogeny of sea turtles using skeletochronology: Methods, validation, and application to conservation, Ph.D. diss., Duke University, Durham, NC, 2002.

Snover, M.L., and Hohn, A.A., Validation and interpretation of annual skeletal marks in loggerhead (*Caretta caretta*) and Kemp's ridley (*Lepidochelys kempii*) sea turtles, *Fish. Bull.*, 102, 682–692, 2004.

Snover, M.L., Avens, L., and Hohn, A.A., Back-calculating length from skeletal growth marks in loggerhead sea turtles *Caretta caretta*. *Endang. Species Res.*, 3, 95–104, 2007a.

Snover, M.L., Hohn, A.A., Crowder, L.B., and Heppell, S.S., Age and growth in Kemp's ridley sea turtles: Evidence from mark-recapture and skeletochronology, in *Biology and Conservation of Ridley Sea Turtles*, P.T. Plotkin (ed.), Baltimore, MD: Johns Hopkins University Press, 89–105, 2007b.

Spotila, J.R., Dunham, A.E., Leslie, A.J., Steyermark, A.C., Plotkin, P.T., and Paladino, F.V., Worldwide population decline of *Dermochelys coriacea*: Are leatherback turtles going extinct?, *Chel. Cons. Biol.*, 2, 209–222, 1996.

Spotila, J.R., Reina, R.D., Steyermark, A.C., Plotkin, P.T., and Paladino, F.V., Pacific leatherback turtles face extinction, *Nature*, 405, 529–530, 2000.

Stone, P.A., and Babb, M.E., A test of the annual growth line hypothesis in *Trachemys scripta elegans*, *Herpetologica*, 61, 409–414, 2005.

Suzuki, H.K., Studies on the osseous system of the slider turtle, *Ann. N.Y. Acad. Sci.*, 109, 351, 1963.

Tucker, A.D., Broderick, D., and Kampe, L., Age estimation of *Eretmochelys imbricata* by schlerochronology of carapacial scutes, *Chel. Cons. Biol.*, 4, 219–222, 2001.

Van Buskirk, J., and Crowder, L.B., Life-history variation in marine turtles, *Copeia*, 1994, 66–81, 1994.

Weems, R.E., Paleocene turtles from the Aquia and Brightseat formations, with a discussion of their bearing on sea turtle evolution and phylogeny, *Proc. Biolo. Soc. Wash.*, 101, 109–145, 1988.

Wilson, D.S., Tracy, C.R., and Tracy, C.R., Estimating age of turtles from growth rings: A critical evaluation of the technique, *Herpetologica*, 59, 178–194, 2003.

Wood, R.C., *Stupendemys geographicus*, the world's largest turtle, *Breviora*, 436, 1–31, 1976.

Zug, G.R., Balazs, G.H., and Wetherall, J.A., Growth in juvenile loggerhead seaturtles (*Caretta caretta*) in the north Pacific pelagic habitat, *Copeia*, 1995, 484–487, 1995.

Zug, G.R., Balazs, G.H., Wetherall, J.A., Parker, D.M., and Murakawa, S.K.K., Age and growth of Hawaiian green sea turtles (*Chelonia mydas*): An analysis based on skeletochronology, *Fish. Bull.*, 100, 117–127, 2002.

Zug, G.R., Chaloupka, M., and Balazs, G.H., Age and growth in olive ridley seaturtles (*Lepidochelys olivacea*) from the North-central Pacific: A skeletochronological analysis, *Mar. Ecol.*, 27, 263–273, 2006.

Zug, G.R., and Glor, R.E., Estimates of age and growth in a population of green sea turtles (*Chelonia mydas*) from the Indian River lagoon system, Florida: A skeletochronological analysis, *Can. J. Zool.*, 76, 1497–1506, 1998.

Zug, G.R., Kalb, H.J., and Luzar, S.J., Age and growth in wild Kemp's ridley sea turtles *Lepidochelys kempii* from skeletochronological data, *Biol. Cons.*, 80, 261–268, 1997.

Zug, G.R., and Parham, J.F., Age and growth in leatherback sea turtles, *Dermochelys coriacea* (Testudines, Dermochelyidae): A skeletochronological analysis, *Chel. Cons. Biol.*, 2, 244–249, 1996.

Zug, G.R., Wynn, A.H., and Ruckdeschel, C., Age determination of loggerhead sea turtles, *Caretta caretta*, by incremental growth marks in the skeleton, *Smith. Contr. Zool.*, 427, 1986.

3 Evolution and Structure of the Turtle Shell

Peter C.H. Pritchard

When you write to Linn., next, pray talk to him about tortoises. There are tortoises whose shells are always open *behind* and *before* 'apertura testae anterior,' as he says himself, 'pro capite et brachiis; posterior pro cauda et femoribus.' These apertures are supported, as it were, by pillars on each side and can never be closed. Of such construction is the shell of Mrs. Snooke's present living tortoise, Timothy. But then there are tortoises whose under shell has a cardo, an hinge, about the middle of their bellies, commanding one lid or flap forward, and one lid backward (like the double-lidded snuff-boxes) which when shut conceal the head and legs and tail of the reptile entirely, and keep out all annoyances. Two such (very small they were) Mrs. Snooke had formerly; and the shells still lie in her room over the hall.

In a letter from Gilbert White to his brother John White, dated September 26, 1774, urging him to tell Linnaeus a thing or two about tortoises

CONTENTS

3.1 ORIGINS

The body form of a turtle, with the shoulder girdles enclosed between the ribs and the plastron, is so peculiar that it could easily be used to justify a fundamental dichotomous division of vertebrates into "turtles" (including tortoises, softshells, and so on) and "non-turtles" (giraffes, lampreys, salamanders, vultures, and so on). Agassiz (1857) was one of the earliest to contemplate this strange body plan, but as a pre-evolutionist his explanation bordered upon the mystical:

> There is hardly any other type in the whole animal kingdom, in which the direct intervention of thought, as the first cause of its characteristic features, can be so fully and so easily illustrated as in the order of Testudinata. In the first place, these animals are so peculiar in their form and in their structure, that they strike, at first sight, every observer as belonging almost to another creation. They have been represented as inverted Vertebrata; and the peculiarity in the position and connection of their limbs has been so magnified, even to the rank of a class character, that very special conditions would seem necessary for their existence; and yet they are so extensively scattered upon the whole surface of the globe, among other animals of entirely different form and structure, upon land, in the fresh waters, and in the ocean, that, unless it can be shown that, besides its known properties, matter possesses also a turtle-making property, it must be granted that there are special thoughts expressed both in their structure and in their forms, and that the plan to which they belong, notwithstanding their striking differences, must have been designed and executed by a thinking being.

Ironically, in addition to his profound interest in turtles Agassiz was also an exceptional embryologist. His engravings of turtle embryos at all stages of development are among the best ever, and his zeal in acquiring turtle eggs at the earliest stages of development is the subject of a brilliant essay, "Turtle Eggs for Agassiz" (Sharp, 1916). Ontogeny, we hope, repeats phylogeny. Twentieth-century authors (Ruckes, 1929; Walker, 1947) have, like Agassiz, looked to ontogeny as revealed by embryology to solve the riddles of turtle evolution. As Zangerl (1969) wrote:

> Ruckes … showed that a great proliferation develops in the mid-dorsal region of the embryonic body, and that this Anlage of the dorsal disc spreads rapidly in a peripheral direction. The cartilaginous ribs do not remain beneath the dermis, but enter and grow longitudinally within the dorsal disc. The rate of growth of this system notably exceeds the rate of growth of the rest of the embryo with the result that the girdle elements (which remain relatively stationary—an observation confirmed by Walker, 1947)—become dorsally and laterally overgrown.

It would be satisfying if a range of really early chelonian fossils were available to confirm and document this transition but, alas, missing links remain missing. The fossil record of turtles is astonishing in its complexity and in the frequency with which genuinely novel offshoots are discovered, even within the relatively conservative pleurodire suborder (Gaffney et al., 2006). But the earliest turtles—the Proganochelyidae of the Triassic—were fully formed, differing from certain extant types mostly in technical osteologic details, and their distribution, which included what are now Germany, Greenland, and Thailand, suggests that turtles attained a worldwide distribution very early in their evolution. As to how they got there, both morphologically and geographically, Zangerl once again sums up the situation succinctly: "The fossil record, at present, provides no clues."

If a living *Proganochelys* were to cross the path of the average *Homo sapiens*, the man would dismiss the animal as a slightly odd-looking snapping turtle; the giants, the truly weird species, and other instantly noteworthy chelonians of the past—*Stupendemys*, *Drazinderetes*, *Meiolania*, *Archelon*, *Sinemys gamera*, *Kallokibotion*, *Nahnsiungchelys*—came much later. The heavily sculptured and tuberculate dorsal scutes, the long tail, and the serrate carapace margin of *Proganochelys* would not look unfamiliar to those on the Suwannee or the Apalachicola who had seen *Macrochelys*. Already in this earliest known turtle, the carapace scutes included a vertebral series, a set of costals on each side, a ring of marginals, and the plastron was virtually identical to that of many modern pleurodires in having an intergular, paired contiguous mesoplastra, and the standard paired scutes (gulars to anals).

Nevertheless, close inspection of the carapace of *Proganochelys* would reveal some differences—for example, unusually wide vertebral scutes, only four in number. There was also a unique double ring of marginals, the outer circle with as many as 17 elements on each side, and the inner circle, the supramarginals, forming a nearly complete ring in sharp contrast to the only living taxon with supramarginals (namely *Macrochelys*) in which there are only two to four of these scutes on each side. Even the casual observer would also be impressed by the *Stegosaurus*-like spines on the long tail, not the mention the ones on the back of the neck.

In the slightly younger *Idiochelys fitzingeri* from the Upper Jurassic of Bavaria, Germany, the vertebral scutes are still very wide and just four in number but the supramarginal scutes have disappeared. Interestingly, the vertebral scutes are also proportionately extremely wide in hatchlings of *Podocnemis* (Figure 3.1), perhaps the most ancient extant turtle genus, although the condition is transient and is no longer shown by individuals that have achieved moderate growth.

It is in the skull that *Proganochelys* departs radically from modern turtles and demonstrates its affinities with truly ancient reptiles such as the pareiasaurs. The presence of teeth (albeit only on the palate) is one of these characteristics linking *Proganochelys* with pre-chelonian reptiles. Others include the presence of a supratemporal bone, lachrymal bone and duct, kinetic basipterygoid

FIGURE 3.1 Hatchling *Podocnemis unifilis*. Note the very wide vertebral scutes.

articulation, and lack of bony walls to the middle ear (Gaffney, 1986). The slightly later *Kayentachelys* (early Jurassic of Arizona, U.S.A.) had lost the basipterygoid articulation but retained the palatal teeth typical of *Proganochelys*. *Kayentachelys* is identifiable as a cryptodire by the typically cryptodiran trochlear device allowing the jaw musculature to slide over the top of the otic chamber (Gaffney et al., 1987). The early existence of the pleurodire lineage is exemplified by the very distinctive *Araripemys* (Early Cretaceous of the Santana formation, Ceará, Brazil), in which the long neck and flattened head suggest adaptation for piscivory. Gaffney et al. (2006) discuss anatomical details.

Once the early turtle ancestors had developed the shell, with both carapace and plastron primarily for purposes of passive defense, this elaborate, literally all-embracing structure turned out to be invaluable for a great many other functions ranging from crypsis (many turtles having highly disruptive carapace patterns and others having uniform earth tones) to thermoregulation, the favorable surface/volume ratio favoring thermal stability, especially in the larger, terrestrial species. The shell also provided an encasement for the voluminous viscera, and especially the bulky digestive tract characteristic of grazing species, notably tortoises. Furthermore, the presence of the shell provided some surprising incidental biochemical advantages. These include the function of providing a buffer for lactic acid during periods of anoxia, permitting protracted survival without access to oxygen, as may occur during hibernation underwater for several months at temperature around 3°C. The bone in the shell may constitute as much as 32% of the body mass of a small turtle such as *Chrysemys picta*, with the rest of the skeleton being an addition 5.5%—together comprising more than three times the percentage found in a small mammal or non-chelonian reptile. Over 99% of the total body calcium, magnesium, and phosphate, over 95% of the carbon dioxide, and over 60% of the body's sodium may reside in the shell and bones (Jackson, 2000). The buffering occurs in two ways: by releasing calcium and magnesium carbonates and by storing and buffering lactic acid within the bone of the shell.

3.2 STANDARD CONFIGURATION AND VARIANT FORMS

3.2.1 Bones

Among extant turtle species, certain aspects of the bony shell mosaic are very conservative whereas others may show extreme variation. The plastron shows less variation in its components than the carapace, although just as much or more variation in shape, and most turtles have four pairs of plastral bones, evolutionarily derived from the clavicles, interclavicle, and the abdominal ribs or gastralia, and perhaps some new bones. They are the epiplastra, hyoplastra, hypoplastra, and xiphiplastra, together with an anterior entoplastron. One extant family (the Pelomedusidae) includes an additional pair of mid-plastral bones, the mesoplastra. These are wide and together span the whole width of the plastron in *Pelusios* (Figure 3.2), whereas in *Podocnemis* they are laterally placed and widely separated (Figure 3.2). Such mesoplastra are also often present in the testudinid cryptodire *Malacochersus* (Figure 3.3). In *Pelomedusa*, they are separated by the large mid-plastral fontanel. Irregular bones present between the hyoplastra and the hypoplastra of many specimens of *Manouria* may also be called mesoplastra, and Pritchard (1966) records their anomalous presence in a specimen of *Lepidochelys olivacea*. Other variant plastral conditions are described later, under Kinosternidae, Trionychidae, Chelydridae, and so on.

The carapace includes a large anterior element, the nuchal, present in all turtles, even in *Dermochelys*. This important bone is the origin of numerous muscles and usually has a sutural connection with the first pair of marginals, the first costals, and the first neural bone. Posterior to the nuchal is a series of neural bones. In the typical condition, each neural bone alternates with the underlying, attached vertebral centrum, and the rib heads lie approximately halfway along the length of the corresponding neural bone. However, throughout the Testudines there is an astonishing amount of variation:

(a)

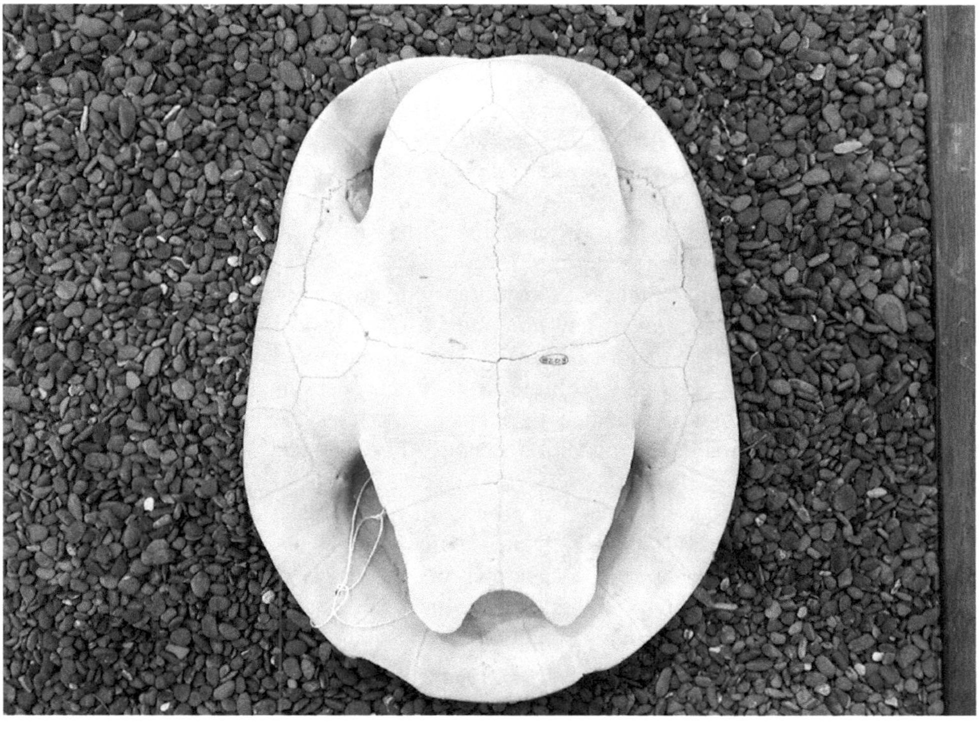

(b)

FIGURE 3.2 Plastron of *Pelusios castaneus*, showing contiguous mesoplastra. Bottom: Plastron of *Podocnemis erythrocephala* showing separated mesoplastra.

FIGURE 3.3 Plastron of *Malacochersus tornieri* showing mesoplastra. Right: Plastron of *Podocnemis erythrocephala* with abnormal absence of mesoplastra.

The presumed ancestral condition of about eight predominantly hexagonal neurals, each of which has the broad end anteriorly directed, may be modified in many ways—by proliferation of elements (to as many as 15), as in *Lepidochelys*; by reduction (usually at the ends of the series) as in many kinosternids and chelids; by total loss of exposed neurals, as in most *Chelodina* and *Emydura*; and by changing in shape—by broadening, as in *Deirochelys*, *Macrochelys*, etc.; narrowing as in *Carettochelys*; reduction to isolated kite-shaped elements as in many *Phrynops gibbus* and *Cycloderma senegalensis*; reversal (to hexagons with the broad end posteriorly directed as in *Rhinoclemmys* and various other smaller batagurids, or to an alternating series of octagonal and quadrilateral elements in most testudinids (Pritchard, 1988).

The pygal bone is usually single but the suprapygals are very variable, the principal variants being complete absence (*Podocnemis lewyana*, most trionychids); single, as in the other *Podocnemis* species; or paired, as in the majority of turtles, with an anterior elongate element and a posterior wider one. The suture between the two is often strongly curved, as in many testudinids and cheloniids. In extreme cases, the anterior element develops posterolateral arms that embrace the posterior element (Figure 3.4), although this condition is not fully developed until maturity is approached (Auffenberg, 1976).

The costal bones number eight pairs in almost all turtles, each costal bearing a rib and corresponding to vertebral II to vertebral IX. Dorsal vertebra 1 thus usually lacks both ribs and costal bones, but Kordikova (2000) reported a specimen of *Pelodiscus sinensis* (PCHP 2771) in which these elements were present, giving a total of nine pairs of costals, and Pritchard (1993) reported a *Dogania* (PCHP 3368) with the same condition on one side. A similar abnormal individual of *P. sinensis* with nine pairs of costals was the type specimen of Heude's *Tortisternum novemcostatum*, and one assumes that his *Tortisternum novemcostatum* also should have had nine costals, although his illustration showed only eight (Heude, 1880). In *Rafetus*, the eighth costals are strongly reduced (Figure 3.5), and this reduction is further advanced in American trionychids, in which the eighth costals are either reduced (Figure 3.6) or completely missing (or sometimes present on one side only). The peripheral bones are also relatively stable, generally numbering 11 pairs, although they

FIGURE 3.4 Variation in suprapygal structure of mature *Eretmochelys imbricata*.

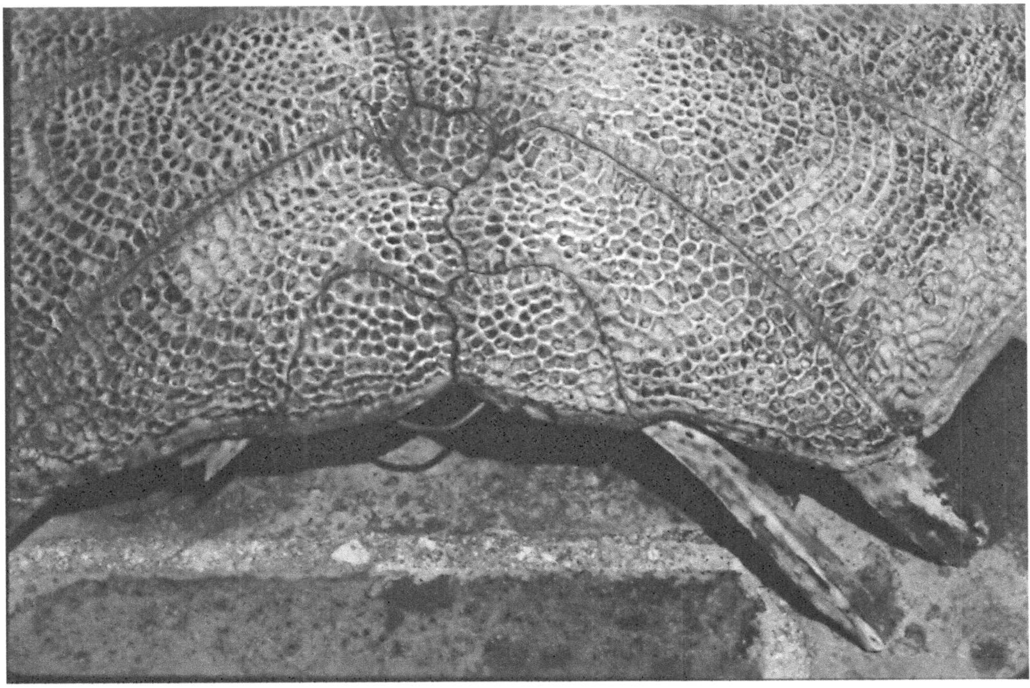

FIGURE 3.5 Rear of bony carapace of *Rafetus swinhoei* showing reduced eighth pleural bones.

(a)

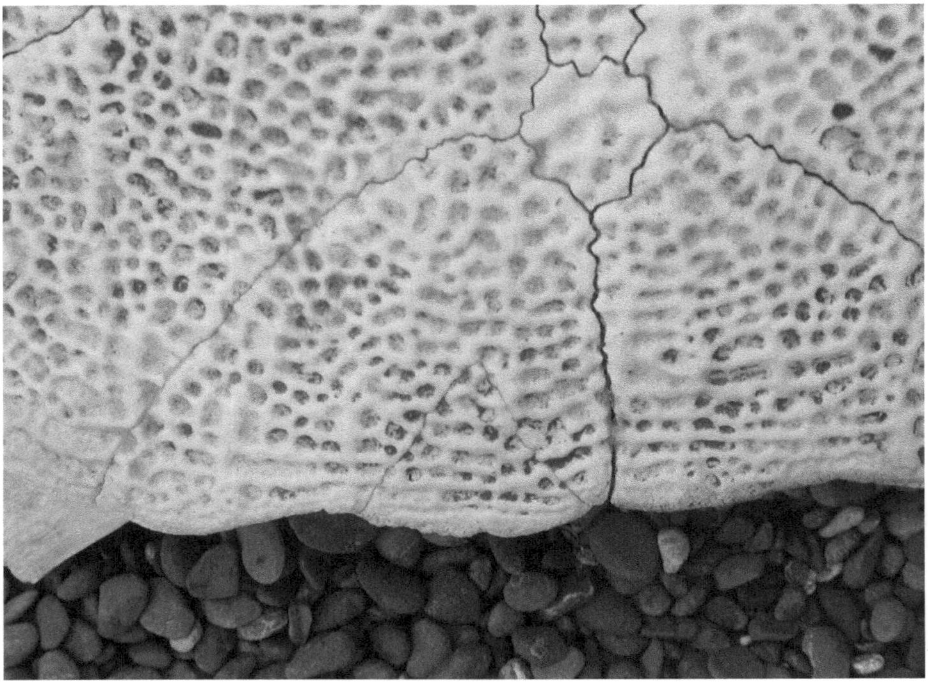

(b)

FIGURE 3.6 Carapace of *Apalone ferox* with a single rudimentary eighth pleural bone.

FIGURE 3.7 Carapace of *Callagur borneoensis* (right) and *Emydura australis* (left) showing extensive inguinal buttressing.

are reduced to 10 pairs in kinosternids and carettochelyids and they disappear completely in dermochelyids and trionychids.

Buttressing between carapace and plastron, as already evident to White in 1774, is very variable among turtles. Extensive shell buttressing is shown by *Callagur*, *Batagur*, *Kachuga*, and also certain sidenecks (e.g., *Emydura australis*), probably associated with defense against predators with powerful jaws, especially crocodilians (Figure 3.7). At the other extreme, buttressing is eliminated in species with hinged plastra, although a remnant of the axillary buttresses not contacting the inner surface of the carapace may be present in *Pelusios*, where they serve as a pair of fixed levers for the muscles that elevate the kinetic anterior plastral lobe (Figure 3.8). In some batagurids in which the posterior plastral lobe shows minor kinesis, the inguinal buttress is not completely eliminated but is shortened and connects to the inner surface of the carapace by means of a pad of fibrous tissue rather than an immovable suture. Buttresses are absent in chelydrids, cheloniids, trionychids, and so on.

3.2.2 Scutes

Whereas some turtle families (Dermochelyidae, Trionychidae, Carettochelyidae) have dispensed entirely with scutes (large external scales that cover the carapace and plastron), they are retained by the vast majority of turtles. Scutes may be remarkably thick (*Eretmochelys*, *Astrochelys yniphora*) or may be paper-thin (*Dermatemys*, *Callagur*). Usually, they are juxtaposed but they are imbricate (overlapping) in most *Eretmochelys*.

Scutes grow by proliferation of keratin around the periphery (as well as the entire undersurface of the scute) as the turtle itself grows, and it is common for this growth to be seasonal, or at least episodic, so that rings or bands may be formed on each scute throughout the area of post-hatchling growth. Typically, the outline of each scute of the hatchling is preserved as an "areola," a small, stippled or shagreened area somewhere near the center of the scute. The growth rings or annuli around the aureolae may often be counted to give an index of the age of the animal (Figure 3.9), although this is reliable only in younger specimens and those that live in temperate latitudes, where winter dormancy occurs. Also, some tropical species (e.g., in the llanos of Venezuela) develop annuli that

FIGURE 3.8 Shell of *Pelusios castaneus* showing axillary buttresses modified into levers for elevation of anterior lobe of plastron.

FIGURE 3.9 Scute details of *Astrochelys radiata* showing aureolae and growth annuli.

reflect an extreme hydric cycle. Moreover, some aquatic turtles (e.g., *Chrysemys*) will routinely lose the outer layers of the scutes through peeling, so that the shell maintains a uniformly smooth texture even in younger specimens. In addition, very old turtles of many species may achieve completely smooth scutes simply as a result of a lifetime of abrasion by vegetation, substrate, and so on, as the animal goes about its life. Nevertheless, certain relatively sessile chelonian species, ranging from the diminutive, terrestrial *Pyxis planicauda* to the very large, aquatic *Macrochelys temminckii*, may retain countable annual rings up to an age of two decades or more, allowing the reconstruction of the entire lifelong growth curve simply by careful measurements of the scutes and the growth annuli of an adult individual.

Extant turtle species, and most extinct ones too, show extraordinary conservatism in the deployment of scutes. For example, the carapace scute patterns of a green sea turtle (*Chelonia*) and a testudinid (tortoise) have remained essentially identical since Mesozoic times, with a normal configuration of five vertebral scutes, four pairs of costals (the large lateral elements), a single, small nuchal scute, 22 small peripheral scutes, and a supracaudal scute at the extreme posterior. The plastron also shows a remarkable lack of diversity, with six pairs of scutes (the gulars, humerals, pectorals, abdominals, femorals, and anals) being the norm for the vast majority of chelonian species.

Nevertheless, some variant scute configurations do occur, even among quite closely related species. Many testudinids have a nuchal scute but many do not, and its presence or absence is one of the few features that distinguishes a South African geometric tortoise (*Psammobates geometricus*) from an Indian star tortoise (*Geochelone elegans*). The chelid genus *Elseya* and the testudinid *Indotestudo* both include some species that have a nuchal scute and some that do not. Furthermore, some species with rather narrow shells have a very narrow nuchal scute (*Geochelone* (*Chelonoidis*) *carbonaria* and *denticulata*), whereas species with wide, flat shells (*Gopherus polyphemus*, *Manouria impressa*) may have a correspondingly broad nuchal.

FIGURE 3.10 Carapace of *Notochelys platynota* showing "extra" vertebral scute.

Nearly all turtles have five vertebral scutes, although anomalies may occur (see examples). Among the few consistent alternatives are the configurations shown by *Notochelys platynota* (a batagurid) with six or sometimes seven, rarely only five, vertebrals (Figure 3.10), and *Lepidochelys olivacea*, a cheloniid, in which five to nine vertebrals may be present. The most striking example of costal scute variation is the additional costal at the front of the series on each side in *Caretta* and *Lepidochelys*, and even though *L. kempii* shows only five pairs of costals, *L. olivacea* may have as many as nine and often shows a different number of costals on each side.

Plastral scute variation is rather limited, except in the families Kinosternidae and Chelydridae (discussed later), but some cryptodires (e.g., cheloniids and dermatemyids), as well as all pleurodires, have an intergular scute. This is small in *Podocnemis* (a pelomedusid) but may be very large in some chelids (*Hydromedusa*, *Pseudemydura*), and in *Chelodina* it is not only enlarged but recessed inward so that it does not form part of the plastral margin (Figure 3.11). Some systematists use the "plastral formula" as an important character in dichotomous identification keys. This formula ranks the lengths of the midline seams between each pair of plastral scutes, e.g., abd > fem > gul > hum > pect > anal. The formula is reasonably stable within a species but may come unstuck when there is significant sexual dimorphism of the posterior lobe and the anal notch.

Fire, trauma, disease, and so on may cause the scutes of a turtle to separate or peel off. This exposes the underlying bone. The outermost layer of bone, no longer protected by scutes, will necrophy, but the dead bone (or sequestrum) remains in place for a long time, and by the time it finally falls away a new layer of keratin will have formed beneath it. However, the original color, patterning, and texture of the scute is not usually recovered.

FIGURE 3.11 Anterior of plastron of *Hydromedusa tectifera* (left) and *Chelodina mccordi* showing enlarged and distinctive intergular scutes.

3.2.3 External Form

Turtle shells vary greatly in shape. Some are almost circular (e.g., adult *Lepidochelys*, *Chelodina steindachneri*, hatchlings of many species) whereas others may be elongate and narrow (e.g., *Chelodina oblonga*; Figure 3.12). In lateral profile, the carapace may be highly elevated or domed (e.g., *Astrochelys yniphora*, *Geochelone pardalis*) or extremely flat (*Malacochersus tornieri*, *Platysternon megacephalum*, *Apalone mutica*). It may be smooth (*A. mutica*) or rough (e.g., *Macrochelys temminckii*, *Chelus fimbriatus*). The carapace may have a single keel (e.g., *Pelusios carinatus*, *Sternotherus carinatus*, *Graptemys* spp., *Pangshura* spp.), two keels (*Platemys platycephala*), three keels (many batagurids, *Staurotypus* spp., many *Kinosternon*; Figure 3.13), or as many as seven keels (*Dermochelys*). A pair of transient plastral keels is present in hatchling *Podocnemis sextuberculata*, and these may also be present in *Eretmochelys imbricata*.

As a first-order generalization, one might observe that terrestrial species (testudinids, Terrapene, and so on) are inclined to be domed and aquatic ones to be flattened. However, there are numerous exceptions. Moreover, "domed" or "flattened" are not the only two options—for example, a number of riverine species of the genera *Graptemys* and *Pangshura*, accustomed to maneuvering in fast currents, have tectiform shells with flat, sloping sides and a high median keel. Some aquatic turtles may have highly elevated shells (east Pacific *Lepidochelys olivacea*, many *Chelodina mccordi*) and some terrestrial ones may be very flat (*Homopus signatus*, *Malacochersus tornieri*), often as adaptations to life under boulders or in rocky cracks.

FIGURE 3.12 Extremes within a genus: *Chelodina steindachneri* (left) and *Chelodina oblonga* (right), both from southwestern Australia.

FIGURE 3.13 Extreme development of carapace keels in an adult *Staurotypus triporcatus*.

Some recent studies by Claude et al. (2003) have sought to establish mathematical interpretations of the extent to which the shell morphology of a turtle is shaped by environmental considerations and how much by the phylogenetic history of the species. This procedure requires that the customary analog descriptions of the turtle shell be replaced by a digital system that uses a series of objective "landmarks" on both the bony and the external shell—for example, a typical "landmark" being the point of intersection of a superficial scute boundary or sulcus and a bony suture line. The data are recorded using a three-dimensional computerized recording system (MicroScribe 3D Digitizing System). It was found that both environment and phylogeny play important roles in shaping the shell of a turtle.

3.2.4 Kinesis

Once having evolved a rigid all-enveloping corselet, turtles were relatively quick to produce forms in which some degree of flexion or kinesis was possible, although a substantial percentage of living taxa, both aquatic and terrestrial, retain a completely rigid shell. The extreme of kinesis is reached by the small trionychid *Dogania subplana* (Pritchard, 1993), in which a bizarre condition known as pankinesis occurs with maturity. In this species, the sutural connections between all adjacent carapace bones break down, producing a flat, floppy shell, with an extensive unossified periphery, apparently ideal for finding refuge under submerged rocks and stones in upland streams.

Another dramatic form of kinesis is developed by members of the testudinid genus *Kinixys*, in which a hinge develops across the rear of the carapace. A hinge across a strongly convex surface would appear to be mechanically untenable, but kinesis is made possible by an infusion of soft tissue between the relevant peripheral bones on each side of the lower edge of the carapace (Figure 3.14).

FIGURE 3.14 Carapace of *Kinixys erosa* showing posterior hinge.

This tissue is absent along the midline, but flexion at this level is made possible by the presence of a "floating" neural bone that is not tightly sutured to any of its neighbors. The hinge is absent in the hatchlings and young but its potential is suggested by the alignment of the relevant bony sutures and scute seams in the carapace and the unusual, rounded shape of the posterior plastral lobe.

The most common form of shell kinesis occurs in the plastron, and several genera (notably *Terrapene* and *Cuora*) have a single transverse hinge across the middle of the plastron that makes possible complete retraction and protection of all the extremities. In other genera (*Emys*, *Cyclemys*, *Emydoidea*, *Notochelys*, and so on), some degree of transverse mid-plastral kinesis is possible, but its primary function would seem to be something other than simple protection.

Other forms of transverse plastral hinging also seem to contribute little toward physical defense. The different conditions found in kinosternids are discussed later; noteworthy also is the small testudinid *Pyxis arachnoides*, in which two of the subspecies have an anterior hinge that is not attached to powerful retractor muscles and that moves primarily as a passive result of locomotion or head retraction/extension (Figure 3.15). In a number of turtle genera in which the eggs are few (even just one) but unusually large (*Homopus*, *Rhinoclemmys*, *Leucocephalon*, and so on), there may be some breakdown of the bridge sutures, as well as kinesis of the posterior plastral lobe, to facilitate oviposition.

In the pleurodires, in which the pelvis is reduced to two separated, rigid pillars extending from plastron to carapace, there is little potential for complete shell protection, but the complex genus *Pelusios*, with over a dozen African species, is distinguished—with the single exception of *P. broadleyi* (Bour, 1986)—by having anterior plastral kinesis that usually offers very effective protection. In *Pelusios gabonensis*, the plastron has some degree of longitudinal (midline) kinesis as well as an angled (as opposed to straight) anterior lobe hinge-line (Figure 3.16). Elevation of the anterior lobe is thus accompanied by the slight projection or bulging of the mid- and hind section of the plastron. Comparably complex modes of kinesis are also shown by certain kinosternids (e.g., *K. s. scorpioides*), in which the hypo-xiphiplastral sutures meet at an obtuse angle, forcing the sides of

FIGURE 3.15 (left to right) Plastra of *Pyxis arachnoides arachnoides*, *P. a. oblonga*, and *P. a. brygooi*. Only *brygooi* has no anterior hinge.

FIGURE 3.16 Angled anterior plastral hinge of *Podocnemis gabonensis*.

the posterior lobe to fold upward slightly when the posterior lobe is raised. This results in exceptionally tight closure of the posterior shell opening.

3.3 SEXUAL DIMORPHISM

Sexual dimorphism in the chelonian shell manifests itself in several ways, most obviously in the form of a size difference between adult males and female. Within the order Chelonia, some species vary little in adult size, but there are many others in which one sex in much larger than the other. Extreme cases of small males and large females are represented by such species as *Graptemys barbouri* and *Hardella thurjii*, where females may have three times the carapace length of males (Figure 3.17). On the other hand, in Galapagos tortoises (*Geochelone nigra*), the males may reach two or three times the weight of adult females.

In terms of shape, the most familiar sexual shell difference relates to the concave plastron in males of many terrestrial species, a detail probably crucial to successful copulation. There are also frequently differences in the general form of the carapace, which may be relatively rounded or domed in females and elongate and more flattened in males (e.g., *Geochelone denticulata*, *Chelonia mydas*). In others (e.g., *Testudo boettgeri*), the posterior of the carapace may be much wider in males than in females. In saddleback tortoises (*Geochelone nigra* ssp.), the degree of saddling increases with size, thus reaching a much more extreme expression in the (large) adult males as opposed to the (much smaller) adult females. One also encounters scute softening and de-ossification of the middle

FIGURE 3.17 Extreme size dimorphism: adult males (top) and adult females (bottom left) of *Graptemys barbouri*. Center right: Subadult female with fontanels.

of the plastral concavity in males of certain unrelated species (*Caretta caretta*; *Phrynops hogei*), which may serve as a frictional "assist" for a mating male to maintain his position. In adults of such species as *Heosemys spinosa*, the adult male has a completely rigid shell with a deep concavity occupying much of the plastral surface, whereas in the female, the concavity is absent and there is extensive kinesis of the posterior lobe (Figure 3.18).

In many turtle species, there may be differences in the shape of the xiphiplastral region of the plastron. This is brought about by the different requirements of males and females as regards the shape of the aperture between the rear of the carapace and the plastron. Females may need to pass impressively large eggs through this aperture, and the space between the xiphiplastra and the pygal bones is rounded accordingly. Males may require a shorter, wider notch to facilitate extension and downcurving of the tail during copulation. In species with extensive plastral mobility, the anal notch is often lacking.

In species with significant size dimorphism, the bony carapace of specimens of small to medium size may reveal the sex of the specimen by the presence or absence of intercostal fontanels. Thus, the carapace of a 10-cm adult male *Graptemys barbouri* shows no fontanels whereas the carapace of a female of the same size has extensive fontanels. This characteristic reflects the capacity for considerable further growth in the female before maturity is reached, at which point the fontanels eventually close and growth ceases or becomes asymptotic. The reverse is true with *Macroclemys temminckii*—males 40 to 50 cm in length have open fontanels, whereas these are closed in females of this size, at which they are already sexually mature (Figure 3.19).

In those tortoise species with gular "horns" (single or double), including some *Gopherus* species, *Astrochelys yniphora*, *Chersina angulata* (Figure 3.20), and *Geochelone sulcata*, the horns are considerably more developed in males than in females, corresponding to the demands of male

FIGURE 3.18 Adult male (left) and female (right) of *Heosemys spinosa*.

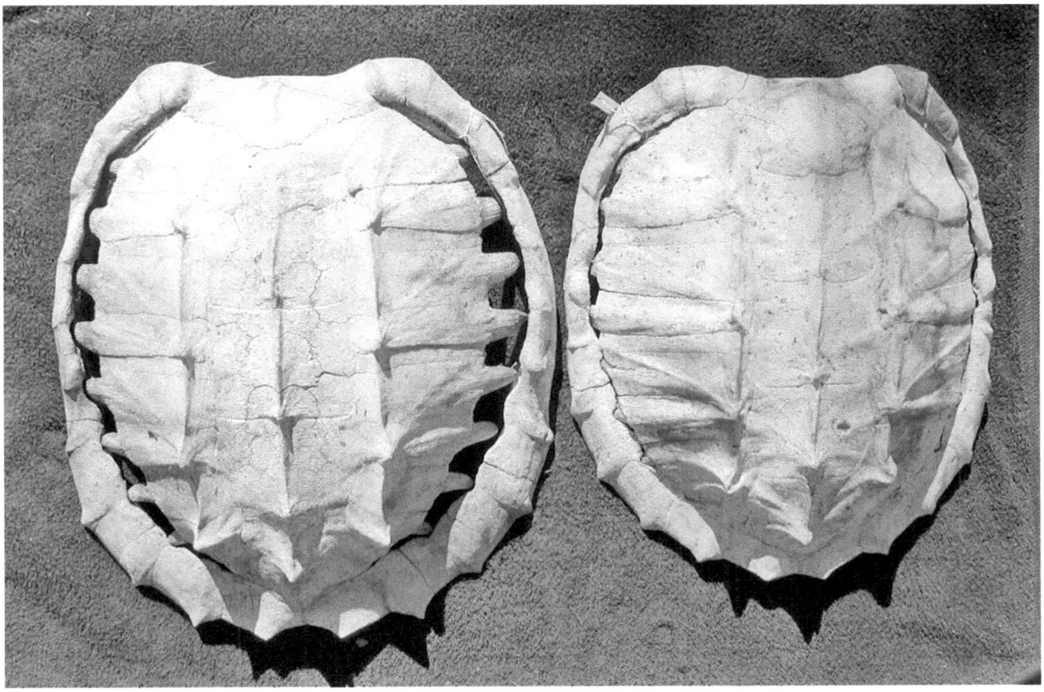

FIGURE 3.19 Adult female with closed fontanels (right) and subadult male with open fontanels (left) of *Macrochelys temminckii.*

combat and "vigorous" courtship of females. Nonetheless, somewhat reduced horns are also present in females.

3.4 ANKYLOSIS

As turtles approach or reach maturity, the shell fontanels usually close at some point between hatching and the attainment of maturity, but in most species the individual shell bones remain distinct. However, in a minority of species shell bones typically fuse with maturity and the sutures disappear. Such ankylosis commonly occurs in several subspecies of *Terrapene carolina* (e.g., *carolina*, *mexicana*, *bauri,* but not in most *T. c. major*), also in some Asian batagurids of the genus *Cuora* (e.g., members of the *C. galbinifrons* complex). Some large river turtles (*Kachuga kachuga*, *Dermatemys mawi*, *Batagur baska*, *Callagur borneoensis*; Figure 3.21), all of which also have very thin shell scutes, also show complete shell ankylosis in adults, and unlike the box turtles this is associated with the disappearance or fusion of the scute seams also. In the smaller, semi-terrestrial batagurid *Heosemys spinosa*, the shell bones may become fully ankylosed in older adults apart from the transverse plastral hinge in adult females. In other, unrelated species (e.g., *Apalone ferox*), carapacial ankylosis may occur in exceptionally large, old individuals. Bony sutures generally become more tightly knit and less obvious with ontogeny and onset of old age in many chelonian species (e.g., in *Geochelone nigra*). In the extinct Mascarene tortoises of the genus *Cylindraspis*, the shell bones, which were very thin, routinely became ankylosed in adults; generalized ankylosis was also evident in the skulls.

(a)

(b)

FIGURE 3.20 Gular "horn" of male *Chersina angulata* (left) and female (right)

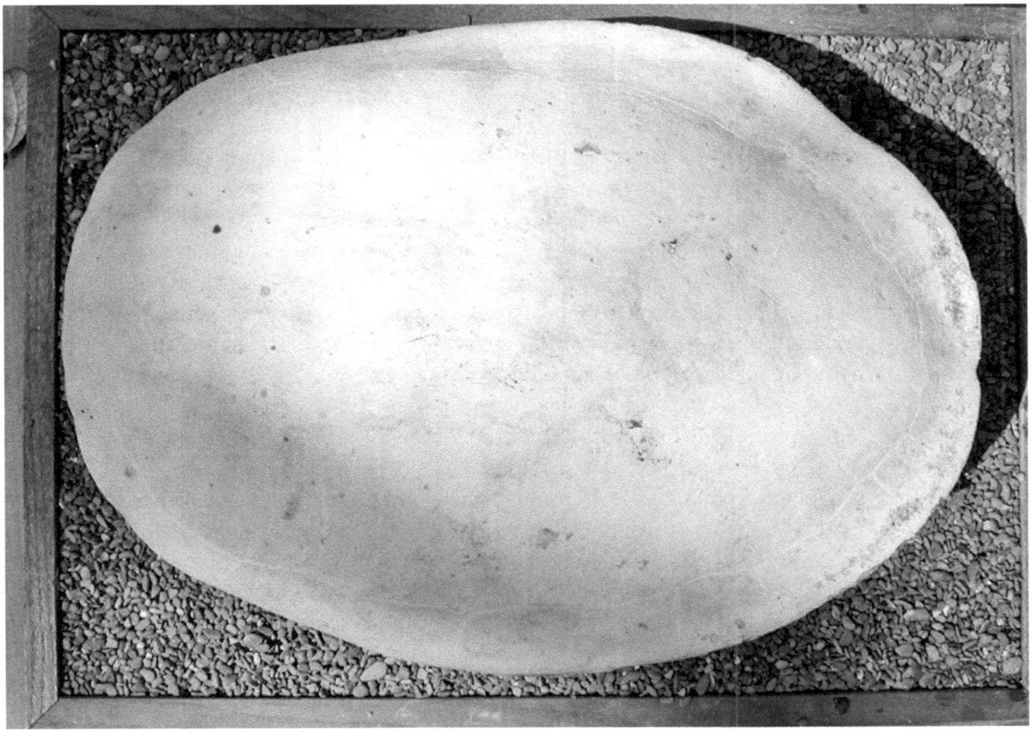

FIGURE 3.21 Carapace of adult male *Batagur baska* showing complete ankylosis.

Advantages and disadvantages of shell ankylosis are somewhat speculative—as are the pros and cons of the shell not ankylosing. The condition could be, in part, simply a reflection or by-product of growth slowing and becoming asymptotic or stopping altogether, and certainly full ankylosis of the shell would seem to preclude further growth. In the terrestrial box turtles (*Terrapene carolina*, and so on), the achievement of shell ankylosis at a rather small size (SCL of less than 15 cm) may correspond to the fact that the moveable shoulder girdle and pelvis of these species may facilitate shell closure but it also severely compromises the weight-bearing capability of these structures (Bramble, 1974). In other words, terrestrial box turtles have to be small or they would not be able to walk, and early shell ankylosis may be the mechanism that ensures that they remain small even after many decades of life. Possibly, this is also the reason that old specimens of *Kinixys homeana* develop fully ankylosed shells.

The relationship between ankylosis and resistance of the shell to fracture has not been examined adequately, and it is complicated by the fact that some turtles with alkylosed shell bones (e.g., *Batagur*, *Callagur*) may indeed have very strong shells, but they also have other features that enhance shell strength, such as heavy buttressing and increased shell thickness. Sutures do not appear to be lines of weakness in living turtles, and examination of traffic-killed specimens of different ages and sizes shows little if any tendency for fracture to occur along sutural lines. Alternatively, Arnold (1979) has suggested that the retention of sutures throughout life may curtail or control the propagation of a crack imposed upon the shell by violent trauma, whereas in a fully ankylosed shell the crack may be unimpeded by sutures and thus spread catastrophically, as in a broken egg.

3.5 FONTANELS

A fontanel in a turtle may be defined as a gap between shell bones, usually filled in life with collagenous tissue, that is typical of young, growing specimens. Fontanels normally eventually close with ontogeny; openings that appear or develop with growth and age are generally called fenestrae.

The typical configuration in hatchling turtles is one in which the neural and peripheral bones are fully developed but with wide open spaces between the exposed ribs, called intercostal fontanels. In the plastron, the edges of the anterior and posterior lobes may be defined by the outer edges of the curved epiplastra and xiphiplastra, but there may be an extensive fontanel (or fontanels) along the midline, from the rear of the entoplastron to the union of the xiphiplastra. Each hyoplastron is sutured to the ipsilateral hypoplastron, but these bones have merely juxtaposed, non-sutural contact with epiplastra and xiphiplasta. The "spikes" of the rear of the entoplastron and the proximal sides of the hyoplastra and hypoplastra may extend into the median fontanel, giving some support to the fibrous tissue therein.

Nonetheless, both neonatal and post-hatchling fontanels are subject to enormous variation. Most remarkable is the condition in the pelomedusid sideneck turtles, in which the hatchlings emerge with a fully plated shell lacking all fontanels, even though in the extreme case (*Podocnemis expansa*) an individual may grow from 30 mm upon hatching to over 80 cm at adulthood. Painted turtles (*Chrysemys picta*) also close the fontanels at an early age. At the other extreme, in *Leucocephalon yuwonoi* and *Heosemys spinosa* the (usually single) young is hatched at 20 to 30% of the adult carapace length but has an extremely undeveloped bony shell structure, especially in the plastron (Figure 3.22), which is barely more developed than in *Dermochelys*. Yet by the time maturity is reached in these forms, the entire shell has become unusually sturdy. In *Notochelys*, there is one intercostal fontanel on each side (at the lower edge, between costal 5 and costal 6; Figure 3.23)

FIGURE 3.22 Bony plastron of juvenile *Leucocephalon yuwonoi*.

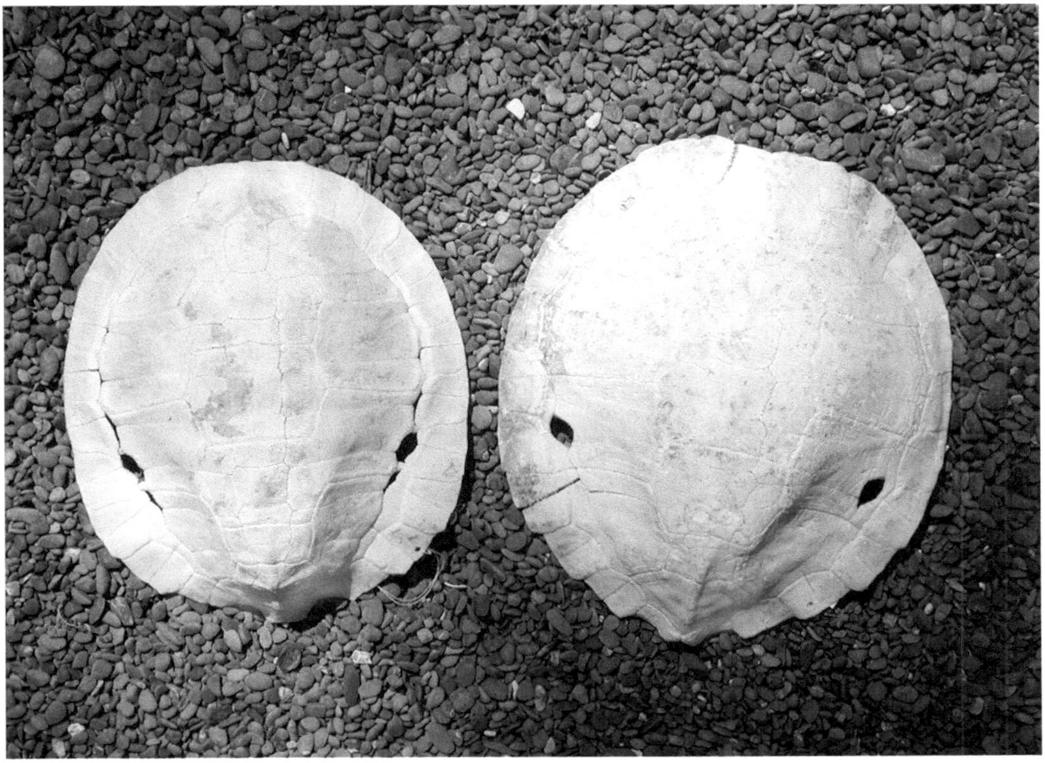

FIGURE 3.23 Carapaces of subadult *Notochelys platynota* showing isolated persistent fontanels.

that is persistent until near-maturity, by which time all the other fontanels have closed. It is located directly above the (somewhat rudimentary) inguinal buttress.

The previous batagurid species share a tropical-forest-floor, leaf-litter, or shallow stream and puddle environment, which seems to correlate with the production of very large, hard-shelled eggs that are not buried but merely concealed beneath a few leaves. In the American tropics, this niche is occupied by the batagurids *Rhinoclemmys punctularia*, *R. diademata*, *R. melanosterna*, and so on.

In several large riverine batagurids (namely *Batagur baska*, *Callagur borneoensis*, *Kachuga kachuga*, *K. dhongoka*, and *K. trivittata*), extensive intercostal fontanels are present in the young but they close progressively with growth and have disappeared in females by the time mature size is reached. Alternatively, in the males (which reach a much smaller adult size than the females), the fontanels do not disappear but the rather sharp, rough inner edges of several of the juvenile fontanels become smooth and rounded, producing the effect of a series of four or five "portholes" that persist even in old animals (Figure 3.24). In some cases, adjacent pairs of these portholes may coalesce, becoming incompletely separated by a bony spike. The function of this strange condition is arguable. Hypothetical explanations must include consideration of the fact that the shell in at least some of these turtles is unusually thick and the internal buttresses developed to a remarkable degree. Furthermore, adults achieve fully ankylosed shells. There is thus a bizarre disparity between the massive overall armoring and the contrasting vulnerability resulting from the persistence of the fontanels, which become permanent openings in the bony shell covered with paper-thin scutes.

The adaptive value must also take note of the presumption that the persistent fontanels are so important that they exist in spite of the apparent need for these turtles to have exceptionally heavy armor. Furthermore, the interpretation must explain why the fontanels are retained in this way only

FIGURE 3.24 Costal "portholes" of adult male *Callagur borneoensis.*

in males. A possible hypothesis is that the lungs, situated in the bony pocket on each side of the shell, protected fore and aft by the heavy buttressing but exposed to vibrations or pressure waves impacting upon the fontanels, serve as a sort of echo chamber to provide exceptional sensitivity to waterborne (or airborne) vibrations. It may be no coincidence that female turtles of the genera *Batagur* and *Callagur* are known as "tuntong" in Malayalam, a word that translates as "thumper." Perhaps there has been great selectivity for the males to be able to detect the thumping as the female closes and conceals her nest, and it may allow such males to be the first to locate and mate with a given female after she returns to the water.

3.6 PLASTRAL REDUCTION

Typically, the plastron covers nearly the entire underside of a turtle, but this is not always the case. The plastron in most species has multiple functions—that of physical protection for the underside of the animal; providing a strong crossbar to prevent splaying of the sides of the carapace if the animal suffers vertical compression; and providing anchorage for the insertion of a wide range of muscles associated with the anterior and posterior limb girdles, as well as those responsible for plastron closure in hinged species.

Nevertheless, in certain turtle species the plastron is reduced to a cruciform condition that would appear to reduce the "protective breastplate" function of the "Bauchpanzer" to a significant degree, and that thus requires some explanation. Extant representatives of such species include the snapping turtles (*Chelydra* and *Macrochelys*), and several kinosternids (*Sternotherus*, *Staurotypus*, *Claudius*). The details of plastral reduction in these species are very varied, but the common driving force for the reduction in all cases appears to be associated with specializations for predation (including cephalic enlargement and strong jaws) and powerful cervical extensor muscles for the predatory "strike." This action involves considerable change in volume of the shell contents as the head and neck are rapidly extended and withdrawn. Plastral reduction is one way of accommodating this rapid visceral displacement.

In *Chelydra* and *Macrochelys*, the plastron is cruciform and strikingly reduced. The bridges are very narrow and the connections to the carapace are achieved through fibrous tissue only, and there is unossified space around the entoplastron. Scute homologies are difficult to interpret, especially in *Macrochelys* (in which the scutes are very variable), but it appears that a prenuchal and an interanal may be present, the abdominals have migrated outward into the bridges, and a single inframarginal is present on each side in addition to the large inguinal and axillary.

In *Staurotypus*, the small anterior plastral lobe is kinetic (Figure 3.25) but has little role in protection, and the main function of the hinge may be to allow the anterior hinge to be depressed by the mandible when the turtle is presenting a threatening gape without extending its head. The bridges are narrow but thickened and are tightly sutured to the sides of the carapace. An entoplastron is present with a persistent fontanel behind it, but the plastron only includes seven scutes—paired humerals, pectorals, and femorals, and an unpaired anal. Axillary and inguinal scutes are very large but no inframarginals are present. The posterior plastral lobe is very narrow and pointed.

In the diminutive *Claudius*, the plastron is extremely reduced but lacks fontanels, and the bridges, just a few millimeters wide, have become reduced to a pair of posterolaterally raked struts, lacking any rigid connection with the carapace. A unique feature is the presence, at all ages, of a single bone replacing the hyoplastron and hypoplastron on each side (Pritchard, 1971). The plastron is not actually kinetic but by being free-floating on the viscera of the underside of the animal, it can

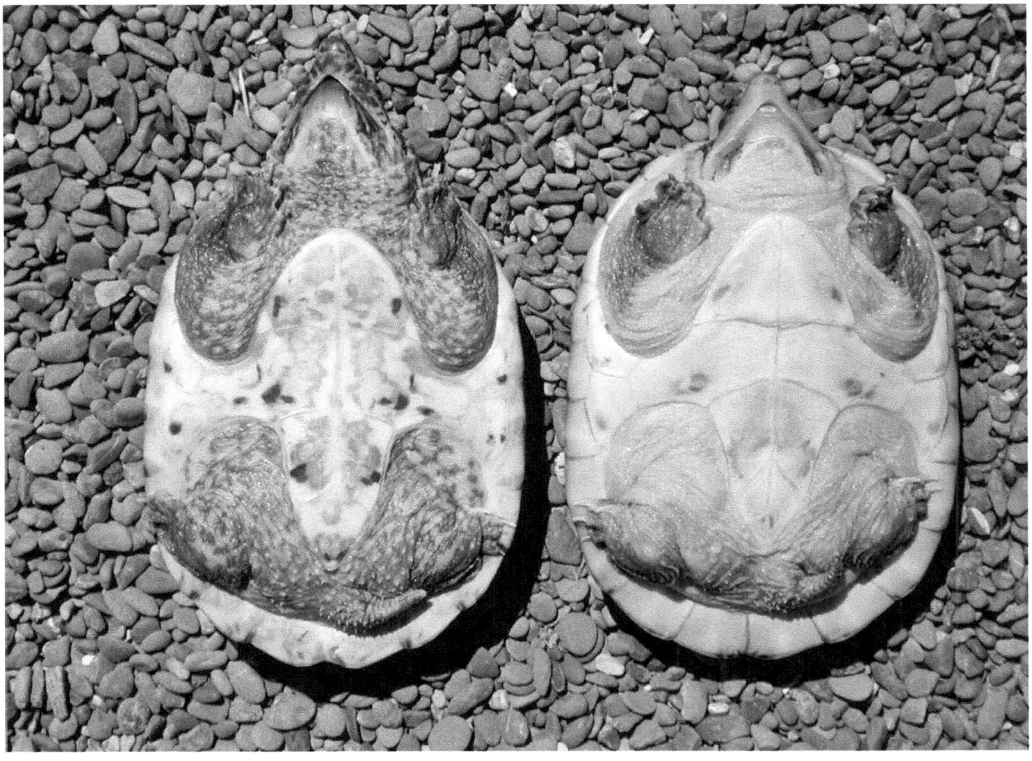

FIGURE 3.25 Reduced, cruciform plastra in *Staurotypus triporcatus* (left) and *S. salvini* (right).

"tilt" to accommodate the depressed mandible during threat display. Scutes are reduced to just four pairs, probably corresponding to the humerals, pectorals, femorals, and anals. Most of the bridge is covered with soft tissue, with a short extension from the pectoral scute covering the proximal part. Axillary and inguinal scutes are lacking.

In the closely related *Kinosternon* and *Sternotherus*, a shared feature is the absence of the entoplastron, an omission otherwise known only in *Dermochelys* among extant turtles. *Kinosternon* includes both species with enlarged plastra as well as those with reduced ones; the plastral kinesis and associated musculature are quite different in the two morphotypes (Bramble et al., 1984). In the small-plastron forms of *Kinosternon* (i.e., *K. subrubrum steindachneri*, *K. angustipons*, *K. herrerai*, *K. chimalhuaca*, and *K. dunni*, with non-overlapping ranges in Florida, Caribbean Costa Rica, eastern Mexico, western Mexico, and northwestern South America, respectively), the plastron presents both anterior and posterior "hinges" although actual kinesis is very modest, with none of the capability for forceful full elevation of both anterior and posterior lobes seen in the large-plastron species. The reduced plastral kinesis is very similar to that of *Sternotherus*, and the two genera are often combined into one, although *Sternotherus* does show some unique features, including a reduction in size of the plastral scutes and proliferation of soft tissue along the scute borders, as well as a long anterior interhumeral seam.

The double kinesis in the *Kinosternon* complex involves an anterior lobe composed of the paired epiplastra, a fixed midsection including the hyoplastra and hypoplastra, and a posterior lobe including the xiphiplastra. The anterior lobe scutation involves a small anterior element preceding two pairs of larger scutes, the midsection is covered by a single pair of scutes, and the posterior lobe bears two further pairs. Homology and thus nomenclature of these scutes is somewhat complex, but the single anterior scute (absent in *Sternotherus carinatus*) is considered to be the gular, followed by paired humerals. The next pair, seemingly the pectorals, are in fact just the anterior parts of those scutes, with the greater part of the pectorals lying posterior to the anterior hinge and covering the midsection of the plastron. The abdominal scutes, large in most turtles, are absent. The scutes covering the posterior lobe are the femorals and anals. Evidence for this nonintuitive interpretation, with the pectorals transversely split by the anterior hinge, are presented by Hutchison and Bramble (1981) and discussed by Pritchard (2003).

3.7 DEFORMITIES

It is not unusual to find tortoises with deformities of shell growth. These are discussed in the following sections.

3.7.1 "PYRAMIDING"

Pyramiding is often seen in tortoises that have been raised under unnatural conditions of humidity and diet, and it results in the vertebral and costal scutes having the central areolae sharply elevated. It also occurs to a mild degree in nature in such tortoises as *Geochelone pardalis*, whereas wild adult females of *Psammobates tentorius tentorius* may show extreme pyramiding (Figure 3.26). The potential for the condition is thus evident even in certain tortoises living in the wild, with an entirely natural diet and environment. Some species, including *Geochelone carbonaria*, may develop carapace pyramiding in captivity but this is never shown in wild specimens. The condition usually makes itself evident during the post-hatchling months, and much more rarely first manifests itself later in life. It has long been believed that the condition resulted primarily from a diet too rich in protein, and an alternative cause was believed to be an unnatural ratio of calcium to phosphate in the diet. Secondary causes were attributed to overfeeding, dietary fiber, temperature, UV light, and so on. However, current belief is that the condition derives from excessively dry conditions, and when hatchlings are raised on a substrate of wet sphagnum, shells will develop normally (Wiesner & Iben, 2003).

FIGURE 3.26 Pathological shell pyramiding in juvenile *Astrochelys radiata* (left) and normal adult female *Psammobates tentorius tentorius* (right).

3.7.2 Kyphosis

This "humpbacked" condition appears to be congenital and shows up occasionally in many species. It is particularly dramatic when it occurs in trionychids, in which the normally flat turtle takes on the appearance of a tall sombrero (Figure 3.27). Lordosis, or swayback, is also seen occasionally and may be the normal condition in old specimens of such specialized, flat-shelled species as *Platysternon megacephalum* and *Malacochersus tornieri.* Scoliosis, or kyphoscoliosis, in which the spinal column forms lateral curves or kinks, may be encountered as a rare condition in many turtle species.

3.7.3 Scute Abnormalities

The standard chelonian configuration described previously that includes five vertebral scutes and four pairs of costals, with about 22 marginals, is remarkably stable but even apart from the minor, species-specific variations mentioned here, it is not unusual for deviations—even drastic ones—to occur. They are more frequent in the carapace than in the plastron, and some of the various configurations that may exist are described by Zangerl (1969), who also observed that a particular set of repetitive variants, each with a definite frequency of occurrence, appears to be characteristic of each species or subspecies. Some scute abnormalities appear to be genetic; others result from accidents, injury, or other trauma during the embryonic stage.

Certain far-reaching derived or abnormal conditions may occur repetitively. An example of this is the replacement of the normal straight series of five vertebral scutes by an asymmetrical zigzag series of eight or nine individual elements, each often reduced to a triangular shape (Figure 3.28).

FIGURE 3.27 A kyphotic adult *Apalone ferox*.

FIGURE 3.28 "Zig-zag" condition of dorsal scutes in *Geochelone carbonaria*.

Usually, such specimens also have an extra costal scute on one side. Such turtles usually show no obvious health impairment but extreme scute abnormality may be associated with reduced viability. Nevertheless, some of the key literature on this phenomenon may be deceptive. For example, Gadow (1899) reported that supernumerary carapace scutes were much more common among hatchling loggerhead turtles than they were in adults, but his hatchlings were the multiscutate *Lepidochelys olivacea*, whereas his adults were the pentacostal *Caretta caretta*, which have a stable count of five vertebrals and five pairs of costals.

3.7.4 Premature Ankylosis of Individual Shell Sutures

Examination of a large series of turtle skeletons reveals that, from time to time, a given suture between two shell bones will fuse completely, leaving other sutures unaffected (Figure 3.29). This may occur in early or in mid-life and results in unilateral or asymmetrical distortion of the carapace; growth continues at the unaffected sutures but is forcibly curtailed at the suture that has ankylosed prematurely. The condition may be found in both wild-caught and captive-raised specimens, and its origins are unknown.

3.8 GEODESICS

Most turtle shells are constructed on some combination of the architectural models of the keystone-stabilized arch and of the cupola. The structure may be highly elevated or shallow according to environmental and ecological constraints, but the carapace is strengthened by its marked convexity and its thickened periphery, and the entire structure is protected from vertical crushing by the strong

FIGURE 3.29 Deformed juvenile *Geochelone nigra* showing extensive premature ankylosis of carapace bones.

FIGURE 3.30 "Geodesic" bony carapace of *Manouria impressa*. Carapace with scutes on right.

bridges connecting the plastron to the peripheral bones. However, a few tortoise species—which happen to run the gamut from the most flattened species of all to the most highly domed—develop very thin or extensively fenestrated shells, with bones of reasonable thickness offering some structural support retained only around the scute sulci.

In *Malacochersus tornieri*, the plastron has a huge central fontanel, and the flattened carapace is extensively fenestrated although the scutes themselves are thin and unspecialized. The structure corresponds to the unique ecology. Rigidity and strength have been sacrificed in favor of literal flexibility; both plastron and carapace can be depressed with light finger pressure, even in adult specimens. This extraordinary degree of shell reduction allows accommodation to the pressures encountered when the tortoise takes refuge in deep, irregular rock fissures, from which few, if any, predators can extract it. Although the tortoise is seemingly highly vulnerable when outside such crevices, it compensates by spending very little time outside.

In *Manouria impressa*, the plastron is fully ossified and unreduced, although the carapace is extensively fenestrated (Figure 3.30). As in *Malacochersus*, the bony carapace is retained at normal thickness only around the scute sulci, but the resulting structure is perfectly rigid as a result of the extraordinary thickening of the dorsal scutes. This thickening does not generate elevation or "pyramiding" of the scutes, but rather the thickening extends inward, with each somewhat lens-shaped scute fitting into a shallow concavity in the bony carapace. The shell in this species is somewhat flattened but not nearly as much as in *Malacochersus*, and it is rigid rather than flexible. *M. impressa* is found at high altitudes in tropical and subtropical southeast Asia but details of its ecology remain lacking. A curious feature is that the costal scute areolae, often clearly defined even in adults, are located near the centers of the scutes. In most turtles, the costal aureolae become increasing displaced toward the vertebrals with growth (*Malaclemys* is an exception).

In *Astrochelys yniphora*, the carapace is unusually high-domed throughout life but the external appearance betrays no sign of unusual internal structure, and the costal scutes do not show centralized areolae as they do in *Manouria impressa*. The carapace is not actually fenestrated but it is extremely thin except along the lines of the scute seams, and the greatly thickened scutes rest in a series of shallow concavities in the bony carapace (Figure 3.31) as in *M. impressa*.

The function or advantage of the thickened scutes and thin bony carapace of *Manouria impressa* and *A. yniphora* remain speculative. Perhaps they simply represent an alternative way of making a rigid shell of adequate thickness and strength—no better and no worse than the standard cupola. It is also conceivable that it is a way of substituting keratin for structural bone in calcium-poor environments.

3.9 SOME TURTLE FAMILIES WITH SPECIALIZED OR UNUSUAL MORPHOLOGIES

3.9.1 CHELONIIDAE

The living hard-shelled marine turtles of the family Cheloniidae show a number of shared trends and morphological features. In the carapace, the genera with the larger heads and broader anterior carapaces (*Caretta*, *Lepidochelys*) have an additional costal scute in the shoulder region on each side, and in ventral view a series of three or four inframarginal scutes is visible on each bridge. The latter is a primitive features also retained in some unrelated turtle species (e.g., *Dermatemys mawi*). An intergular scute is also typically present.

The bony shell shows some interesting specializations for marine life, including general streamlining (assisted by the short, thick neck), and the plastron includes significant unossified sections—especially behind the entoplastron and along the midline, also along the bridges on both sides (Figure 3.32). Whereas the main plastral bones are quite thick and presumably offer good protection

FIGURE 3.31 Bony carapace of adult male *Astrochelys yniphora* showing concavities and elevated sulci.

FIGURE 3.32 Bony plastron of a young green sea turtle, *Chelonia mydas*.

to the viscera, the plastron remains flexible along the midline and at the bridges throughout life, as it does in trionychids. This feature allows the chest to be pressed inward or distended in response to rapid inhalation of air during brief surfacings, or as a result of exposure to high pressure during deep dives.

In juvenile cheloniids the hyoplastra and hypoplastra have a "sunburst" appearance with numerous long spines, especially on the proximal edge. The epiplastra and xiphiplastra are somewhat curved and rather narrow, and they make mutual contact only at the extreme front and rear of the plastron. The sutural line between hyoplastra and hypoplastron on each side is tight and well sutured but is relatively short because of extensive fontanel development both mesially and laterally. The entoplastron is dagger-shaped, only the "handle" making (non-sutural) contact with other bones (i.e., the epiplastra).

Changes with ontogeny include progressive spreading of the bone in the hyoplastra and hypoplastra, so that the transverse sutures become relatively longer as the fontanels close, and the long radiating spines become progressively buried in a bony matrix and eventually may (almost) disappear. Shell ossification and fontanel closure is most complete in the two species of *Lepidochelys*—the smallest cheloniid turtles, fully adult specimens showing enlargement of the xiphiplastra and progressive spread of the hyoplastra and hypoplastra, finally eliminating all fontanelles and attainment of a fully ossified plastron. The plastron of *Natator depressus* is unique among both living and fossil cheloniids in retaining large medial fontanels into full maturity while completely closing the lateral ones (Zangerl et al., 1988).

In the carapace, intercostal fontanels are well developed in young and half-grown specimens. In *Caretta* and *Lepidochelys*, the fontanels may achieve full closure in old adults. In adult *Chelonia*, the anterior and posterior fontanels may progressively close but almost always some trace of fontanels is present toward the middle of the series. In *Eretmochelys*, fontanel development is variable; in some adults, there may be retention of extensive intercostal fontanels but in others, these become very narrow, although still open.

3.9.2 Dermochelyidae

The single living dermochelyid species, the enormous *Dermochelys coriacea*, displays a degree of skeletal neotony more advanced than that of any other turtle. Only the nuchal bone, which has no sutural contacts with any other bones, is relatively intact. The costal bones, each of which, in other turtles, generally forms together with a rib in a fused composite bony structure, are absent. The ribs are thin, even flimsy, and have the appearance of those of an embryonic hard-shelled turtle. The peripheral bones are also absent. In the plastron, only the entoplastron is actually lacking, but the remaining elements (epiplastra, hyoplastra, hypoplastra, and xipihiplastra) are reduced to narrow, splint-like structures that together form a ring around an enormous central fontanel, open throughout life.

The shell thus derives very little rigidity from its core bony structure, but this is not the important point. *Dermochelys* has unusual habits for any turtle: it undertakes rapid, deep dives to depths of over 1000 m, and is able to survive—even thrive—in sub-polar waters. It appears that the shell, and especially the plastron, in which ossification is reduced to vestiges, can deform substantially to accommodate the extreme pressures encountered at great depths. Additionally, the advanced degree of homeothermy shown by adult leatherbacks is made possible not only by gigantothermy but also by various structural modifications such as the development of counter-current heat exchangers in the bases of the limbs. In addition, both the diving and the insulation functions are facilitated by the same thick layer of oily connective or fibrous tissue that forms over both the plastral and the carapacial bones, constituting a sort of stiff insulating blubber layer. Finally, the animal is able to achieve some sort of stability in its overall shape by means of a continuous, neomorphic layer of mosaic bones just beneath the skin of the carapace. These bones are irregular and number in the thousands, and form a sort of pseudo-carapace that has significant flexibility but which is stabilized by the formation of seven longitudinal ridges in which the individual mosaic elements are markedly enlarged.

3.9.3 Trionychidae

The Trionychidae are an ancient (Cretaceous to Recent) family with extremely derived shell configurations—e.g., the loss of scutes (replaced by soft or leathery skin)—as well as certain non-shell features such as three claws on each limb, a nasal tube that may be long or short, and very elongated cervical vertebrae. The unique shell features include elimination of the peripheral and suprapygal bones, although neomorphic analogs of the posterior peripheral elements are present in one genus (*Lissemys*). The plastron includes many oddities, including frequent fusion of the hyoplastra and hypoplastra on each side, the characteristically boomerang-shaped entoplastron, and the elongate, narrow epiplastra. Loveridge and Williams (1957) proposed a different interpretation of the softshell plastron, the boomerang-shaped bone being identified as the fused epiplastra, the entoplastron being missing, and the "epiplastra" of other authors being neomorphic "preplastra." However, Cherepanov (1995) has indicated that the entoplastron is formed from a pair of bony anlages in all turtles, and the only way in which the trionychid entoplastron is unusual is the lack of the posterior spike. Loveridge and Williams' interpretation was also rejected by Meylan (1987).

The softshell carapace is only "soft" in the posterior flap (which sometimes extends to 50% of the bony carapace length) and to a lesser extent at the sides, although these are generally supported by the tips of the ribs extending into the leathery margin. Indeed, in a large softshell, the bony carapace may be very thick and offers substantial protection for the viscera below. The ratio between the size of the bony hard carapace disk and the soft "floppy" disk is variable; for example, in *Cycloderma frenatum* and *Cyclanorbis aubryi* the soft disk is only slightly larger than the hard disk, which in turn is sufficiently developed in adults for the (usually exposed) rib tips to be concealed. Alternately, in many other forms (*Apalone*, *Dogania*, and others) the flexible flap of the soft disk extends considerably beyond the edges of the bony shell. The surface of the carapace bone in trionychids is extensively pocked with hundreds or thousands of shallow pits (or even finer shagreen-like asperities), sometimes coalescing into vermiform ramblings, a feature that probably corresponds to the requirements for attachment of the smooth epidermis that replaces the scutes of other turtles. In *Rafetus*, the eighth pleural bones are reduced, and in *Apalone*, the eighth pleurals may be small or even lost on one or both sides, although the eighth rib on each side is retained as a elongate, thin, flexible structure that extends posterolaterally and that may serve as a partial "stiffener" for the posterior flap.

However, the most remarkable feature of the trionychid plastron is the presence of bony callosities (the "epithecal skeleton," or intermembranous sesamoid bones) in addition to the nine standard plastral bones. These callosities are astonishingly variable in both degree of development and actual deployment over the plastron, and they often show striking ontogenetic change. It has been argued that the callosities are merely roughened areas on the dermal plastral bones, and this may indeed sometimes be the case. For example, in *Rafetus* and in *Dogania*, as well as in *Cyclanorbis elegans*, the callosities are very feebly developed throughout life, consisting at most of a slightly roughened area on each hyoplastral-hypoplastral area. But callosities may not correspond on a one-to-one basis with the dermal bones, and neomorphic, free-standing callosities (the "prenuchal" or the "peripherals" of certain cyclanorbines, or the anterior plastral elements of *Cyclanorbis senegalensis*), unsupported by dermal bones, may develop. The latter species is especially unusual in having the anterior part of the plastron heavily festooned with callosities whereas the xiphiplastra show no trace of callosity formation. A more standard configuration, such as is seen in the giant softshells *Chitra* or *Pelochelys*, offers substantial callosities on each hyoplastral-hypoplastral combination as well as extensive development of a xiphiplastral callosity on each side but with the entoplastron and epiplastron remaining simple, narrow, and without callosities throughout life. Such a configuration is also seen in much smaller forms (*Pelodiscus*, *Palea*).

The previously mentioned genera, large and small, show full development of the callosities very early in life, indeed shortly after the hatchling stage is passed. Alternatively, species such as *Apalone ferox* hatch without traces of callosities, although early in development the hyoplastra and hypoplastra become completely fused. Initially, the hyoplastra, hypoplastra, and xiphiplastral elements are narrow and show elaborate elongate spikes (some sharp, some blunt) extending from the corners and extremities. However, early in life a superficial callosity develops on each of these bones and spreads like thick molasses until it reaches and indeed surpasses the edges of the original bones; eventually, the bones have a plate-like shape with the elaborate outline of the original bone completely concealed (except as a sort of shadow or "ghost" in visceral view). In extreme old age, *Apalone ferox* may show a single massive plastral bone on each side corresponding to the hyoplastron, hypoplastron, and xphiplastron, and a deep groove forms along the distal edge of the hypoplastron as growth of the edges of the bone and of the attached callosity extend in divergent planes. Despite these elaborate changes, the entoplastron and epiplastra remain simple and smooth-surfaced throughout life.

In some cyclanorbines, the plastral callosities may show remarkable development. In *Lissemys scutata*, the initially small, circular entoplastral callosity shows spectacular enlargement in adulthood, and in *Cycloderma aubryi* the plastral callosities become so huge that they virtually constitute a "recovered" complete bony plastron (Figure 3.33). The situation is even more extreme in *Lissemys punctata*, in which the hyoplastra-hypoplastral callosities extend dorsolaterally so that they offer an entire bony bridge on each side, and continue dorsally to ossify the lateral edges of the carapace in the areas where peripheral bones would be present in hard-shell turtle species—a classic case of evolution reversing itself to create analogous but not homologous structural forms. By contrast, in the two largest cyclanorbine species, *Cycloderma frenatum* (Figure 3.34) and *Cyclanorbis elegans* (Figure 3.35 and Figure 3.36), callosity development remains minimal throughout life.

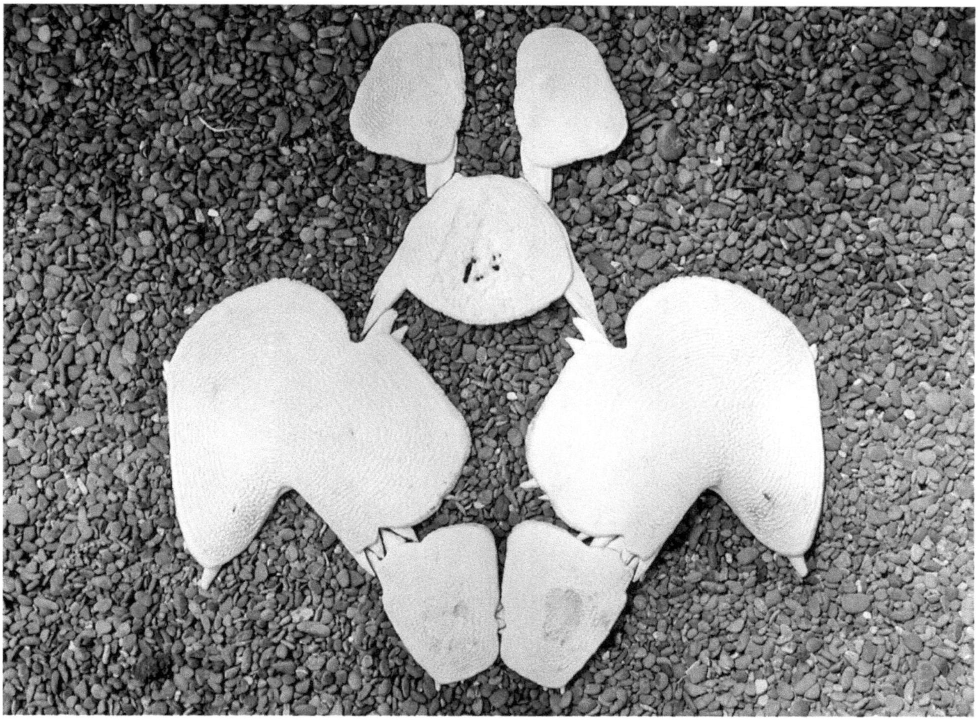

FIGURE 3.33 Bony plastron cf adult *Cycloderma aubryi*.

FIGURE 3.34 Bony plastron cf adult *Cycloderma frenatum*.

FIGURE 3.35 Bony plastron of adult *Cyclanorbis senegalensis.*

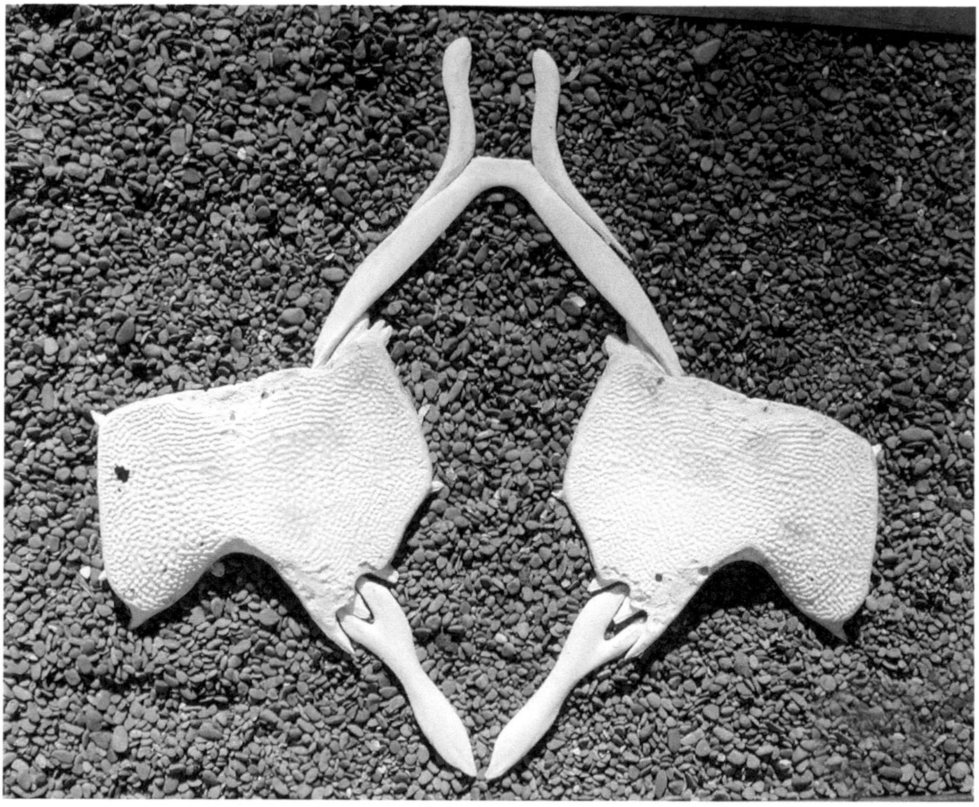

FIGURE 3.36 Bony plastron of adult *Cyclanorbis elegans.*

Figure 3.37 shows the bony carapace of an extremely old, record sized (42 kg in life) *Apalone ferox* with complete ankylosis of shell elements. Figure 3.38 shows the bony carapace of a juvenile *Trionyx triunguis* (CL = 21 cm) with a bizarre and complete ankylosis of costal I through costal IV on each side and including the associated neural bones, these elements being fused into a huge, continuous plate; the nuchal bone and the posterior half of the carapace are unaffected. The specimen was starting to show a degree of lordosis at the time of death.

ACKNOWLEDGMENTS

Many thanks to Jeanette Wyneken for sending reprints of relevant source materials and for her patience with my numerous lapses from approved format, not to mention her good-humored tolerance of my tardy submissions. Erin Dougherty and Simona Ceriani kindly donated their time and skills to the preparation of the photographic illustrations for this chapter.

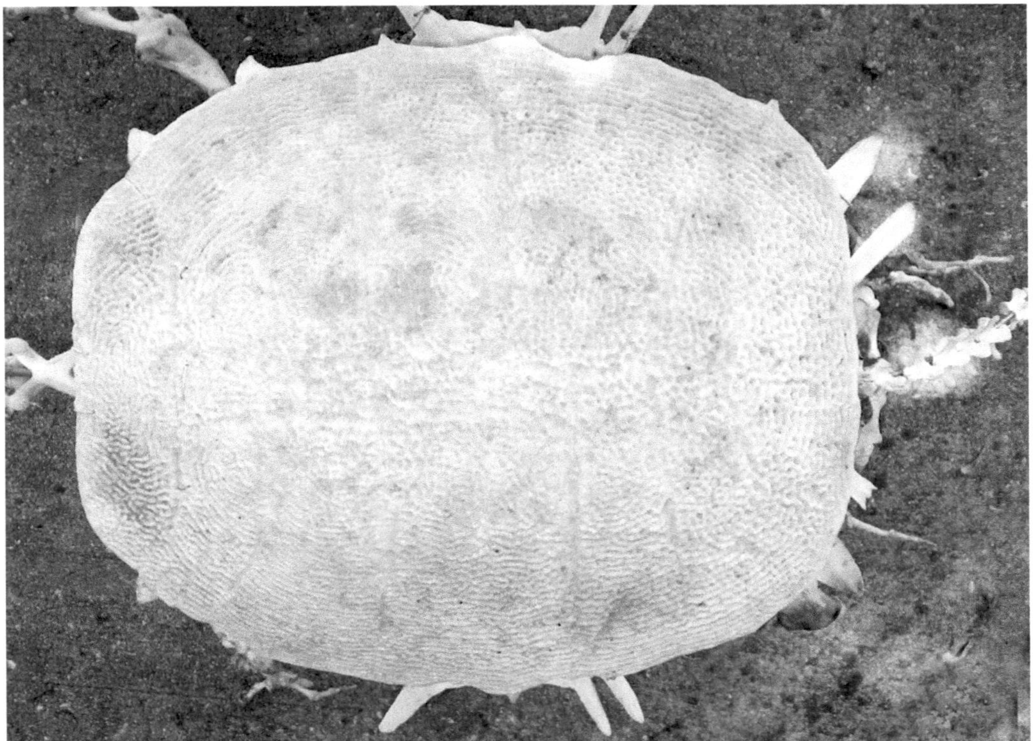

FIGURE 3.37 The bony carapace of an extremely old *Apalone ferox* of record size (42 kg in life), showing the complete ankylosis of shell elements.

FIGURE 3.38 The bony carapace of a juvenile *Trionyx triunguis* (CL = 21 cm) shows complete ankylosis of costal I through costal IV on each side, as well as neural 1 to neural IV as a single block. The nuchal bone and the posterior half of the carapace retain all of their sutures.

REFERENCES

Agassiz, L., *Contributions to the Natural History of the United States of America, Vol. 1*, Boston: Little, Brown & Company, 1857, 452.

Arnold, E.N., Indian Ocean giant tortoises: Their systematics and island adaptations, *Phil. Trans. Roy. Soc., London, B.*, 286, 127–145, 1979.

Auffenberg, W., The genus *Gopherus* (Testudinidae): Pt. 1. Osteology and relationships of extant species. *Bull. Florida State Mus., Biol. Sci.* 20, 2, 47–110, 1976.

Bour, R., Note sur *Pelusios adansonii* (Schweigger, 1812) et sur une nouvelle espèce affine du Kenya (Chelonii, Pelomedusidae), *Studia palaeocheloniologica*, Salamanca, 2, 2, 23–54, 1986.

Bramble, D.M., Emydid shell kinesis: Biomechanics and evolution, *Copeia*, 3, 707–727, 1974.

Bramble, D.M., Hutchison, J.H., and Legler, J.M., Kinosternid shell kinesis: Structure, function, and evolution, *Copeia*, 2, 456–475, 1984.

Cherepanov, G.O., Ontogenetic development of the shell in *Trionyx sinensis* (Trionychidae, Testudinata) and some questions on the nomenclature of bony plates, *Russian J. Herpetol.*, 2, 2, 129–133, 1995.

Claude, J., Paradis, E., Tong, H., and Auffray, J.-C., A geometric morphometric assessment of environment and cladogenesis on the evolution of the turtle shell, *Biol. J. Linn. Soc.*, 79, 485–501, 2003.

Gadow, H., Orthogenetic variation in the shells of Chelonia, *Willey's Zool. Results*, Pt. III, 207–262, 1899.

Gaffney, E.S., Triassic and early Jurassic turtles, in *The Beginning of the Age of Dinosaurs*, K. Padian (ed.), Cambridge, UK: Cambridge University Press, 1986, 183–186.

Gaffney, E.S., Hutchinson, J.H., Jenkins, F.A., Jr., and Meeker, L.J., Modern turtle origins: The oldest known cryptodire, *Science (Amer. Soc. Adv. Sci.)*, 237, 289–291, 1987.

Gaffney, E.S., Tong, H., and Meylan, P.A., Evolution of the side-necked turtles: The families Bothremydidae, Euraxemydidae, and Araripemydidae, *Bull. Am. Mus. Nat. Hist.*, 300, 1–698, 2006.

Heude, P-M., Memoire sur les *Trionyx. Memoires concernant l'histoire naturelle de l'Empire Chinois par des Pères de la Companie de Jesus, Chang-Hai, imprimérie de la Mission Catholique à l'orphelinat de Tou-sè-wè'*, 1, 1–45 + pls., 1880.

Holt-White, R., *The Life and Letters of Gilbert White of Selbourne*, Vol. I, London: John Murray, 1901, 330.

Hutchison, J.H., and Bramble, D.M., Homology of the plastral scales of the Kinosternidae and related turtles, *Herpetologica*, 37, 2, 73–85, 1981.

Jackson, D.C., How a turtle's shell helps it survive prolonged anoxic acidosis, *News. Physiol. Sci.*, 15, 4, 181–185, 2000.

Kordikova, E., Paedomorphosis in the shell of fossil and living turtles. *N. Jb. Geol. Paläont. Abh.*, 218, 3, 399–446, 2000.

Loveridge, A., and Williams, E.E., Revision of the African tortoises and turtles of the suborder Cryptodira, *Bull. Mus. Comp. Zool.*, 115, 6, 163–557 + pls., 1957.

Meylan, P.A., Phylogenetic relationships of soft-shelled turtles, *Bull. Am. Mus. Nat. Hist.*, 186, 1–101, 1987.

Pritchard, P.C.H., Occurrence of mesoplastra in a cryptodiran turtle, *Lepidochelys olivacea*, *Nature*, 210, 5036, 652, 1966.

Pritchard, P.C.H., Numerical reduction of bony plastral elements in the kinosternid turtle *Claudius angustatus*. *Copeia*, 1, 151–152, 1971.

Pritchard, P.C.H., A survey of neural bone variation among recent chelonian species, with functional interpretations. Festshrift for Marian Mlynarski. *Acta Zool. Cracov.*, 31, 26, 626–686, 1988.

Pritchard, P.C.H., Carapacial pankinesis in the Malayan softshell turtle, *Dogania subplana*, *Chel. Cons. Biol.* 1, 1, 31–36, 1993.

Pritchard, P.C.H., Akinesis and plastral scute homologies in *Sternotherus* (Testudines: Kinosternidae), *Chel. Cons. Biol.*, 4, 3, 671–674, 2003.

Ruckes, H., The morphological relationships between the girdles, ribs and carapace, *Ann. N.Y. Acad. Sci.*, 31, 81–120, 1929.

Sharp, D.L., Turtle eggs for Agassiz, *Atlantic Classics*, Atlantic Monthly Company, 23–44, 1916.

Walker, W.F. Jr., Development of the shoulder region of the turtle *Chrysemys picta marginata*, with special reference to the primary musculature, *J. Morph.*, 80, 195–249, 1947.

Wiesner, C.S., and Iben, C., Influence of environmental humidity and dietary protein on pyramidal growth of carapaces in African spurred tortoises (*Geochelone sulcata*), *J. Anim. Physiol. Anim. Nutr.*, 87, 66–74, 2003.

Zangerl, R., The turtle shell, in *Biology of the Reptilia, Vol. 1: Morphology,* C. Gans, A. d'A. Bellairs, and T.C. Parsons (eds.), New York: Academic Press, 1969, 311–339.

Zangerl, R., Hendrickson, L.P., and Hendrickson, J.R., Redescription of the flatback sea turtle, *Natator depressus*, *Bishop Mus. Bull. Zool.*, 1, Honolulu, 69, 1988.

4 Long Bone Allometry in Tortoises and Turtles

Gustavo A. Llorente, Xavier Ruiz, Adrià Casinos, Ignacio Barandalla, and Carles Viladiu

CONTENTS

4.1 INTRODUCTION

Chelonians are in many ways unique among the tetrapods. Above all, the presence of the shell, with girdles that lie within the rib cage and limbs emerging laterally from the anterior and posterior apertures of the shell (Webb et al., 1978), affects various aspects of the biology and morphology of all turtles. Various chelonian species inhabit a variety of marine, freshwater, and terrestrial habitats. Morphologically, a turtle's leg design corresponds to its habitat: marine chelonians have paddles, whereas terrestrial and freshwater turtles have more typical tetrapod limbs. At the current time, the plesiomorphic limb design remains unknown. Extant marine turtles appear to be derived from an amphibious ancestor but it is possible that other extinct groups of marine turtles existed that had different derivations (Renous, 1995; Pritchard, 1979). In cryptodires, it appears that the oldest turtles were amphibious (Romer, 1968; Gaffney et al., 1987). In this case, both marine and terrestrial conditions would be apomorphic. There are important differences in the long bones of marine and non-marine turtles, mainly in proximal bones (humerus and femur); these bones tend to have curved diaphyses in non-marine forms, whereas they are straighter in marine forms (Romer 1968). Alternatively, there is a lengthening of both metacarpus and metatarsus in the autopods of marine turtles (Guibé, 1970).

In terms of locomotion, turtles are a special group within tetrapods. In terrestrial species, gaits are extraordinarily slow. This is not surprising, as terrestrial turtles are herbivorous and consequently do not actively pursue prey and their shells provide protection against predators, thus it seems that there is no reason for natural selection to have favored faster walking speeds (Alexander, 2003). In contrast, freshwater species—both bottom-walkers and swimmers—are able to engage in a moderate speed of movement with a dominant diagonal limb sequence (Renous, 1995; see also Zug, 1971, and Walker, 1972, on this point).

It is assumed that the slower an animal walks or moves, the greater it can tolerate departures from the equilibrium. Normally, in tetrapods low speeds mean that even with only three feet on the ground, the animal can achieve and maintain a stable equilibrium. However, despite their slow gaits some tortoises display normal stride stages with only two feet on the ground at any one time, suggesting that they do not always maintain stable equilibrium (Jayes & Alexander, 1980; Zug, 1971). The lateral sequence of limbs in this case is dominant.

The most likely reason for the existence of slow locomotor speeds in tortoises is that their muscles are remarkably slow. Slow muscles are not able to produce abruptly changing forces. In mammals, the maximum possible isometric stress of leg muscles is about 0.3 MPa, although these high values are apparently required only in very strenuous activities. During walking, tortoise leg muscles exert maximum stresses between 0.1 MPa and 0.2 MPa (Van Leeuwen & Jayes, 1981). These maximum values of turtle legs are still lower than the lowest values recorded in mammals. For tortoises, one possible advantage of having these particular muscles is the ability to maintain tension at low metabolic costs, which in turn facilitates higher energy efficiency (Alexander, 1982).

There are two parameters that illustrate the extent to which walking in tortoises is different from other tetrapods. The first is duty factor, which is the percentage of the stride duration for which the corresponding foot is on the ground. Tortoises walk with duty factor values of 0.75 or more, which are unusually high relative to duty factor values observed in mammals (Jayes & Alexander, 1980; Hildebrand, 1976). Consequently, the forces that tortoises exert while walking are not much higher than those exerted while standing. For mammals to exhibit similarly high duty factor values, they would have to maintain three feet on the ground at all times to maintain adequate equilibrium. In contrast, tortoises often walk with as many as two feet off the ground. Experiments with force platforms have shown that the gait used by tortoises is close to optimal for such slow muscles.

The second parameter is the equilibrium number, equal to

$$\frac{g}{2f^2h} \tag{4 1}$$

where g is the gravitational acceleration, f is the stride frequency, and h is the length of the leg (Alexander, 1993). This dimensionless parameter can be used as an index of the animal's necessity to preserve equilibrium when walking. Assuming that $1/f$ is the duration of a stride, an animal submitted to gravitational acceleration g will fall a distance $g/2f^2$. Remember that frequency is the inverse of time, and the general expression for distance as function of time (t) and any acceleration (a) is $\frac{1}{2}at^2$. The distance that it will fall without hitting the ground will be a little less than h, the length of the leg. In mammals, the highest value of the equilibrium number observed is about 5, corresponding to a dog that is slowly walking. In tortoises, much higher values have been recorded, up to 200 for *Geoemyda* (Alexander, 2003).

According to Romer (1968), the peculiarities of the chelonian limb structure appear to be more related to the presence of the shell rather than to the nature of the habitat frequented by a particular turtle species. This is exceptional relative to other tetrapods (e.g., rodents), where habitat and the corresponding locomotion required in that habitat exert an important selective pressure on limb structure, either for the bones or muscles (Bou et al., 1987, 1990; Casinos, 1994; Castiella & Casinos, 1990). Moreover, Romer (1968) reported that in amphibious reptiles limb length is positively related to the active role that the limbs play as propulsive organs: smaller limbs tend to be associated with a tail that is the primary propulsive element in swimming. Again, this appears to be exceptional among tetrapods, where limb scaling is probably in accordance with geometrical similarity criteria (Alexander, 1985), except in the case of adaptive variations (Casinos et al., 1993).

In this chapter, we compare bone diameters and lengths across various families of turtles to test whether limb scaling in chelonians follows the patterns described previously for other tetrapods and whether there are differences in limb scaling between aquatic and terrestrial species—which may be related to locomotor adaptations, as previously reported in mammals.

4.2 MATERIALS AND METHODS

The total length and the antero-posterior diameter at mid-diaphysis of the long bones (humerus, ulna, femur, and tibia) were measured as straight-line distances, independently of bone curvature. The same was done with the longest metacarpal and metatarsal. In total, 125 specimens were studied, although not all bones were available from each specimen. Specifically, the following bones were examined: 123 humeri, 124 ulnae and femora, 122 tibiae, 42 metacarpals, and 46 metatarsals. The material studied belonged to nine families, of which two were pleurodires (Pelomedusidae and Chelidae) and seven were cryptodires (Testudinidae, Trionychidae, Chelydridae, Emydidae, Kinosternidae, Cheloniidae, and Dermochelyidae) (Webb et al., 1978). All the bones belonged to the collections of the former Laboratoire d'Anatomie Comparée of the Muséum National d'Histoire Naturelle of Paris (Table 4.1).

The lengths of the bones were regressed to diameters, and diameters set as the independent variable. In addition, the relationships between lengths of various bones were compared using regression analysis. We compared:

- The length of the humerus to the length of the ulna
- The length of the femur to the length of the tibia
- The length of the humerus to the length of the femur
- The length of the ulna to the length of the tibia
- The length of the metacarpal to the length of the metatarsal
- The length of the humerus plus ulna to the length of the femur plus tibia
- The length of the humerus, ulna, and longest metacarpal to the length of the femur, tibia, and the longest metatarsal

In these comparisons, either the distal bone length or the hindlimb length was used as the independent variable. All length data were log-transformed for the analyses. Model II (Sokal & Rohlf, 1995) was used for all the regressions. The statistical software used (ALOMETRA; Casinos et al., 1993) calculated confidence intervals for the y-intercept (a) and exponent (b) at 95% of the standard allometric equation (Alexander 1993):

$$y = ax^b \tag{4.2}$$

Length/diameter relationships of the stylopod, zeugopod, and autopod were examined to test whether these parameters were scaled among each other according to geometrical similarity, as it is the case for most tetrapods already studied or, alternatively, according to elastic similarity (Alexander, 1985). Geometrical similarity predicts the following relationship for the scaling of length (l) against diameter (d):

$$l \propto d \tag{4.3}$$

However, if bones are elastically similar, the expected proportion is

$$l \propto d^{2/3} \tag{4.4}$$

4.3 RESULTS

When all bones were considered independently, there was a significant negative allometry, except in the case of the metacarpal and the metatarsal (Table 4.2). Although the metacarpal scaled with a

TABLE 4.1
List of Species and Samples Used in This Study

		Femur	Tibia	Humerus	Ulna	Metacarpal	Metatarsal
Chelidae	*Chelodina expansa*	1	1	1	1	—	—
	Hydromedusa tectifera	1	1	1	1	—	—
Cheloniidae	*Caretta caretta*	1	1	1	1	1	1
	Eretmochelys imbricata	1	1	1	1	1	1
Chelydridae	*Chelydra serpentina*	5	5	5	5	1	1
	Macroclemmys temminckii	1	1	1	1	1	1
Dermochelyidae	*Dermochelys coriacea*	1	1	1	1	1	1
Emydidae	*Clemmys guttata*	3	3	3	3	2	2
	Chrysemys picta	2	2	2	2	—	1
	Emys orbicularis	9	9	9	9	6	7
	Graptemys pseudogeographica	2	2	2	2	1	1
	Malaclemys terrapin	1	1	1	1	—	—
	Terrapene carolina	5	5	5	5	—	2
	Terrapene ornata	1	1	1	1	1	1
	Trachemys dorbigni	1	1	1	1	1	1
	Trachemys scripta	4	4	4	4	4	4
Geoemydidae	*Cuora amboinensis*	3	3	3	3	2	2
	Chinemys reveesii	2	2	2	2	2	2
	Geoclemmys hamiltonii	1	1	1	1	—	—
	Heosemys grandis	1	1	1	1	—	—
	Hieremys annandalii	1	1	1	1	—	—
	Malayemys subtriuga	3	3	3	3	1	1
	Mauremys caspica	3	3	3	3	2	1
	Melanochelys trijuga	1	1	1	1	1	1
	Rhinoclemmys punctularia	2	2	2	2	1	1
	Siebenrockiella crassicolis	1	1	1	1	—	—
Kinosternidae	*Kinosternon leucostomum*	2	2	2	2	—	—
	Kinosternon subrubrum	2	2	2	2	—	—
Pelomedusidae	*Pelusios castaneus*	1	1	1	1	—	—
	Pelusios castanoides	1	1	1	1	—	1
	Pelusios gabonensis	1	1	1	1	—	1
	Pelusios subniger	1	1	1	1	—	—

TABLE 4.1
List of Species and Samples Used in This Study (continued)

		Femur	Tibia	Humerus	Ulna	Metacarpal	Metatarsal
Podocnemididae	*Erymnochelys madagascariensis*	3	3	3	3	3	3
	Podocnemis expansa	1	1	1	1	1	—
	Podocnemis lewyana	1	1	1	1	—	1
Testudinidae	*Dipsochelys arnoldi*	1	1	1	1	1	1
	Dipsochelys dussumieri	—	—	—	1	—	—
	Geochelone carbonaria	8	8	8	8	—	—
	Geochelone pardalis	3	2	3	3	1	2
	Geochelone radiata	10	10	9	10	—	1
	Geochelone sulcata	2	2	2	2	—	—
	Homopus aerolatus	2	2	2	2	—	—
	Indotestudo elongata	3	3	3	3	—	—
	Kinixys belliana	1	1	1	1	—	—
	Kinixys erosa	1	1	1	1	—	—
	Testudo graeca	11	11	11	11	1	—
	Testudo hermanni	1	1	1	1	—	—
	Testudo kleinmanni	1	1	1	1	—	—
	Testudo marginata	4	3	4	3	—	—
Trionychidae	*Aspideretes gangeticus*	2	2	2	2	2	1
	Cycloderma aubryi	1	1	1	1	1	1
	Chitra indica	1	1	1	1	1	1
	Lissemys punctata	1	1	1	1	—	1
	Trionyx triunguis	1	1	1	1	1	1

TABLE 4.2
Regressions of Length to Diameter for the Different Bones Studied; Confidence Interval (CI) for *a* and *b* Parameters of the Regression Line Are Given

Bone	Equation	n	r	CI *a*	CI *b*
Humerus	y = 10.28x0.843	123	0.960	9.536–11.082	0.801–0.885
Ulna	y = 7.275x0.860	124	0.919	6.619–7.995	0.800–0.920
Metacarpal	y = 3.267x1.307	41	0.722	2.446–4.364	1.014–1.599
Femur	y = 11.608x0.819	124	0.965	10.381–12.383	0.780–0.857
Tibia	y = 11.332x0.787	122	0.956	10.689–12.014	0.745–0.829
Metatarsal	y = 6.795x0.908	46	0.826	5.375–7.860	0.754–1.062

TABLE 4.3
Regressions of Length to Length between Bones or Addition of Bones; Confidence Interval (CI) for *a* and *b* Parameters of the Regression Line Are Given

Correlation	Equation	*n*	*r*	CI *a*	CI *b*
Humerus/ulna	$y = 0.508x^{1.031}$	122	0.974	0.433–0.596	0.989–1.073
Tibia/femur	$y = 0.667x^{1.020}$	122	0.984	0.590–0.754	0.988–1.053
Humerus/femur	$y = 0.686x^{1.103}$	123	0.966	0.567–0.831	1.052–1.154
Ulna/tibia	$y = 0.538x^{1.117}$	122	0.954	0.437–0.663	1.056–1.177
Metacarpal/metatarsal	$y = 0.303x^{1.312}$	36	0.936	0.198–0.464	1.152–1.472
Humerus+ulna/femur+tibia	$y = 0.591x^{1.104}$	121	0.969	0.478–0.731	1.054–1.153
Humerus+ulna+metacarpal/ femur+tibia+metatarsal	$y = 0.468x^{1.132}$	36	0.984	0.339–0.647	1.061–1.202

high exponent (1.307), it was not significantly different from 1. Regarding the criteria for elasticity (McMahon, 1975), the predicted exponent (0.67) was always outside of the confidence intervals.

When the equations derived from regressions among lengths of different bones were analyzed (Table 4.3), none of the slopes were significantly different from 1 with the exception of the regression of the longest metacarpal to the longest metatarsal, which showed significant positive allometry. In terms of the relationship between the length and the diameter of the ulna (Figure 4.1), the data from several families stood out by virtue of their position above the regression line (Trionychidae, Pelomedusidae, Chelidae, and Podocnemididae) or below the regression line (Dermochelyidae).

For the relationship between length and diameter in the longest metacarpals, there was a wide dispersion of values as reflected by the lowest correlation coefficient derived from all equations calculated. Individuals from three families (Cheloniidae, Trionychidae, and Dermochelyidae) placed above the regression line displayed a relatively greater length of the metacarpals, whereas two families (Testudinidae and Chelydridae) had values below the regression line, corresponding to relatively short metacarpals (Figure 4.2). For the relationship between the length and the diameter

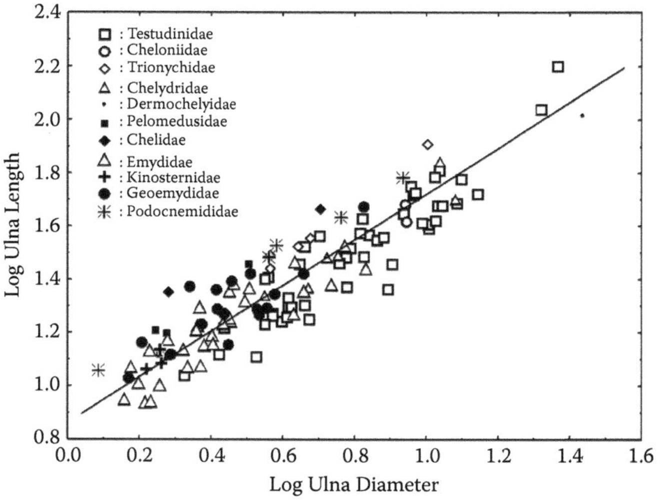

FIGURE 4.1 Scaling of length to diameter of ulna bone.

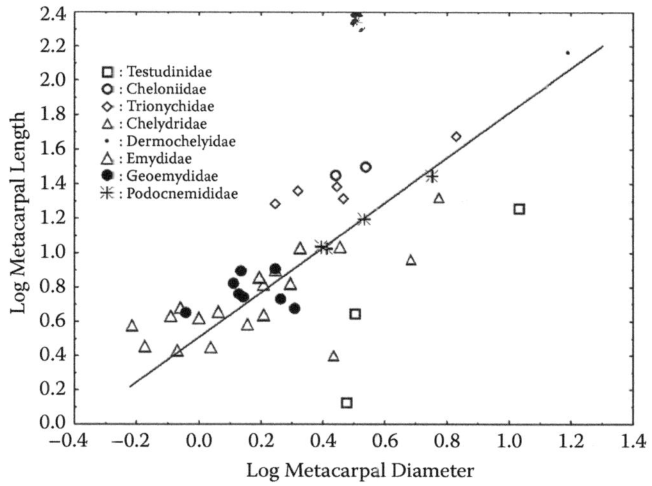

FIGURE 4.2 Scaling of length to diameter of longest metacarpal.

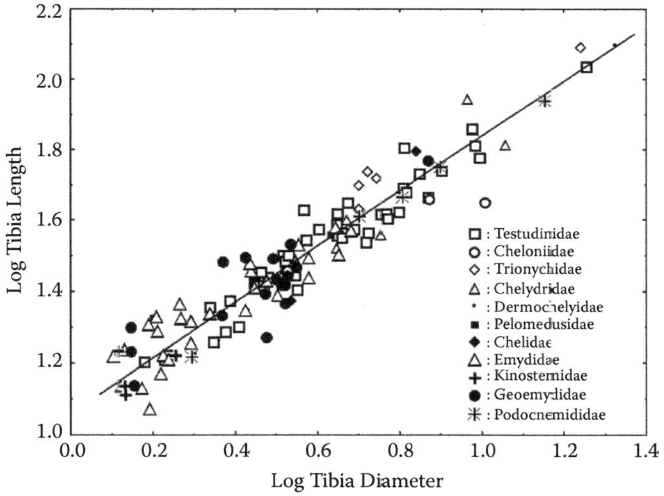

FIGURE 4.3 Scaling of length to diameter of tibia bone.

of the tibia, the values from the Trionychidae and Chelidae were clearly placed above the regression line, whereas those from the Cheloniidae were below (Figure 4.3).

Although the pattern of the scaling of the longest metatarsal (Figure 4.4) also displayed a relatively wide dispersion of values, there were some differences from the patterns seen in the analyses of the longest metacarpal. In particular, the values from the Cheloniidae and Chelydridae were distributed on both sides of the regression line. Only values corresponding to the Dermochelyidae and, to a lesser extent, the Podocnemididae were placed above the regression line, whereas values from the Testudinidae were clearly below. In the case of the Emydidae, values corresponding to terrestrial species were all above the regression line. The pattern of ulna length versus humerus length

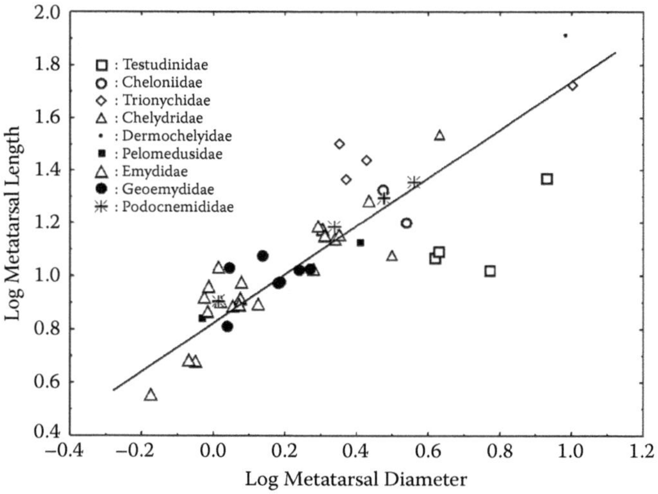

FIGURE 4.4 Scaling of length to diameter of longest metatarsal.

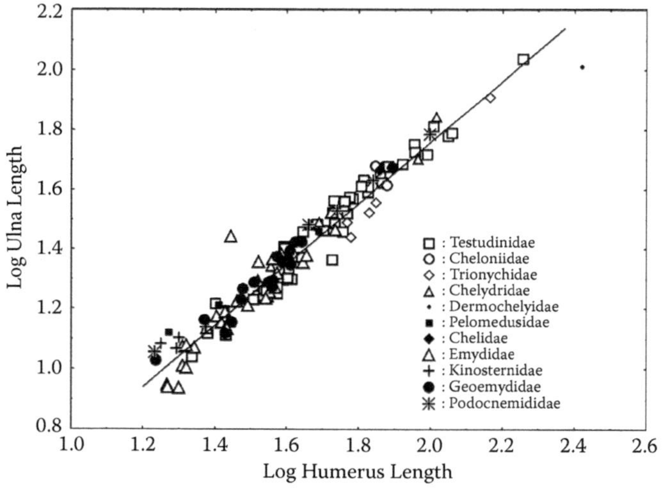

FIGURE 4.5 Scaling of ulna length to humerus length.

was more uniform than for metatarsal or metacarpal (Figure 4.5). However, some families tended to have longer ulnae than humeri (Pelomedusidae, Kinosternidae, and Podocnemididae) or vice versa (Trionychidae and Dermochelidae).

The length of the forelimbs relative to the hind limbs was evaluated by plotting the plot of length of humerus plus length of the ulna against the length of femur plus the length of tibia (Figure 4.6). Although the dispersion of values was tight, specimens from some families were nearly always above the regression line (Testudinidae, Cheloniidae, Dermochelyidae, and Kinosternidae), showing a tendency to have longer long bones in the forelimbs than in the hind limbs. Points corresponding to some other families (Trionychidae, Chelidridae, Pelomedusidae, Chelidae, and Podocnemididae) were placed below the regression line, indicating longer bones in the hindlimbs. In the Emydidae,

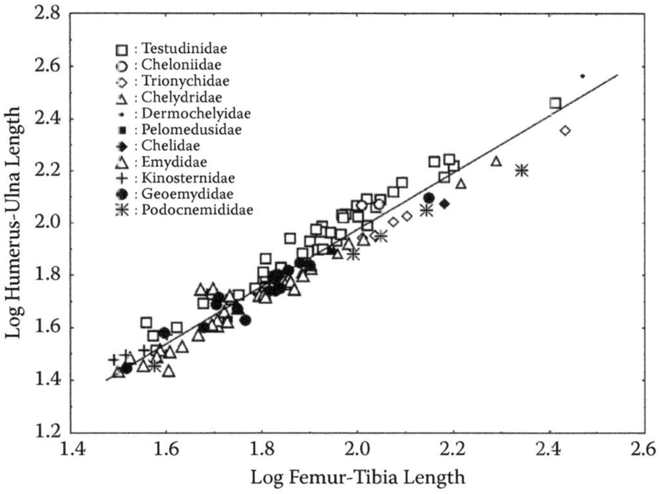

FIGURE 4.6 Scaling of humerus–ulna length to femur–tibia length.

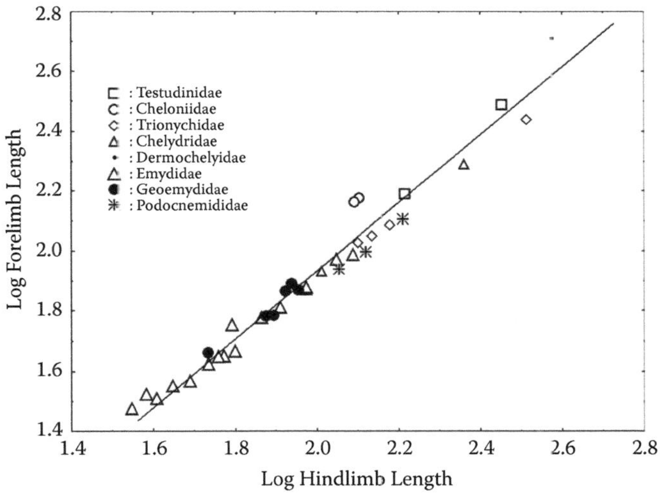

FIGURE 4.7 Scaling of forelimb length to hindlimb length.

terrestrial specimens placed above the regression line, whereas the aquatic forms of this family were distributed on both sides of the regression line.

Finally, we compared the total length of both limbs (Figure 4.7). Individuals from Testudinidae, Cheloniidae, and Dermochelidae displayed relatively longer forelimbs, whereas those from Trionychidae, Chelydridae, and Podocnemididae showed longer hind limbs. These results suggest that groups with longer hind limbs use them mainly to generate forward thrust. However, according to Pace et al. (2001) in the special case of Trionychidae different strategies of thrust generation—involving in some cases forelimbs—seem to exist.

4.4 DISCUSSION

With the exception of metapodial bones, all the regressions produced exponents significantly less than 1 (the predicted outcome of the geometrical similarity hypothesis) but not as low as $^2/_3$ (the predicted outcome of the elastic similarity hypothesis). On the basis of these results, it appears that turtle bones have more in common with the shape of avian hind limb bones (Olmos et al., 1996) than with the long bones of normal quadrupedal mammals (Alexander et al., 1979). The relationship between chelonian locomotion and limb bone length/diameter proportions deserves attention in future research.

Based on research of limb proportions in terrestrial mammals, Raich and Casinos (1991) reported that most of the locomotor specialization took place in the autopod (metacarpals and metatarsals) rather than in the long bones. Our present results suggest a similar situation in turtles. The scaling of length to diameter of the longest metacarpal and metatarsal bones revealed a greater dispersion of values than that observed when other appendicular bones were studied, or even when both long bones of a limb together were considered. In general, it appears that chelonian families that are highly adapted to an aquatic environment have metacarpals and metatarsals that are relatively long and narrow—for example, the Trionychidae and Dermochelyidae, or only long and thin metacarpals, like the Cheloniidae. Note that these three turtle families have flippers. However, families with metacarpals and metatarsals that tend to be short or wide are more difficult to characterize. None of them have flippers, and among these families the most remarkable shape is present in the Testudinidae, the family that includes the most terrestrial turtle forms, with both extremely short metacarpals and metatarsals. When the scaling of total lengths of the forelimb to hind limb was analyzed, the pattern was very different; although the dispersion of the cluster was not great, the Testudinidae, together with two marine families (the Cheloniidae and Dermochelyidae) display rather longer fore limbs.

There remain basic questions about bone size as related to mechanical behavior. For instance, is parameter (1) an adequate index of the speed exhibited by a turtle, i.e., the lower the value, the faster the animal? Renous (1995) showed that aquatic turtle forms are able to attain moderate running speeds. Combined with low values of parameter (1), this information predicts that aquatic animals should have longer limb bones than terrestrial species. However, the data presented above reveal that limb length is mainly a function of the length of the metacarpal or metatarsal. It has been observed that aquatic turtle families have the longest autopodial bones, in accordance with the previous observation by Guibé (1970). Moreover, the Testudinidae—which are the most terrestrial chelonians—have the shortest bone lengths. Interestingly, within the Emydidae terrestrial species tended to have shorter bones whereas aquatic species tended to have longer bones, concordant with the predicted adaptation to aquatic or terrestrial environments exhibited by different species in this family. In general, we conclude that locomotor adaptation is an important factor acting on appendicular bone shape as has been observed in mammals, although the typology does not appear to be the same. Casinos (1994) showed that long bones in aquatic rodents tended to be relatively slender. This could be true for metapods also, based on the results described previously concerning the Dermochelyidae. However, this has not been observed in the ulna and tibia bones as they tend to be rather robust. One possible explanation for this difference is that while swimming, rodents propel themselves with their body and marine turtles produce thrust by beating their flippers (Alexander, 2003).

The results reported in this chapter represent a first pass on the subject. More work is needed to fully understand the role of locomotor adaptation on appendicular shape in turtles, particularly within families that include both terrestrial and aquatic forms, such as the Emydidae. The influence of other factors also deserves attention—for example, geographic variations or sexual dimorphism, whose importance has been highlighted in other types of anatomical studies (Fairbairn, 1997; Kamezaki & Matsui, 1995; Wyneken et al., 1999). Also, methods used to study other species such as birds can inform this type of research. For instance, in turtles it would be useful to study cross-sectional parameters with mechanical significance, such as the area of the cross sections, the second moment, and the polar moment (Cubo & Casinos, 1998). Similarly, more work on bone curvature is needed, given its importance in terrestrial turtles. Although the mechanical meaning of bone curvature is currently under discussion, there is an indication that the sagittal long bone curvature in birds increases very quickly with body mass (Cubo et al., 1999). From this, it can be predicted that curvature must also increase in a similar manner with any other size parameter, such as the total length of the animal.

ACKNOWLEDGMENTS

Funds from program BOS2000-0997 (Ministerio de Educación y Ciencia of Spain) awarded to A. Casinos made this research possible.

REFERENCES

Alexander, R.M., *Locomotion of Animals*, Glasgow: Blackie, 1982.

Alexander, R.M., Body support, scaling, and allometry, in *Functional Vertebrate Morphology*, M. Hildebrand, et al. (eds.), Cambridge, MA: Belknap Press of Harvard University Press, 1985, 26–37.

Alexander, R.M., Gaits of mammals and turtles, *J. Robot. Soc. Jap.*, 11, 314–319, 1993.

Alexander, R.M., *Principles of Animal Locomotion*, Princeton, NJ: Princeton University Press, 2003.

Alexander, R.M. et al., Allometry of the limb bones of mammals from shrews (Sorex) to elephant (Loxodonta), *J. Zool. Lond.*, 189, 305–314, 1979.

Bou, J., Casinos, A., and Ocaña, J., Allometry of the limb long bones of insectivores and rodents, *J. Morphol.*, 192, 113–123, 1987.

Bou, J. et al., Multivariate analysis and locomotor morphology in insectivores and rodents, *Zool. Anz.*, 225, 287–294, 1990.

Casinos, A., Is locomotor typology in rodents possible?, *Pol. Ecol. Stud.*, 20, 375–381, 1994.

Casinos, A., Quintana C., and Viladiu, C., Allometry and adaptation in the long bones of a digging group of rodents (Ctenomyinae), *Zool. J. Linn. Soc.*, 107, 107–115, 1993.

Castiella, M.J., and Casinos, A., Allometry of leg muscles in insectivores and rodents, *Ann. Sci. Nat. Zool.*, 11, 165–178, 1990.

Cubo, J., and Casinos, A., Biomechanical significance of cross-sectional geometry of avian long bones, *Eur. J. Morphol.*, 38, 112–121, 1998.

Cubo, J., Menten, L., and Casinos, A. Sagittal long curvatura in birds, *Ann. Sci. Nat. Zool.*, 20, 153–159, 1999.

Fairbairn, D.J., Allometry for sexual size dimorphism: Pattern and process in the coevolution of body size in males and females, *Ann. Rev. Ecol. Syst.*, 28, 659–687, 1997.

Gaffney, E.S. et al., Modern turtle origins: The oldest known cryptodire, *Science*, 237, 289–291, 1987.

Guibé, J., Le squelette du tronc et des members, in *Traité de Zoologie*, P.P. Grassé, Vol. XIV, Fasc. II, (ed.), Paris: Masson, 1970, 33–77.

Hildebrand, M., Analysis of tetrapod gaits: General considerations and symmetrical gaits, in *Neural Control of Locomotion*, R.M. Herman et al. (eds.), New York: Plenum Press, 1976, 203–236.

Jayes, A.S., and Alexander, R.M., The gaits of chelonians: Walking techniques for very low speeds, *J. Zool. Lond.*, 191, 353–378, 1980.

Kamezaki, N., and Matsui, M., Geographic variation in skull morphology of the green turtle, *Chelonia mydas*, with a taxonomic discussion, *J. Herpetol.*, 29, 51–60, 1995.

McMahon, T.A., Using body size to understand the structural design of animals: Quadrupedal locomotion, *J. Appl. Physiol.*, 39, 619–627, 1975.

Olmos, M., Casinos, A., and Cubo, J., Limb allometry in birds, *Ann. Sci. Nat. Zool.*, 17, 39–49, 1996.

Pace, C.M., Blob, R.W., and Westneat, M.W., Comparative kinematics of the forelimb during swimming in red-eared slider (*Trachemys scripta*) and spiny softshell (*Apalone spinifera*) turtles, *J. Exp. Biol.*, 204, 3261–3271, 2001.

Pritchard, P.C.H., *Encyclopedia of Turtles*, Hong Kong: T.F.H. Publications, 1979.

Raich, J., and Casinos, A., Limb proportions in terrestrial mammals, *Bel. J. Zool.*, 121, 295–314, 1991.

Renous, S., The ontogeny of locomotion in marine turtles, in *Scientia Herpetologica*, G.A. Llorente et al. (eds.), Barcelona: Asociación Herpetológica Española, 112–119, 1995.

Romer, A.S., *Osteology of the Reptiles*, Chicago: University of Chicago Press, 1968.

Sokal, R.R., and Rohlf, F.J., Biometry. *The Principles and Practice of Statistics in Biological Research*, New York: W.H. Freeman & Company, 1995.
Van Leeuwen, J.L., and Jayes, A.S., Estimates of mechanical stresses in tortoise leg muscles during walking, *J. Zool., Lond.*, 195, 53–69, 1981.
Walker, W.F., Jr., Body form and gait in terrestrial vertebrates, *Ohio J. Sci.*, 72, 177–183, 1972.
Webb, J.E., Wallwork, J.A., and Elgood, J.H., *Guide to Living Reptiles*, London: Macmillan Press, 1978.
Wyneken, J. et al., Site differences in hind limbs and carapaces of hatchling green turtles (*Chelonia mydas*) from Hawaii and Florida, *Chel. Cons. Biol.*, 3, 491–495, 1999.
Zug, G.R., Buoyancy, locomotion, morphology of the pelvic girdle and hind limb and systematics of cryptodiran turtles, *Misc. Publ. Mus. Zool. Univ. Michigan*, 142, 1–98,1971.

5 Evolution of Locomotion in Aquatic Turtles

Sabine Renous, France de Lapparent de Broin, Marion Depecker, John Davenport, and Vincent Bels

CONTENTS

5.1 INTRODUCTION

Chelonians exemplify an ancient and well-differentiated group of vertebrates that has extensively diversified in both number of lineages and habitat use. In spite of this extensive diversification, turtles have maintained a relatively conserved basic body plan.

In vertebrate evolution, the changes required to move from aquatic to terrestrial life were profound. In contrast, it is assumed that the reverse move from land to water was "easier" and "faster," occurring several different times, especially in chelonians (Gaffney et al., 1991; Shaffer et al., 1997). These secondary swimmers presumably capitalized on benefits associated with the aquatic biome, such as enhanced possibilities for dispersal and migration, as well as new access to food. Concurrently, they may also have escaped terrestrial predators and come to play the role of new predators themselves. The reinvasion of the aquatic realm by some species of turtles also necessitated the development of new characters of locomotor performance, such as high speed, extended endurance, deep diving, plus alternative kinetics for acceleration and maneuverability.

Applied to fossils, the relationship between limb morphology and habitat preferences (Joyce & Gauthier, 2004) observed among extant turtles provides evidence supporting the hypothesis that turtles first evolved in a terrestrial habitat and only later moved into aquatic environments. It is consistent with similar observations based on the evolution of the morphology of the turtle skeleton and other external anatomical characters (Gaffney, 1990; Lapparent de Broin, 2001). Studying forms of the turtle shell of the Testudinoidae, Claude et al. (2003) hypothesized that the diversity of shell variation in terrestrial species relative to aquatic species indicates that the turtle shell is less functionally constrained in terrestrial environments than it is in aquatic environments, largely due to more complex and heterogeneous habitats and niches in the former (Schluter, 2000). In the terrestrial environment, the weights of the shell and viscera (not buffered by the buoyancy of water) become a principal physical constraint. A heavy body, which can impede rapid progress on land, is far less constraining in water. However, a variety of physical constraints on the turtle shell are associated with the aquatic environment. Each type of aquatic environment possesses its unique suite of characteristics of salinity, pressures, current velocity, and sediment type, all of which impose limitations on locomotion. Water density and viscosity impose great resistance to the motion of the body. Turtles must reduce this water resistance and propel themselves in the relatively dense and incompressible environment with different systems of force exchange. They must also control their position and body equilibrium. An aquatic habitat also implies other adaptations, including control of body temperature and modification of respiratory, circulatory, and osmotic physiology.

Our objective is to examine whether a direct relationship exists among morphological variation of the locomotor systems, their function, and the diversity of the habitats. Our hypothesis is that locomotor performance depends on structural characters. Associated with slight modulations of behavior, locomotor performance has an effect on fitness in a given habitat. In return, this fitness reflects selective pressures acting back at the structural level of the organism (Dominici, 2003).

Although a large number of earlier studies have made at least brief descriptions of movement in turtles (Barth, 1962; Casamiquela, 1964; Deraniyagala, 1930; Lessertisseur, 1955; Magne de la Croix, 1929, 1933; Mertens, 1960; Oliver, 1955, 1959; Peabody, 1959; Reed, 1957; Sukhanov, 1964, 1974; Webb, 1962; Woods, 1945) a fairly limited number of studies have focused specifically on turtle locomotion. The primary foundation for research on turtle locomotion was laid through the studies of Walker (1971a, 1971b, 1972, 1973, 1979) and Zug (1971, 1972) in the 1970s, which compared locomotor patterns and anatomy in a wide range of species. Subsequently, Jayes and Alexander (1980), Van Leeuwen et al. (1981) and Zani et al. (2005) focused on walking mechanics of terrestrial forms and also on the effect of loads and slopes on locomotion (Claussen et al., 2002; Muegel & Claussen, 1994; Wren et al., 1998; Zani & Claussen, 1994). Davenport and collaborators compared swimming marine and freshwater species (Davenport et al., 1984) and investigated several marine species including *Caretta caretta* (Davenport & Clough, 1986), *Dermochelys coriacea* (Davenport, 1987), and *Lepidochelys olivacea* (Davenport & Pearson, 1994). Renous and her co-workers were interested in the terrestrial locomotion used by marine turtles (Renous, 1988; Renous & Bels, 1991; Renous et al., 1989) and in comparison with swimming activity (Renous & Bels, 1991, 1993). Syntheses of research on sea turtle locomotion were produced by Wyneken (1997) and Renous et al. (1999). More recently, Pace et al. (2001) compared the kinematics of swimming in a freshwater emydid (*Trachemys scripta*) and a highly aquatic trionychid (*Apalone spinifera*), whereas Rivera et al. (2006) examined the maneuverability of *Chrysemys picta* during swimming. Willey and Blob (2004) studied the role of the tail in bottom-walking and several studies focused on muscle activity in freshwater turtles (Earhart & Stein, 2000; Gillis & Blob, 2001; Stein, 2003).

5.2 SYNTHESIS OF KNOWLEDGE OF TURTLE LOCOMOTION

5.2.1 Environmental Constraints and the Origin of Aquatic Turtles: Paleontological Data

The habitat preferences of the principal taxa of turtles (families and isolated genera) described from the Triassic to the present are illustrated in Figure 5.1, showing the Pleurodira to the left of one of the most primitive turtles, *Proganochelys*, and the Cryptodira to the right. The oldest and most primitive forms appear at the bottom of the figure, with more modern forms appearing further up the *x*-axis. As a whole, modern forms present a higher number of derived characters. This arrangement largely conforms to known phylogenies (e.g., Gaffney et al., 1991, 1998; and others), but the relationships given among the groups and taxa are not completely resolved because of the

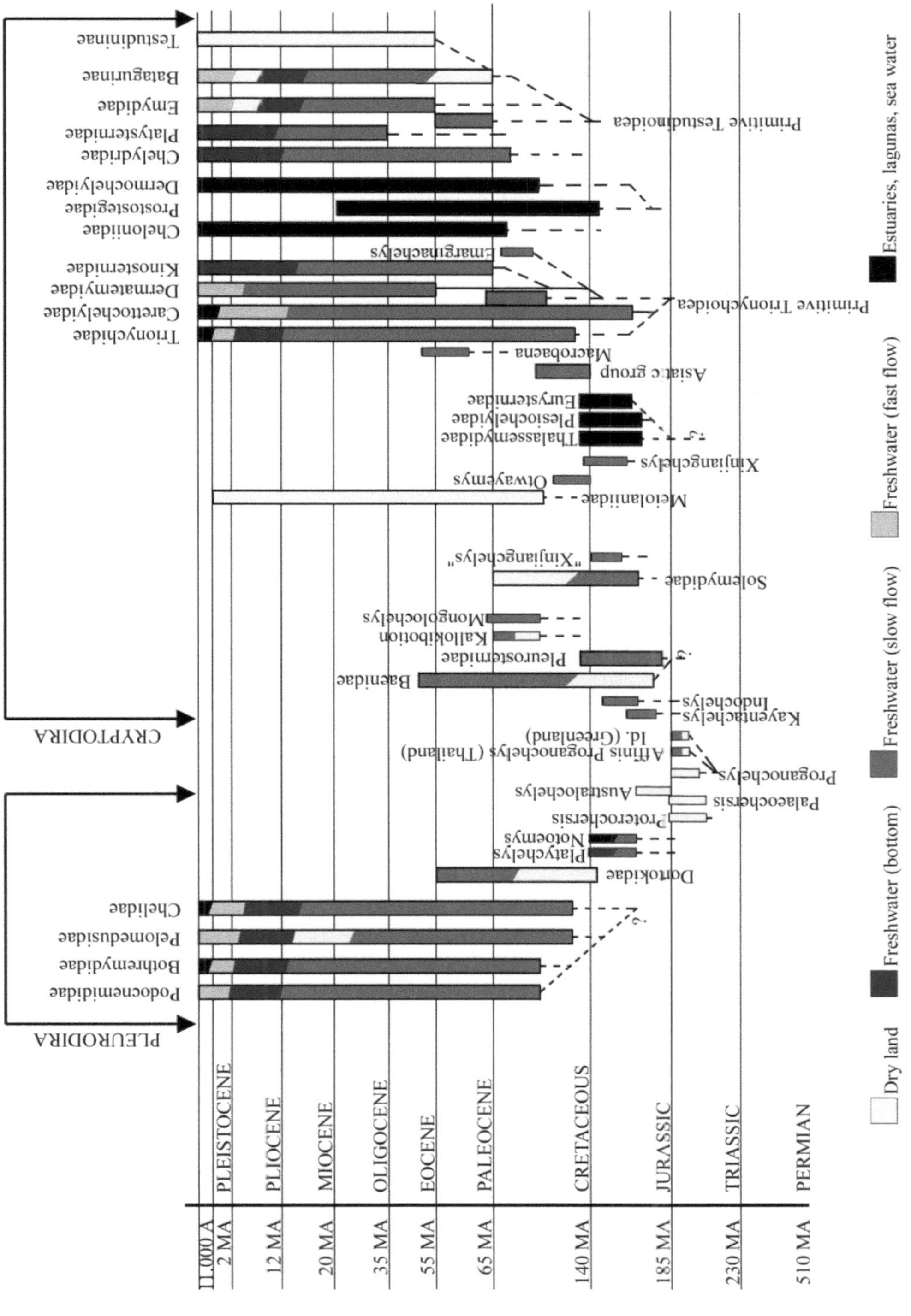

FIGURE 5.1 Preferential habitats of the major fossil and extant groups of Chelonians. Categories of habitats are indicated by grey gradients.

possibility of homoplasy in the characters used to generate the phylogeny. Most of these characters correspond to common trends in the evolution of turtles and are often linked together. They also principally concern our present problem: the evolution of locomotion, particularly in water. These characters are linked to body modifications relative to *Proganochelys* (Gaffney, 1990) (Figure 5.2): shell, neck, and tail (formation of condyles and cotyles), modifications to facilitate better mobility (implying loss of neck ribs and chevrons), lightening of transverse apophyses, skull emargination, and the loss of osteoderms on neck, limb, and tail, including the tail club. Some of these last features may have been subsequently reversed following the reacquisition of a terrestrial habitat. The characters also may be linked to the acquisition of novel prey in the new habitats and therefore some are also linked with characters of the skull, such as palatal, maxillary, and jaw modifications.

The evolutionary history of the chelonians (given here in deliberately non-cladistic format) probably began during the late Triassic with heavy armored terrestrial forms (Gaffney et al., 1987). This presentation in Figure 5.1 emphasizes the habitat preferences (terrestrial versus aquatic) of the groups to demonstrate the division between environments of the fossil and extant turtles. *Proganochelys* from the late Triassic of Germany (Gaffney, 1985, 1990; Gaffney et al., 1987)—which is not the oldest known fossil turtle (Broin, 1985; Lapparent de Broin, 2000a, 2001)—likely represents the fossil most closely related to primitive ancestors. The geographic distribution of the earliest chelonians was extensive in Pangea (Lapparent de Broin, 2000b, 2001). Besides *Proterochersis* in Germany, different species attributed to the genus *Proganochelys* (Lapparent de Broin, 2000a) are also known from Greenland (Jenkins et al., 1994) and Thailand (Broin, 1985), whereas *Palaeochersis* (Rougier et al., 1995) was found in Argentina. The heavy structural characters of *Proganochelys* from Germany (Gaffney, 1990) support the thesis of robustness as terrestrial animals. There is no evidence that they were also aquatic, even though the sediments in which they were found were located near riverside areas. The form from Thailand (Broin, 1985) already had aquatic features, as evidenced by the associated aquatic fossil fauna, as well as its own slender structural aspect. The third most robust form from Greenland (Jenkins et al., 1994) could be interpreted as a semiaquatic animal, as it was found in association with aquatic as well as continental animals. *Palaeochersis*

FIGURE 5.2 Reconstruction of the Triassic turtle *Proganochelys* and a lateral view of its skeleton showing the connections between the girdles, the limbs, and the different parts of the carapace (used with permission from Gaffney, 1990).

is considered to have lived a completely terrestrial existence (Rougier et al., 1998). Overall, it is certain that the "aquatic turtle niche" has been filled since the Triassic period, despite the absence of the distal portions of chelonian limbs in most of the fossil deposits. From very early in turtle evolution, they had already attained a significant ecological diversity including terrestrial and semi-aquatic freshwater forms.

Gaffney et al. (1991) have confirmed the monophyly of the two lines of chelonians: pleurodires and cryptodires. *Proganochelys* was presented as the sister group of all other turtles. The oldest pre-Cretaceous pleurodires are *Proterochersis* from the late Triassic of Germany, *Notoemys* from the Upper Jurassic of Cuba (Oxfordian) and Argentina (Tithonian), and *Platychelys* from the Upper Jurassic (Kimmeridgian-Tithonian) of Switzerland and Germany. *Kayentachelys*, from the early Jurassic of North America, is the sister-group to all other cryptodires (other forms of Middle Jurassic age were found in Morocco and China). Note that there is still some debate about the true phylogenetic relationships of chelonians. For example, Rougier et al. (1995) proposed the exclusion of *Proterochersis* from the pleurodires, considering it as pre-dating the pleurodire-cryptodire separation. *Palaeochersis* might be considered the oldest pleurodire by virtue of its link of the pelvis (very primitive) to the shell, in addition to the shape of the carapacial posterior notch, and based on revised definitions of the characters used in the analyses and taking into account possible homoplasies. Its age in relation to the age of the other Triassic turtles, particularly *Proterochersis* and *Proganochelys*, is not precisely known. *Proganochelys* appears in the Germanic beds after *Proterochersis*, as well as together in the youngest beds, and thus it is not possible to state definitively if *Proganochelys* is the most primitive taxon (Gaffney & Kitching, 1994).

We consider the early Cretaceous as the period of expansion of the pleurodires in Gondwana. Several pleurodiran families were evident from the late Triassic onward, showing a great and early diversification that increased during the Jurassic and Cretaceous periods. The diversification of cryptodires occurred at the same time, principally in Laurasia. By comparison with the terrestrial *Palaeochersis* and *Proterochersis*, *Platychelys* had features suggesting that it was a freshwater or a littoral turtle because it was associated with aquatic forms in the sediment. It was probably not a good swimmer but rather a bottom-walker. The pleurodiran *Notoemys* (De la Fuente & Fernandez, 1989; Fernandez & de la Fuente, 1994; Lapparent de Broin et al., in press) of the same period in Argentina was found in a littoral environment, and the form of the carapace and the structure of the limbs suggests that it was an excellent swimmer in running water. Other Notoemydidae are now known that present the same carapace characteristics: *Caribemys* (de la Fuente & Iturralde-Vinent, 2001) (Oxfordian of Cuba) and *Notoemys zapatocaensis* (Cadena Rueda & Gaffney, 2005) (early Cretaceous of Colombia). The European Cretaceous-Paleocene Dortokidae (Lapparent de Broin & Murelaga, 1996, 1999; Lapparent de Broin et al., 2004) appear to have been semiaquatic forms, whereas the primitive Chelidae from the Cretaceous (de la Fuente et al., 2001; Lapparent de Broin & de la Fuente, 2001) were substantially more aquatic freshwater turtles that had body forms similar to extant forms, such as *Phrynops* (Lapparent de Broin et al., 1998; Lapparent de Broin & de la Fuente, 1999). The group was already characterized by a great diversity of aquatic habitats and probably exhibited an associated wide diversity of aquatic modes of locomotion. Some of the extant Chelidae live in estuaries and can reach islands close to the continents.

Among the extant species, some seem to prefer the slower moving freshwater of slow rivers, ponds, pools, and marshes in forests. They often can bury themselves in mud. Other species prefer running water and can emerge from rivers to rest on the banks. Yet other species live in ponds by day but can walk on land by night to reach other ponds. For this reason, the Chelidae include poor as well as very good swimmers, bottom-walkers, and effective land walkers. In other groups of pleurodires—including the Pelomedusidae, Bothremydidae, and Podocnemididae—we note similar evolutionary diversity. Turtles from massive rivers can also bury themselves into mud. Some Pelomedusidae can occasionally walk in terrestrial habitats. In the fossil record, from the end of the early Cretaceous to the early-middle period of that epoch that encompassed the radiation of the Bothremydidae (Broin, 1988; Lapparent de Broin, 2000a; Lapparent de Broin & Werner, 1998), several littoral African and Mediterranean forms returned to the freshwater environment in Europe. Among the littoral forms, few limbs have been preserved and these do not indicate a strong adaptation to swimming in deep seas, except with respect to the more robust humeri as known in one form. Within the pleurodires, more than 20 fossil forms of Bothremydidae were adapted to littoral conditions (Gaffney et al., 2006; Lapparent de Broin & Werner, 1998) and of extant forms, more than 10 chelid species have acquired secondary adaptations for littoral habitats. However, none has limb apparatus adapted to marine locomotion.

Evolution of the cryptodires resulted in more highly adapted marine forms. These marine cryptodires appeared during the early Jurassic, although they must have already diversified during the Triassic given the similarly earlier diversification of the pleurodires. The cryptodires included forms that invaded terrestrial, freshwater, and also marine habitats (Hirayama, 1998). In their tests of phylogeny by molecular, morphological, and paleontological approaches, Shaffer et al. (1997) estimated that the major radiation of these cryptodiran lineages occurred between 120 and 90 million years ago. However, it is evident that marine-littoral forms were present before, at least during the Lias (Schleich, 1984), and that the first important radiation occurred during the late Jurassic, 160 million years ago.

Kayentachelys (early Jurassic) and *Indochelys* (possibly early to middle-late Jurassic) (Datta et al., 2000) of India were probably freshwater turtles as suggested by the structure of their carapace and the type of sediment in which they were found being associated with freshwater fauna rather than terrestrial. During the Jurassic period, the remains of unidentifiable turtles (not presented in Figure 5.1, see Lapparent de Broin, 2001, for Europe) showed possible littoral adaptations. The Baenidae were possibly semiaquatic, similar to *Kallokibotion*, though perhaps with a greater terrestrial potential as indicated by their heavy body construction (Gaffney, 1972; Gaffney & Meylan, 1992). The Pleurosternidae were freshwater swimmers. *Mongolochelys* was possibly amphibious. Likewise, the Solemydidae probably included semiaquatic forms, but *Solemys* showed adaptations to terrestrial life (Lapparent de Broin & Murelaga, 1999) similar to the Meiolaniidae, in spite of "their bizarre appearance" (Gaffney, 1996). Other probable freshwater forms in Asia include the *Xinjiangchelys* complex from the Jurassic of Asia (Peng & Brinkman, 1993) and *Otwayemys* from the Cretaceous of Australia (Gaffney et al., 1998). During this period in Asia, four other forms (*Sinemys*, *Ordosemys*, *Dracochelys*, and *Hangaïemys*), in addition to *Macrobaena* from the Paleocene, were examples of lines of turtles adapted to freshwater environments, described as "chelydroid" in their appearance (Sukhanov, 2000) because they resemble modern snapping turtles. All of these fossils correspond more or less to related groups without recognized relationships with the modern turtles.

In the highly diversified stock of cryptodires, several turtles progressively adopted characters designed for a marine life. They were probably first bottom-walkers in littoral ecosystems or in lagoons. During the late Jurassic, several families of small littoral turtles, especially the Plesiochelyidae and the Thalassemydidae, spread into European seas. Around 17 genera in the three families (Plesiochelyidae, Thalassemydidae, and Eurysternidae) were involved (Broin, 1994; Lapparent de Broin, 2001; Lapparent de Broin et al., 1996). For example, the genera *Plesiochelys*, *Craspedochelys*, *Tropidemys*, and *Thalassemys* were littoral forms, whereas *Idiochelys*, *Eurysternum*, the *Euryapis–Solnhofia* group, *Hydropelta*, *Achelonia*, and other eurysternids were lagoon species. The Eurysternidae are interesting because they clearly show hand morphology that represents a small paddle, intermediate between the hand of the freshwater forms and the paddle of the open sea forms; this may be considered as an adaptation to the shores and lagoons (Lapparent de Broin, personal observation). A further evolutionary move into the marine environments occurred during the late Jurassic of Argentina with *Neusticemys* (probably the first protostegid) (Fernandez & de la Fuente, 1993). Subsequently, three groups can be distinguished, the Protostegidae including *Santanachelys* (Hirayama, 1994) (early Cretaceous, Brazil), the Cheloniidae, and the Dermochelyidae, these last two families being extant today. By the end of the Cretaceous, some marine turtles had become huge animals, as exemplified by *Archelon* (with a 4-m flipper span) from North America and Allopleuron from Europe (carapace = 1.50 m long; flipper span ca. 2 m; Mulder, 2003).

Within a stock of semi-aquatic to highly aquatic turtles exists numerous extant forms:

- The Trionychidae and Carettochelyidae live in habitats that range from river mud to deep waters of large rivers and lakes, from brackish to sea water (Hughes, 1979; in Scott Thompson's http://www.carettochelys.com and at http://www.chelodina.com; Cann, 1998; Pritchard, 1979). *Carettochelys* is particularly interesting because of its modification of fore limbs as flippers, whereas it retains mobile articulations between the forelimb elements. It also has an ability to use a synchronous movement of the forelimbs, similar to that used by marine forms.
- The Dermatemytidae and Kinosternidae (and their fossil relatives) that can both walk on the bottom and emerge from rivers or lakes to climb in the vegetation of the banks.
- The Chelydridae, bottom-dwellers that are more or less swimmers (Zug, 1971).
- The Platysternidae that live in mountain torrents and emerge by gripping and climbing on rocks and branches.
- The Testudinoidae that present a suite of adaptive designs adopted under the selective pressures of terrestrial or aquatic environments that is difficult to unravel.

While some aquatic forms may have had a terrestrial ancestor, some aquatic forms may have secondarily returned to terrestrial conditions. The Emydidae—which live in ponds, rivers, and lakes—are also found in marshes and brackish waters of lagoons, estuaries, and the coastal marine environment. They include more terrestrial forms in humid forests and some even inhabit dry areas. A terrestrial emydid from the middle Miocene suggests that this evolution toward the terrestrial habitat appeared later in the group (Holman, 1987), and probably several times in the Emydinae, whereas all species of the Deirochelyinae remained aquatic. The Geoemydidae (or Bataguridae) (Spinks et al., 2004) contain both terrestrial and aquatic species, also indicating several occurrences of an environmental change in the habitat preference of these turtles. Only the Testudinidae, which are derived from freshwater forms, are strictly terrestrial.

Finally, we must emphasize that the oldest chelonians possessed distinctly terrestrial habitat preferences and hence locomotion adapted to gravitational constraints. Their girdles were short and stout and the limbs were short with robust elements. The phalange number may be reduced to two (*Proganochelys*, *Palaeochersis*) with the claws being stout, as in extant relatively terrestrial forms such as *Pelomedusa* and in tortoises. The early invasion of the freshwater environment suggests that there was a concurrent rapid secondary adaptation to the new physical constraints. Because egg laying and embryonic development remained associated with the terrestrial habitat, the turtles repeatedly developed the ability to move in both terrestrial and aquatic environments. Around 200 million years ago, they adapted to water and colonized numerous niches in this large and varied environment but always retained the ability to return, at least temporarily, to land. This capacity for reversion makes it challenging to elucidate accurately the functional history of the group.

5.2.2 Structural Constraints on the Chelonian Body Plan

5.2.2.1 Basic Pattern

Vertebrates show a structural organization with a visceral skeleton, an axial skeleton composed of a skull, ribs, and articulated vertebrae, and also an appendicular skeleton composed of pectoral and pelvic girdles and associated paired appendages located outside the rib cage (Figure 5.3). The basic chelonian body plan differs from the typical vertebrate pattern in many features, particularly the presence of a shell and the location of the girdles within the rib cage. The value of such a pattern is proved by the fact that the basic turtle bauplan has remain relatively unchanged throughout

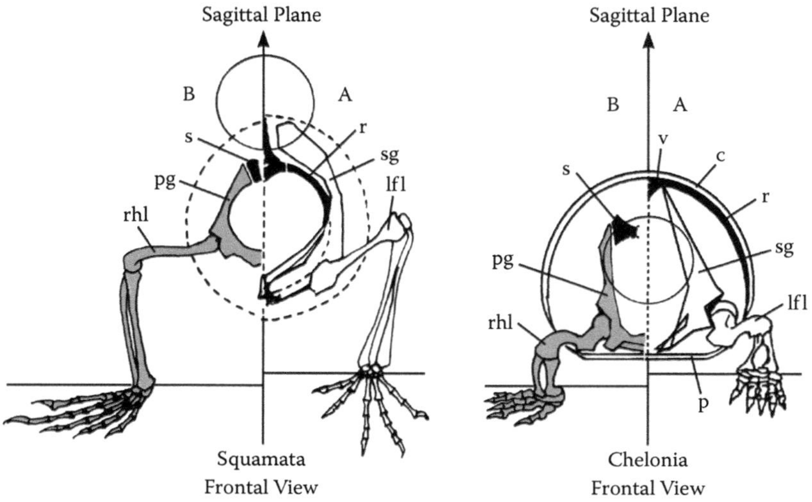

FIGURE 5.3 Comparison of the body organization in two categories of extant reptiles: Chelonia and Squamata. The body of each is crossed by a sagittal plane that separates a right part (A) and a left part (B) corresponding to a frontal view of the body at the levels of the shoulder (in white) and the pelvic (in grey) girdles. The head is indicated only by a projection of a circle on the other structures. c, carapace; lfl, left fore limb; p, plastron; pg, pelvic girdle; r, rib; rhl, right hind limb; s, sacrum; sg, shoulder girdle; v, thoracic vertebra.

evolutionary history, its modifications corresponding to simple adjustments, and also its adaptability because chelonians have dispersed into many different environments. Even if it does not touch the substratum during locomotion, the presence of a shell is largely responsible for the slow speed and peculiarities of limb movements used to maintain balance in heavy animals.

The rigid shell forms a solid box composed of two parts: a dorsal carapace and a ventral plastron. It is clearly established that the carapace is formed (Gaffney & Meylan 1988; Zangerl, 1969) from costal bones with fused ribs, neural bones with fused thoracic vertebrae, and marginal bones, including nuchal and pygal bones. The plastron is formed from the interclavicle and five paired bones (including clavicles) sutured together. The carapace and the plastron are articulated at the lateral margins to enclose the shoulder and pelvic girdles. Although the presence of extensive external armor is a defining characteristic of chelonians, other vertebrates—extant and fossil—also display different forms of external armor. However, the relationship between the ribs and dermis in turtles is unique, as is the position of the girdles within the shell, rather than outside as is the case with the other armored vertebrates (see Chapter 1); this is an autapomorphy (a uniquely derived character). The triradiate pectoral girdle (Lee, 1996a) is different from all other tetrapods; for chelonians, it is composed of a long scapula, an acromion process, and a coracoid process. The acromion anchors the girdle to the plastron, allowing it to pivot during locomotion, whereas the scapula articulates on the internal surface of the carapace. In the absence of a clavicle, these two elements constitute a bony arm between dorsal and ventral parts of the shell. This was reinforced by the dorsal epiplastral processes in Triassic turtles such as *Proganochelys*. By comparison with *Captorhinus*, the dorsal epiplastral process is considered to be homologous with the clavicle (Gaffney, 1990), assuming that the coracoid represents a posterior coracoid and the epiplastron an anterior coracoid. However, the location of this dorsal epiplastral process is variable in the Triassic turtles (Broin, 1985).

According to Walker (1973), in modern chelonians the scapula and acromion constitute a truss that is resistant to transverse compression when the forelimb is retracted. The swinging of the glenoid cavity, produced by the rotation of the pectoral girdle, would amplify the excursion of the limb, adjusting this displacement to that of the longer hindlimb. The exceptional forward protraction of the forelimb during locomotion (and also withdrawal into the shell) would result from the orientation of the articular surface of the humeral head (close to the longitudinal axis of the bone) and axial rotation of the dorsally arching shaft. The shorter radius and ulna are essentially in the same plane. The pelvic girdle shows three components: ventral pubis and ischium separated by a fairly large (in modern forms) thyroidian fenestra, supporting a dorsal vertical ilium that articulates with the sacrum. The pectineal process of the pubis rests upon the xiphiplastron. The ischium also contacts the xiphiplastron, posteriorly to the pubis. However, during evolution this last contact appears only when the posterior extension of the xiphiplastron replaces the hypoischium. The acetabulum articulates with a dorsally arched femur that proximally has a deep fossa between two trochanters and distally hosts a large tibial condyle.

This body bauplan required a complete rearrangement of the basic elements of the vertebrate skeleton, especially the axial elements. Burke (1989) showed the shell to be the result of an epithelial-mesenchymal interaction (as with feathers or limb buds) in the body wall of the embryo of turtles. The analogue of the carapace margin, the carapacial ridge, is composed of mesenchyme of the dermis and overlying ectoderm, dorsally to the ectodermal boundary between somatic and lateral plate mesoderm. The ectoderm of the carapacial ridge is thickened into a pseudostratified epithelium that overlies a condensing in the mesenchyme of the dermis. As in other reptiles, the dermal component of the chelonian carapace is capable of producing ossifications. However, the epithelial-mesenchymal interactions responsible for these dermal ossifications occur late in development when endochondral condensations of the skeleton are already established. Burke (1989) proposed that the interaction that initiates dermal outgrowth influences presumptive costal cells of the somite, and hence has a causal connection to rib placement. Burke and collaborators (Gilbert et al., 2001) showed that the distal aspects of the ribs are affected by remodeling, indicating initiation centers for dermal ossification of costal bones. The authors suggested that the carapacial ridge, which is responsible for this novel morphology, could be generated by a slight alteration of timing in one of the morphogenetic events in the sequence of development. An earlier publication suggested a similar non-gradualist hypothesis to explain quick selection for limbless squamates (Gasc & Renous, 1989), as limb-reduced forms are poorly represented in the fossil record.

The fusion of the rib cage into a dermal armor changes many functions because the viscera can occupy only a restricted volume. For example, this body plan has complicated the turtle's breathing (Gans & Hughes, 1967; Gaunt & Gans, 1969). On land, as in water, turtles exclusively use their appendicular system in locomotion, in association with the rigid trunk volume. Limitations imposed by the shell cause greater restriction of limb movement than they affect coordination of the limbs in different gaits (Zug, 1971). For example, retraction of the forelimbs and protraction of the hindlimbs are restricted by the shell bridge.

This basic armor arrangement is unique and has remained virtually unchanged over 200 million years. The shell probably represents an adaptation for protection against attacks, covering the turtles' vulnerable viscera. *Proganochelys* even had the presence of bony knoblike armor along the top of its neck and tail and had osteoderms on its limbs (Gaffney, 1990). The earliest turtles already had a highly integrated bauplan. Different scenarios have been proposed to explain the development of the dermal ossifications of the shell and the location of the pectoral girdle under the carapace within the rib cage (Lee, 1993). Lee (1996a, 1996b) analyzed the stages of a speculative evolution from pareiasaurs to turtles based on complex interconnections of traits, with a change of one influencing the evolution of the others. However, his theory of the homology of the formation of the shell in turtles and in pareiasaurs is not justified—compare Lee's theory (1996b) and Burke (1989). In turtles, the ontogenetic development reveals the formation of the pleural disc of the carapace. Dermal bone progressively includes each thoracic rib (dorsoventrally) from the vertebra up to the lateral extremity of the rib. Included in dermal bone, the ribs suture between them but do not fuse. This is in opposition to Lee's assertion that the shell is formed by the agglomeration of juxtaposed osteoscutes arranged in transversal rows. Therefore, the synapomorphies for the two groups, namely some of the postcranial characters linked to the presence of armor, may be homoplasies. However, shared characters may be retained, and pareiasaurs arguably appear to be the closest relatives to the chelonians (Rieppel, 1996). Pareiasaurs (Gregory, 1946; Lee, 1993, 1996a, 1996b, 1997) are not unique candidates for ancestors of turtles. Procolophonids (Reisz & Laurin, 1991; Laurin & Reisz, 1995), captorhinids (Gaffney & McKenna, 1979), diapsids (crocodiles, lepidosauromorph diapsids) (de Braga & Rieppel, 1997; Hedges & Poling, 1999; Platz & Conlon, 1997; Rieppel, 1995, 2000; Rieppel & de Braga, 1996; Rieppel & Reisz, 1999; Zardoya & Meyer, 2001) have all been presented as possibilities but none are convincing for the following reasons:

1. Some theories are based on extant forms and many derived characters without taking into account the plesiomorphic states of the Triassic turtles and their evolution.
2. Some of them present plesiomorphic characters as derived states, implying unlikely reversions.
3. Some characters are considered as synapomorphies whereas they are not strictly homologous. The question of ancestry is therefore still debatable.

As in the evolution of birds, turtles probably followed a "narrow pathway" determined only by a small number of characters. For birds, flight was definitive; for turtles, the definitive character appears to be their structural protection. This definition has greatly impacted the design of the turtle body. It is difficult to understand the biological roles of all of the particular chelonian characters because several roles can often be played by a single, unique structure. This probably explains why a structure can persist when one of its roles, often the most evident, has disappeared during evolution. For turtles, shell initially provided protection against predators but today serves additional roles, such as providing a highly efficient hydrodynamic shape (e.g., *Dermochelys*) or even acting as a respiratory organ (some Tryonychidae; Davenport & Wong, 1992) under conditions when environmental physical conditions are constraining. The turtle body bauplan has shown great temporal stability and has offered a novel source of diversity.

5.2.2.2 Variation of the Basic Pattern within the Aquatic Environment

In spite of the constraints imposed by the basic pattern, chelonians show a wide adaptive radiation in terrestrial, semi-aquatic, and highly aquatic forms in fresh and marine waters. This adaptation to an aquatic life clearly appears at two levels of the body plan: the shell and the locomotor appendages. In general, changes correspond to an adaptation of the shell shape and modifications of the fore- and hindlimbs as paddles and flippers that are efficient for swimming in the aquatic environment. Various studies have quantified the correlations associated with the transformations affecting the form of the shell together with the form of the pectoral and pelvic girdles and also the humerus and femur.

5.2.2.2.1 Modification of Shell Shape in Extant Turtles

We note that in water, an incomplete and reduced pedomorphic shell (Lapparent de Broin et al., 1996; Lapparent de Broin, 2000b) seems to be favored. However, considering the physical constraints of the aquatic environment (hydrostatic pressure and hydrodynamic forces), there is no clear relationship between a lighter shell and the aquatic life, except for the flattening of the shell, which is more hydrodynamically efficient. It is possible that the flattening associated with a more streamlined shape has sometimes been facilitated by pedomorphic lightening. Many aquatic forms have solid shells (e.g., *Carettochelys*), a secondarily strengthened armor with hard skin (cheloniid marine turtles, forms with fontanelles), or a mosaic of plates (Dermochelyidae), or secondary bony callosities (plastron of Trionychidae, Carettochelyidae) for protection. Additionally, the plastral buttresses remain well developed in many freshwater turtles. The shell, with pedomorphic fontanels and cartilaginous links between plastron and carapace or between pleural disc and peripheral border, is not necessarily linked to an aquatic life, such as in *Malacochersus*, the African pancake tortoise; a derived carapace, relative to primitive stages such as the heavy shell in *Proganochelys*, appears first as an adaptation for better agility. When aquatic turtles became amphibious and began to walk in terrestrial habitat, the lighter shell that was derived initially for the aquatic environment also proved useful for moving about more easily on land. Lighter shells also appeared under terrestrial conditions in tortoises, which are characterized by greatly arched shells that show a reduction in thickness of the bony plates of the carapace such as in *Astrochelys*. However, in the more terrestrial forms of the Testudinidae, there was an increased carapace vault, primarily achieved by the raising of the plates of the bridge between the dorsal and the ventral parts of the shell (Lapparent de Broin, 2002, 2003, personal observation). These plates play a role of support for the carapace roof, when the buttresses were reduced. In water, the forces selecting for a streamlined form must be favored over forces selecting for reduced body weight because in water, gravitational influences are minimized. Two adaptive directions seem to be evident in aquatic turtles (Figure 5.4):

- Structural features that facilitated streamlining in swimming species. These include flattening (with lack of plastral buttresses), cordiform, or oblong shape, with or without anterior nuchal protrusion, posterior tapered, or expanded borders, keels, smooth surface (freshwater as well as marine species), and flexible margins (Trionychidae).
- The development of larger limb apertures to allow greater freedom of movement of the limbs (powered by a more substantial musculature acting at the shoulder and hip) during bottom-walking or swimming. These liberated apertures tend to result in a reduced,

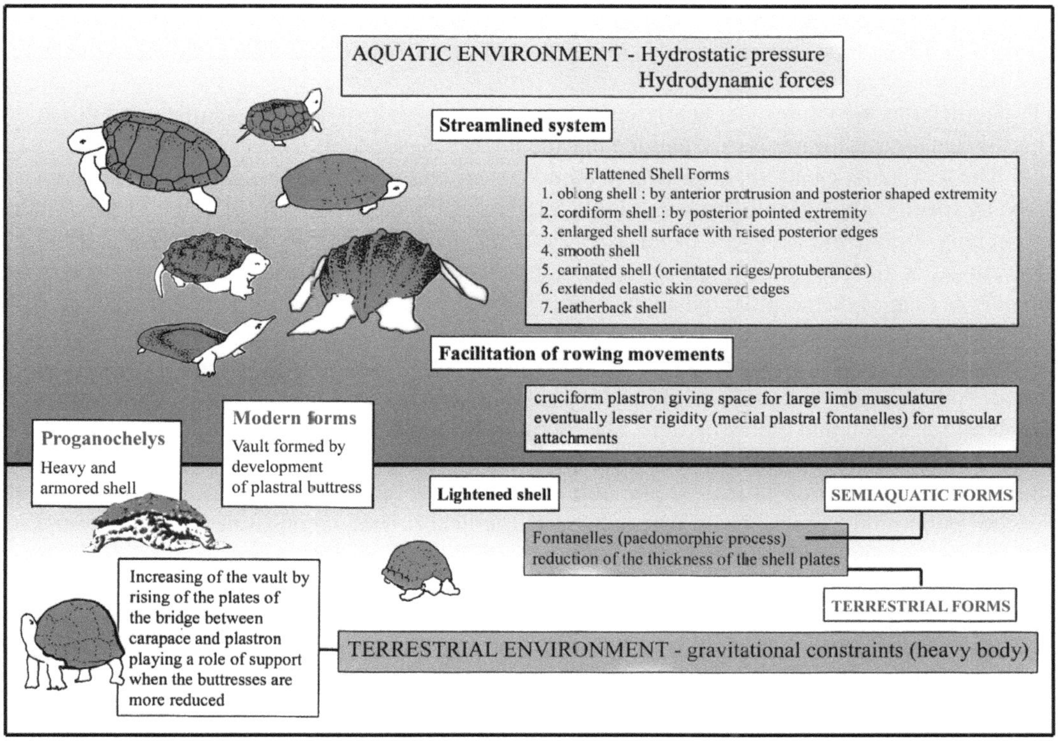

FIGURE 5.4 Modulation of the shell shape of extant turtles producing structures favoring performances in aquatic conditions and comparison with the shape adopted by the terrestrial forms. The upper part of the figure corresponds to the aquatic conditions and the lower part to the terrestrial conditions. The semiaquatic and terrestrial forms are presented as two adaptive strategies to lighten the body under gravitational constraints.

cruciform plastron (Chelydridae, some fossil Kinosternidae, fossil Laurasiatic Carettocheyidae, Asiatic Cretaceous and Paleocene cryptodires). The elements of both types of adaptation combined to select the general pattern of the shell.

To better understand the effects of environment on turtle shell morphology, Claude et al. (2003) conducted a study of the Testudinoidea, the most taxonomically and ecologically diversified group. The results showed that mean shell size did not differ significantly between terrestrial and aquatic forms. However, the authors noted greater shell variation for the terrestrial forms; for instance, species living in rocky crevices have a strongly serrated carapace, whereas species living under leaf litter have a flat carapace. In comparison, box turtles, which live in forested areas and leaf litter, have domed shells with a hinged plastron to provide the animals with more complete protection (Feldman & Parham, 2002). Such variation was related to a more complex adaptive process in the terrestrial environment that provides multiple niches. The flat shell morphology of many aquatic forms, a design that enhances hydrodynamics, was considered to be more homogeneous and more functionally constrained than in the terrestrial environment (Figure 5.5).

5.2.2.2.2 Paddles and Flippers in Extant Turtles

Significant variations in limb morphology are seen in species specialized for particular types of aquatic life. Three types of swimming appendages have arisen:

- Small-footed paddles consisting of mobile digits bearing long claws and short webs (found in most freshwater turtles, with differing amounts of webbing between the digits of fore and hind feet)
- Long mobile digits connected by an expanded web with reduced claws (Trionychidae, Carettochelyidae)
- Long rigid flippers (anterior edge longer than posterior edge (Cheloniidae, Dermochelyidae)

These different structures classically correspond to three subpatterns: the semiaquatic forms, the highly aquatic freshwater forms, and the marine forms (Figure 5.6).

The semiaquatic pattern, occurring in numerous species in various lineages of turtles characterized by specific phylogenetic features (Table 5.1), seems the most difficult to define. The Emydidae, especially *Trachemys scripta*—listed as *Pseudemys* in Walker (1973) and *Chrysemys scripta* in Davenport et al. (1984)—is often chosen to exemplify it. In this pattern, the coracoid is approximately as long as the scapula, the angle between the scapula and acromion is 90°, and the femur is a little longer than the humerus. Protraction and abduction of the limbs are governed by powerful muscles, for example, at the forelimb, the common mass of the teres major and latissimus dorsi, the deltoideus, and the triceps brachii with a smaller scapular part. Both parts of the biceps act in humeral retraction and elbow flexion. The forearm extensors and flexors are all present and complex; however, the intrinsic muscles of the digits are less complex than in better swimmers such as the highly aquatic Trionychidae. In the hindlimb, the dorsal puboischiofemoralis internus and the ventral puboischiofemoralis externus, as well as the flexor tibialis, constitute powerful complexes of muscles, whereas the puboischiotibialis is vestigial or absent. The shank extensors and flexors are numerous and complex. The phalangeal formula is usually 2-3-3-3-3. All of the digits bear long claws, which are typically longer in males for use in courtship. There is some information on overall animal locomotor performance (Walker, 1962, 1963, 1973; Zug, 1971; Davenport et al., 1984; Wyneken, 1997). However, to date there has been no mechanical approach to the study as with some terrestrial forms (Van Leeuwen et al., 1981). Electromyographic data have been recently published (Earhart & Stein, 2000; Gillis & Blob, 2001; Stein, 2003).

The highly aquatic pattern is characterized by the more elaborate paddles (Zug, 1971; Walker, 1973) of the Trionychidae. In the pectoral girdle, the dorsal scapula and ventral acromion form an angle of approximately 65°, whereas the coracoid is elongated. At the pelvic girdle, the pectineal

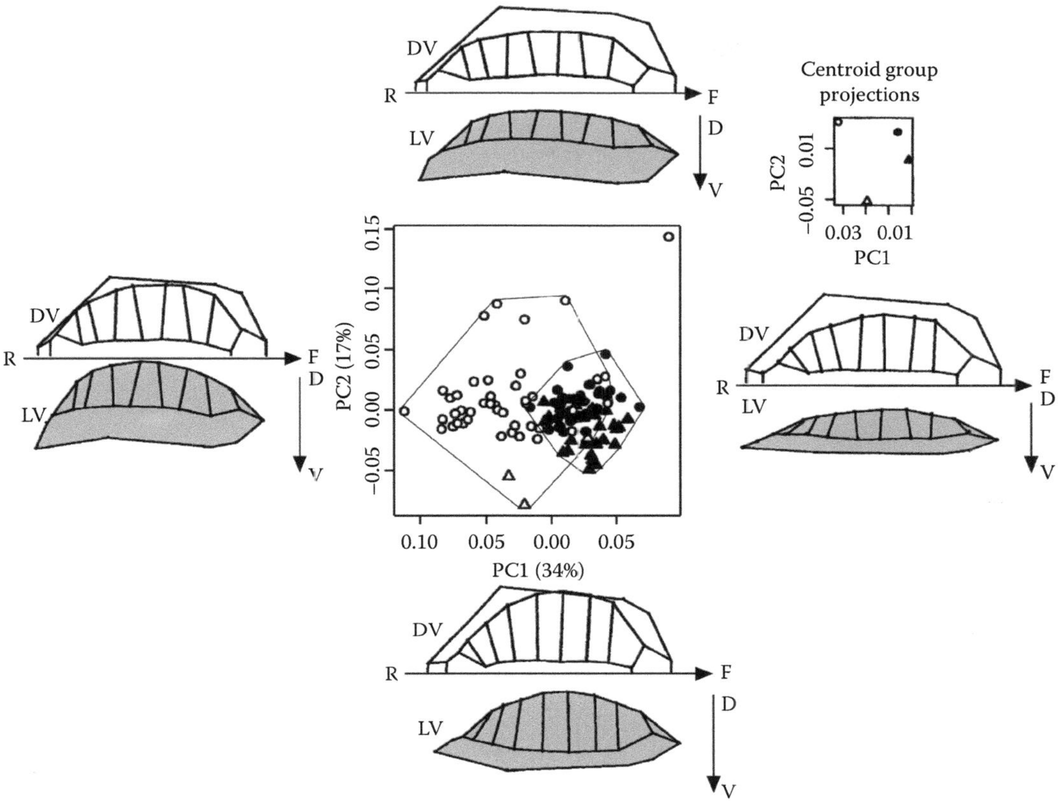

FIGURE 5.5 Morphological changes in the carapace of species of Testudinoidae (which showed the largest variation in size) and Emydidae (which showed the lowest variation in size) between aquatic and terrestrial environments. Principal component analysis was conducted on coordinates for the bony carapace of the species. Diagrams around the graph are amplified shape, in hemi-carapace in dorsal (DV) and lateral (LV) views, derived from the eigenvectors along the first two principal components (PC1 and PC2). Arrows indicate midbody on the dorsal view and opposite body sides on the lateral view. On the graph, areas are delimited for aquatic species (filled symbols) and terrestrial species (open symbols). Circles are Testudinoidae and triangles are Emydidae. The graph in the upper right shows the projections of the mean for each group. D, dorsal; F, front; R, rear; V, ventral (used with permission from Claude et al., 2003).

process of the pubis is exceptionally large. As with the semiaquatic pattern, the humerus is still shorter than the femur. The segments of the front and rear flippers are also elongated, and this is associated with hyperphalangy (the phalangeal formula ranges from 2-3-3-4-3 to 2-3-3-6-4 for the hand and from 2-3-3-4-2 to 2-3-3-5-3 for the foot). Only the first three digits bear claws. According to Walker (1973), adaptation for an extreme protraction of the humerus is shown by corresponding muscular features: large teres major clearly separated from a slender latissimus dorsi, large triceps scapularis, and large deltoideus clavicularis. Humeral retractors (supracoracoideus, coracobrachialis magnus, and subscapularis) are well developed, as are the elbow and hand flexors. In the same way, at the pelvic level different components of the puboischiofemoralis externus (e.g., large origin and subdivision) give control of protraction and retraction. The large insertion of the flexor tibialius internus on the tibia completes this retraction and adds to shank flexion. The great mobility of the fingers and webbing can be explained by the origin of the flexor carpi ulnaris, crossing the palm to reach the flexor tendons of several digits, and also the well developed abductors of the first and fifth

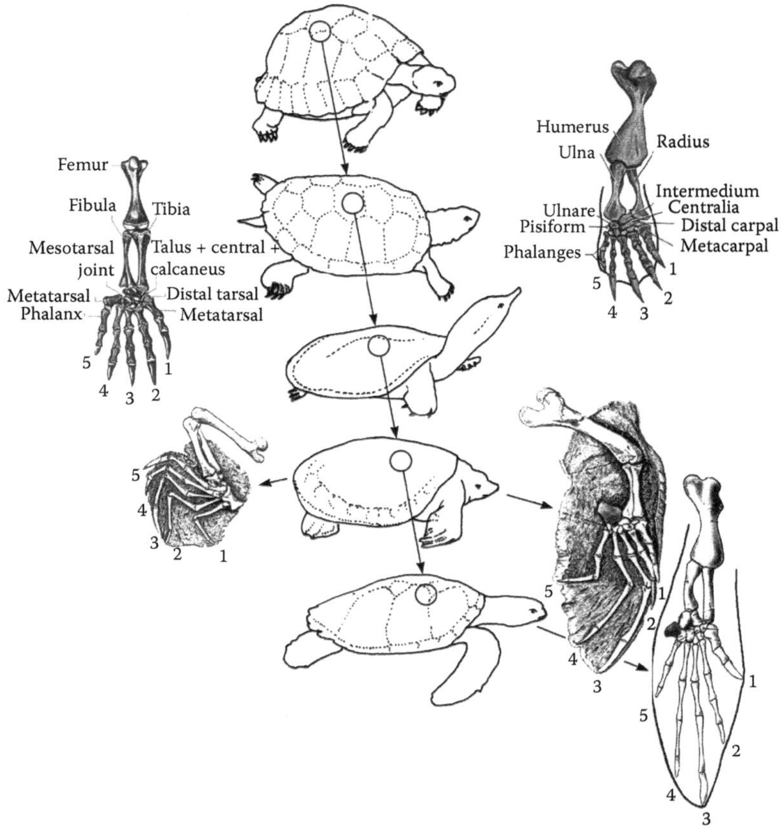

FIGURE 5.6 Variation of the limbs in extant turtles. From top to bottom: an extant terrestrial testudinid, a freshwater form, a highly aquatic trionychid, a highly aquatic carettochelid, and a marine turtle. Some representation of the skeleton of the fore and the hind limbs illustrate the modification in a natatory paddle and a flipper. Numbers correspond to the different digits of both limbs. A circle underlines the forward displacement of the center of mass.

fingers. The fibers of the abductor digiti minimi and lumbricales enter the web, enabling greater control over web form. The lumbricales play the same role in the foot.

The elongated coracoid and the increasing area of attachment of associated muscles—including the supracoracoideus, coracobrachialis magnus, and biceps—seem to be clearly related to aquatic locomotion. However, the most important features concern the structure and the functioning of the paddles. The complex relationship and division of the extensors and flexors of the forearm, and especially the hand, suggest a great mobility of the digits and an accurate movement of each of them in the paddle. To assume this movement, muscles have changed their orientations and insertion points to connect new areas.

The case of *Carettochelys*, the only non-marine turtle with true flippers, merits special consideration. According to previous descriptions (Waite, 1905; Walther, 1921; Pritchard, 1979), the animal has hypertrophied forelimbs used synchronously to swim by flapping as occurs in extant sea turtles. However, the flippers show a structure comparable with that of the flippers of Trionychidae, despite some modifications, mainly of the humerus, similar to marine forms. Although *Carettochelys* undoubtedly demonstrates synchronized flapping of its flippers, there appears to be no convincing demonstration yet that this flapping results in lift-based rather than drag-based

TABLE 5.1a
Examples of Structural Features Characterizing Different Extant or Fossil Species Possessing Distinct Habitats in Terrestrial or Aquatic Environments: Summary of Characters Implicated in Locomotion*

Habitat	Terrestrial	Terrestrial (presumed)	Freshwater Amphibious		
Species	*Chelonoidis*	*Proganochelys*	*Erymnochelys*	*Hydromedusa*	*Macroclemmys*
Fore/hind limb length	short	short	short	short	short
Angle scapula-acromion	obtuse	right	slightly acute ca right	slightly acute ca right	acute ca right
Coracoid	relatively short distally widened	short wide	short distally expanded	short distally expanded	elongated distally expanded
Autopodial articulations	present	present	present	present	present
Humerus	± curved	few curved	slightly curved	slightly curved	few curved
	diverging trochanters at obtuse angle	post. diverging trochanters at obtuse angle	diverging trochanters at obtuse angle	diverging trochanters at obtuse angle	at right angle
	deep long trochanteric fossa	radial weakly oblique inferiorly toward the body axis	radial obliquely oriented inferiorly toward the body axis	radial obliquely oriented inferiorly toward the body axis	radial slightly oblique inferiorly
	ulnar much longer than condyle	larger ulnar not beyond the condyle	larger ulnar not beyond the condyle	larger ulnar not beyond the condyle	ulnar longer than condyle
Humerus/femur length	shorter	shorter	shorter	shorter	shorter
Ectepicondylar canal	reduced sulcus ± closed	canal ventrally open	open sulcus	open sulcus	open sulcus
Radius/ulna	radius < ulna	radius < ulna	radius < ulna	radius < ulna	radius < ulna
Radius-ulna/tibia-fibula	1 < 2	subequal	1 < 2	1 < 2	1 < 2
Hand	shortened digits, elephantine legs, no paddle	interdigital webs, no paddle	interdigital webs, no paddle	interdigital webs, no paddle	interdigital webs, no paddle
Metacarpals	shortened	short	short	short	short
Carpals	short (except 1st medial) 2-2-2-2-2	short 2-2-2-2-2	short 2-3-3-3-3	short 2-2-2-2-1	short 2-3-3-3-3
Phalanges	shortened	shortened	relatively short	short	short
Anterior claws	5 oval, flattened	5 ovaloids	5 long, sharp	4 long, sharp	5 long, sharp
Sacrum	ligamentar attach with ilium, iliac blade knocking against a process, raised behind thoracic rib 10	strong, sutured with dorso-medial ilium part reinforced	S1 integrated in shell behind thoracic rib 10 against the iliac suture, free sacral rib 2	S1 integrated in shell behind thoracic rib 11 against the iliac suture, free sacral rib 3	ligamentar attach with ilium

TABLE 5.1a (continued)
Examples of Structural Features Characterizing Different Extant or Fossil Species Possessing Distinct Habitats in Terrestrial or Aquatic Environments: Summary of Characters Implicated in Locomotion*

Habitat	Terrestrial	Terrestrial (presumed)	Freshwater Amphibious		
Species	*Chelonoidis*	*Proganochelys*	*Erymnochelys*	*Hydromedusa*	*Macroclemmys*
Thyroid fenestrae	2 medium well separated	2 small well separated	1 large vertical	2 large vertical	2 medium separated by cartilage
Pectineal process	long, distally expanded	short, anteriorly directed	short oval sutured on xiphiplastron	short oval sutured on xiphiplastron	moderate anteriorly directed
Tarsals	4 short, 5th enlarged and flattened	short, 5th slightly curved	short, 5th enlarged, flattened, T shaped	short, 5th enlarged, flattened, T shaped	short, 5th enlarged, flattened, T shaped
Metatarsals	4 elongated(1-4)	1 and 5 short 2-4 longer	5 rather long	6 rather long	long
Phalanges	1st shortened 2-2-2-2 (1-4)	1st short 2-2-2-2-2	rather long, 5th reduced 2-3-3-3-3	short 2-2-2-2-1	long 2-3-3-3-3
Posterior claws	4 oval, flattened	ovaloids	4 long, sharp	4 long, sharp	5 long, sharp
Reinforcement of body armor	osteoderms on limbs and tail	osteoderms on neck, limbs, tail, tail club	no	no	protuberances or 3 carenae
Retractile head	very	no	yes	yes	yes
Head posteriorly emarginated	yes	armored forms	no	yes	partly emarginated
Carapace	■	■ heavy, short posterior lobe	▼4,5	▼2±,6±	●

TABLE 5.1b
Examples of Structural Features Characterizing Different Extant or Fossil Species Possessing Distinct Habitats in Terrestrial or Aquatic Environments: Summary of Characters Implicated in Locomotion*

Habitat	Riverine	Riverine	Lagoonal littoral	Marine	Marine
Species	*Chitra*	*Carettochelys*	*Eurysternidae*	*Eretmochelys*	*Dermochelys*
Fore/hind limb length	short	short	short	long	long
Angle scapula-acromion	acute	acute	right ca obtuse	obtuse	obtuse
Coracoid	very elongated distally widened	very elongated	elongated distally widened	very elongated narrow	very elongated narrow
Autopodial articulations	present	present	present	absent	absent

TABLE 5.1b (continued)

Habitat	Riverine	Riverine	Lagoonal littoral	Marine	Marine
Humerus	few curved	very few curved	few curved	flattened	flattened
	diverging trochanters at obtuse angle	at acute angle	diverging trochanters at obtuse angle	distally widened trochanters at acute angle	distally widened trochanters at acute angle
Species	*Chitra*	*Carettochelys*	*Eurysternidae*	*Eretmochelys*	*Dermochelys*
	radial slightly oblique inferiorly	radial oblique inferiorly opposite to the body axis	radial oblique inferiorly toward the body axis	radial inferiorly reduced and directed opposite to the body axis	radial inferiorly reduced and directed opposite to the body axis
	ulnar longer than condyle	ulnar much longer than condyle	ulnar slightly longer than condyle	ulnar much longer than condyle	ulnar much longer than condyle
				deltopectoral crest of radial anteriorly expanded on medial edge	deltopectoral crest separated from condyle transversally expanded
Humerus/femur length	shorter	slightly longer	shorter	longer	longer
Ectepicondylar canal	open sulcus	canal ventrally open	wide open sulcus	wide open sulcus	wide canal ventrally open
Radius/ulna	radius > ulna	radius < ulna	radius > ulna	radius > ulna	radius > ulna
Radius-ulna/tibia-fibula	1 < 2	1 < 2	1 < 2	1 > 2	subequal
Hand	moderately elongated interdigital webs paddle from metacarpals	interdigital webs paddle form fingers 3,4,5	elongated (max 4) interdigital webs pseudo-paddle	elongated (max 2,3,4) interdigital webs paddle from carpals	elongated (max 2,3,4) interdigital webs paddle from carpals
Metacarpals	short	short	short	flattened, first widened	elongated 2 to 4
Carpals	short, medium sized pisiform, 2-3-3-4-3	short, large pisiform, 2-3-3-3-3	short, short pisiform, 2-2-(or 3)-3-3-3	most large & flat large pisiform, 2-3-3-3-2	most large & flat large pisiform, 2-3-3-3-2
Phalanges	elongated	elongated 3,4,5	elongated 2,3,4,5	2-4 elongated & flattened	elongated and rounded
Anterior claws	3 (1,2,3) sharp	2 (1,2) sharp	5 long, sharp	1 (1) much reduced, sharp	0
Sacrum	ligamentar attach with ilium	ligamentar attach with ilium	ligamentar attach with ilium	ligamentar attach with ilium	ligamentar attach with ilium
Thyroid fenestrae	1 large	1 large	1 wide	1 wide with thin median cartilage	2 small well separated
Pectineal process	large diverging	moderate diverging enlarged medial pubis	short, anteriorly directed	large, diverging	large, diverging

TABLE 5.1b (continued)
Examples of Structural Features Characterizing Different Extant or Fossil Species Possessing Distinct Habitats in Terrestrial or Aquatic Environments: Summary of Characters Implicated in Locomotion*

Habitat	Riverine	Riverine	Lagoonal littoral	Marine	Marine
Species	*Chitra*	*Carettochelys*	*Eurysternidae*	*Eretmochelys*	*Dermochelys*
Tarsals	short, 5th enlarged, flattened, T shaped	short, 5th enlarged, flattened, T shaped	short, 5th enlarged, flattened, quadratic	flat, 5th flattened curved	normal size
Metatarsals	long	long	long	flattened, slender	1-4 elongated, rounded 5th curved, flattened
Phalanges	long 2-3-3-3-2	long 2-3-3-3-2	long 2-3-3-3-3 or 2-3-3-3-2	elongated, flattened 2-3-3-3-2	elongated, rounded 3-3-3-3-3
Posterior claws	3 long, sharp	2 long, sharp	5 long, sharp	1st very short, sharp	0
Reinforcement of body armor	no	no	no	no	no
Retractile head	yes	yes	no	yes	yes
Head posteriorly emarginated	yes	no	no	no	no
Carapace	■ 1, 5, 7	■ 2a, 4, 6a	■ 1, 5	■ 2a, b, 4, 5	■ 1, 6a+

Notice the problem of homology of the alleged "hooked fifth metatarsal" in Chelonii, here considered as a element of the tarsus; see text for more discussion of this issue. In Erymnochelys, the autopod 5 has one independent quadrangular, flattened and hooked in T tarsal element as in other turtles with a "hooked tarsal," besides one metatarsal and three phalanges; the hooked bone cannot be a tarso-metatarsal element. As in all the other turtles, the element following the hooked bone of the finger 5 has the typical morphology of a metatarsal and not of a phalange. In Cheloniidae, there is a semi-hooked tarsal 5 and a flattened and expanded metatarsal 1. Carapace: terrestrial type, ■; aquatic type, summary of hydrodynamic characters of the shell, ▼ (1, flattened shell; 2, oblong shell by anterior protrusion (a) and posterior sharpened extremity (b), 3, cordiform, 4, enlarged surface of the shell and raising posterior edges, 5, smooth surface, 6, carinated shell with oriented ridges (s), 7, elastic skinned edges. Facilitation of rowing movements: ●, cruciform plastron giving space for limb musculature.

propulsion. Interestingly, members of the Trionychidae sometimes show brief bursts of synchronized forelimb movement when digging in sediment (Davenport, unpublished).

The marine pattern of extant turtles (Hirayama, 1994; Hirayama & Chitoku, 1996; Manlius, 1996; Walker, 1973), the Cheloniidae and Dermochelyidae, is characterized by forelimbs modified into wing-like rigid flippers and hindlimbs into semi-rigid paddles or rudders. In contrast to the selection of a large flexible skinned surface with mobile digits, as in the Trionychidae, the coracoid is elongated in relation to a powerful locomotor musculature and an angle greater than 90° exists between acromion and scapula. The rigidity of the flippers stems from a loss of joint mobility in the digits and metapodial elements, and an associated reduction, loss, or replacement of muscular units of the hand and foot by connective tissue (fascialization). In the pelvic girdle, the pectineal process of the pubis is exceptionally large. The humerus is stouter and flatter than in non-marine taxa, with a proximally shifted ulnar process. The intertrochanteric fossa is reduced by displacement of the deltoid crest toward the opposite trochanter. The three central digits in the foreflipper are greatly elongated by increasing the length of the metacarpals and phalanges rather than by hyperphalangy. An enlarged pisiform supports the ulnar border of the flipper. The femur, which is similar in length to the humerus, is flat, as are the tibia and fibula. The intertrochanteric fossa also is reduced.

As in the other aquatic turtles, the muscles, latissimus dorsi, subscapularis, pectoralis, and supracoracoideus are large but in marine turtles the deltoideus scapularis and coracobrachialis magnus are also enlarged. The flexors of the elbow and hand are well developed, except the flexor brevis superficialis and the lumbricales. The adductors of the fingers are absent. The supinator manus is vestigial, the extensores radialis profundus are absent, and the extensor digitorum communis is distally reduced to a connective tissue fascia. At the hindlimb, parts of the puboischiofemoralis show a more extensive origin, whereas units of the shank and foot are reduced to a fascia (extensor digitorum communis) or are absent (flexor digitorum communis sublimis) (Manlius, 1996; Walker, 1973; Wyneken, 2001, 2003).

5.2.2.2.3 Structure of the Posterior Zeugopod: The Hooked Fifth Tarso-metatarsus or Fifth Tarsus

Chelonians are characterized by the presence of a hooked bone at the angle of the foot. There is some debate about whether there is a homologous bone equivalent found in other reptiles (Guibé, 1970; Romer, 1956; Schaeffer, 1941; Rieppel & Reisz, 1999). The hooked bone is considered to be the fusion of either tarsal 4 and tarsal 5 (a hooked tarsal), or tarsal 5 with metatarsal 5 (Romer, 1956; Schaeffer, 1941; Rieppel & Reisz, 1999). It is apparently also present in some Squamata and all crocodiles. Its ontogenetic formation and morphology have not been adequately studied in any group where it is present. This bone is not similar morphologically across the different groups of reptiles where it appears. In crocodilians, it is a long bone that resembles a metatarsal curved slightly proximally. In turtles, there are considerable differences in shape: it can either be a rectangular bone with a small prolongation in the middle of the distal face, articulating with a bone similar in shape to the metatarsal, or it may be a more quadratic bone. Its homology, ontogenetically and morphologically, is not known and will have to be studied employing a cladistic approach, including every form with and without a hooked bone and considering its shape. In terms of its role in locomotion, it appears to act as a pisiform-like bone that supports the back of the foot and increases the surface area of the limb. The bone that distally articulates with the "hooked bone" has clearly a metatarsal and not a phalange shape. Besides, fossil taxa such as Eurysternids (Figure 5.6) and extant forms such as the pleurodiran *Erymnochelys* (Table 5.1) clearly present a rather quadratic hooked bone, distally followed by four bones, a metatarsal, and three phalanges.

5.2.2.2.4 Shells and Limbs in Aquatic Fossils

The fossils of Trionychidae and Carettochelyidae had limbs with elongation and flattening of well-articulated elements, phalangeal multiplication, and reduction in the number of claws. Similarly, new anatomical developments are seen in other fossil forms. The late Jurassic families of littoral turtles, especially the Plesiochelyidae, Thalassemydidae, and Eurysternidae, which spread into the European seas, were primarily characterized by a flattened or carinated shell with a relatively light dermal skeleton, pedomorphic in the two latter families (Lapparent de Broin et al., 1996; Lapparent de Broin 2000b; see the various processes of pedomorphosis in Kordikova, 2000). The autopods were enlarged, when known—i.e., in the Eurysternidae. The radius was curved and longer than the ulna. Digits two to five were particularly elongated, producing a small paddle, but they still retained distinct mobile finger and toe elements, as well as inter-digital webbing. The hindlimb was longer than the forelimb primitively. Eurysternidae, especially known by their skull (*Solnhofia* group: Lapparent de Broin, 2001), are characterized—as in marine turtles—by a descending parietal process much reduced in length and inferiorly notched, at the rear of the inter-orbital foramen. This notch is thought to have housed hypertrophied lachrymal glands adapted for salt excretion (Hirayama, 1998). This feature could conceivably have been acquired in a non-marine hypersaline environment, such as a tropical lagoon subject to evaporation, and subsequently proved appropriate to life in the marine environment. In the second wave of "marine" cryptodires, early Cretaceous *Santanachelys* (Protostegidae from the freshwater lake of Araripe) had the same skull notch particularity and similar limb characters as the Eurysternid European forms (Figure 5.7). However, a

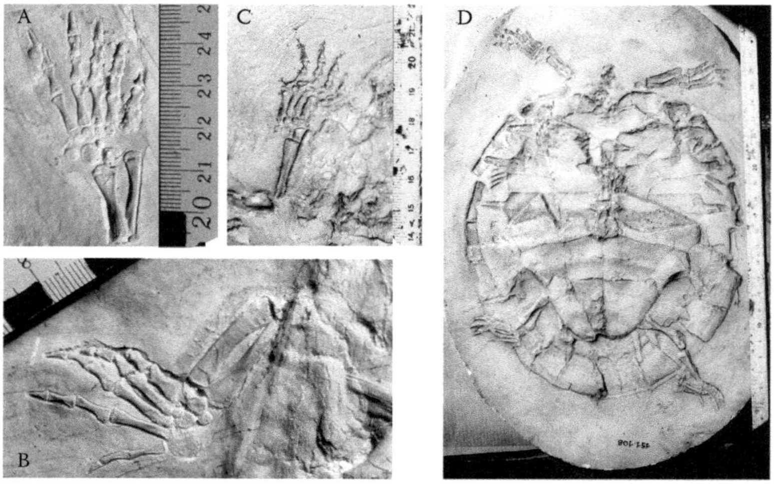

FIGURE 5.7 Structure of the fore (A) and hind (B) limb of *Eurysternum wagleri,* Aff. *Solnhofia* sp. (C) and *Idiochelys fitzingeri* (D) giving an idea of the Jurassic turtles of Cerin (France) living in aquatic conditions. Prints of webs are often visible between and around the digits suggesting the existence of paddles. (With thanks to R. Alberdsdörfer, H.V. Karl, and M. Maisch for their images and material, in particular for the unpublished Eurysternids from the Jurassic of Bavaria from the private collection of R. Albersdörfer.)

proportion of the fingers were still mobile and allowed articulations (Hay, 1908) but the others were flattened and immobile, having lost the ability to articulate, as is the case in modern marine turtle flippers. *Santanachelys* also had a flattened and lighter shell with a long bridge but with a protruded anterodorsal part, probably facilitating more extensive movement of the forelimbs. However, the littoral *Neusticemys* (skull unknown), a possible first protostegid from the late Jurassic of Argentina, already had a similar carapace and also an elongated forelimb (longer than the hindlimb) and flattened fingers with completely reduced articulations.

The complete transformation of the limbs into flippers in marine turtles occurred during the late Cretaceous. This modification included many other changes: the forelimb became much longer than the hindlimb, the humerus was flattened with an elongated ulnar trochanter, and the position of the deltopectoral crest was modified, becoming anteriorly directed. There was torsion of the elongated radius, flattened fingers that lost articulations between the elements, elongated fingers (particularly 2 to 4), reduction of the claws, skin enveloping all the limbs, and modification of the pectoral girdle (elongated coracoid, obtuse angle between scapula and acromion, and so on). In Cheloniidae, the deltopectoral crest is still linked to the humerus condyle, whereas in Dermochelyidae that link is lost and the crest develops transversely and the distal extremity is more expanded and widened; the flippers are much longer relative to other marine turtle species.

For comparison, it is interesting to consider the structural features of the Testudinidae, the single exclusively terrestrial turtle group. The structural pattern of these species includes a rather domed shell and more or less elephantine features of the short fore- and hindlimbs. In particular, the short forelimb results from reduced lengths of the carpus and fingers segments, and also a reduction of the phalangeal formula (2-2-2-2-2 or 1), whereas the hindlimbs have short yet massive toes, the phalangeal formula being 2-2-2-2. There is no connecting web of tissue between the digits, which bear heavy claws. Proximally on the humerus, the length of the intertrochanteric fossa is longer, whereas on the femur the ventral fusion of the trochanters translates into smaller size and the depth of the intertrochanteric fossa. Concerning the locomotor muscles, these forms exhibit a reduced development of the pectoralis, a single biceps, a loss of pronator profundus, and a clear division of the extensor carpi radialis at the forelimb. Walker (1973) proposed that this last feature facilitates the placement of the hand in a dorsi-flexed position without a simultaneous extension of the forearm. At the pelvis, the flexor tibialis complex is reduced to a single muscle. Reduction of the extrinsic muscles of the forelimb (flexor brevis superficialis, abductors of the fist and fifth digits, adductor digiti minimi, and lumbricales) and the hindlimb (flexor digitorum communis sublimis) suggests weak independent action of the fingers.

5.2.2.2.5 The Geometric Morphometry of the Shoulder Girdle and Humerus

Morphological traits of the limbs can be used to distinguish morphofunctional groups of turtles. Recently, geometrical morphometry based on the Procrustes method (Bookstein, 1991; Penin, 2003; Penin et al., 2002; Berge & Penin, 2004) was used to analyze the size and shape variations of two skeletal elements (the humerus from 69 species of 11 extant and 2 extinct families, and the shoulder girdle from 57 species of 13 families of extant adults specimens) and to understand the relationships of these structures aquatic and terrestrial locomotor adaptations (Depecker et al., 2006a, 2006b). According to Bookstein's method (1991), 21 reference marks were placed on the humerus and 16 reference marks on the pectoral girdle (Figure 5.8). The data for the shoulder girdle (Depecker et al., 2006a) reveal that the first two principal components axes of Principal Component Analyses (PCA) represent 73.2% of the total variance. They allow the identification of four main groups, independent of size (Figure 5.9A):

Extant terrestrial turtles (Testudinidae)
Eight families of freshwater turtles (Geoemydidae or Bataguridae, Chelidae, Chelydridae, Emydidae, Kinisternidae, Pelomedusidae, Platysternidae and Podocnemididae)
Two other families of freshwater turtles (Carettochelyidae and Trionychidae)
Two families of marine turtles (Cheloniidae and Dermochelyidae).

Data for the humerus (the first two principal components of PCA representing 51.2% of the total variance) also reveal similar distinct groups (Depecker et al., 2006b) (Figure 5.9B):

The extant terrestrial Testudinidae, including the supposed terrestrial fossil *Ptychogaster*
The same eight families of freshwater turtles
The extant, highly aquatic, Trionychidae and the terrestrial fossil Meiolaniidae
The extant highly aquatic Carettochelyidae
The Cheloniidae
The Dermochelyidae

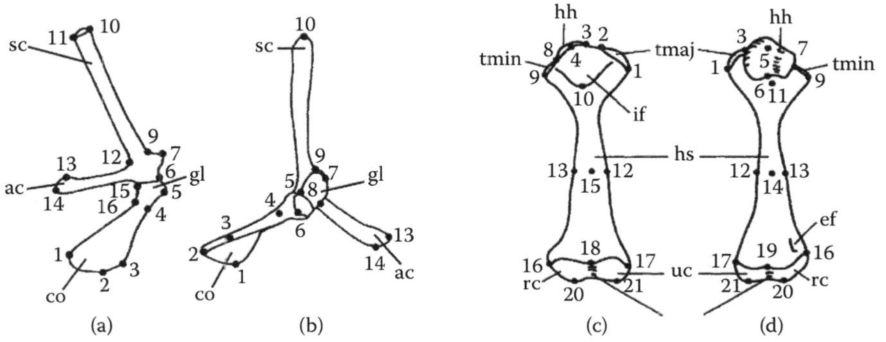

FIGURE 5.8 Position of 16 landmarks on the right shoulder girdle, dorsal posterior view (a) and lateral view (b), and 21 on the right humerus, ventral view (c) and dorsal view (d) in a semi-aquatic freshwater turtle (*Chelydra serpentina*). a, acromial process; co, coracoid; ef, ectepicondylar foramen; gl, glenoid cavity; hh, humeral head; hs, humeral shaft; sc, scapular prong; rc, radial condyle; tch, trochlee; tmaj, trochanter major; tmin, trochanter minor; uc, ulnar condyle (used with permission from Depecker et al., 2006a, 2006b).

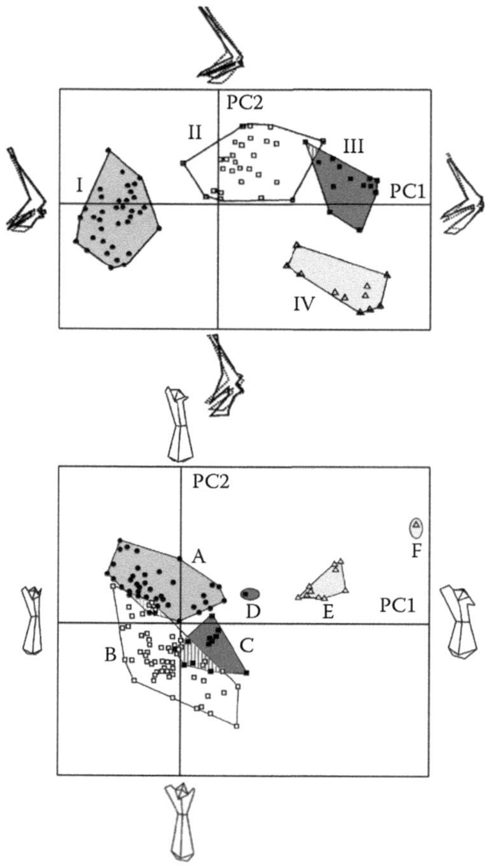

FIGURE 5.9 Principal Component Analyses of 88 shoulder girdles (top graph) and 122 humerus (bottom graph) onto the plane of the first principal component PC1 and PC2. Pleurodira and Cryptodira are not distinguished in the analysis. Filled circles correspond to groups I (top) and A (bottom) (Testudinidae). Open squares correspond to groups II (top) and B (bottom) (Emydidae, Geoemydidae or Bataguridae, Chelydridae, Platysternidae, Kinosternidae, Chelidae, Pelomedusidae, Podocnemididae). Filled squares correspond to group III (top) (Trionychidae and Carettochelidae) and the groups C (bottom) (Trionychidae and Meiolaniidae), D (bottom) (Carettochelyidae). Open triangles correspond to group IV (top) (Cheloniidae and Dermochelyidae) and groups E (bottom) (Cheloniidae) and F (bottom) (Dermochelyidae). Dorsal views of line diagrams of the humerus for the two extremities of PC1 and PC2 are presented. The line diagram of the shoulder girdle also changes along PC1 and PC2 (solid lines) with respect to the consensus (mean shape) in dotted line (used with permission from Depecker et al., 2006a, 2006b).

Based on a multivariate regression performed to calculate the correlation between humerus shape and the degree of specialization to aquatic habitat, the authors demonstrated relatively few differences in shape between terrestrial and semiaquatic turtles but greater differences between them and truly aquatic turtle species (Figure 5.10). In this latter group, the humerus shape (including robustness, larger and dorsoventrally flattered shaft, proximally developed medial trochanter, distally developed lateral trochanter, laterally enlarged radial condyle) appears directly related to the constraints associated with the group's specific environments or habitats.

It is generally acknowledged that modern terrestrial and marine turtles may have evolved independently from less-specialized freshwater turtles (Gaffney, 1985; Gaffney & Meylan, 1988; Gaffney et al., 1991; Shaffer et al., 1997). For this reason, the authors used these freshwater turtles as a

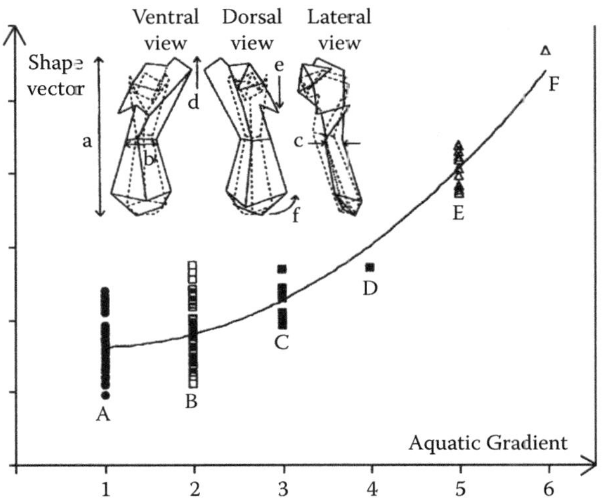

FIGURE 5.10 Correlation between shape changes of the humerus (Vreg) and specialization for the aquatic habitat (aquatic gradient). Curve formula: $y = 0.0144x^2 + 0.0032x - 0.0377$ ($R^2 = 0.81$) Symbols are the same as in Figure 5.9. Ventral, dorsal, and lateral views of line diagrams of the humerus between both extremities of Vreg: dashed lines, semiaquatic turtles; full lines, marine turtles. Shape changes observed: a, longer humerus; b, larger shaft; c, flatter shaft; d, developed trochanter major; e, distal trochanter minor; f, enlarged radial condyle (used with permission from Depecker et al., 2006).

reference to analyze the shape of the shoulder girdle for each group of terrestrial and marine turtles (Figure 5.11). Unlike semiaquatic freshwater turtles, terrestrial forms exhibit a more massive girdle with a larger glenoid cavity, a wider angle between scapula and acromial process, a longer scapula (vertically oriented), and a shorter and broader (medially longer) coracoid. The highly aquatic freshwater turtles exhibit a more developed pectoral girdle ventrally (coracoid and acromial process) relative to the dorsal scapula. The coracoid is longer, broader, and more curved. The scapula is shorter, forming an acute angle with the acromial process. Comparatively, the marine turtles show a more massive girdle. The coracoid, which has a straight posterior border, is longer, whereas the acromial process and the scapula are shorter. The scapula forms an obtuse angle with the acromial process. The shoulders of terrestrial and marine turtles appear to be highly specialized. Large freshwater species such as the trionychids and carettochelyids also have a specialized pattern that is reminiscent of marine turtles.

Allometric investigation has showed that the shape of the shoulder girdle gradually changes with size: the coracoid becomes proportionally longer, whereas the scapula and the acromial process become proportionally shorter (Figure 5.12). The smallest turtle species occur among the terrestrial forms and the generalized freshwater turtles, whereas the largest species are terrestrial and marine turtles. The largest highly aquatic freshwater turtles (*Pelochelys*, some trionychine species) are the same size as the smallest marine turtles (*Lepidochelys*). When size increases, the coracoid becomes longer in marine turtles and both longer and broader in highly aquatic freshwater turtles.

Locomotor adaptations of aquatic and terrestrial species could explain these distinct groups. On land, the limbs must provide both support and propulsion (Williams, 1981). They act like beams to support the weight of the shell (Walker, 1962, 1971a, 1973). Jayes and Alexander (1980) and Van Leeuwen et al. (1981) noted that the vertical components of muscular forces in walking must be much larger than the horizontal components to counterbalance body weight. In contrast, for aquatic turtles the girdles do not play a supporting role, whereas propulsive forces are generated by the limbs acting against the water. The streamlined shape of the body of marine turtles is associated with hypertrophied flapping forelimbs that give an advantage in terms of locomotor efficiency

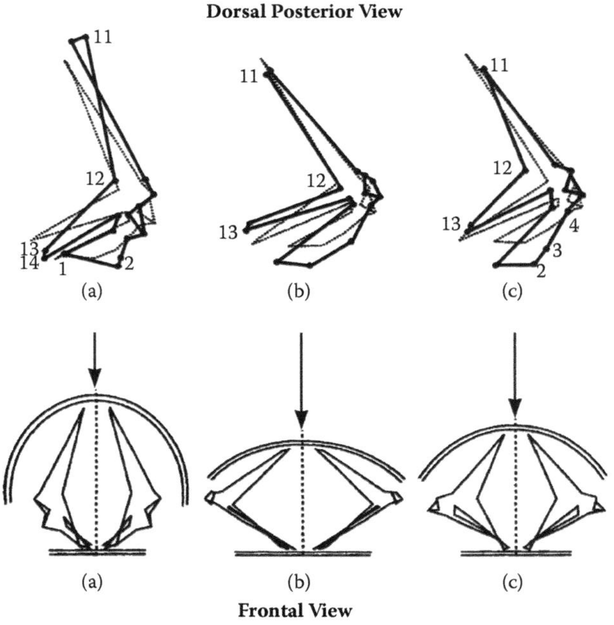

FIGURE 5.11 Discriminant traits of the shoulder girdle (dorsal posterior view), the semiaquatic freshwater turtles (dotted lines) being used as a reference to compare with other groups (solid lines), terrestrial forms (a) highly aquatic freshwater forms (b) and marine turtles (c). The shape and the orientation of the girdle are also given in relation to the shape of the shell (frontal view) in the same categories of turtles. The numbers indicate selected remarkable landmarks: 1 and 2, respectively medial anterior and posterior angles of the coracoid; 3 and 4, posterior border of the coracoid at maximal curvature and collar; 11, medial dorsal angle of the scapular prong; 12, limit between scapula and acromial process; 13 and 14, respectively medial anterior and posterior angles of the acromial process (used with permission from Depecker et al., 2006).

(Davenport et al., 1984). Walker (1973) suggested that there is a relationship between the shape of the shoulder girdle and the configuration of the turtle shell. Aquatic turtles possess a long coracoid, which gives them a mechanical advantage in swimming in terms of lever arms and muscular attachments. In this case, pectoral morphology is directly linked to the streamlined shape of the shell (Figure 5.11). In terms of terrestrial turtles, the humerus of an extinct member of the Meiolaniidae (representing a primary adaptation, i.e., the primary condition) is distinct from those of the Testudinidae, that represent a secondary return to terrestrial lifestyle from freshwater ancestors.

5.3 LOCOMOTION OF THE AQUATIC TURTLES

A highly specific structure, the turtle shell generates biomechanical limitations to the physical constraints of the terrestrial and aquatic environments and to body functioning. The inflexibility of the body axis obliges the turtles to use limb movements alone for propulsion. Significant variations in limb morphology are found in species that are specialized for particular patterns of locomotion (Walker, 1973). Shell mass and body geometry play an important role on land and also in shallow water. To demonstrate the importance of weight, Zani and Claussen (1995) experimented with increased extrinsic loads; they found a correlation between increased loads and a decrease of speed and stride length during terrestrial locomotion. However, weight and size are generally not limiting factors at sea and are probably favorable to other functions, such as high fecundity, gigantothermy, prolonged diving, oxygen storage, and resistance to heat loss (James et al., 2006). The greater

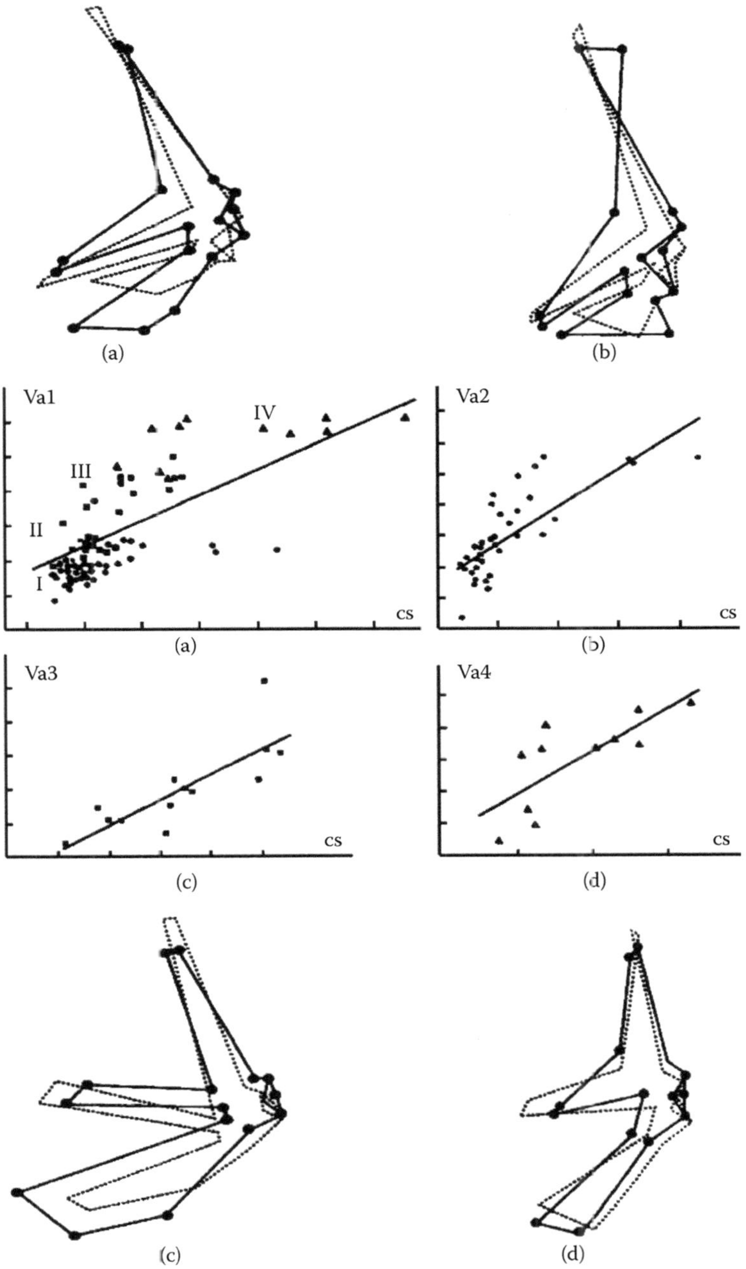

FIGURE 5.12 Allometric shape vectors calculated in all turtles Va1 (a), in terrestrial turtles Va2 (b), Va3 in highly aquatic freshwater turtles (c), in marine turtles Va4 (d). CS, centroid size; the groups I, II, III, and IV correspond to the graph A of Figure 5.9. Line diagrams corresponding to each graph show the variation between the smallest specimens (dotted lines) and the largest (solid lines) ones (used with permission from Depecker et al., 2006).

absolute body size of marine turtles could incorporate a larger pectoral musculature that is able to generate a greater swimming power. For this reason, the forward displacement of the center of mass, functionally related to stable swimming in freshwater aquatic turtles, becomes even more marked in the marine turtles (Davenport et al., 1984). Finally, as reported by Sukhanov (1974) it is important to note that some aquatic turtles, such as *Platysternon* or *Chelus*, do not swim well but often move on the bottom of lakes, streams, and so on. In addition, terrestrial as well as aquatic turtles (e.g., bottom-walkers like Chelydridae and Kinosternidae) walk on land, and the highly aquatic *Apalone*, for example, is able to cover large distances on land. The speed of progression varies from 0.2 kph (*Gopherus*, *Clemmys*) to 1.7 kph (*Chrysemys*) and 3.2 kph (*Chelonoidis nigra* of Galapagos) in relationship to the body size.

5.3.1 Different Function of the Limbs

In water, the mechanics of thrust production by vertebrates varies markedly among different groups (Fish, 1996). One general approach corresponds to a typically asynchronous rowing action (limbs oriented in a horizontal plane), where the limbs use the water resistance to generate thrust by a retraction action or a paddling (limbs oriented in a vertical parasagittal plane). During rowing, retraction corresponds to a power phase of a cyclic movement in which protraction is the recovery phase. The power phase generates a large pressure drag that provides anterior thrust by reaction of the water (Figure 5.13). For this reason, some authors use the expression "drag-based propulsion" to characterize this system (Davenport et al., 1984; Fish, 1996; Wyneken, 1997). For maximum efficiency, there must be a maximum extension of area (by abduction of the digits and web extension) during the power phase. All anatomical structures that augment paddle area contribute to greater mechanical efficiency (for example, enlargement of fingers, flattening of claws, interdigital webbing, and so on). Conversely, the recovery phase—which repositions the limb—must ideally be a non-thrust generating stage. To avoid or to limit drag production during recovery, the paddles benefit from an adduction of the digits to reduce the total area, and also a change in timing to temporarily reduce the relative limb velocity.

A second approach corresponds to a more typically derived synchronous swimming mode when a lift force is used to generate the thrust. The limb becomes a hydrofoil and lift-based thrust is produced by its oscillation at a controlled angle of attack (Figure 5.14) relative to the effective water flow over it (Davenport et al., 1984). To enhance the lift and decrease hydrodynamic drag, the angle of attack should be small enough to avoid separation of the water flow from the hydrofoil surface. However, thrust is the resultant of lift and drag forces on the hydrofoil. Broadly speaking, the greatest thrust will be produced before the hydrofoil stalls, when both lift and drag are high. Thrust may be generated either during a specific phase of the cycle movement of the hydrofoil (upstroke or downstroke) or continuously during its entire duration of the cycle according to the direction of the general movement of the limb, the direction of the flow, and the angle of attack. During non-thrust generating phases, the angle of attack is reduced close to zero, reducing both lift and drag. The efficiency of propulsion is also a function of the shape of the hydrofoil.

5.3.2 Different Patterns of Propulsion

As a group, aquatic turtles use both types of mechanical systems. In extant forms, the limbs of freshwater animals show rowing strokes during slower swimming in semiaquatic species (Figure 5.13) as well as in the faster swimming of the highly aquatic Trionychidae. Movement of the limbs is characterized by dominant antero-posterior movements, combined with a rotation of the limb to become perpendicular to the water flow during the retraction phase that generates thrust but parallel to flow during protraction (recovery phase). Fore- and hindlimbs show similar movements, but the studies of Zug (1971) and Walker (1973) indicate that the hindlimbs predominate in generating thrust. Comparative kinematics of the forelimb during the swimming of slider (*Trachemys scripta*)

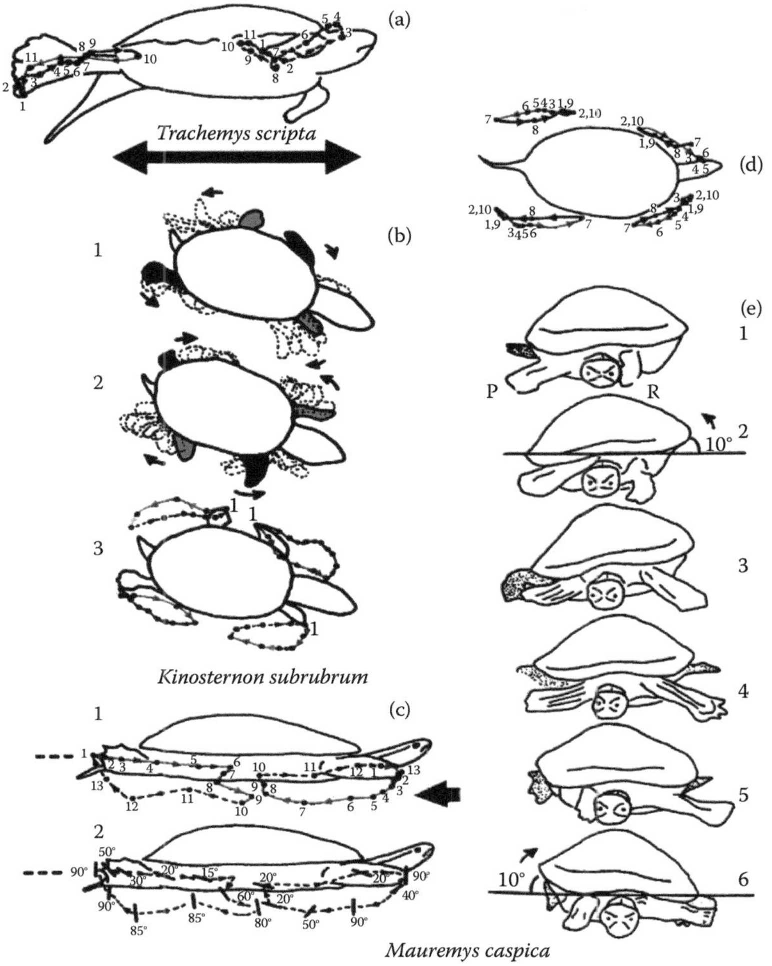

FIGURE 5.13 Kinematics of swimming of three freshwater turtles: *Trachemys scripta*, *Kinosternon subrubrum*, and *Mauremys caspica*. The dorsal views of *K. subrubrum* (b, 1, 2, 3) and *M. caspica* (d) show the synchronization of the limb movements and their lateral amplitude. The lateral views of *T. scripta* (a) and *M. caspica* (c) underline the up and down amplitude of these movements and the limb inclination, whereas the front views of *M. caspica* (e) indicate the degree of alternate body incline according to the swimming phases. Numbers are stages of the coordinated limb movements (used with permission from Davenport et al., 1984).

and soft-shell (*Apalone spinifera*) turtles (Pace et al., 2001) give additional insight (Figure 5.15). In *Trachemys*, the forelimbs are employed primarily for balance and orientation rather than propulsion. However, the long claws can assist in propulsion in highly vegetated habitats. In *Apalone*, greater elbow extension and lesser wrist flexion combine to straighten and extend the forelimb more during the power stroke, which in turn maximizes the volume of water accelerated to generate thrust. The increase of flow velocity over the limb leads to an increase of the propulsive forces (proportional to the square of flow velocity). Drag is minimized during the recovery phase. This demonstrates that there are different styles of forelimb rowing among freshwater turtles.

Pace et al. (2001) also note that the substantial elevation and depression of the humerus of the slider turtle bears a kinematic resemblance to the dorsoventral movements of the same bone in sea

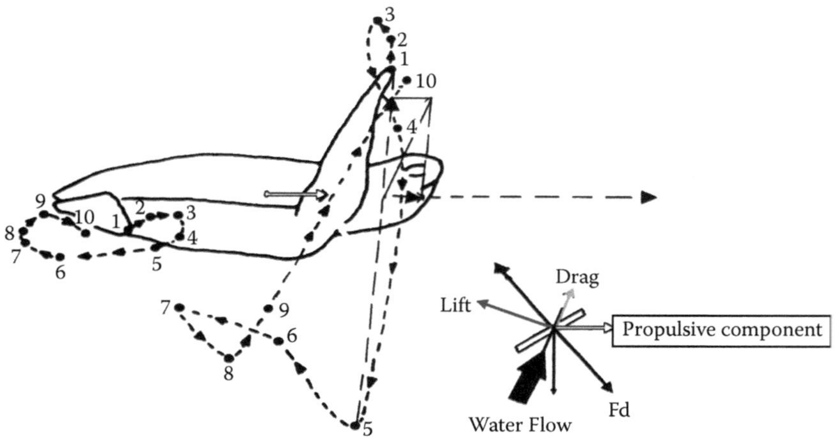

FIGURE 5.14 Kinematics of the swimming of *Dermochelys coriacea* showing the trajectory of the forelimbs extremity during their dorso ventral flapping, which could generate lift and propulsive components. An example of the forces exchanged is indicated to illustrate the hydrofoil function of the limb and the production of the propulsive force. Numbers are stages of the coordinated limb movements (used with permission from Davenport, 1987).

turtles. We must expect a great diversity of limb rowing mechanisms—in terms of their efficiency—according to phylogenetic differences among the freshwater turtles. In the absence of information on the movement of the fore- and hindlimbs in diverse species of turtles ranging from the semi-aquatic to highly aquatic freshwater turtles, it is difficult to draw general conclusions on the diversity of patterns used for propulsion. The diagonal mode of limb coordination used in these animals might lead to expectations of an equivalent role of fore- and hindlimbs in thrust production and thrust intensity. However, there is more extensive webbing between the toes of the hind feet relative to the forefeet (Walker, 1973) in emydid turtles, which spend time in both terrestrial and aquatic habitats (Gibbons, 1970; Bodie & Semlitsch, 2000), producing a broader paddle-like foot that has a predominant role in generating drag-based thrust in water (Pace et al., 2001). However, extensive webbing between the toes of front and rear limbs (Webb, 1962) in softshell turtles (Trionychidae) suggests that both hind- and forelimbs are sources of drag-based thrust in water—although softshells often dig in soft mud when seeking prey, so the webbed forelimbs may have a dual function.

The role of the tail in turtle propulsion was studied by Willey and Blob (2004) in juvenile common snapping turtles, *Chelydra serpentina*, during aquatic walking (Figure 5.16). Long muscular tails are characteristic of Chelydridae and Platysternidae and can facilitate the climbing inclines (Finkler & Claussen, 1997), in contrast to the vast majority of other extant turtles that possess reduced tails that play a negligible/limited role in locomotion (Pace et al., 2001; Walker, 1973; Zug, 1971). In the long-tailed forms, the tail may leave a twisted trail on the ground. Common snapping turtles move their tails cyclically during aquatic locomotion. This tail motion does not appear to be a passive effect of limb retractions but is potentially controlled by caudi-iliofemoralis contractions on the same side of the body. Lateral compression of the tail also suggests that lateral tail oscillations may conceivably accelerate water volume posteriorly and thus contribute to aquatic thrust. However, the functional effects of these tail movements demand greater study, particularly as such tail-based propulsion represents the only known involvement of the axial system in thrust generation in turtles.

Many publications have described the dominant dorsoventral flapping of the foreflippers acting as hydrofoils in extant sea turtles, the Cheloniidae and Dermochelyidae (Figure 5.14). The hindlimbs function as combined rudders and elevators (Davenport et al., 1984; Davenport & Clough, 1986; Davenport, 1987; Renous, 1988; Wyneken, 1988a, 1988b, 1997; Renous & Bels, 1991, 1993; Davenport & Pearson, 1994). Use of the foreflippers in combination with the streamlined body confers

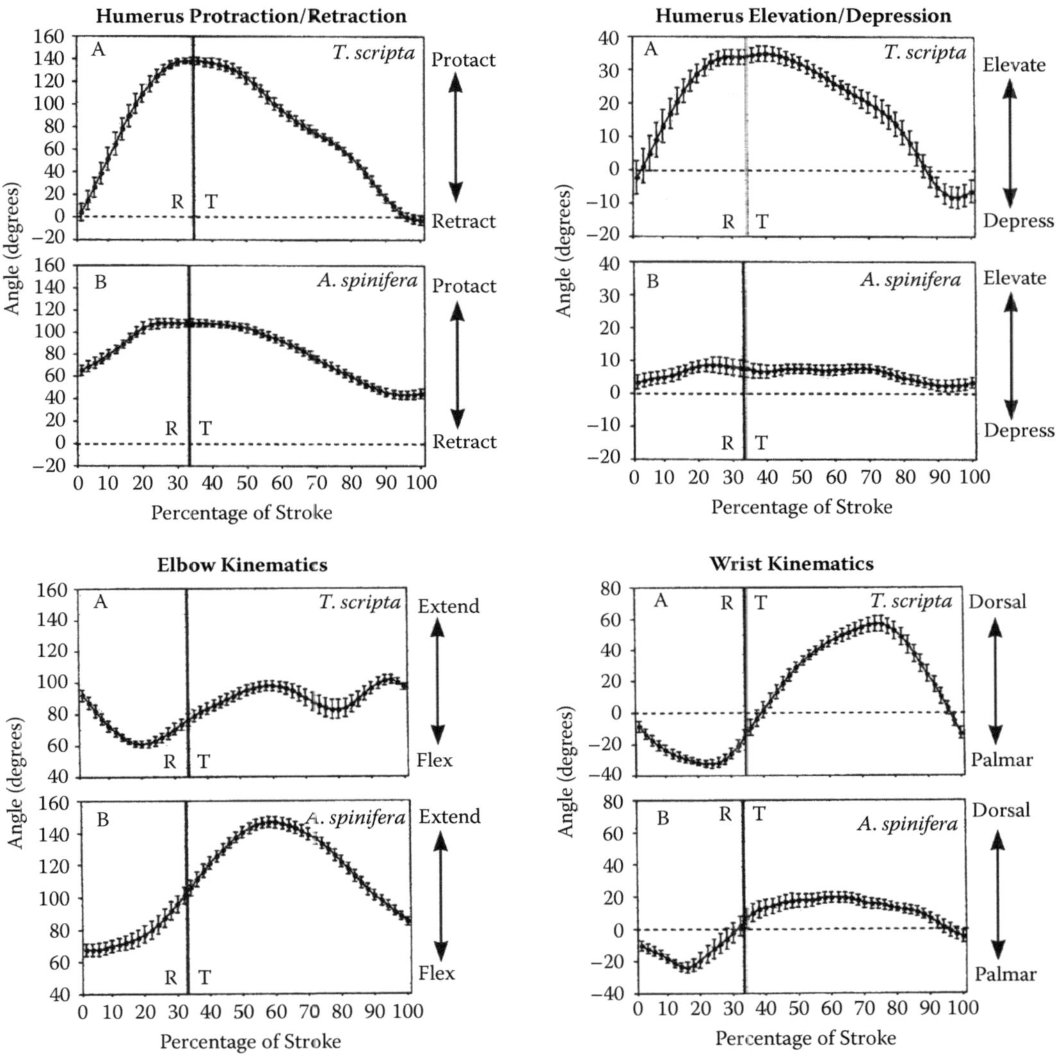

FIGURE 5.15 Mean kinematic profiles of humerus protraction and retraction (angle from the transverse plane), humerus elevation and depression (angle from the horizontal plane), elbow extension and flexion (true angle in three dimensions), and wrist dorsiflexion and palmar flexion (true angle in three dimensions) during the swimming of *Trachemys scripta* (A) and *Apalone spinifera* (B). Variations of mean values of kinematic angles ($N = 15$ trials) (with standard errors) are plotted. A bold vertical line in each plot demarcates recovery phase (R) from thrust phase (T). Humerus is perpendicular to the anteroposterior midline of the turtle, and also horizontal for angles of 0°. In protraction/retraction, the distal end of this bone is directed posteriorly for negative values and inversely for positive values. In elevation/depression, it is elevated above the proximal end for positive values, and depressed below the proximal end for negative values. Larger values indicate greater elbow extension. 0° corresponds to a straight wrist, the positive values showing a dorsiflexion and negative values a palmar flexion (used with permission from Pace et al., 2001).

advantages such as enhanced speed. Comparing the swimming of marine and freshwater turtles, Davenport et al. (1984) demonstrated that juvenile *Chelonia mydas* could swim approximately six times as fast as *Mauremys caspica*, the fastest semiaquatic form studied (the range of speed for *Kinosternon subrubrum* was 5 to 128 cms^{-1} and 1.82 BLs^{-1} for the maximal speed; *Trachemys*

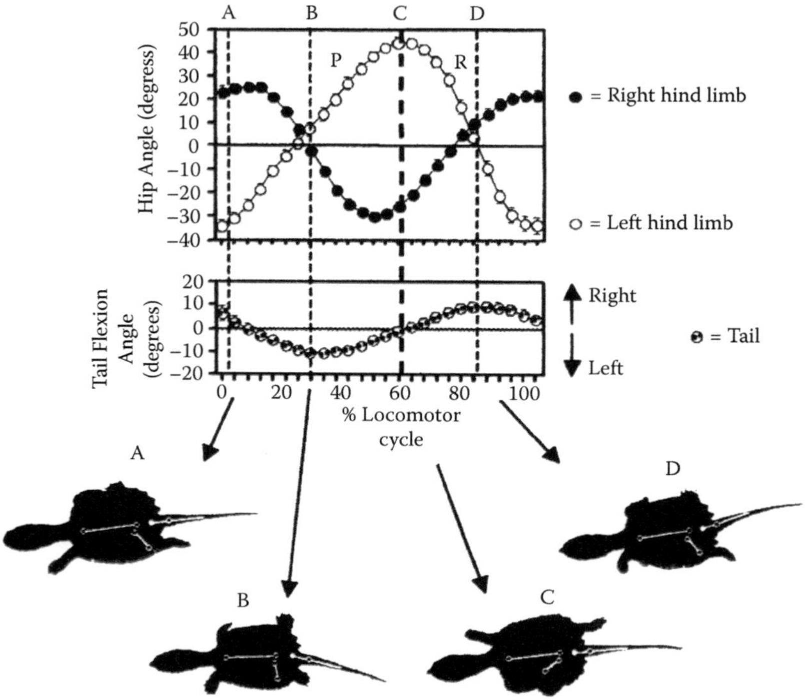

FIGURE 5.16 Mean kinematic profiles for the two hind limbs and tail of juvenile *Chelydra serpentina* during aquatic walking. A angle of 0° for the hip angle of each limb indicates a femur perpendicular to the sagittal plane of the body. A positive slope indicates a protraction of the limb and a negative slope a retraction. For each limb, a bold vertical line separates the protraction (P) from the retraction (R). This line is labelled C for the left limb. A straight tail is indicated by an angle of 0°, a flexion to the left by a negative value, and a flexion to the right by a positive value. The time A corresponds to the maximal retraction of the left limb, B to the maximal tail flexion toward the left, C to the maximal protraction of the left limb, and D to the maximal tail flexion toward the right. Solid points and lines are drawn or superimposed on the video images to calculate position of the sacrocaudal joint (used with permission from Willey & Blob, 2004).

scripta elegans, 6.8 to 13.3 cms^{-1} and 1.11 BLs^{-1} for the maximal speed; *Mauremys caspica*, 10 to 22.7 cms^{-1} and 2.27 BLs^{-1} for the maximal speed; *Chelonia mydas*, 10 to 143 cms^{-1} in vigorous swimming and 13 BLs^{-1} for the maximal speed, the routine swimming being lower than 10 cms^{-1}; "BL" signifies body length, and the thrust generated was higher, during a period of 5 s, for approximately the same body mass: *Chelonia mydas* developed 2.44 N (average forward thrust for all body masses = 0.89 N) and *Mauremys caspica* 1.04 N (average forward thrust for all body masses = 0.46 N). *Apalone spinifera* (Pace et al., 2001) reached speeds of 3.44 ± 0.48 BLs^{-1}. We have no swimming speed data for *Carettochelys*, a "highly aquatic" freshwater species with flipper-shaped forelimbs similar to those of sea turtles. In marine turtles, locomotion involves the simultaneous beating of the forelimbs (Davenport et al., 1984). Thrust, generated during the stance phase of the limb cycle in terrestrial locomotion and the retraction phase in swimming (in freshwater forms), is created by the upstroke or the downstroke, or both, as is the case with the largest marine turtles, *Chelonia mydas* and *Dermochelys coriacea* (Davenport et al., 1984; Davenport, 1987). Davenport et al. (1987, 1997) were also able to quantify the advantage of simultaneous forelimb beating within a single marine species. By observing young green turtles (*Chelonia mydas*) in a flume and gradually increasing the current velocity, they noticed that the turtles swam with a drag-based alternate-limb dogpaddle action similar to that of freshwater turtles, used when diagonally opposite limbs moved together in a contralateral synchronization (Davenport et al., 1984). At higher current velocities, they employed simultaneous foreflipper beating. At 25°C, the transition between swimming modes occurred at a sustained speed of 1.32 BLs^{-1} and the mean maximum sustained swimming speed using the flapping foreflippers was 3.31 BLs^{-1}. In this case, the synchronized lift-based flapping gave a 2.5-fold advantage over dogpaddle (Davenport et al., 1984), and the morphological configuration (shell shape, streamlining) was identical in the two conditions.

According to Pace et al. (2001), the evolution of aquatic specialization appears to have followed two distinct paths. Good swimming ability is correlated with forelimb hypertrophy and the use of flapping limb kinematics in sea turtles and carettochelyids. In contrast, in softshell turtles good swimming ability is correlated with rowing limb kinematics, in which the thrust-generating surface area of the limbs is maximized and drag minimized. In comparison to these specialized paths, the more functionally and morphologically generalized freshwater turtles, such as emydids, can be predicted to exhibit systems of propulsion with a greater capacity for both aquatic and terrestrial performance.

5.3.3 Coordination of the Limbs in Aquatic Locomotion

There are important differences in the limb coordination of turtles during aquatic locomotion. Zug (1971) analyzed sequences of limb movement during swimming of 15 species of cryptodires, mainly including semiaquatic or highly aquatic forms of four families: Chelydridae, Kinosternidae, Emydidae, and Trionychidae. He showed that all animals adopted in-water gaits equivalent to fast walks or moderate runs when the retraction phase of rowing was considered as equivalent to the stance phase of vertebrate terrestrial locomotion (they both are the propulsive stage of the limb cycle). In several cycles of swimming of each species, calculation of the percentage of stride during which the left fore-footfall follows the left hind foot, according to Hildebrand's (1966) method of symmetrical gait analysis, enabled us to characterize swimming as equivalent to a diagonal sequence of walk or running trot (Figure 5.17). In these gaits, in the displacement of a diagonal couplet the beginning of retraction of a hindlimb occurs before the beginning of retraction of the diagonally opposed forelimb. Zug (1971) estimated that this overlapping diagonal sequence would produce a minimal amount of yaw. This reinforces the idea that generation of consistent amounts of thrust can be produced from diagonally opposed limb propulsive phases operating more-or-less at the same time. In swimming, the variation of timing of limb movement is great, caused by variation in the ipsilateral pairs. In contrast, in swimming adult sea turtles (Figure 5.18) the foreflippers flap simultaneously. In hatchlings, this simultaneous movement is not perfect, but most swimming results from near-synchronous beating of the foreflippers. However, hatchling marine turtles can also swim with a type of rowing, termed "dogpaddling," occurring when diagonally opposite limbs move together, (Davenport et al., 1997). This is similar to the swimming gaits of freshwater turtles except that the limbs move in the horizontal and vertical planes to a greater extent. Marine turtle hatchlings (*Caretta caretta*) can also adopt a propulsive rear flipper kicking by simultaneously beating the hindlimbs; in this case, the forelimbs are tucked against the shell and take no part in locomotion (Davenport & Pearson, 1994).

Differences also appear when semiaquatic or highly aquatic turtle forms walk on land and are therefore forced to move under gravitational constraints. Zug (1971) studied the terrestrial locomotion of aquatic species and noticed a basic change (Figure 5.17): the slower gaits fall, in Hildebrand's analysis system (1966), between the walk and the trot of the lateral sequence. In each diagonal couplet, the beginning of retraction of the forelimb started before the beginning of retraction of the opposite hindlimb. In gaits characterized by this lateral sequence, the footfall of the hindlimb is followed by the footfall of the ipsilateral forelimb. In the diagonal sequence, the hindlimb footfall is, in contrast, followed by the diagonally opposite forelimb footfall.

The gait of terrestrial forms (Testudinidae) is characterized by this lateral sequence alone (Figure 5.19). These data signify that semiaquatic and highly aquatic freshwater turtles (such as Trionychidae) that have "discovered" a diagonal sequence of moving when they progress in water have

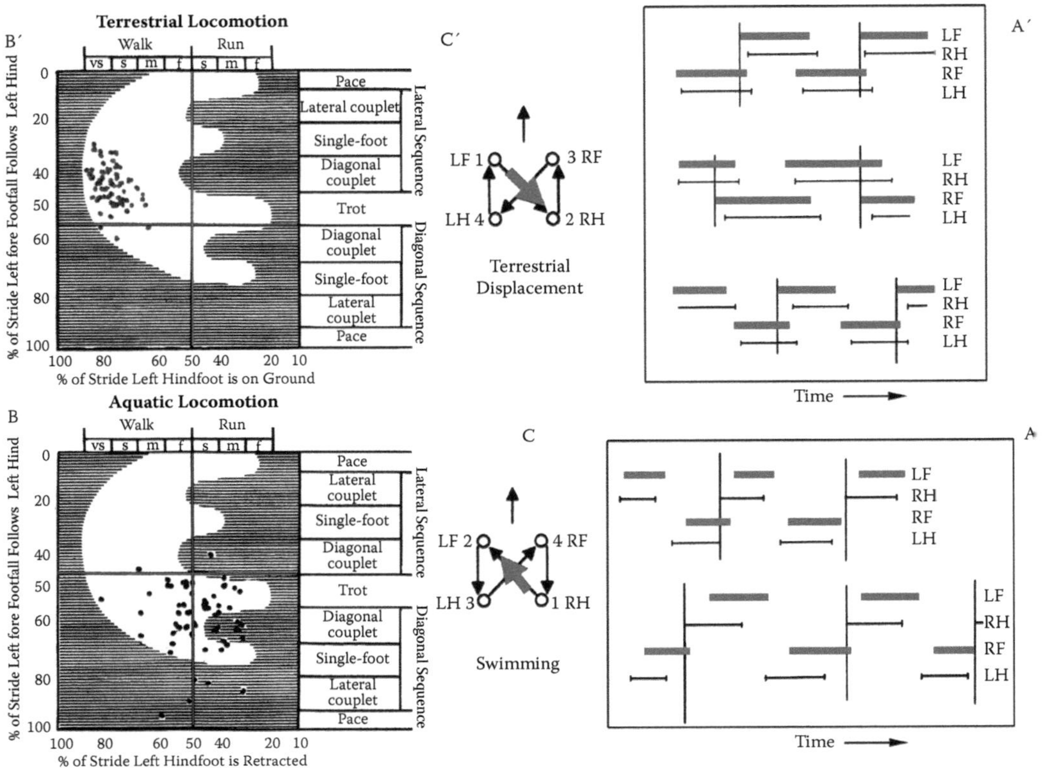

FIGURE 5.17 Gaits used by the freshwater cryptodires turtles to swim and walk. Gait diagrams in aquatic locomotion (A) showing the adoption of a trotting walk or a trot characterized by a diagonal sequence (B). In addition (C) the order of limbs succession is given: a hind limb (1) followed by the diagonally opposed fore limb (2), the hind limb on the same side of the body (3) followed by the diagonally opposed fore limb (4) and the hind limb on the same side of the body (1 again), cyclically during the progression. Gait diagrams in terrestrial locomotion (A′) showing the adoption of a trotting walk or a trot characterized by a lateral sequence (B′). In addition (C′) the order of limbs succession is given: a fore limb (1) followed by the diagonally opposed hind limb (2), the fore limb on the same side of the body (3) followed by the diagonally opposed hind limb (4) and the fore limb on the same side of the body (1 again), cyclically during the progression. LF, left fore limb; LH, right fore limb; RF, right fore limb; RH, right hind limb. f, fast walk; m, moderate walk; s, slow walk; sw, very slow walk (adapted from Zug, 1971).

never lost the typical gaits of the terrestrial forms: a slow walking trot with a lateral sequence giving the most stable dynamic basis (via maximized tripodal support) that is often used by other slow vertebrates. A recent test conducted with *Emys orbicularis* has supported Zug's conclusions (Depecker, personal communication). We have no data for *Carettochelys.*

When sea turtles leave the ocean to crawl up beaches for egg laying, some cheloniids (*Caretta*, *Eretmochelys*, *Lepidochelys*) use typical alternate ("terrestrial") movements of their flippers to move on sand and produce thrust. The terrestrial locomotion of the cheloniid *Natator* is less clear because it can employ either alternate or simultaneous movements of its four limbs. In contrast, simultaneous limb movements on land are exclusively adopted by the largest marine forms, the cheloniid *Chelonia* and dermochelyid *Dermochelys* (Renous et al., 1989; Renous & Bels, 1991, 1993; Wyneken, 1997). While slow, this gait does not involve raising the entire body; instead, the forelimbs essentially produce the lifting and forward displacement of the body as if the animal was "walking on crutches" (Figure 5.20). The hindlimbs help to raise the proximal part of the forelimbs

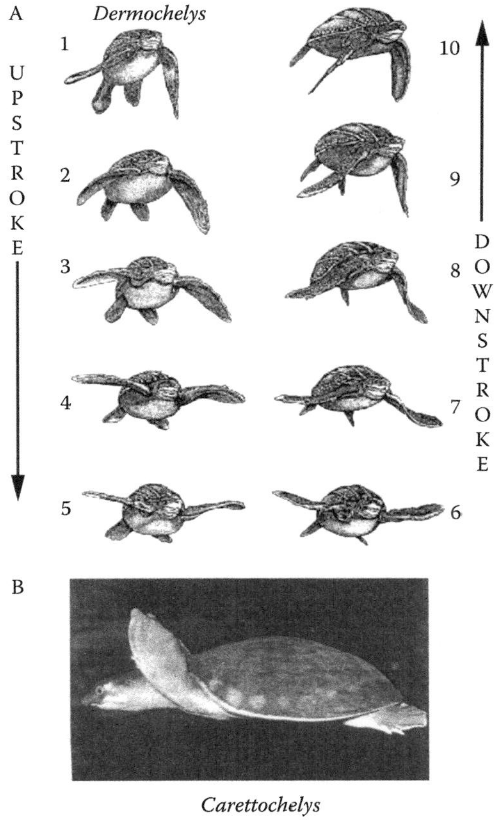

FIGURE 5.18 Simultaneous beating of the fore limbs during the swimming of *Dermochelys coriacea* (A) and suggestion of the same limb synchrony in *Carettochelys*. (Used with permission from Pritchard, 1979.) (B) The numbers indicate successive stages of the two phases of the forelimb cycle: upstroke (from 1 to 5) and downstroke (from 6 to 10) (used with permission from Renous & Bels, 1993).

and contribute to the equilibrium of the body. These turtles have lost the basic terrestrial gait and have acquired, by minimal adjustment of the swimming mode, a type of energy-saving system appropriate for moving a heavy mass forward under gravity.

5.3.4 A History of Limb Coordination

Patterns of limb coordination are vital to understand the evolution of locomotion in aquatic turtles and the selection of ecological strategies (Hendrickson, 1980). We believe that coordination guides the structural modifications of the limbs. Particularly interesting are the questions posed by the existence of synchronous foreflipper action in locomotion in diverse extant groups, some of which have rigid flippers (Cheloniidae, Demochelyiidae), others not (Carettochelyiidae). Even among the still living species, it seems that rigid foreflippers are not a prerequisite for synchronization.

Recently, a study of large tetrapod track ways with exceptional fossilization (Figure 5.21) from Cerin (Upper Jurassic, France) (Bernier, 1985; Roman et al., 1994) suggests that giant aquatic turtles used simultaneous movements of contralateral limbs (Gaillard et al., 2003). These extinct turtles moved in shallow waters of a lagoon (though we do not know whether they also swam in deeper waters). Interpretation of the footprints suggests that the limbs differed from extant marine turtles in the presence of free claws and an interdigital web. The fossilized animals did not possess true rigid

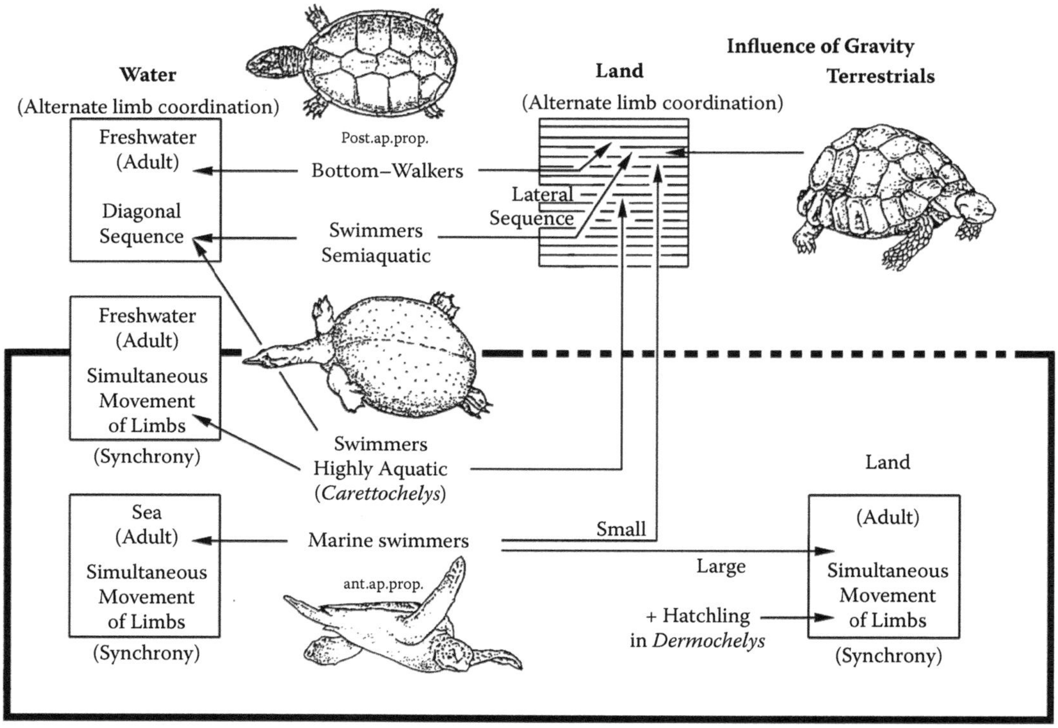

FIGURE 5.19 Patterns of propulsion and limb coordination to progress in water and on land in adult extant chelonians (adapted from Renous et al., 1999) ant.ap.prop., anterior appendicular propulsion; post.ap.prop., posterior appendicular propulsion (which can be shared with the anterior limb in the better swimmers). The marine swimmers are divided in small and large forms. The white rectangle indicates the synchronous periodic movements of the limbs introducing a novelty in the swimming adaptation. The grey rectangle indicates the action field of gravity.

flippers and so must have acquired the synchronous pattern of limb coordination before the selection of highly modified rigid fins. This important difference raises questions about the locomotor mechanisms of primitive groups of Eurysternidae and allied forms also present at the site, as well as primitive Protostegidae. Besides, the hindlimbs ought to have been longer than the forelimbs. Synchronous movement of the limbs evidently appeared several times as far as the Jurassic forms of Cheloniidae-Dermochelyiidae and Carettochelyidae are concerned. As we can deduce from Zug (1971) for non-marine and non-carettochelyid species, no phylogenetic relationships between these groups can be resolved from their limb coordination patterns.

So what drove the evolution of synchronous foreflipper action? Large size does not seem an adequate explanation; large individuals of riverine species (e.g., *Batagur*, *Podocnemis expansa*) may overlap in body size with small individuals of extant marine turtles but employ the diagonal pattern associated with rowing in most aquatic turtles, which is probably retained for its flexibility that permits maneuverability and regular use of the terrestrial habitat.

We suggest that synchronous forelimb movement is the primary feature that allows a true pelagic marine life in modern forms—it is seen in its most highly developed form in those large turtles that, as adults, carry out extensive high speed oceanic migrations (*Chelonia*, *Dermochelys*). Synchronous limb movement was already used by other species during the Jurassic (Cerin and probably Bavaria and Canjuers), possibly also by the first South American protostegids. At the same time, it was found used with in freshwater environments (as suggested by the extant *Carettochelys*;

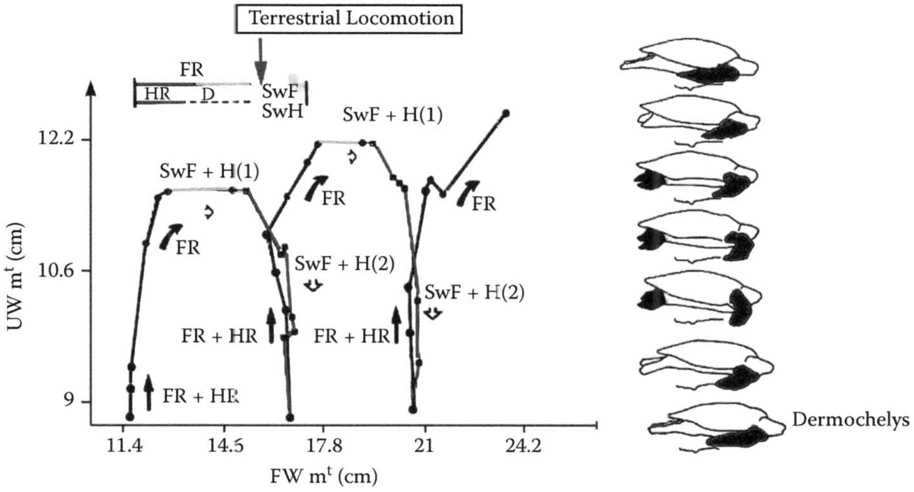

FIGURE 5.20 Terrestrial locomotion of *Dermochelys coriacea* displacement of the cranial section of the body (anterior limit of the dorso-lateral hull) of a juvenile of 11 kg in a vertical plane. FR, fore limb retraction; FR+HR, simultaneous retraction of the fore- and hind limbs; FWm^t, forward movement; HR, hind limb retraction; SwF and SwH, fore- and hind limbs swings; SwF+H(1) and (2), first and second parts of the fore and hind limbs swing; UWm^t, upward movement; D, dragging. The arrows underline the resulting movement. A gait diagram documents the duration of the respective phases of the fore and the hind limbs during a complete revolution cycle of the four limbs. Drawings of the stages of the progression are also presented (used with permission from Renous 1988, Renous & Bels, 1993).

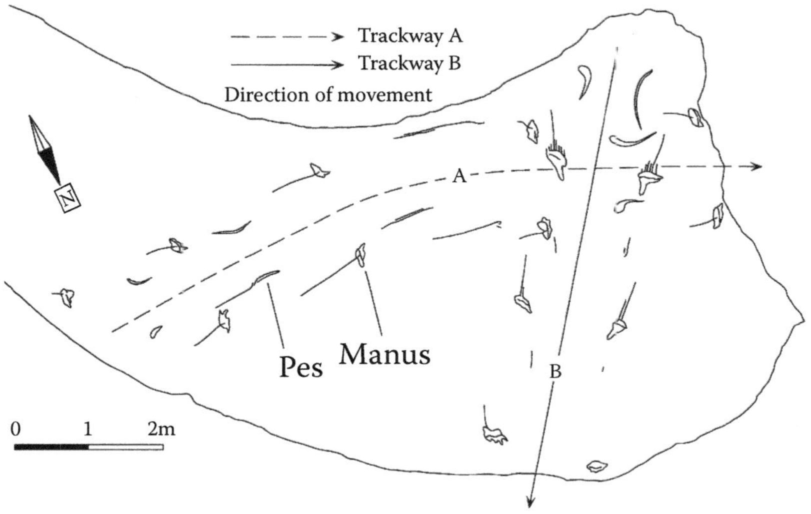

FIGURE 5.21 Footprints of a Jurassic fossil turtle from Cerin (Cerin, Ain, France). Two tracks, A and B, are visible on the sediment and show footprints assigned to hands and feet (the former being out of the latter along the middle line of each track) (used with permission from Gaillard et al., 2003).

the oldest recorded Carettochelyidae are either from the late Jurassic or early Cretaceous (Young & Chow, 1953) but they were probably already well present at Jurassic times, according to their wide distribution during the Lower Cretaceous (Lapparent de Broin, 2004) and new discoveries in the Cretaceous from France. This implies an early separation from the Trionychidae, their sister group, probably during the late Jurassic. The carettochelyid humerus is known from the early Cretaceous from Laos and complete limbs are described from the Eocene (Harrassowitz, 1922); they are identical to those of the extant forms. The carettochelyid paddle had therefore characterized all members of the family at its appearance and was probably associated to synchronous movement of the limbs. The ability of extant Carettochelyidae to penetrate estuaries and salt waters is known in New Guinea and Australia, thus it is not impossible that the track ways from Cerin was made by a carettochelyid, if not a protostegid or an eurysternid.

The challenge is to identify the environmental or life history features that have selected for synchronous forelimb movement in non-pelagic situations and at relatively small sizes. The very shallow environment of the Cerin trackways and the occasional synchronous action of forelimbs during mud digging in some extant *Trionyx* species may give a clue; perhaps synchronous limb action aids semi-terrestrial movement in shallow water overlying extensive soft substrata? More study is clearly required.

5.4 CONCLUSION

In chelonians, phylogenetic studies and cladograms constructed with combined molecular and morphological data have revealed satisfactory correlations between the great diversity of habitats occupied by turtles and the major patterns of locomotion that they exhibit. This is an important first step toward improving our understanding of the evolutionary processes of this terrestrial group of vertebrates. In reality, all chelonians are "terrestrial" animals by virtue of their major physiological functions, especially respiratory physiology and principal behaviors (notably, egg-laying). At regular periods in their lives, at least females of all turtle species must return to land. In spite of secondary adaptations to aquatic habitats that introduced changes in the mode of coordination of limb movements (from lateral sequences on land under the influence of gravity to diagonal sequences and synchronous sequences in water), they all must use some kind of limb-based locomotion to move on land.

Different evolutionary scenarios and locomotor strategies have been selected in the chelonians. It is likely that the basic structural "turtle pattern" arose in terrestrial forms solely in the context of a terrestrial life—most likely to provide maximum body protection. Subsequent adaptations to a semiaquatic or highly aquatic way of life were rapidly adopted, allowing the development of a mode of locomotion that was able to work within the constraints of the two environments, aquatic and terrestrial. This mode retained an alternating limb function, even if it underwent some further adjustments when the environment changed. In this context, the morphological-functional features of turtles cannot be solely attributed to the constraints of a single environment. However, we must take into account the fact that the constraints of the terrestrial environment are strong, with animals simultaneously benefiting from and being compromised by these constraints (e.g., structural integrity of the body versus lightening of the body). Nevertheless, even in the context of this "polyvalent" adaptation requiring a diagonal coordination of the limbs, as more time is spent swimming in the aquatic environment the more turtles exhibit true "challenges" when returning to the terrestrial environment (even to lay eggs). The result is a decrease of speed and other locomotor parameters and the induction of greater energetic cost. The use of synchronous limb movement was probably adopted very early before the forelimbs were highly modified into rigid flippers (e.g., Jurassic turtles from Cerin), whereas the hindlimbs were flexible elements, similar to those of some extant forms (Carettochelyidae). This pattern of coordination, present in Cerin (an environment distinct from the sea and characterized by shallowness, great evaporation, and high salinity, hence selecting for hypertrophied salt glands), for example, was the first that permitted the eventual full conquest of the marine environment. Subsequently, the differentiation of rigid fore- and hindlimbs, longer in the case of the foreflippers, has favored a more complete adaptation to a pelagic marine life.

It is probable that the synchronous movement of the forelimbs in truly marine locomotion was the result of exaptation: perhaps performance enhancement that allowed greater thrust and speed, in prey capture or predator avoidance. The development of rigid limbs and their synchronous functioning in marine turtles facilitated the coverage of great distances in open sea, either to move between far-flung benthic food resources (coral reefs, sea grass) or to seek out ephemeral pelagic food concentrations (e.g., jellyfish). These morphofunctional parameters strongly influence efficiency on land when the female turtles come back to lay eggs with a large body that is very heavy under the influence of gravity. Alternate functioning of the limbs could introduce added instability and greater energy expenditure. In contrast, the alternate limb movements exhibited by the great majority of semi-aquatic and highly aquatic turtles found their optimum efficiency in the exploitation of the great diversity of fresh water habitats, yet also allowed more extensive use of the terrestrial environment.

REFERENCES

Barth, R., Observacoes sobre a grande tertaruga marinha, *Chelonia mydas* L., feitas en Ilha de Trinadade, *An. Acad. Brasil. Cienc.*, 34, 405, 1962.

Berge, C., and Penin, X., Ontogenetic allometry, heterochrony and interspecific differences in the skull of African apes, using tridimensional Procrustes analysis, *Am. J. Phys. Anthropol.*, 124, 124, 2004.

Bernier, P., *Une Lagune Tropicale au Temps des Dinosaures*, Paris: CNRS Paris-Muséum Lyon Eds., 1985, 136.

Bodie, J.R., and Semlitsch, R.D., Spatial and temporal use of floodplain habitats by lentic and lotic species of aquatic turtles, *Oecologia*, 122, 138, 2000.

Bookstein, F.L., *Morphometrics Tools for Landmarks Data: Geometry and Biology*, Cambridge, UK: Cambridge University Press, 1991.

Braga, M. de, and Rieppel, O., Reptile phylogeny and the interrelationships of turtles, *Zool. J. Linn. Soc.*, 120, 281, 1997.

Broin, F. de, *Proganochelys ruchae* n. sp., Chélonien du Trias supérieur de Thaïlande, *Stud. Geol. Salm., Stud. Palaeochel.*, 1, 87, 1985.

Broin, F. de, Les Tortues et le Gondwana: Examen des rapports entre le fractionnement du Gondawana et la dispersion géographique des Tortues pleurodires à partir du Crétacé, *Stud. Geol. Salm., Studia Palaeochel.*, 2, 103, 1988.

Broin, F. de, Données préliminaires sur les Chéloniens du Tithonien inférieur de Canjuers (Var, France), Table Ronde Internationale Calcaire Lithographique, Lyon, Juillet 1991, *Geobios*, 16, 168, 1994.

Burke, A.C., Epithelial-mesenchymal interactions in the development of the Chelonian Bauplan, in *Trends in Vertebrate Morphology*, E. Spechtana and H. Hilgers (eds.), Stuttgart: Fischer Verlag, 1989, 206.

Cadena Rueda, E.A., and Gaffney, E.S., *Notoemys zapatocaensis*, a new side-necked turtle (Peurodira: Platychelyidae) from the Early Cretaceous of Colombia, *Amer. Mus. Nov.*, 3470, 19, 2005.

Cann, J., *Australian Freshwater Turtles*, Singapore: J. Cann & Beaumont Publishing, 1998, 292.

Casamiquela, R.M., Estudios icnologicos: Problemas y metodos de la icnologia con aplication al studio de pisadas mesozoicas (Reptilia, Mammalia) de la Patagonia, Gobierno de la Provincia de Río Negro, Ministerio de Asuntos Sociales, 229pp. Buenos Aires, 1964.

Claude, J., Paradis E., Tong, H., and Auffray, J.-Ch., A geometric morphometric assessment of the effects of environment and cladogenesis on the evolution of the turtle shell, *Biol. J. Linn. Soc.*, 79, 485, 2003.

Claussen, D.L., Lim, R., Kurz, M., and Wren, M., Effects of slope, substrate, and temperature on the locomotion of the ornate box turtle, *Terrapene ornate*, *Copeia*, 2, 411, 2002.

Datta, P.M., Manna, P., Ghosh, S.C., and Das, D.P., The first Jurassic turtle from India, *Palaeontology*, 43, 99, 2000.

Davenport, J., Locomotion in hatchling leatherback turtles *Dermochelys coriacea*, *J. Zool. Lond.*, 212, 85, 1987.

Davenport, J., and Clough, W., Swimming and diving in young loggerhead sea turtles (*Caretta caretta* L.), *Copeia*, 53, 1986.

Davenport, J., Monks, S.A., and Oxford, P.J., A comparison of the swimming in marine and freshwater turtles, *Proc. R. Soc. Lond. B*, 220, 447, 1984.

Davenport, J., and Pearson, G.A., Observations on the swimming of the Pacific ridley, *Lepidochelys olivacea* (Eschscholtz 1829): Comparisons with other sea turtles, *Herpetol. J.*, 4, 60, 1994.

Davenport, J., de Verteuil, N., and Magill, S.H., The effects of current velocity and temperature upon swimming in juvenile green turtles *Chelonia mydas* L., *Herpetol. J.*, 7, 143, 1997.

Davenport, J., and Wong, T.M., Aerial and aquatic respiration in the black-rayed softshell turtle *Amyda cartilaginea* (Boddaert), *Herpetol. J.*, 2, 115, 1992.

De la Fuente, M.S., and Fernandez, M.S., *Notoemys laticentralis* Cattoi & Freiberg, 1961, from the Upper Jurassic of Argentina: A member of the Infraorder Pleurodira (Cope, 1868), *Studia Geol. Salm. Studia Paleochel.*, 3, 25, 1989.

De la Fuente, M.S., and Iturralde-Vinent M., A new pleurodiran turtle from Jagua Formation (Oxfordian) of Western Cuba, *J. Paleontol.*, 75, 860, 2001.

De la Fuente M., Lapparent de Broin, F. de, and Manera de Bianco, T., The oldest and first nearly complete skeleton of a chelid, of the Hydromedusa group (Chelidae, Pleurodira), from the Upper Cretaceous of Patagonia, *Bull. Soc. Geol. Fr.*, 172, 237, 2001.

Depecker, M., Berge, C., Penin, X., and Renous, S., Geometric morphometry of the shoulder girdle in extant turtles (Chelonii), *J. Anat.*, 208, 35, 2006a.

Depecker, M., Renous, S., Berge, C., and Penin, X., Procrustes analysis: A tool to understand shape changes of the humerus in turtles (Chelonii), *C.R. Palevol.*, 5, 509, 2006b.

Deraniagala, P.E.P., The testudinata of Ceylon, *Ceylon J. Sci. B.*, 16, 43, 1930.

Dominici, P., Habitat, body design and the swimming performance of fish, in *Vertebrate Biomechanics and Evolution*, V.L. Bels, J.P. Gasc, and A. Casinos (eds.), Oxford: BIOS Scientific Publishers, 2003,137.

Earhart, G.M., and Stein, P.S.G., Step, swim, and scratch motor patterns in the turtle, *J. Neurophysiol.*, 84, 2181, 2000.

Feldam, C.R., and Parham, J.F., Molecular phylogenetics of emydine turtles: Taxonomic revision and the evolution of shell kinesis, *Mol. Phylogenet. Evol.*, 22, 388, 2002.

Fernandez, M.S., and de la Fuente, M.S., Las tortugas casiquelidias de las calizas litograficas titonianas del area Los Catutos, Neuquén, Argentina, *Ameghiniana*, 30, 283, 1993.

Fernandez, M.S., and de la Fuente, M.S., Redescription and phylogenetic position of *Notoemys*: The oldest Gondwanian pleurodiran turtle, *Neues Jahrb. Geol. P.A.*, 193, 81, 1994.

Finkler, M.S., and Claussen, D.L., Use of the tail in terrestrial locomotor activities of juvenile *Chelydra serpentine*, *Copeia*, 884, 1997.

Fish, F.E., Transitions from drag-based to lift-based propulsion in mammalian swimming, *Am. Zool.*, 36, 628, 1996.

Gaffney, E. S., The systematics of the North American Family Baenidae (Reptilia, Cryptodira), *Bull. Mus. Nat. Hist.*, 147, 241, 1972.

Gaffney, E.S., The shell morphology of the Triassic turtle *Proganochelys*, *N.J. Geol. Palont. Abh.*, 170, 1, 1985.

Gaffney, E.S., The comparative osteology of the Triassic turtle *Proganochelys*, *Bull. Am. Mus. Nat. Hist.*, 194, 1, 1990.

Gaffney, E.S., The postcranial morphology of *Meiolania platyceps* and a review of the Meiolaniidae, *Bull. Am. Mus. Nat. Hist.*, 229, 166, 1996.

Gaffney, E.S., Hutchinson, J.H., Jenkins, F.A., and Meeker, L.J., Modern turtle origins: The oldest known Cryptodire, *Science*, 237, 289, 1987.

Gaffney, E.S. and Kitching, J.W., The most ancient African turtle, Letters to Nature, *Nature*, 369, 55, 1994.

Gaffney E.S., Kool, L., Brinkman, D.B., Rich, T., and Vickers-Rich, P., *Otwayemys*, a new Cryptodiran turtle from the early Cretaceous of Australia, *Am. Mus. Novit.*, 3233, 28, 1998.

Gaffney, E.S., and McKenna, M.C., A late Permian captorinid from Rhodesia, *Am. Mus. Novit.*, 2688, 15, 1979.

Gaffney, E.S., and Meylan, P.A., A phylogeny of turtles, in *The Phylogeny and Classification of the Tetrapods, Vol. 1: Amphibians, Reptiles, Birds*, Systematics Association Special Volume 35A, M.J. Benton (ed.), Oxford: Clarendon Press, 1988, 5, 157.

Gaffney, E.S., and Meylan, P.A., The Transylvanian turtle, *Kallokibotion*, a primitive Cryptodire of Cretaceous Age, *Amer. Mus. Novit.*, 3040, 1, 1992.

Gaffney, E.S., Meylan, P.A., and Wyss, A.R., A computer assisted analysis of the relationships of the higher categories of turtles, *Cladistics*, 7, 313, 1991.

Gaffney, E.S., Tong, H., and Meylan P.A. Evolution of the side-necked turtles: The families Bothremydidae, Euraxemydidae, and Araripemydidae, *Bull Amer. Mus. Nat. Hist.* 300, 700, 2006.

Gaillard, C., Bernier, P., Barale, G., Bourseau, J.P., Buffetaut, E., Ezquerra, R., Gall, J.C., de Lapparent de Broin, F., Renous, S., and Wenz, S., A giant upper Jurassic turtle revealed by its trackways, *Lethaia*, 36, 315, 2003.

Gans, C., and Hughes, G.M., The mechanism of lung ventilation in the tortoise, *Testudo graeca*, *J. Exp. Biol.*, 47, 1, 1967.

Gasc, J.P., and Renous, S., Les données ontogénétiques peuvent-elles expliquer l'origine évolutive des Squamates apodes?, in *Ontogénèse et Evolution*, Coll. Intern. CNRS, *Géobios*, Mém. Sp., 12, 173, 1989.

Gaunt, A.S., and Gans, C., Mechanics of respiration in the snapping turtle, *Chelydra serpentina* (Linné), *J. Morphol.*, 128, 195, 1969.

Gibbons, J.W., Terrestrial activity and population dynamics of aquatic turtles, *Am. Midl. Nat.*, 83, 404, 1970.

Gilbert, S.F., Loredo, G.A., Bruckman, A., and Burke, A., Morphogenesis of the turtle shell: The development of a novel structure in tetrapod evolution, *Evol. & Dev.*, 3, 47, 2001.

Gillis, G.B., and Blob, R.W., How muscles accommodate movement in different physical environments: Aquatic vs. terrestrial locomotion in vertebrates, *Com. Biochem. Phys. A.*, 131, 61, 2001.

Gregory, W.K., Pareiasaurs versus placodonts as near ancestors to the turtles, *Bull. Amer. Mus. Nat. Hist.*, 86, 276, 1946.

Guibé, J., Le squelette du tronc et des membres, in, *Traité de Zoologie*, Grassé, P.P. (ed.), Paris: Masson & Co, 1970, 14, 33.

Harrassowitz, H., Die Schildtrötengattung ANOSTEIRA von Messel bei Darmstadt und ihre stammeesgeschichliche Bedeutung, *Abh. Hess. Geol. Land. Darmstadt*, 6, 238,1922.

Hay, O.P., The fossil turtles of North America, *Pub. Carnegie Inst.*, 75, 568, 1908.

Hedges, S.B., and Poling, L., A molecular phylogeny and the interrelationships of turtles, *Science*, 283, 898, 1999.

Hendrickson, J.R., The ecological strategies of sea turtles, *Am. Zool.*, 20, 597, 1980.

Hildebrand, M., Analysis of the symmetrical gaits of Tetrapods, *Folia Biotheor.*, 6, 9, 1966.

Hirayama, R., Phylogenetic systematics of chelonioid sea turtles, *The Island Arc*, 3, 270, 1994.

Hirayama, R., Oldest known sea turtle, *Nature*, 392, 705, 1998.

Hirayama, R., and Chitoku, T., Family Dermochelyidae (Superfamily Chelonioidea) from the Upper Cretaceous of north Japan, *Trans. Proc. Palaeont. Soc. Japan, N.S.*, 184, 597, 1996.

Holman, J.A., Herpetofauna of the Egelhoff site (Miocene: Barstovian) of north-central Nebraska, *J. Vert. Paleon.*, 7, 109, 1987.

Hughes, B., Occurrence of the freshwater trionychid turtles *Cyclanorbis elegans* and *C. senegalensis* in Ghana, *Bull. Inst. Fond. Afr. Noire*, 41, 193, 1979.

James, M.C., Davenport, J., and Hays, G.C., Expanded thermal niche for a diving vertebrate: A leatherback turtle diving into near-freezing water, *J. Exp. Mar. Biol. Ecol.*, 2006.

Jayes, A.S., and Alexander, R.M., The gaits of Chelonians: Walking techniques for very low speeds, *J. Zool. Lond.*, 191, 353, 1980.

Jenkins, F.A. Jr., Shubin, N.H., Amaral, W.W., Gatesy, S.M., Schaff, C.R., Clemmensen, L.B., Downs, W.R., Davidson, A.R., Bonde, N., and Osbaeck, F., Late Triassic continental vertebrates and depositional environments of the Fleming Fjord Formation, Jameson Land, East Greenland, Meddelelser om Grönland, *Bioscience*, 32, 2, 1994.

Joyce, W.G., and Gauthier, J.A., Palaeoecology of Triassic stem turtles sheds new light on turtle origins, *Proc. R. Soc. Lond.*, 271, 1, 2004.

Kordikova, E.G., Paedomorphosis in the shell of fossil and living turtles, *Neues Jahrb. Geol. P.A.*, 218, 399, 2000.

Lapparent de Broin, F. de, African chelonians from the Jurassic to the Present. A preliminary catalog of the African fossil chelonians, *Palaeontol. Africana*, 36, 43, 2000a.

Lapparent de Broin, F. de, The oldest Podocnemidid turtle (Chelonii, Pleurodira), from the early Cretaceous, Ceará state, Brasil, and its environment, *Treballs Mus. Geol. Barcelona*, 9, 43, 2000b.

Lapparent de Broin, F. de, The European turtle fauna from the Triassic to the Present: A short review, *Dumerilia*, 4, 155, 2001.

Lapparent de Broin F. de, A giant tortoise from the late Pliocene of Lesvos Island (Greece) and its possible relationships, *Ann. Géol. pays Héllén. 1re Sér.*, 39, A, 99, 2002.

Lapparent de Broin F. de, The Miocene chelonians from the southern Namibia, in *Faunas from the Southern Namibia,* Senut, B., and Pickford, M. (coord.), *Mem. Geol. Surv. Namibia*, 19, 67, 2003.

Lapparent de Broin F. de, A new Shachemydinae (Chelonii, Cryptodira) from the Lower Cretaceous of Laos: Preliminary data, *C.R. Palevol. Paris*, 3, 387, 2004.

Lapparent de Broin F. de, and de la Fuente, M.S., Particularidades de la fauna continental de tortugas del Cretacico de Argentina, Res. Com. XIV Jornadas Argentinas de Paleontología de Vertebrados, 18–20 Mai 1998, Neuquén, *Ameghinina*, 36, 104 1999.
Lapparent de Broin, F., and de la Fuente, M.S., Oldest world Chelidae (Chelonii, Pleurodira), from the Cretaceous of Patagonia, *C.R. Acad. Sc. Paris*, II, 333, 463, 2001.
Lapparent de Broin, F., de la Fuente, M.S., and Calvo, J.O., Presencia de los mas antiguos quélidos (tortugas pleurodiras) en el Cretácico inferior de "El Chocón," provincia del Neuquen, Argentina, XIII Jornadas Argentinas de Paleontología de Vertebrados, La Rioja, 11–15 Mai 1997, *Ameghiniana*, 34, 538, 1998.
Lapparent de Broin, F., de la Fuente, M.S., and Fernandez, M., *Notoemys* (Chelonii, Pleurodira), late Jurassic of Argentina: New examination of the anatomical structure and comparisons, *Rev. Paléobiol. Genève*, in press.
Lapparent de Broin, F. de, Lange-Badré, B., and Dutrieux, M., Nouvelles découvertes de tortues dans le Jurassique supérieur du Lot (France) et examen du taxon Plesiochelyidae, *Rev. Paleobiol.*, 15, 533, 1996.
Lapparent de Broin, F. de, and Murelaga, X., Une nouvelle faune de Chéloniens dans le Crétacé supérieur européen, *C.R. Acad. Sc., Paris*, 2a, 323, 729, 1996.
Lapparent de Broin, F. de, and Murelaga, X., Turtles from the Upper Cretaceous of Laño (Iberian Peninsula), in *Geology and Palaeontology of the Upper Cretaceous Vertebrate-bearing Beds of the Laño Quarry (Basque-Cantabrican Region, Iberian Peninsula)*, H. Astibia et al. (eds.), *Estud. Mus. Cienc. Nat. Alava*, 14 (Num. Esp. 1), 135, 1999.
Lapparent de Broin, F. de, Murelaga Bereikua, X., and Codrea, V., Presence of Dortokidae (Chelonii, Pleurodira) in the earliest Tertiary of the Jibou Formation, Romania: Paleobiogeographical implications, *Acta Paleont. Rom.*, 4, 203, 2004.
Lapparent de Broin, F. de, and Werner, C., New late Cretaceous turtles from the Western Desert, Egypt, *Ann. Pal.*, 84, 131, 1998.
Laurin, M., and Reisz, R.R., A reevaluation of early amniote pylogeny, *Zool. J. Linn. Soc.*, 113, 165, 1995.
Lee, M.S.Y., The origin of the turtle body plan: Bridging a famous morphological gap, *Science*, 261, 1716, 1993.
Lee, M.S.Y., The homologies and early evolution of the shoulder girdle in turtles, *Proc. R. Soc. London B*, 263, 111, 1996a.
Lee, M.S.Y., Correlated progression and the origin of turtles, *Nature*, 379, 812, 1996b.
Lee, M.S.Y., Pareiasaur phylogeny and the origin of turtles, *Zool. J. Linn. Soc.*, 120, 197, 1997.
Lessertisseur, J., Traces fossils d'activité animale et leur signification paléobiologique, *Mem. Soc. Geol. France*, 34, 74, 1, 1955.
Magne de la Croix, P., Filogenia de las locomociones cuadrupedal y bipedal en los vertebrados y evolution de la forma consecutiva de la evolution de la locomocion, *Ann. Soc. Cient. Argentina*, 108, 383, 1929.
Magne de la Croix, P., Evolution de la locomocion terrestre en los vertebrados, *Rev. Med. y Veterinar.*, 15-19, 2, 1933.
Manlius, N., Organisation de la ceinture pectorale et du membre antérieur de la tortue luth, *Dermochelys coriacea* Vandelli, 1761 (Chelonia, Dermochelyidae) et comparaison avec quatre autres espèces de cheloniens, *Bull. Soc. Herpetol. Fr.*, 73-74, 1, 1996.
Mertens, R., *The World of Amphibians and Reptiles*, New York: McGraw-Hill, 1960.
Muegel, L.A., and Claussen D.L., Effects of slope on voluntary locomotor performance in the turtle, *Terrapene carolina carolina*, *J. Herpetol.*, 28, 6, 1994.
Mulder, E.W.A., On latest Cretaceous tetrapods from the Maastrichtian type area, *Publ. Naturhist. Genootschop Limburg*, XLIV, 1, 2003.
Oliver, J.A., *The Natural History of the North American Amphibians and Reptiles*, Princeton, NJ: D. Van Nostrand, 1955.
Oliver, J.A., The Matamata: Surely all turtles can swim?, *Anim. King.*, 62, 167, 1959.
Pace, C.M., Blob, R.W., and Westneat, M.W., Comparative kinematics of the forelimb during swimming in red-eared slider (*Trachemys scripta*) and spiny softshell (*Apalone spinifera*) turtles, *J. Exp. Biol.*, 204, 3261, 2001.
Peabody, F.E., Trackways of living and fossil Salamanders, *Univ. Calif. Publs. Zool.*, 63, 1, 1959.
Peng, J.H., and Brinkman, D.B., New material of *Xinjiangchelys* (Reptilia: Testudines) from the Late Jurassic Qigu Formation (Shishugou Group) of the Pingfengshan locality, Junggar basin, Xinjiang, *Can. J. Earth Sc.*, 30(10-11), 2013, 1993.
Penin, X., APS: A Procrustes Software, http:/www.cpod.com/monoweb/aps, 2003.

Penin, X., Berge, C., and Baylac, M., Ontogenetic study of the skull in modern humans and common chimpanzees: Neotenic hypothesis reconsidered with tridimensional Procrustes analysis, *Am. J. Phys. Anthropol.*, 118, 50, 2002.

Platz, J.E., and Conlon, J.M., Reptile relationships turn turtle … and turn back again, *Nature*, 389, 246, 1997.

Pritchard, P.C.H., *Encyclopedia of Turtles*, Neptune, NJ: T.F.H. Publications, 1979, 895.

Reed, C.A., Non-swimming water turtles in Iraq, *Copeia*, 1, 51, 1957.

Reisz, R.R., and Laurin, M., Owenetta and the origin of turtles, *Nature*, 349 (6307), 324, 1991.

Renous, S., Retentissement de l'adaptation à la vie pélagique de la tortue luth (*Dermochelys coriacea*) sur sa locomotion terrestre, *Mésogée*, 48, 79, 1988.

Renous, S., and Bels, V., Etude cinématique de la palette natatoire antérieure de la tortue luth, *Dermochelys coriacea* (Vandelli, 1761), au cours de sa locomotion terrestre intermittente, *Can. J. Zool.*, 69, 495, 1991.

Renous, S., and Bels, V., Comparison between aquatic and terrestrial locomotions of the leatherback sea turtle (*Dermochelys coriacea*), *J. Zool., Lond.*, 230, 357, 1993.

Renous, S., Bels, V., and Davenport, J., Locomotion in marine chelonian: Adaptation to the aquatic habitat, *Hist. Biol.*, 14, 1, 1999.

Renous, S., Gasc, J.P., Lescure, V., and Bels, V., Intervention des membres dans la locomotion terrestre et le creusement du nid chez la tortue luth (*Dermochelys coriacea*), *Amphib. Reptilia*, 10, 355, 1989.

Rieppel, O., The genus Placodus: Systematics, morphology, paleobiogeography and paleobiology, *Fieldiana (Geol.), N.S.*, 31, 1, 1995.

Rieppel, O., Testing homology by congruence: The pectoral girdle of turtles, *Proc. R. Soc. Lond. B.*, 263, 1395, 1996.

Rieppel, O., Turtles as diapsid reptiles, *Zool. Scr.*, 29, 199, 2000.

Rieppel, O., and de Braga, M., Turtles as diapsid reptiles, *Nature*, 384, 453, 1996.

Rieppel, O., and Reisz, R.R., The origin and early evolution of turtles, *Annu. Rev. Ecol. Syst.*, 30, 1, 1999.

Rivera, G., Rivera, A.R.V., Dougherty, E.E., and Blob, R.W., Aquatic turning performance of painted turtles (*Chrysemis picta*) and functional consequences of a rigid body design, *J. Exp. Biol.*, 209, 4203, 2006.

Roman, J., Atrops, F., Arnaud, M., Barale, G., Barrat, J.-M., Boullier, A., de Broin, F., Gill, G.A., Michard, J.-G., Taquet, P., and Wenz, S., Le gisement tithonien inférieur de Canjuers (Var, France): État actuel des connaissances, Table Ronde Internationale: Calcaire Lithographique, Lyon, Juillet 1991, *Geobios*, M.S. 16, 126, 1994.

Romer, A.S., *Osteology of the Reptiles*, Chicago: University Chicago Press, 1956, 772.

Rougier, G.W., de la Fuente, M.S., and Arcucci, A.B., Late Triassic turtles from South America, *Science*, 268, 855, 1995.

Rougier, G.W., de la Fuente, M.S., and Arcucci, A.B., L'evolution des tortues, Pour la Science, 249, 42, 1998.

Schleich, H.H., Ein neuer Schildkrötennachweis aus Lias-Sedimenten Deutschlands (Testudines: Chelonioidea: ? Thalassemydidae), *Naturw. Zeitschr. F. Niederbayern*, 30, 56, 1984.

Schaeffer, B., The morphological and functional evolution of the Tarsus in amphibians and reptiles, *Bull. Amer. Mus. Nat. Hist.*, 78, 395, 1941.

Schluter, D., *The Ecology of Adaptive Radiation*, Oxford: Oxford University Press, 2000.

Shaffer, H.B., Meylan, P., and McKnight, M.L., Tests of turtles phylogeny: Molecular, morphological, and paleontological approaches, *Syst. Biol.*, 46, 235, 1997.

Spinks, P.Q., Shaffer, H.B., Iverson, J.B., and McCord, W.P., Phylogenetic hypotheses for the turtle family Geoemydidae, *Mol. Phylogenet. Evol.*, 32, 164, 2004.

Stein, P.S.G., Neuronal control of turtle hindlimb motor rhythms, *J. Comp. Physiol. A*, 191, 213, 2003.

Sukhanov, V.B., Locomotion in lizards and its place in the general system of locomotion in terrestrial vertebrates, in *Problems of Herpetology*, University of Leningrad, 1964, 67.

Sukhanov, V.B., *General System of Symmetrical Locomotion of Terrestrial Vertebrates and Some Features of Movement of Lower Tetrapods*, New Delhi: Amerind Publishing, 1974, 1.

Sukhanov, V.B., Mesozoic turtles of Middle and central Asia, in *The Age of Dinosaurs in Russia and Mongolia*, M.J. Benton, M.A. Shishkin, D.M. Unwin, and E.N. Urochkin (eds.), Cambridge, UK: Cambridge University Press, 2000, 17, 309.

Van Leeuwen, J.L., Jayes, A.S., and Alexander, R.M., Estimates of mechanical stresses in tortoise leg muscles during walking, *J. Zool. Lond.*, 195, 53, 1981.

Waite, R., The osteology of the New Guinea Turtle (*Carettochelys insculpta* Ramsay), *Rec. Austr. Mus.*, 6, 110, 1905.

Walker, W.F., Aspects of the functional anatomy of the chelonian pectoral girdle and limb, *Am. Zool.*, 566 (abstract), 1962.

Walker, W.F., An analysis of force developed at the feet of turtles in walking, *Am. Zool.*, 3, 488, 1963.

Walker, W.F., A structural and functional analysis of walking in the turtle, *Chrysemys picta marginata*, *J. Morph.*, 90, 263, 1971a.

Walker, W.F., Swimming in sea turtles of the family Cheloniidae, *Copeia*, 2, 229, 1971b.

Walker, W.F., Body form and gait in terrestrial vertebrates, *Ohio J. Sci.*, 72, 177, 1972.

Walker, W.F., The locomotor apparatus of Testudines, in *Biology of the Reptilia, Vol. 4: Morphology*, C. Gans and T.S. Parsons (eds.), London: Academic Press, 1973.

Walker, W.F., Locomotion, in *Turtles Perspective and Research*, Harless, M., and Morlock, H. (eds.), New-York: J. Wiley & Sons, 1979, 435.

Walther, W.G., Die Neu-Guinea-Schildkröte *Carettochelys insculta* Ramsay, *Nova-Guinea*, 13, *Zool.*, 607, 1921.

Webb, R.G., North American recent soft-shelled turtles (Family Trionychidae), *Univ. Ks. Mus. Nat. Hist.*, 13, 431, 1962.

Willey, J.S., and Blob, R.W., Tail kinematics of juvenile common snapping turtles during aquatic walking, *J. Herpetol.*, 38, 360, 2004.

Williams, T.L., Experimental analysis of the gait and frequency of locomotion in the tortoise, with a simple mathematical description, *J. Physiol.*, 310, 307, 1981.

Woods, G.T., Rate of travel of the wood turtle, *Copeia*, 1, 49, 1945.

Wren, K., Claussen, D.L. and Kurz, M., The effects of body size and extrinsic mass on the locomotion of the ornate box turtle, *Terrapene ornate*, *J. Herp.*, 32, 144, 1998.

Wyneken, J., Functional motivations in swimming analyses of sea turtle locomotor patterns, *Amer. Zool.*, 28(A), 13A, 1988a.

Wyneken, J., Comparative and functional considerations of locomotion in turtles, Ph.D. diss., University of Illinois, Urbana-Champaign, IL, 1988b.

Wyneken, J., Sea turtle locomotion: Mechanisms, behaviour and energetics, in *The Biology of Sea Turtles*, P.L. Lutz and J.A. Musick (eds.), Boca Raton, FL: CRC Press, 1997, 165.

Wyneken, J., The Anatomy of Sea Turtles, US Department of Commerce: NOAA Technical Memorandum NMFS-SEFSC-470, 2001, 172.

Wyneken, J., The external morphology, musculoskeletal system, and neuroanatomy of sea turtles, in *The Biology of Sea Turtles, Vol. 2*, P.L. Lutz, J.A. Musick and J. Wyneken (eds.), Boca Raton, FL: CRC Press, 2003, 455.

Young, C.C., and Chow, M-C., New fossil reptiles from Szechuan China, *Acta Pal. Sin.*, 1, 87, 1953.

Zani, P.A., and Claussen, D.L., Voluntary and forced terrestrial locomotion in juvenile and adult painted turtles, *Chrysemys picta*, *Copeia*, 2, 466, 1994.

Zani, P.A., and Claussen, D.L., Effects of extrinsic load locomotion in painted turtles, *Copeia*, 735, 1995.

Zani, P.A., Gottschall, J.S., and Kram, R., Giant Galapagos tortoises walk without inverted pendulum-energy exchange, *J. Exp. Biol.*, 208, 1489, 2005.

Zangerl, R., The turtle shell, in *Biology of the Reptilia, Vol. 1*, C. Gans (ed.), New York: Academic Press, 1969, 311.

Zardoya, R., and Meyer, A., The evolutionary position of turtles revised, *Naturwissenschaften*, 88, 193, 2001.

Zug, G.R., Buoyancy, locomotion, morphology of the pelvic girdle and hind limb and systematics of cryptodiran turtles, *Misc. Publ. Mus. Zool. Univ. Michigan*, 142, 1, 1971.

Zug, G.R., A critique of walk pattern analysis of symmetrical quadrupedal gaits, *Anim. Behav.*, 20, 436, 1972.

6 Hindlimb Function in Turtle Locomotion: *Limb Movements and Muscular Activation across Taxa, Environment, and Ontogeny*

Richard W. Blob, Angela R.V. Rivera, and Mark W. Westneat

CONTENTS

6.1 INTRODUCTION

Turtles possess one of the most distinctive body plans among vertebrates (Burke, 1989). The presence of a bony shell in turtles affords them physical protection from a wide range of predators and environmental hazards, providing opportunities for turtles to use distinctive life history and survival strategies (Cagle, 1950; Jayes & Alexander, 1980; Gibbons, 1990; Jackson, 2000, 2002, 2004; Ultsch, 2006). However, in addition to the benefits that they convey, the shells of turtles also place unusual constraints on their locomotor systems. Because the carapace of the shell is fused to the dorsal vertebrae, the body axis of turtles is inflexible except for the neck and tail (Zangerl, 1969;

Ernst et al., 1994). Although the tail plays an important role in the locomotion of many vertebrates (Snyder, 1962; Fish, 1984; Gatesy, 1990; Lauder, 2000), in almost all extant turtles the tail is highly reduced and thus contributes little to the production of locomotor forces (Zug, 1971; Walker, 1973; Pace et al., 2001; Willey & Blob, 2004). As a result, turtles must rely exclusively on movements of the limbs to propel themselves (Gillis & Blob, 2001; Pace et al., 2001; Rivera et al., 2006).

Despite the potential constraints that the turtle body plan places on their locomotion (Rivera et al., 2006), turtles have diversified into a wide range of both aquatic and terrestrial habitats, with many species spending substantial amounts of time in, and moving through, both types of environments (Cagle, 1944; Bennett et al., 1970; Gibbons, 1970; Zug, 1971; Davenport et al., 1984; Gillis & Blob, 2001). The physical conditions characterizing these habitats differ substantially. For example, the limbs of animals on land must support their body weight against gravity, whereas buoyancy afforded by water largely eliminates this need in aquatic habitats (Denny, 1993; Vogel, 1994). However, the high viscosity of water may require limbs to exert greater forces during some parts of the locomotor cycle than they would when surrounded by less resistive air while moving over land (Gillis & Blob, 2001). As a consequence of these different physical conditions, the locomotor systems of turtles specialized primarily for life in aquatic habitats will likely be exposed to different functional demands than those of turtles living mainly in terrestrial habitats; moreover, turtles spending substantial time in both habitats will have to use their limbs to accommodate a range of diverse functional demands (Biewener & Gillis, 1999; Gillis & Blob, 2001).

One of the primary ways in which animals accommodate changes in the demands placed on their locomotor system is by adjusting the activity patterns of their locomotor muscles (Zernicke & Smith, 1996; Biewener & Gillis, 1999; Gillis & Blob, 2001; Reilly & Blob, 2003). Muscles affecting limb function have the potential to be activated in different combinations, for varying durations of time, and with different degrees of intensity. The regular use of habitats that differ in physical conditions is likely to make such modulations of muscular motor patterns an important aspect of limb function for many turtle species. In addition, the restriction of turtle propulsive structures to the limbs makes them an excellent general model for studies of how animals adjust the function of their muscles to perform a variety of tasks, because it allows comparisons of motor function under different conditions to be focused on a specific anatomical region, without complications such as the addition of, or shift to, other propulsive structures (e.g., body axis and tail) under particular conditions.

In this chapter, we outline our studies of hindlimb function during aquatic and terrestrial locomotion in turtles. The studies we discuss have taken advantage of differences in habits and locomotor performance among turtle species to perform functional comparisons that address broader questions about the modulation of neuromotor control. First, we will describe our comparisons of aquatic and terrestrial patterns of hindlimb kinematics and muscle activation (EMG) between adults of generalized, semiaquatic sliders (*Trachemys scripta*) and specialized, highly aquatic spiny softshell turtles (*Apalone spinifera*). The differing degrees of aquatic specialization in these species allow us to experimentally test not only how limb function is modulated between environments, but also whether locomotor function in freshwater turtles might be subject to tradeoffs such that specialization for one habitat impedes performance in others. As an extension of these comparisons, we will present a complementary study comparing the modulation of limb motor function between juvenile turtles from these same two species. These studies provide data on the development of the locomotor system in turtles and can be further examined as a basis for understanding the contrasts in function observed in adults. These studies, along with distinctive features of several other turtle lineages, indicate that studies of turtle locomotion can not only provide insight into the biology of these fascinating animals, but also provide a promising direction for broader studies of functional adaptation and evolution.

6.2 MOTOR CONTROL OF SWIMMING AND WALKING IN ADULT SLIDER AND SOFTSHELL TURTLES: TESTING FUNCTIONAL CORRELATES OF HABITAT SPECIALIZATION

Vertebrate limbs are called upon to perform diverse tasks that subject them to multiple, sometimes conflicting demands. How do animals use the restricted number of muscles in their limbs to perform all the different functions required for survival? Three general ways that animals might use their muscles to perform different behaviors can be outlined (Biewener & Gillis, 1999; Gillis & Blob, 2001). First, muscles might not exhibit any changes in their activity patterns between behaviors. A lack of versatility in motor function could impair the performance of some behaviors (Biewener & Gillis, 1999) but might be found if central pattern generator input were the dominant source of control for the muscles in question (Buford & Smith, 1990; Pratt et al., 1996). Second, the same muscles might be activated to produce different behaviors, but the intensity or timing of their activity might change (Gruner & Altman, 1980; Macpherson, 1991; Roy et al., 1985, 1991; Johnston & Bekoff, 1996; Kamel et al., 1996; Gillis & Biewener, 2000; Reilly & Blob, 2003). In some cases, general patterns of muscle coactivation might be maintained between different behaviors with only moderate differences in the timing or intensity of muscle activity (Gruner & Altman, 1980; Johnston & Bekoff, 1996), though in others the timing of muscle activity might change so drastically between behaviors that muscles acting as synergists in one task could act as antagonists in another (Walker, 1973; Buchanan et al., 1986). Third, different muscles or sets of muscles might be activated to execute different behaviors. In some cases, this might involve the recruitment of particular muscles only during the performance of specific tasks (Gatesy, 1997). Given the high degree of muscular redundancy in most vertebrate limbs, with multiple muscles capable of contributing to each direction of movement at each limb joint, the three possibilities outlined here are not mutually exclusive (Biewener & Gillis, 1999; Gillis & Blob, 2001). However, the potential for these three patterns does raise a more general question as to whether modifications of muscle function between behaviors are generally widespread among muscles that participate in the behaviors, or whether functional modifications should only be expected in a small subset of key muscles.

Turtle species that make use of both aquatic and terrestrial habitats can provide insight into such questions about how muscles accommodate competing functional demands because the physical differences between these habitats allow tests of specific predictions for how the activity of limb muscles might change to optimize limb function in each habitat. For example, because turtle limbs must support the weight of the body on land, the duration or intensity of activity might be expected to increase for limb extensor muscles that counter the force of body weight and propel the body forward during terrestrial locomotion (Gillis & Blob, 2001). Based on similar principles, the high viscosity of water may require a greater duration or intensity of activity among limb flexor muscles during the recovery phase of aquatic locomotion versus the swing phase of terrestrial locomotion because the limbs must overcome greater drag moving through water than they encounter moving through air on land (Gillis & Blob, 2001). If modulation patterns such as these are common for turtles, it would suggest that the physical characteristics of the external environment can be a primary influence governing motor output (Zernicke & Smith, 1996).

Variation in the degree to which different turtle species use aquatic and terrestrial habitats provides a further opportunity to examine questions about the diversity of motor function and factors that might affect locomotor performance. The habitat preferences of turtles range from almost exclusively terrestrial among tortoises to almost exclusively aquatic among sea turtles (Ernst et al., 1994). Between these extremes, numerous species divide their time between both habitats (Bennett et al., 1970). Comparison of function between such habitat "generalists" with species specialized for life in particular habitats provides insight into the functional consequences of specialization, such as whether optimizing for performance in one habitat leads to a sacrifice of performance in others (Gilchrist, 1995).

To examine these questions, we compared hindlimb kinematics and motor patterns for major hindlimb muscles between sliders (*Trachemys scripta*) and spiny softshell turtles (*Apalone spinifera*) during aquatic and terrestrial locomotion. Sliders are a morphologically generalized member of the emydid turtle lineage that swim proficiently but also spend considerable time traveling over land (Gibbons, 1970; Ernst et al., 1994; Bodie & Semlitsch, 2000). In contrast, softshell turtles are members of the trionychid lineage that are highly specialized for life in water and travel over land only infrequently (Webb, 1962; Zug, 1971; Plummer et al., 1997; Pace et al., 2001). Although trionychids possess extensive interdigital webbing in the manus that results in their forefeet forming broader paddles than in lineages like emydids (Pace et al., 2001), they still move both sets of limbs during swimming with a stroke that is fundamentally similar to that of emydids, using rowing or "drag-based" (Vogel, 1994) motions in which the primary range of motion is antero-posterior (Zug, 1971; Pace et al., 2001). The hindlimbs are also thought to provide the majority of propulsive force in both groups (Zug, 1971). This differs from the aquatic specialist species of sea turtles, in which the forelimbs are extensively hypertrophied relative to the hindlimbs, and in which the majority of thrust is produced through dorsoventral flapping movements that make use of fundamentally different ("lift-based") mechanisms of thrust production (Renous & Bels, 1993; Wyneken, 1997). Thus, comparisons of hindlimb movements and motor function between sliders and softshells are highly suitable for examining how muscles function differently between behaviors with different demands, and avoid complications that would arise in comparisons involving highly aquatic lineages such as sea turtles in which the role of the hindlimbs is substantially altered.

Our comparisons of sliders and softshells focus on two questions. First, do sliders and softshells show different ways of adjusting limb function between swimming and walking? Second, is terrestrial locomotor performance impeded in aquatic specialist softshells compared with more generalized emydids like sliders?

6.2.1 Experimental Methods

Locomotor data were collected from similarly sized adult individuals of each species collected from Union and Alexander counties, Illinois, USA (*T. scripta*, $N = 2$, mean ± SD: body mass = 985 ± 303 g, carapace length = 197 ± 14 mm; *A. spinifera*, $N = 4$, mean ± SD: body mass = 668 ± 78 g, carapace length = 174 ± 19 mm). Kinematic data were collected simultaneously in lateral and ventral views at 60 Hz using two digitally synchronized high-speed digital video cameras (Redlake Motionscope PCI 1000S). A custom-built flow tank with a transparent glass side and bottom was used as the filming arena for both aquatic and terrestrial trials. Ventral views were provided by a mirror placed at 45° under the transparent flow tank bottom. Because locomotor speed was difficult to control in both species, turtles were allowed to select their own preferred speed (i.e., the speed at which they naturally moved when stimulated). For aquatic trials, the tank was filled with water and flow was adjusted to the point at which forward swimming behavior was elicited (Pace et al., 2001). For terrestrial trials, water was drained from the arena and turtles were coaxed to walk using lures of food or gentle pinching of the tail. The speed of each locomotor trial was then measured from the video by measuring the displacement of the turtle over a known period of time and, for aquatic trials, adding the flow speed of the water.

To analyze locomotor kinematics, the locations of landmarks on the shell, hip, knee, ankle, and tips of digits 1, 3, and 5 were digitized from every second frame of each video view (producing an effective framing rate of 30 Hz) using QuickImage® software for Macintosh, a modification (written by J. Walker) of public domain NIH Image software developed by the U.S. National Institutes of Health (available online at http://www.usm.maine.edu/~walker/software.html). The shells of the turtles occasionally obscured some joint positions for a few frames in either lateral or ventral view but never both views for the same frame. The position of an obscured joint in a specific view was evaluated based on the measured lengths of the limb segments and the point of intersection of lines extended in the video frame along the visible portions of the limb segments meeting at the joint

(Pace et al., 2001). Kinematic variables (three-dimensional joint angles and the angles of limb segments relative to specific planes) were calculated from the three-dimensional coordinate data generated for each trial using a custom MATLAB® (The Mathworks Inc.) program. The program QuickSAND®, available online at http://www.usm.maine.edu/~walker/software.html (Walker, 1998), then was used to fit a quintic spline function to the values of each kinematic variable for each trial, smoothing the data and allowing all of the trials to be normalized to the same duration prior to comparisons between locomotor habitats or species. These procedures helped to clarify patterns of limb movements by reducing variation due to minor errors in locating anatomical landmarks on video frames during the use of the digitizing software. They also allowed us to evaluate the timing of landmark kinematic events as a percentage of total cycle duration, allowing us to perform comparisons across runs with differences in cycle duration. These smoothed and normalized data were then used to calculate average kinematic profiles and standard errors for each variable through the course of walking and swimming cycles.

Hindlimb cycles for both species were defined to begin with the first forward movement of the femur (protraction) and end with the extension of the femur to its most rearward position (retraction) in a single limb. Thus, swimming cycles were divided into an initial recovery phase (as the leg moved forward) followed by a thrust phase (as the leg moved backward), which was the propulsive portion of the stroke. Similarly, walking cycles were divided into an initial swing phase (as the leg moved forward through the air) followed by a stance phase (as the leg moved backward) in which the foot contacted the ground and pushed off to propel the body forward. For angles of femoral protraction and retraction, a line coming out of the hip socket perpendicular to the anteroposterior axis of the turtle was designated as the 0° axis, with positive angles indicating a protracted position and negative angles a retracted position (Pace et al., 2001). For femoral elevation and depression, a 0° angle indicated a horizontal femur, with positive angles indicating femoral elevation and negative angles indicating femoral depression (Pace et al., 2001). For the knee, larger angles indicate greater extension; thus, 0° would indicate a completely flexed knee and 180° would indicate a fully straightened knee (Pace et al., 2001). Foot kinematics were calculated as the angle between a vector pointing forward along the anteroposterior midline of the turtle and a vector emerging from the plantar surface of a plane defined by the ankle and the tips of digits 1 and 5, transformed by subtracting 90° from each value (Pace et al., 2001). Thus, a feathered (flat) orientation of a foot is indicated by an angle of 0°, whereas a high-drag orientation (perpendicular to oncoming flow) is indicated by an angle of 90°.

Motor patterns for turtle limb muscles were evaluated following standard methods (Loeb & Gans, 1986). Turtles were anesthetized (ketamine HCl intramuscular injection, 30 to 60 mg/kg base dose with supplements as needed) to allow the implantation of fine-wire (insulated stainless steel bifiler, California Fine Wire Co.), bipolar electrodes through the skin into target muscles using hypodermic needles. Electrode wires exiting each turtle were allowed a few centimeters of slack before they were bundled together and glued into a single cable that was sutured to the underside of the rear portion of the carapace, protecting the wires without hindering limb movement. Turtles were then allowed to recover overnight prior to data collection. Motor patterns were measured on the day following electrode implants by relaying electromyographic (EMG) signals from each muscle to preamplifiers (Grass P511or AM Systems 1700) for amplification (5000 or 10,000 times) and filtering (60 Hz notch filter, 30 Hz to 10kHz bandpass). Analog EMG signals then were converted to digital data collected at 5000 Hz and analyzed on a personal computer. Electrode positions were verified anatomically upon the completion of data collection (Westneat & Walker, 1997). Custom LabView® software routines were used to identify muscle activity bursts and evaluate variables, including EMG onset and offset times, EMG burst duration, and EMG intensity (mean amplitude) for each limb cycle. Measurements of these variables allowed us to determine when and to what degree muscles were active during particular behaviors, allowing us to test for differences in muscle action under different conditions (e.g., water versus land, juvenile versus adult, between species). Because minor variations in electrode construction can affect signal strength, burst intensities were

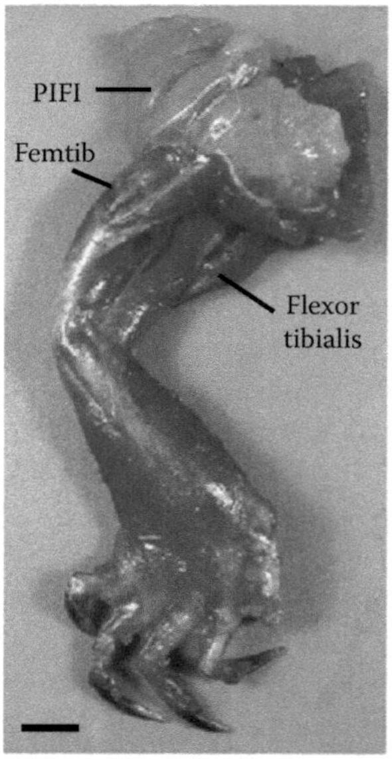

FIGURE 6.1 Dorsolateral view of hindlimb muscles from the left side of adult *Trachemys scripta*, indicating muscles of primary focus for EMG data collection. Legend: Femtib = femorotibialis (a knee extensor); Flexor tibialis = flexor tibialis complex (a limb retractor); PIFI = puboischiofemoralis internus (a limb protractor).

not directly compared across individuals; however, data collected from the same electrode are consistent within an individual, allowing patterns of intensity change (e.g., consistency of higher or lower intensity in one behavior) to be compared across individuals. EMG data were synchronized with kinematic data using a signal generator that simultaneously sent signal pulses visible in both video and EMG records.

Our comparisons of aquatic and terrestrial motor function between *T. scripta* and *A. spinifera* focused on three hindlimb muscles (Figure 6.1). The first was femorotibialis, which originates from the proximodorsal aspect of the femur and inserts on the proximal tibia via a robust tendon (Walker, 1973). The second was the flexor tibialis complex, which originates from the pelvis and inserts on the flexor surface of the proximal tibia. The third was puboischiofemoralis internus (PIFI), which originates from the pelvis and inserts on the dorsal aspect of the proximal femur. Based on their anatomical positions, femorotibialis has been hypothesized to act as a knee extensor, flexor tibialis has been hypothesized to act as a knee flexor and femoral retractor, and PIFI has been hypothesized to act as a femoral protractor (Walker, 1973; Earhart & Stein, 2000; Gillis & Blob, 2001; Stein, 2003).

6.2.2 Results: Aquatic versus Terrestrial Performance, Hindlimb Kinematics, and Muscle Activation in Slider and Softshell Turtles

Comparisons of hindlimb function between *T. scripta* and *A. spinifera* focus primarily on three factors: locomotor performance, hindlimb movement patterns (i.e., kinematics), and motor output of major hindlimb muscles (EMGs).

6.2.2.1 Locomotor Performance

To compare locomotor performance between sliders and softshells, we evaluated the preferred velocities of these species both in water and on land. Not surprisingly, in water the preferred swimming velocity of aquatic specialist softshells (2.03 ± 0.14 body lengths/s, mean ± 1 SE) was significantly greater than that of sliders (1.00 ± 0.04 body lengths/s, mean ± SE; Mann–Whitney U-test, $P < 0.001$; Figure 6.2a). However, limb cycle frequency did not differ significantly between the species during swimming (1.53 ± 0.08 Hz, mean ± SE for *T. scripta*, 1.56 ± 0.08 Hz mean ± 1 SE for *A. spinifera*; Figure 6.2a). Thus, softshells swim faster than sliders not because they stroke more often or more quickly; instead, they must be propelling themselves more forcefully per stroke than sliders.

On land, softshell turtles frequently walk faster than sliders, though the difference in preferred terrestrial velocity between these species is not significant (Mann–Whitney U-test, $P = 0.063$; Figure 6.2b). At the least, it appears that softshell terrestrial performance is no worse than that of sliders. In contrast to swimming, limb cycle frequency on land is significantly faster in softshells

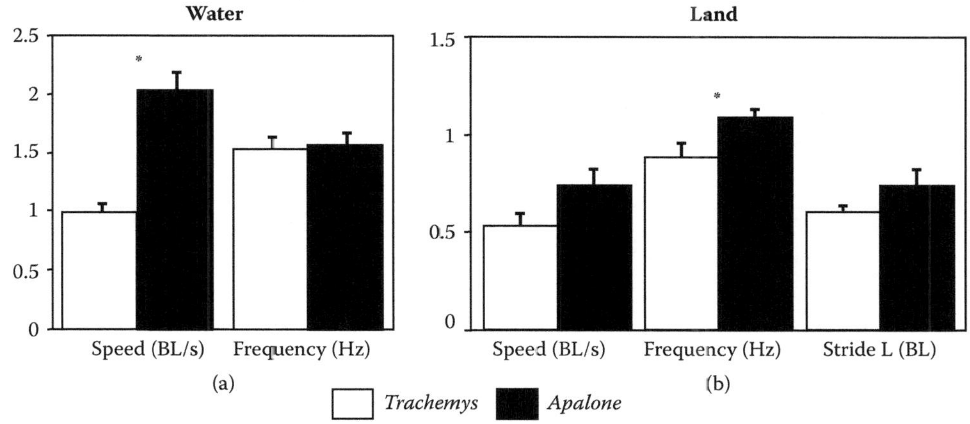

FIGURE 6.2 Comparison of aquatic (a) and terrestrial (b) locomotor performance between slider (*Trachemys scripta*, unshaded bars) and spiny softshell (*Apalone spinifera*, shaded bars) turtles. Error bars represent 1 SE. Comparisons marked with asterisks (*) indicate significant differences ($P < 0.05$, Mann–Whitney U-test). $N = 10$ trials for *T. scripta* and 13 trials for *A. spinifera* for each comparison. Abbreviations: L, length; BL, body lengths.

than in sliders (Mann–Whitney U-test, $P = 0.019$; Figure 6.2b); however, stride lengths (the distance traveled between footfalls of the same foot) do not differ significantly between the species (Figure 6.2b). Thus, to achieve terrestrial speeds as fast as those of sliders, softshell turtles must move their legs more quickly.

6.2.2.2 Hindlimb Kinematics

Patterns for some kinematic variables differed between sliders and softshells with little variation between aquatic and terrestrial habitats. For example, the femur showed a similar magnitude of anteroposterior angular excursion (i.e., protraction and retraction) between aquatic and terrestrial locomotion in each species (Figure 6.3 & Figure 6.4; Table 6.1). However, the location in which this excursion occurred was shifted between the two species. *T. scripta* moves the femur between a maximally protracted position of 74.4° ± 3.6° and maximally retracted position of –7.0° ± 11.5° in water, and a maximally protracted position of 71.4° ± 6.6° and maximally retracted position of –8.8° ± 6.2° on land (values mean ± 1 SD). In contrast, *A. spinifera* moves the femur between a maximally protracted position of 57.2° ± 9.1° and maximally retracted position of –30.6° ± 9.0° in water, and between a maximally protracted position of 48.1° ± 8.7° and maximally retracted position of –33.0° ± 6.2° on land. Thus, in *T. scripta* femoral motion was located primarily anterior to the hip joint, whereas in *A. spinifera* femoral protraction and retraction were more evenly divided between regions anterior and posterior to the hip (Mann-Whitney U-tests, $P < 0.001$ for interspecies comparisons of maximal protraction and retraction angles in the same habitat; Table 6.1). Similar patterns are evident when aquatic and terrestrial angles of femoral elevation and depression are compared between these species: although neither species shows significant differences in femoral depression between swimming and walking, *T. scripta* depress the femur over twice as much as *A. spinifera* (17 to 20° versus 8°) in both aquatic and terrestrial habitats (Mann-Whitney U-tests, $P < 0.001$ for interspecies comparisons of maximal femoral depression angle and dorsoventral excursion in the same habitat; Figure 6.5 and Figure 6.6; Table 6.1). As result, sliders raise the body higher off the ground than softshells while traveling over land, and sliders would be expected to generate thrust upward, as well as forward, while swimming in water.

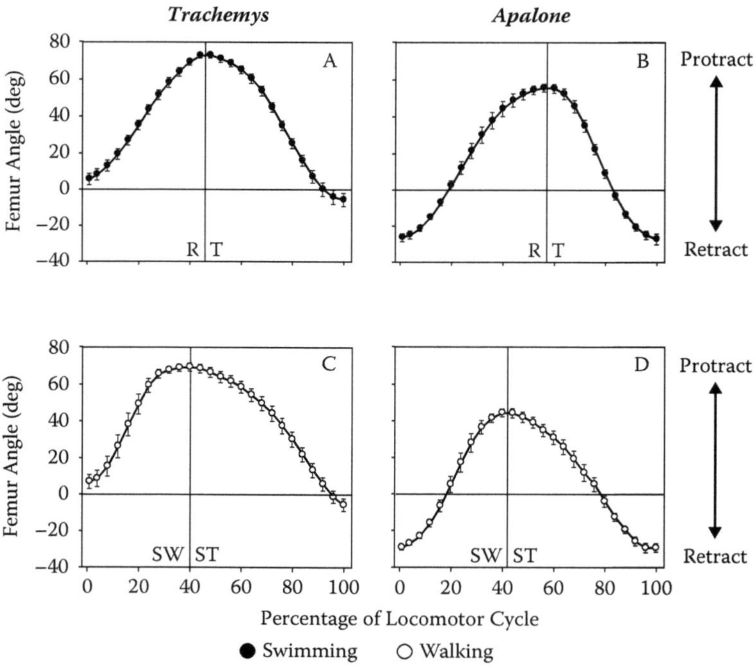

FIGURE 6.3 Mean kinematic profiles of femoral protraction and retraction (i.e., angle from the transverse plane) for (a) *Trachemys scripta* swimming, (b) *Apalone spinifera* swimming, (c) *T. scripta* walking, (d) *A. spinifera* walking. Each kinematic trial was normalized to the same duration, and values for kinematic angles were interpolated for 100 equally spaced increments through the locomotor cycle, allowing mean angles and standard errors to be calculated for each 1% increment through each cycle type in each species. Angle values ± 1 SE are plotted for every fourth increment (every 4% through the cycle); $N = 10$ trials for *T. scripta* swimming, 11 trials for *T. scripta* walking, and 13 trials for *A. spinifera* in both swimming and walking. Angles of 0° indicate that the femur is perpendicular to the anteroposterior midline of the turtle, with negative values indicating that the distal end of the femur is directed posteriorly and positive values indicating the distal end of the femur is directed anteriorly. The vertical line in each panel divides the cycle into recovery (R) and thrust (T) phases for swimming, or swing (SW) and stance (ST) phases for walking.

In contrast to kinematic patterns for femoral motion at the hip, patterns of knee motion showed changes between water and land that were similar in both sliders and spiny softshells. In water, both species show a major phase of knee flexion during the recovery portion of the stroke, followed by a phase of knee extension that progresses into the thrust phase of the stroke (Figure 6.7A,B). Some brief periods of flexion may occur during the extension phase but these are minor excursions compared to the main phase of knee flexion at the beginning of the aquatic limb cycle. Terrestrial knee movements differ significantly from aquatic knee movements in both species. On land, both species show two cycles of knee flexion and extension: the first occurs during femoral protraction and the second occurs during femoral retraction (Figure 6.7C,D). Thus, during swing phase (femoral protraction) the knee initially flexes as the leg comes forward but then extends as the foot reaches out to contact the ground. After the foot reaches the ground, the knee flexes again at the beginning of limb retraction (as the limb begins to support the weight of the body) and then extends a second time as the limb pushes off the ground and propels the body forward. Both species show similar magnitudes of knee flexion during swing phase and stance phase. However, despite the similarities in adjustments of knee kinematics between water and land in these species, sliders and spiny softshells also showed differences in their knee kinematics. In particular, the overall range of angles

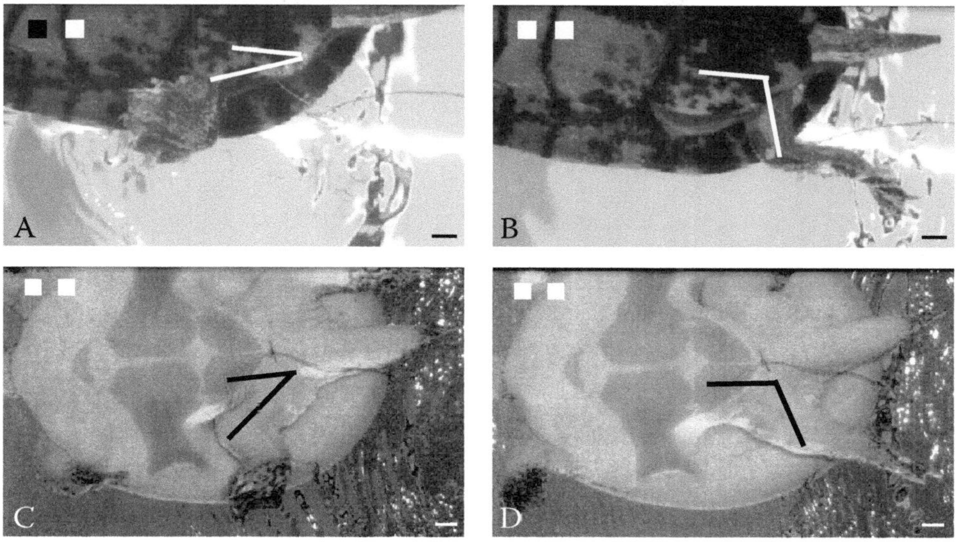

FIGURE 6.4 Still images from ventral view video footage of turtle hindlimbs during swimming, comparing femoral protraction and retraction between species. (A) *Trachemys scripta*, early thrust phase, (B) *T. scripta*, late thrust phase, (C) *Apalone spinifera*, early thrust phase, (D) *A. spinifera*, late thrust phase. Scale = 1 cm. Lines on limb highlight the position of the femur relative to the anteroposterior axis of the turtle.

through which the knee moves on land differs substantially between sliders and spiny softshells: in the *A. spinifera* we examined, the knee was always held in a more extended position (approximately 70 to 125°) than it was in *T. scripta* (approximately 40 to 110°) during terrestrial locomotion (Mann–Whitney U-tests, $P < 0.02$ for interspecies comparisons of extreme angles of knee flexion and extension on land; Figure 6.7C,D; Table 6.1).

Comparisons of foot kinematics between *T. scripta* and *A. spinifera* show a different pattern of interspecies differences than those exhibited by the hip and knee. On land, turtles use a plantigrade foot posture (Walker, 1971; Zug, 1971), and we did not attempt to quantify differences between species in this behavior. However, in water the hindfoot rotates into a flat (feathered) orientation during the initial recovery portion of the stroke, then rotates into a high drag orientation (perpendicular to oncoming flow) during the thrust phase of the stroke (Figure 6.8A,B), producing a pattern of motion much like the rowing of a boat oar (Vogel, 1994; Walker & Westneat, 2000). However, unlike comparisons of hip and knee movement in these species, kinematic patterns of hindfoot orientation are almost identical in sliders and spiny softshells, with both species moving the hindfoot into an orientation nearly parallel with the horizontal plane (and oncoming flow) during recovery, and both rotating to a peak angle of approximately 60° from horizontal during thrust phase (Figure 6.8A,B; Table 6.1). This similarity in aquatic hindfoot movement patterns between *T. scripta* and *A. spinifera* stands in sharp contrast to the highly divergent kinematic patterns exhibited by the forefeet in these species (Pace et al., 2001).

6.2.2.3 Motor Patterns of Hindlimb Muscles

Standard patterns of the timing (onset, offset, and duration) and intensity of EMG bursts for the three focal muscles (femorotibialis, flexor tibialis, and PIFI) during aquatic and terrestrial locomotion in adult *T. scripta* and *A. spinifera* are illustrated in Figure 6.9. Patterns for *T. scripta* (Gillis & Blob, 2001) are generally similar to those observed in other studies of this species (Earhart & Stein, 2000; Stein, 2003). There is little change in burst timing for either PIFI or flexor tibialis between

TABLE 6.1
Values (mean ± SD) of Kinematic Variables Measured from the Hindlimbs of Adult Slider (*T. scripta*) and Spiny Softshell (*A. spinifera*) Turtles during Aquatic (Water) and Terrestrial (Land) Locomotion.

	T. scripta		*A. spinifera*					
Variable	**Water**	**Land**	**Water**	**Land**	P_1	P_2	P_3	P_4
Maximum femur protraction	74.4° ± 3.6°	71.4° ± 6.6°	57.2° ± 9.1°	48.1° ± 8.7°	0.398	0.026*	< 0.001*	< 0.001*
Maximum femur retraction	–7.0° ± 11.5°	–8.8° ± 6.2°	–30.6° ± 9.0°	–33.0° ± 6.2°	0.481	0.270	< 0.001*	< 0.001*
Antero-posterior femur excursion	81.5° ± 9.6°	80.2° ± 9.8°	87.8° ± 16.1°	81.2° ± 10.1°	0.573	0.522	0.495	0.839
Maximum femur elevation	0.4° ± 5.3°	0.4° ± 3.9°	–0.5° ± 2.6°	–0.5° ± 2.5°	0.833	0.939	0.072	0.401
Maximum femur depression	–20.6° ± 6.7°	–17.4° ± 3.3°	-8.3° ± 1.6°	–8.3° ± 3.4°	0.181	0.898	< 0.001*	< 0.001*
Dorso-ventral femur excursion	20.9° ± 6.9°	17.8° ± 3.2°	7.9° ± 2.8°	7.8° ± 3.3°	0.190	0.858	< 0.001*	< 0.001*
Minimum knee flexion	38.4° ± 7.6°	38.7° ± 10.4°	61.6° ± 7.0°	72.4° ± 16.9°	0.888	0.522	< 0.001*	< 0.001*
Maximum knee extension	120.5° ± 8.1°	108.4° ± 11.5°	121.4° ± 6.9°	125.1° ± 17.2°	0.014*	0.270	0.420	0.019*
Angle of hindfoot paddle from horizontal plane	59.0° ± 11.9°	N/A	58.1° ± 9.8°	N/A	N/A	N/A	0.901	N/A

The P values reported are for Mann-Whitney U-tests of the following comparisons in each row: P_1 = *T. scripta*, water vs. land; P_2 = *A. spinifera*, water vs. land; P_3 = water, *T. scripta* vs. *A. spinifera*; P_4 = land, *T. scripta* vs. *A. spinifera*.

* *Significant at* $P < 0.05$.

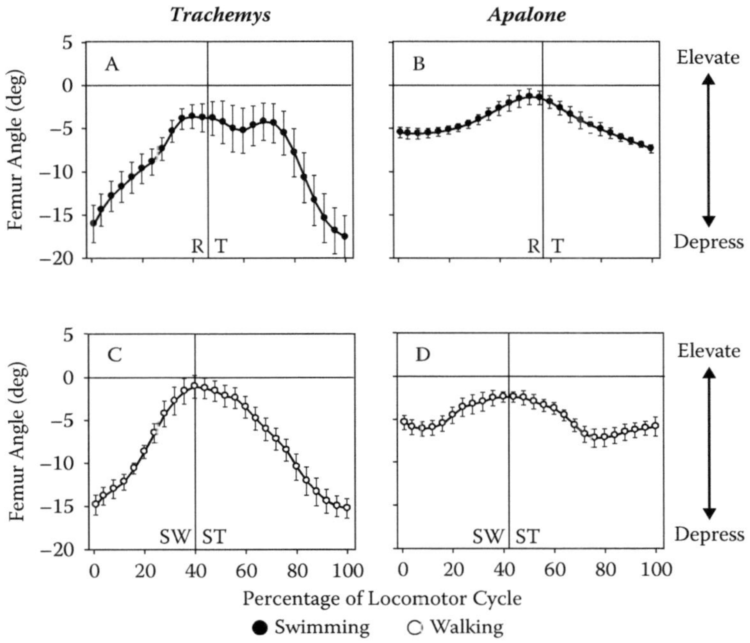

FIGURE 6.5 Mean kinematic profiles of femoral elevation and depression (i.e., angle from the horizontal plane) for (A) *Trachemys scripta* swimming, (B) *Apalone spinifera* swimming, (C) *T. scripta* walking, (D) *A. spinifera* walking. Figure format and method of profile calculation are the same as in Figure 6.3. Angles of 0° indicate a horizontal femur, with negative values indicating that the distal end of the femur is depressed below the proximal end, and positive values indicating the distal end of the femur is elevated above the proximal end.

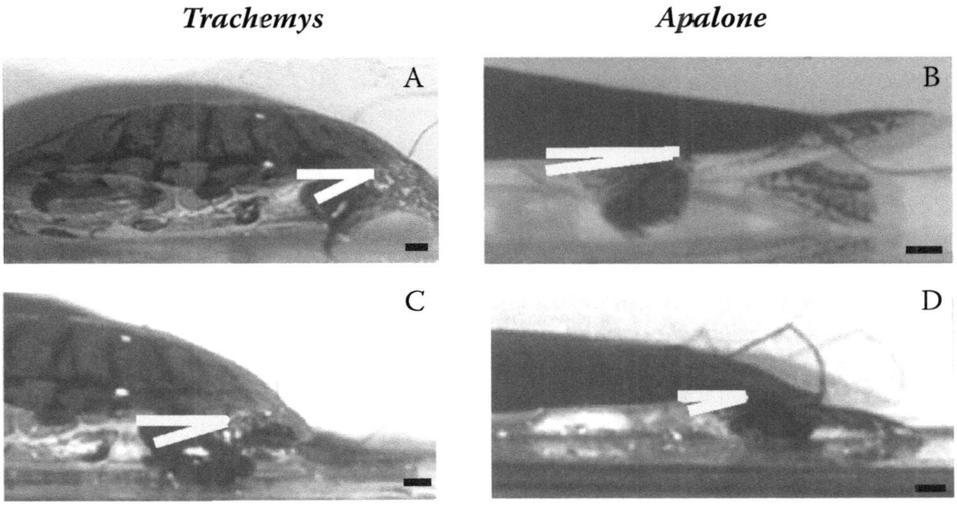

FIGURE 6.6 Still images from lateral view video footage of turtle hindlimbs during aquatic and terrestrial locomotion, comparing femoral depression between species. (A) *Trachemys scripta*, swimming, (B) *Apalone spinifera* swimming, (C) *T. scripta* walking, (D) *A. spinifera* walking. Scale = 1 cm. White lines highlight the position of the femur relative to the horizontal plane.

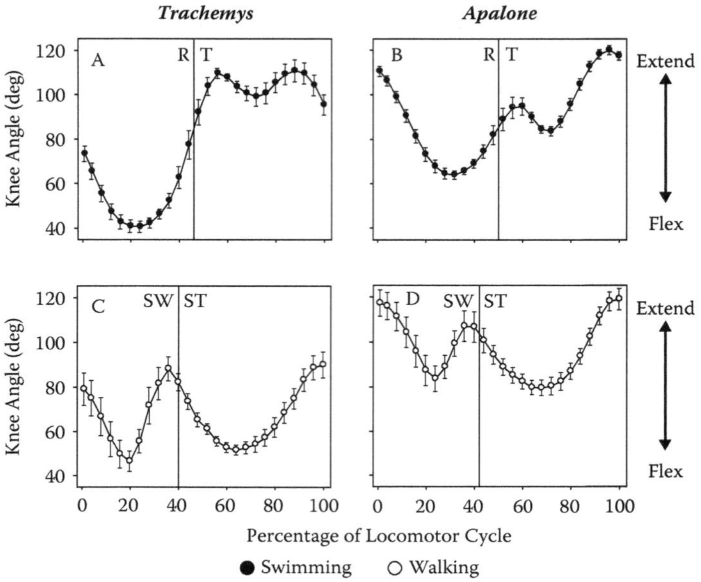

FIGURE 6.7 Mean kinematic profiles of knee flexion and extension for (A) *Trachemys scripta* swimming, (B) *Apalone spinifera* swimming, (C) *T. scripta* walking, (D) *A. spinifera* walking. Figure format and method of profile calculation are the same as in Figure 6.3. Larger angles indicate greater knee extension; 180° would indicate a straight knee joint.

swimming and walking in *T. scripta*. PIFI activity extends through much of femoral protraction in both behaviors but its onset begins prior to the beginning of the recovery phase (Figure 6.9), suggesting its initial contraction might serve to brake limb retraction at the end of the thrust phase in anticipation of the limb protraction to occur in the following recovery phase. Flexor tibialis is active primarily during limb retraction, and during swimming in particular its activity does not substantially overlap with knee flexion, suggesting that, at least in swimming, its primary function may be limb retraction rather than knee flexion. However, unlike PIFI and flexor tibialis, femorotibialis shows a dramatic change in activation timing between swimming and walking with a second burst during terrestrial locomotion that corresponds to the second phase of knee extension exhibited by this species while walking on land (Figure 6.9). Although only femorotibialis showed dramatic differences in burst timing among the muscles examined in *T. scripta*, all three muscles examined show significant differences in burst intensity between habitats (Mann-Whitney U-tests, $P < 0.05$), with PIFI showing higher amplitude bursts during walking, and femorotibialis and flexor tibialis showing higher amplitude bursts during swimming (Figure 6.9).

EMG patterns for *A. spinifera* show several similarities to those for *T. scripta*. As in *T. scripta*, in *A. spinifera* activity of PIFI begins before the start of limb protraction and continues through much of the duration of limb protraction with little change in burst timing between swimming and walking (Figure 6.9). Also similar to *T. scripta*, the single flexor tibialis burst in *A. spinifera* occurs later in the locomotor cycle during swimming than during walking, coinciding with limb retraction, rather than knee flexion, during aquatic locomotion. Both PIFI and flexor tibialis show minor differences in burst timing between the species (e.g., PIFI bursts onset earlier before the start of limb protraction in *T. scripta* than in *A. spinifera*); however, these small differences do not suggest substantial changes in the functional roles of these muscles between these species.

Although *T. scripta* and *A. spinifera* show many similarities in their hindlimb motor patterns, significant differences also emerge between these species. As in *T. scripta*, in *A. spinifera* the

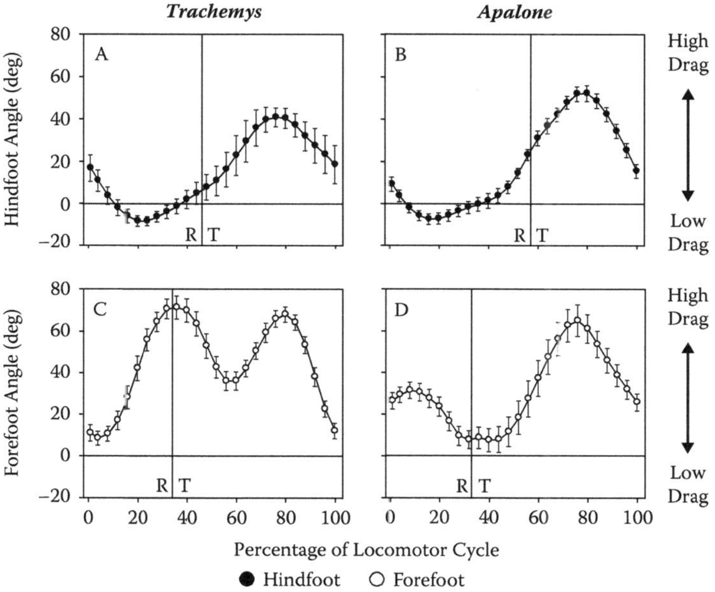

FIGURE 6.8 Mean kinematic profiles of paddle (i.e., foot) orientation during swimming for (A) *Trachemys scripta* hindfoot, (B) *Apalone spinifera* hindfoot, (C) *T. scripta* forefoot, (D) *A. spinifera* forefoot. Figure format and method of profile calculation are the same as in Figure 6.3. The angle plotted is the angle between a vector pointing forward along the anteroposterior midline of the turtle and a vector emerging from the plantar surface of a plane defined by the ankle (hindfoot) or wrist (forefoot) and the tips of digits 1 and 5, transformed by subtracting 90° from each value. Thus, a low-drag orientation of a foot (ideal feathering) is indicated by an angle of 0°, whereas a high-drag orientation (with increased thrust-generating potential during limb retraction) is indicated by an angle of 90°. Data for forefeet were replotted for comparison from Pace et al. (2001).

femorotibialis exhibits a single burst of activity during swimming but two bursts of activity during walking, each of which correspond to one of the two phases of knee extension during terrestrial locomotion (Figure 6.9). *A. spinifera* femorotibialis EMG bursts also show greater mean amplitudes during swimming than during walking, as in *T. scripta* (Mann-Whitney U-test, $P < 0.05$). However, during swimming in *T. scripta* the femorotibialis burst occurs during the recovery phase of the stroke, whereas in swimming *A. spinifera* this burst occurs during the thrust phase of the stroke (Figure 6.9). A further distinction in motor patterns between the species is that, unlike *T. scripta*, in *A. spinifera* mean EMG burst amplitudes do not differ between water and land for either flexor tibialis or PIFI (Figure 6.9). In addition, differences in terrestrial knee kinematics between sliders and spiny softshells suggest the potential for a further contrast in hindlimb muscle function between these species. Because spiny softshells walk with the knee more extended than it is in sliders, the femorotibialis muscle of *A. spinifera* could be contracting when it is already highly shortened. As a result, in aquatic specialist softshell turtles, the femorotibialis might be active at a length that is suboptimal for force generation (Schmidt-Nielsen, 1990).

6.2.3 Conclusions from Comparisons of Hindlimb Function in Adult Slider and Softshell Turtles: Functional Consequences of Habitat Specialization

Sliders and spiny softshells show largely similar ways of adjusting limb function between aquatic and terrestrial habitats. These adjustments can be widely distributed in the locomotor system (Gillis & Blob, 2001), though they are not necessarily present among all muscles (e.g., *A. spinifera* PIFI).

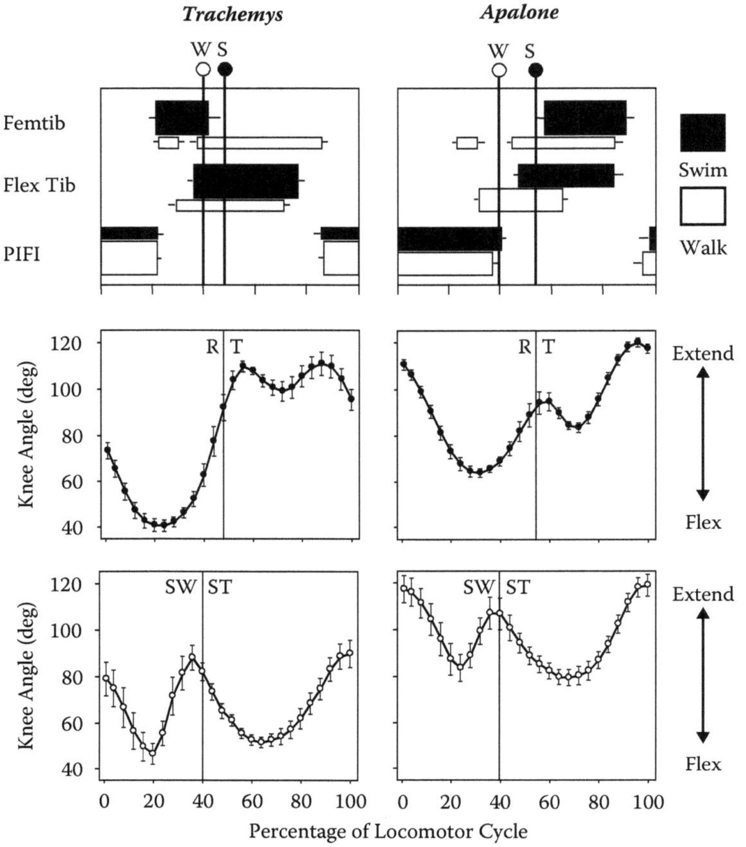

FIGURE 6.9 Representative mean activity patterns for three focal hindlimb muscles during swimming (shaded bars) and walking (unshaded bars) for *Trachemys scripta* (left column) and *Apalone spinifera* (right column), synchronized with mean kinematic profiles of knee kinematics for both behaviors in each species (replotted from Figure 6.7). Within a muscle, greater bar thickness in swimming or walking indicates significantly greater mean EMG amplitude for that behavior (Mann–Whitney U-test, $P < 0.05$). Vertical lines demarcate cycles into recovery (R) and thrust (T) phases for swimming and swing (SW) and stance (ST) phases for walking; these demarcations are both indicated with appropriate labels on muscle activity graphs. All error bars indicate 1 SE.

However, even with largely similar modulation patterns between *T. scripta* and *A. spinifera*, some prominent contrasts in hindlimb function can be identified between these species. For example, sliders depress the femur twice as much as softshells during both swimming and walking. Sliders also flex their knee significantly more than softshells during walking (i.e., softshells walk with straighter hindlegs). In addition, although the femorotibialis is active during knee extension in both species, during swimming this activity occurs during the recovery phase of the stroke in sliders but during the thrust phase of the stroke in spiny softshells. *A. spinifera* also do not show all of the changes in burst amplitude between swimming and walking that are exhibited by *T. scripta*.

The changes in motor function between swimming and walking exhibited by sliders and softshell turtles provide some broader insights into how muscular function can be modulated to perform a variety of tasks. First, both activation timing and intensity (e.g., EMG burst amplitude) can be modulated independently between behaviors (Gillis & Blob, 2001), and these patterns may differ for a muscle even between closely related species. For example, in femorotibialis both the timing and intensity of EMG bursts changed between aquatic and terrestrial locomotion in both species; however, in PIFI burst intensities differed significantly between water and land without a change in burst timing in *T. scripta*, and neither burst timing nor intensity changed significantly between water and land in *A. spinifera* (Figure 6.9). Second, differences in motor function between locomotor environments might not always be well-predicted by differences in the physical characteristics of those environments. For example, the need for the limbs to bear the weight of the body on land, in addition to moving the body, led us to predict that limb extensor muscles would show greater mean burst amplitudes on land than in water (Gillis & Blob, 2001). However, in both *T. scripta* and *A. spinifera* just the opposite was the case: the knee extensor femorotibialis showed bursts of greater mean amplitude in water (Figure 6.9). Similarly, the high fluid viscosity of water led us to expect that protractor muscles would show greater burst intensity during swimming than during walking. However, the femoral protractor PIFI showed no change in mean burst amplitude between swimming and walking in *A. spinifera*, and actually showed higher mean amplitude bursts during walking in *T. scripta* (Figure 6.9). In the context of these patterns, factors besides gravitational load and fluid viscosity must contribute substantially to the control of motor pattern modulation between environments (Gillis & Blob, 2001).

The performance implications of the differences in hindlimb kinematics and motor function observed in sliders and spiny softshells are challenging to assess. Preferred swimming speed is markedly greater in the aquatic specialist *A. spinifera* than in the generalist *T. scripta* (Figure 6.2A)—though this may, at least in part, be related to differences in forelimb function between these species (Pace et al., 2001). Terrestrial locomotor speed does not appear to be impeded in the aquatic specialist *A. spinifera* relative to generalist species like sliders, as these species are able to achieve similar speeds over land. However, though terrestrial performance of aquatic specialist softshell turtles may be comparable to that of generalist species, it may be achieved with some higher costs. For example, *A. spinifera* produce their speed over land by moving their limbs at a higher frequency than that used by *T. scripta*, which could require increased energy expenditure. In addition, knee extensors in *A. spinifera* may operate over a range of the muscular force-length curve that is not optimal for force production. It is possible that further consequences of habitat specialization for softshell turtles may be manifested in other aspects of locomotor performance. For example, the limited femoral depression of *A. spinifera* (coupled with the sensitive skin covering the plastron of softshells) might impede the ability of this species to negotiate rough or uneven substrates. Thus, although some aspects of terrestrial performance might not be impeded in aquatic specialists, possible performance sacrifices may become evident with further examinations of muscle function, such as measurements of muscle length change (Olson & Marsh, 1998; Gillis & Biewener, 2000; Gillis & Blob, 2001) and the study of locomotion over a wider range of conditions (Biewener & Gillis, 1999; Reilly & Blob, 2003).

6.3 ONTOGENY OF MOTOR FUNCTION IN TURTLE HINDLIMBS: CAN INTERSPECIFIC DIFFERENCES IN ADULT MOTOR PATTERNS BE TRACED TO JUVENILES?

Adult *T. scripta* and *A. spinifera* exhibit a number of differences in hindlimb function, at least some of which (e.g., reduced femoral depression in *A. spinifera*) may be related to the differing degrees of specialization for aquatic habitats in these species. When in the lives of these animals do such differences first appear? Can functional contrasts between adult turtle species be traced back to juveniles, or might all juveniles show similar patterns of hindlimb function distinct from those of adults?

In vertebrates with altricial young, motor activity can appear similar across species because the fairly uncoordinated movements of young with limited development are typically highly variable, obscuring possible distinctions in species patterns (Cazalets et al., 1990; Westerga & Gramsbergen, 1993). However, turtles have fairly precocial young that are typically capable of proficient locomotion soon after hatching (Cagle, 1950; Miller et al., 1987; Zani & Claussen, 1994). Even for young showing coordinated locomotion, patterns of hindlimb function might be similar across species of juvenile turtles for two primary reasons. First, hatchling turtles are typically less than 25% adult carapace length. Because small animals tend to use high frequency limb movements (Heglund & Taylor, 1988), there may be a limited number of ways for juvenile turtles to move their limbs very quickly (Carrier, 1996), causing all juvenile turtles to use similar motor control patterns. Second, for *T. scripta* and *A. spinifera* in particular, juveniles are more similar in habits than adults. Juveniles of both sliders and spiny softshells are generally predators of evasive aquatic prey (typically small invertebrates); though *A. spinifera* maintain predatory habits into adulthood, *T. scripta* tend to shift to a diet dominated by aquatic plants as they grow larger in size (Cagle, 1950; Dalrymple, 1977; Hart, 1983). Thus, differences in locomotion among adults that could be associated with the requirements for acquiring different types of food (evasive versus non-evasive) might not be expected among juveniles that face similar locomotory demands.

6.3.1 Experimental Methods

To test if contrasts in limb function between adults of different turtle species can be traced to their juveniles, we collected high-speed video and hindlimb EMGs during swimming and walking from juvenile *T. scripta* and *A. spinifera* (8 to 15 weeks posthatching, 18 to 25% adult length, $N = 2$ individuals per species) purchased from a commercial supplier (Turtletown, Waukegan, Illinois, U.S.A.). Patterns observed in juveniles were then compared to those exhibited by adults. Methods of video and EMG data collection were the same as those described previously for adults of these species and targeted the same three muscles (femorotibialis, flexor tibialis complex, and PIFI), with the exception that filming was frequently conducted at 125 Hz rather than 60 Hz. Kinematic analyses (digitizing of limb movements, smoothing of kinematic variables) and EMG analyses (calculation of burst variables) also proceeded as described previously for adults, standardizing procedures to facilitate comparisons between the groups.

6.3.2 Results: Ontogeny of Hindlimb Function in Slider and Softshell Turtles

Limb cycle frequencies in juvenile sliders and softshells for both swimming (*T. scripta*: 3.81 ± 0.13 Hz; *A. spinifera*: 3.00 ± 0.12 Hz; mean ± SE) and walking (*T. scripta*: 2.47 ± 0.22 Hz; *A. spinifera*:

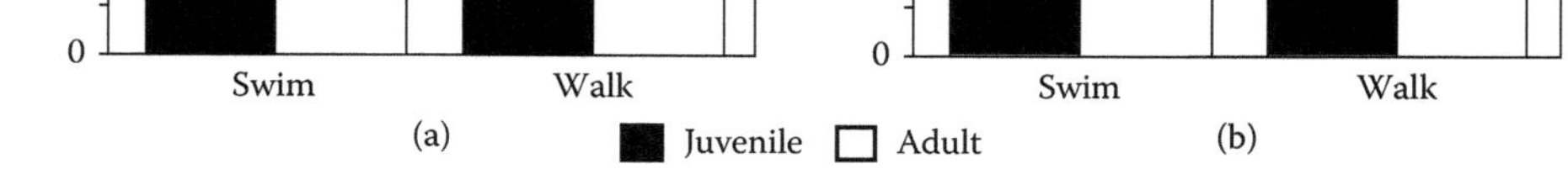

FIGURE 6.10 Comparison of locomotor cycle frequencies for juvenile versus adult turtles during swimming (shaded bars) and walking (unshaded bars) for (a) *Trachemys scripta* and (b) *Apalone spinifera*. Data for adults replotted from Figure 6.2; $N = 10$ cycles for *T. scripta* juveniles in both swimming and walking, 7 cycles for *A. spinifera* juveniles in swimming, and 5 cycles for *A. spinifera* juveniles in walking. Error bars indicate 1 SE. *All comparisons of juvenile to adult of the same species for locomotion within a specific environment are significantly different (Mann–Whitney U-tests, $P < 0.01$).

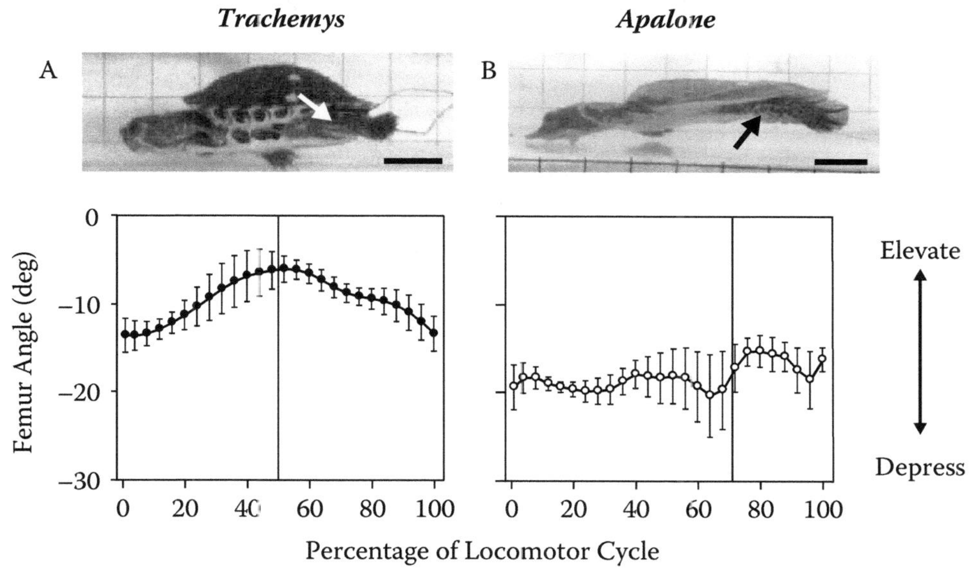

FIGURE 6.11 Comparison of mean kinematic profiles for femoral depression during swimming in juvenile (A) *Trachemys scripta* and (B) *Apalone spinifera*. Lateral view still images from high-speed video recordings accompany each plot; arrows indicate location of the femur. Format for kinematic plots follows that in Figure 6.3. Error bars indicate 1 SE. Scale = 1 cm.

1.85 ± 0.09 Hz; mean ± SE) are significantly higher than those for adults of both of these species (Mann–Whitney U-tests, $P < 0.01$; Figure 6.10). This result supports the possibility that juvenile turtles of different species might show similarities of hindlimb function related to their shared small size.

Kinematic comparisons of juvenile sliders and spiny softshells indicate that some of the main contrasts in hindlimb function found between adults of these species are not evident early in ontogeny. For example, in contrast to patterns observed in adults, juvenile *A. spinifera* actually depress the femur more (−26.4° ± 6.6°) than juvenile *T. scripta* (−17.2° ± 1.3°) during swimming (Mann-Whitney U-test, $P < 0.01$; Figure 6.11; Table 6.2), potentially increasing the upward component of thrust in *A. spinifera*. Similarly, in contrast to patterns observed among adults, juvenile *A. spinifera* flex the knee more (74.1° ± 10.9°) than juvenile *T. scripta* (93.4° ± 14.2°) during terrestrial walking (Mann-Whitney U-test, $P < 0.01$; Figure 6.12; Table 6.2).

Comparisons of EMG burst patterns between juvenile *T. scripta* and *A. spinifera* show that a major contrast in muscle activation between adults of these species is present in juveniles: as in adults, the femorotibialis is active during the recovery phase of swimming in juvenile sliders but during the thrust phase of swimming in juvenile spiny softshells (Figure 6.13). However, juveniles of these species also show a number of contrasts in motor patterns beyond those seen in adults. For example, in both species femorotibialis shows one EMG burst during swimming cycles in adults but shows two EMG bursts during swimming cycles in juveniles (Figure 6.13). In addition, despite the similar EMG burst patterns shown by PIFI both between species and between locomotor environments in adults, juvenile *T. scripta* and *A. spinifera* show very different PIFI burst patterns both between species and between water and land (Figure 6.13).

TABLE 6.2
Values (mean ± SD) of Kinematic Variables Measured from the Hindlimbs of Juvenile Slider (*T. scripta*) and Spiny Softshell (*A. spinifera*) Turtles during Aquatic (Water) and Terrestrial (Land) Locomotion

	T. scripta		*A. spinifera*					
Variable	**Water**	**Land**	**Water**	**Land**	P_1	P_2	P_3	P_4
Maximum femur protraction	43.1° ± 6.1°	47.7° ± 4.9°	34.7° ± 4.5°	50.2° ± 8.5°	0.221	0.005*	0.047*	0.597
Maximum femur retraction	–29.0° ± 9.1°	–43.6° ± 6.4°	–60.0° ± 3.9°	–41.6° ± 10.5°	0.014*	0.003*	0.009*	0.880
Maximum femur elevation	–2.9° ± 1.6°	–4.7° ± 5.0°	–13.3° ± 2.8°	–6.0° ± 2.2°	0.270	0.002*	0.009*	0.496
Maximum femur depression	–17.2° ± 1.3°	–27.2° ± 2.5°	–26.4° ± 6.6°	–20.4° ± 6.9°	0.002*	0.086	0.009*	0.016*
Minimum knee flexion	92.2° ± 5.6°	93.4° ± 14.2°	88.0° ± 5.6°	74.1° ± 10.9°	0.327	0.270	0.076	0.005*
Maximum knee extension	142.5° ± 8.9°	136.7° ± 10.6°	132.7° ± 5.0°	125.0° ± 12.2°	0.903	0.037*	0.175	0.041*

The P values reported are for Mann-Whitney U-tests of the following comparisons in each row: P_1 = *T. scripta*, water vs. land; P_2 = *A. spinifera*, water vs. land; P_3 = water, *T. scripta* vs. *A. spinifera*; P_4 = land, *T. scripta* vs. *A. spinifera*.

* *Significant at* $P < 0.05$.

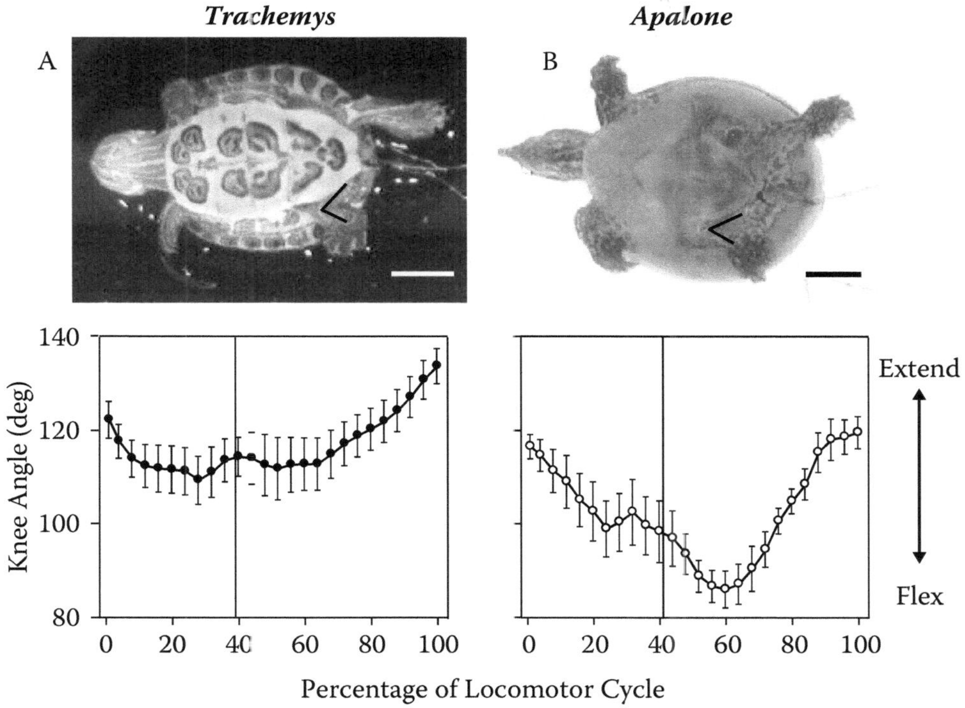

FIGURE 6.12 Comparison of mean kinematic profiles for knee flexion and extension during walking in juvenile (A) *Trachemys scripta* and (B) *Apalone spinifera*. Ventral view still images from high-speed video recordings accompany each plot; black lines highlight the knee joint. Format for kinematic plots follows that in Figure 6.3. Error bars indicate 1 SE. Scale = 1 cm.

6.3.3 Conclusions from Comparisons of Hindlimb Function in Juvenile Slider and Softshell Turtles

Although the complex array of differences and similarities in hindlimb function between juvenile and adult *T. scripta* and *A. spinifera* makes interpretation challenging, some general conclusions can be drawn. First, interspecific differences in adult limb function are often not evident in juveniles of those species. Thus, even if functional tradeoffs related to specialization for particular habitats are evident across adults of different species, some of those tradeoffs might only appear later in the lives of those animals. Second, juveniles may show some kinematic and motor similarities to each other, potentially related to having small body sizes (Carrier, 1996), but they also can show species-specific functional specializations that are not seen in adults. Evaluating potential functional implications for such differences could provide substantial insight into the survival of animals at a crucial stage of life history.

6.4 Future Directions for Studies of Turtle Locomotion

The studies we have presented indicate how research on turtle locomotion and hindlimb function can contribute not only to our understanding of the functional morphology, behavior, and ecology of species in the turtle lineage, but also more broadly to studies in areas ranging from neuromotor control to functional development. Research on turtle locomotion is therefore critical for understanding how the diverse species of these distinctive animals survive in a variety of habitats. Moreover,

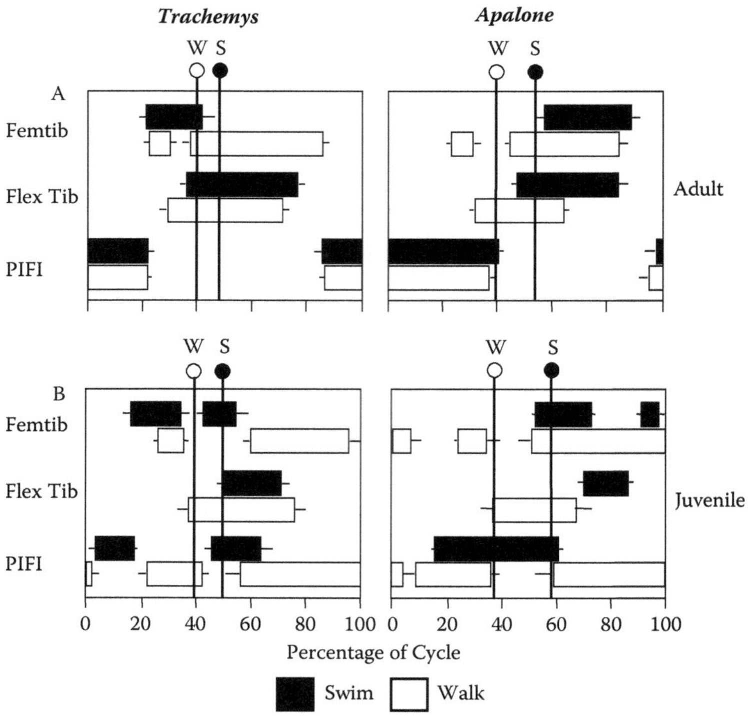

FIGURE 6.13 Comparison of representative mean activity patterns for three focal hindlimb muscles during swimming (shaded bars) and walking (unshaded bars) for *Trachemys scripta* (left column) and *Apalone spinifera* (right column) for (A, replotted from Figure 6.9) adults and (B) juveniles. Vertical lines demarcate cycles into recovery (R) and thrust (T) phases for swimming and swing (SW) and stance (ST) phases for walking; these demarcations are both indicated with appropriate labels on muscle activity graphs. All error bars indicate 1 SE.

studies of turtle locomotion can provide a model for workers on other animal systems who are seeking to understand general principles of how the musculoskeletal system works and how its function can diversify through the course of evolution.

Several future directions for research on turtle locomotion are suggested by our results. First, measurement of EMG patterns from additional muscles could test for distinctions in the roles of muscles that appear—based on anatomical position—to have redundant functions across particular joints. Such studies could evaluate potential correlations of muscle motor patterns with distinctions in, for example, muscle fiber types (Laidlaw et al., 1995) or moment arms (Kargo & Rome, 2002), providing further insight into the versatility of motor control in turtle limbs despite the potential constraints placed on locomotion by the shell. Second, some proposed functions of muscles could be examined through direct measurement of muscle length change using sonomicrometry (Olson & Marsh, 1998; Gillis & Biewener, 2000). Among other questions, such measurements could address the potential for limb muscles to be operating at suboptimal lengths during terrestrial locomotion in aquatic specialist species. A third area warranting study is the production of propulsive forces during both terrestrial and aquatic environments. Although force platform studies have provided insight into the maintenance of slow speeds and the locomotor energetics of turtles (Jayes & Alexander, 1980; Zani et al., 2005), the implications of external locomotor forces for muscular force production and loading of the limb skeleton (Biewener, 1983; Blob & Biewener, 2001) remain to be examined. Efforts to measure the net force produced by all limbs during swimming have been made in turtles (Davenport et al., 1984), but measurements of forces produced by individual limbs—for example, using particle image velocimetry (Drucker & Lauder, 1999; Blob et al., 2003)—could inform evaluations of functional divergence between the fore and hindlimbs. A fourth potential area of focus for future research on turtle limb function is an expansion of the locomotor situations in which limb function is evaluated. Additional data on limb function in turtles under a variety of conditions, including variations in temperature, external load, and locomotor slope (Muegel & Claussen, 1994; Wren et al., 1998; Claussen et al., 2002, 2004), and in different behaviors, such as aquatic maneuvering (Rivera et al., 2006), could aid understanding of the full locomotor capacity of these animals and the impact it has on their survival (Biewener & Gillis, 1999; Reilly & Blob, 2003). Efforts to further examine forelimb function (Pace et al., 2001; Blob et al., 2004) and interlimb coordination (Renous & Bels, 1996; Blob et al., 2003) could also substantially improve understanding of turtle functional morphology, evolution, and ecology.

Examinations of any of these issues in single turtle species would be worthwhile, but because the turtle lineage includes species with a diverse range of habits and reasonably well resolved phylogenetic relationships, comparative studies of limb function in turtles hold especially high promise. Despite containing less than 300 species, turtles include species with diverse locomotor habits, including multiple evolutions of highly terrestrial groups (tortoises, box turtles), different types of highly aquatic groups (bilateral forelimb flapping sea turtles and carettochelyids, forelimb rowing trionychids, bottom walking kinosternids and chelydrids), and species (e.g., many emydids) with intermediate habits (Zug, 1971; Walker, 1973; Ernst & Barbour, 1989). The proliferation of recent research on evolutionary relationships of turtle species has provided an important framework for studies of this diversity (Shaffer et al., 1997; Van der Kuyl et al., 2002; Stephens & Wiens, 2003; Engstrom et al., 2004; Krenz et al., 2005). With data from a sufficiently broad ecological and phylogenetic range of species, it should be possible to evaluate primitive and derived character states for limb function in turtles, allowing tests of adaptive hypotheses through the application of phylogenetic comparative methods (Westneat, 1995; Butler & King, 2004; Garland et al., 2005). Thus, with continued expansion from efforts made to date, studies of turtle locomotion should provide an important model for future research on functional diversity, adaptation, and evolution.

ACKNOWLEDGMENTS

We are grateful to the editors for their invitation to contribute to this volume, and to J. Wyneken and anonymous reviewers for helpful comments on an early version of this chapter. We also thank Cinnamon Pace and Erin Scanga for assistance with data collection for many of the experimental results reported in this chapter, Jeff Walker for access to QuickImage® and QuickSAND® software used for kinematic analysis, and Stephen Gosnell for pointing out helpful literature. Field collection of animals for this work was authorized by Illinois Scientific Permit No. A99.0550. This research was supported by NIH (1F32NS10813-01 and 5F32NS10813-02 to RWB) and the Office of Naval Research (N000149910184 to MWW and J. Walker). Support during final analyses and manuscript preparation was provided by NSF (IOB 0517340, to RWB), NIH (2 R01 DC005063-06A1 to E. Peterson, Ohio University; subaward UT10853 to RWB), and the Clemson Department of Biological Sciences.

REFERENCES

Bennett, D.H., Gibbons, J.W., and Franson, J.C., Terrestrial activity in aquatic turtles, *Ecology,* 51, 738–740, 1970.

Biewener, A.A., Locomotory stresses in the limb bones of two small mammals: The ground squirrel and chipmunk, J. Exp. Biol., 103, 131–154,1983.

Biewener, A.A., and Gillis, G.B., Dynamics of muscle function during locomotion: Accommodating variable conditions, *J. Exp. Biol.*, 202, 3387–3396, 1999.

Blob, R.W., and Biewener, A.A., Mechanics of limb bone loading during terrestrial locomotion in the green iguana (*Iguana iguana*) and American alligator (*Alligator mississippiensis*), *J. Exp. Biol.*, 204, 1099–1122, 2001.

Blob, R.W., Willey, J.S., and Lauder, G.V., Swimming in painted turtles: Particle image velocimetry reveals different propulsive roles for the forelimb and hindlimb, *Integr. Comp. Biol.*, 43, 985, 2003.

Blob, R.W., Willey, J.S., and Wyneken, J., Limb muscle activation patterns during locomotion in loggerhead sea turtles, *J. Morphol.*, 260, 279, 2004.

Bodie, J.R., and Semlitsch, R.D., Spatial and temporal use of floodplain habitats by lentic and lotic species of aquatic turtles, *Oecologia*, 122, 138–146, 2000.

Buchanan, T.S., Almdale, D.P.J., Lewis, J.L., and Rymer, W.Z., Characteristics of synergic relations during isometric contractions of human elbow muscles, *J. Neurophysiol.*, 56, 1225–1241, 1986.

Buford, J.A., and Smith, J.L., Adaptive control for backward quadrupedal walking II. Hindlimb muscle synergies, *J. Neurophysiol.*, 64, 756–766, 1990.

Burke, A.C., Development of the turtle carapace: Implications for the evolution of a novel bauplan, *J. Morphol.*, 199, 363–378, 1989.

Butler, M., and King, A., Phylogenetic comparative analysis: A modeling approach for adaptive evolution, *Am. Nat.*, 164, 683–695, 2004.

Cagle, F.R., Home range, homing behavior, and migration in turtles, *Misc. Publ. Mus. Zool. Univ. Mich.*, 61, 1-34, 1944.

Cagle, F.R., Life history of the slider turtle, *Pseudemys scripta troosti* (Holbrook), *Ecol. Monogr.*, 20, 33–54, 1950.

Carrier, D.R., Ontogenetic limits on locomotor performance, *Physiol. Zool.*, 69, 467–488, 1996.

Cazalets, J.R., Menard, I., Crémieux, J., and Clarac, F., Variability as a characteristic of immature motor systems: An electromyographic study of swimming in the newborn rat, *Behav. Brain Res.*, 40, 215–225, 1990.

Claussen, D.L., Lim, R., Kurz, M., and Wren, K., Effects of slope, substrate, and temperature on the locomotion of the ornate box turtle, *Terrapene ornata*, *Copeia*, 2002, 411–418, 2002.

Claussen, D.L., Snashall, J., and Barden, C., Effects of slope, substrate and temperature on forces associated with locomotion of the ornate box turtle, *Terrapene ornata*, *Comp. Biochem. Physiol. A*, 138, 269–276, 2004.

Dalrymple, G.H., Intraspecific variation in the cranial feeding mechanism of turtles of the genus *Trionyx* (Reptilia, Testudines, Trionychidae), *J. Herpetol.*, 11, 255–285, 1977.

Davenport, J.S., Munks, A., and Oxford, P.J., A comparison of the swimming of marine and freshwater turtles, *Proc. R.S. Lond. B*, 220, 447–475, 1984.

Denny, M.W., *Air and Water*, Princeton, NJ: Princeton University Press, 1993.

Drucker, E.G., and Lauder, G.V., Locomotor forces on a swimming fish: Three-dimensional vortex wake dynamics quantified using digital particle image velocimetry, *J. Exp. Biol.*, 202, 2393–2412, 1999.

Earhart, G.M., and Stein, P.S.G., Step, swim, and scratch motor patterns in the turtle, *J. Neurophysiol.*, 84, 2181–2190, 2000.

Engstrom, T.N., Shaffer, H.B., and McCord, W.P., Multiple data sets, high homoplasy, and the phylogeny of softshell turtles (Testudines: Trionychidae), *Syst. Biol.*, 53, 693–710, 2004.

Ernst, C.H., and Barbour, R.W., *Turtles of the World*, Washington, DC: Smithsonian Institution Press, 1989.

Ernst, C.H., Lovich, J.E., and Barbour, R.W., *Turtles of the United States and Canada*, Washington, DC: Smithsonian Institution Press, 1994.

Fish, F., Kinematics of undulatory swimming in the American alligator (*Alligator mississippiensis*), *Copeia*, 1984, 839–843, 1984.

Garland, T. Jr., Bennett, A.F., and Rezende, E.L., Phylogenetic approaches in comparative physiology, *J. Exp. Biol.*, 208, 3015–3035, 2005.

Gatesy, S.M., Caudofemoral musculature and the evolution of theropod locomotion, *Paleobiology*, 16, 170–186, 1990.

Gatesy, S.M., An electromyographic analysis of hindlimb function in *Alligator* during terrestrial locomotion, *J. Morph.*, 234, 197–212, 1997.

Gibbons, J.W., Terrestrial activity and the population dynamics of aquatic turtles, *Am. Midl. Nat.*, 83, 404–414, 1970.

Gibbons, J.W., *Life History and Ecology of the Slider Turtle*, Washington, DC: Smithsonian Institution Press, 1990.

Gilchrist, G.W., Specialists and generalists in changing environments. I. Fitness landscapes of thermal sensitivity, *Am. Nat.*, 146, 252–270, 1995.

Gillis, G.B., and Biewener, A.A., Hindlimb extensor muscle function during jumping and swimming in the toad (*Bufo marinus*), *J. Exp. Biol.*, 203, 3547–3563, 2000.

Gillis, G.B., and Blob, R.W., How muscles accommodate movement in different physical environments: Aquatic vs. terrestrial locomotion in vertebrates, *Comp. Biochem. Physiol. A*, 131, 61–75, 2001.

Gruner, J.A., and Altman, J., Swimming in the rat: analysis of locomotor performance in comparison to stepping, *Exp. Brain Res.*, 40, 374–382, 1980.

Hart, D.B., Dietary and habitat shift with size of red-eared turtles (*Pseudemys scripta*) in a southern Louisiana population, *Herpetologica*, 39, 285–290, 1983.

Heglund, N.C., and Taylor, C.R., Speed, stride frequency, and energy cost per stride. How do they change with body size and gait?, *J. Exp. Biol.*, 138, 301–318, 1988.

Jackson, D.C., Living without oxygen: Lessons from the freshwater turtle, *Comp. Biochem. Physiol. A*, 125, 299–315, 2000.

Jackson, D.C., Hibernating without oxygen: The painted turtle, *J. Physiol. Lond.*, 543, 731–737, 2002.

Jackson, D.C., Surviving extreme lactic acidosis: The role of calcium lactate formation in the anoxic turtle, *Resp. Physiol. Neurobiol.*, 144, 173–178, 2004.

Jayes, A.S., and Alexander, R.M., The gaits of chelonians: Walking techniques for very slow speeds, *J. Zool.*, 191, 353–378, 1980.

Johnston, R.M., and Bekoff, A., Patterns of muscle activity during different behaviors in chicks: Implications for neural control, *J. Comp. Physiol. A.*, 179, 169–184, 1996.

Kamel, L.T., Peters, S.E., and Bashor, D.P., Hopping and swimming in the leopard frog, *Rana pipiens*: II. A comparison of muscle activities, *J. Morph.*, 230, 17–31, 1996.

Kargo, W.J., and Rome, L.C., Functional morphology of proximal hindlimb muscles in the frog *Rana pipiens*, *J. Exp. Biol.*, 205, 1987–2004, 2002.

Krenz, J.G., Naylor, G.J.P., Shaffer, H.B., and Janzen, F.J., Molecular phylogenetics and evolution of turtles, *Mol. Phylogenet. Evol.*, 37, 178–191, 2005.

Laidlaw, D.H., Callister, R.J., and Stuart, D.G., Fiber-type composition of hind limb muscles in the turtle, *Pseudemys (Trachemys) scripta elegans*, *J. Morphol.*, 225, 193–211, 1995.

Lauder, G.V., Function of the caudal fin during locomotion in fishes: Kinematics, flow visualization, and evolutionary patterns, *Am. Zool.*, 40, 101–122, 2000.

Loeb, G.E., and Gans, C., *Electromyography for Experimentalists*, Chicago: University of Chicago Press, 1986.

Macpherson, J.M., How flexible are muscle synergies?, in *Motor Control: Concepts and Issues*, D.R. Humphrey and H.J. Freund (eds.), Chichester, UK: John Wiley & Sons, 1991, 33–47.

Miller, K., Packard, G.C., and Packard, M.J., Hydric conditions during incubation influence locomotor performance of hatchling snapping turtles, *J. Exp. Biol.*, 127, 401–412, 1987.

Muegel, L.A., and Claussen, D.L., Effects of slope on voluntary locomotor performance in the turtle, *Terrapene carolina carolina*, *J. Herpetol.*, 28, 1994.

Olson, J.M., and Marsh, R.L., Activation patterns and length changes in hindlimb muscles of the bullfrog *Rana catesbeiana* during jumping, *J. Exp. Biol.*, 201, 2763–2777, 1998.

Pace, C.M., Blob, R.W., and Westneat, M.W., Comparative kinematics of the forelimb during swimming in red-eared slider (*Trachemys scripta*) and spiny softshell (*Apalone spinifera*) turtles, *J. Exp. Biol.*, 204, 3261–3271, 2001.

Plummer, M.V., Mills, N.E., and Allen, S.L., Activity, habitat and movement patterns of softshell turtles (*Trionyx spiniferus*) in a small stream, *Chel. Cons. Biol.*, 2, 514–520, 1997.

Pratt, C.A., Buford, J.A., and Smith, J.L., Adaptive control for backward quadrupedal walking: V. Mutable activation of bifunctional thigh muscles, *J. Neurophysiol.*, 75, 832–842, 1996.

Reilly, S.M., and Blob, R.W., Motor control of locomotor hindlimb posture in the American alligator (*Alligator mississippiensis*), *J. Exp. Biol.*, 206, 4341–4351, 2003.

Renous, S., and Bels, V., Comparison between aquatic and terrestrial locomotions of the leatherback sea turtle (*Dermochelys coriacea*), *J. Zool. Lond.*, 230, 357–378, 1993.

Renous, S., and Bels, V.L., Swimming behaviour during first immersion in hatchling leatherback turtles (*Dermochelys coriacea*), *Ann. Sci. Nat. Zool. Paris*, 17, 25–38, 1996.

Rivera, G., Rivera, A.R.V., Dougherty, E.E., and Blob, R.W., Aquatic turning performance of painted turtles (*Chrysemys picta*) and functional consequences of a rigid body design, *J. Exp. Biol.*, 209, 4203–4213, 2006.

Roy, R.R., Hirota, W.K., Kuehl, M., and Edgerton, V.R., Recruitment patterns in the rat hindlimb muscle during swimming, *Brain Res.*, 337, 175–178, 1985.

Roy, R.R., Hutchinson, D.L., Pierotti, D.J., Hodgson, J.A., and Edgerton, V.R., EMG patterns of rat ankle extensors and flexors during treadmill locomotion and swimming, *J. Appl. Physiol.*, 70, 2522–2529, 1991.

Schmidt-Nielsen, K., *Animal Physiology: Adaptation and Environment*, New York: Cambridge University Press, 1990.

Shaffer, H.B., Meylan, P., and McKnight, M.L., Tests of turtle phylogeny: Molecular, morphological, and paleontological approaches, *Sys. Bio.*, 46, 235–268, 1997.

Snyder, R.C., Adaptations for bipedal and quadrupedal locomotion in lizards, *Am. Zool.*, 2, 191–203, 1962.

Stein, P.S.G., Neuronal control of turtle hindlimb motor rhythms, *J. Comp. Physiol. A*, 191, 213–229, 2003.

Stephens, P.R., and Wiens, J.J., Ecological diversification and phylogeny of emydid turtles, *Biol. J. Linn. Soc.*, 79, 577–610, 2003.

Ultsch, G.R., The ecology of overwintering among turtles: Where turtles overwinter and its consequences, *Biol. Rev.*, 81, 339–367, 2006.

Van der Kuyl, A.C., Ballasina, D.L.P., Dekker, J.T., Maas, J., Willemsen, R.E., and Goudsmit, J., Phylogenetic relationships among the species of the genus *Testudo* (Testudines: Testudinidae) inferred from mitochondrial 12S rRNA gene sequences, *Mol. Phylogenet. Evol.*, 22, 174–183, 2002.

Vogel, S., *Life in Moving Fluids*, Princeton, NJ: Princeton University Press, 1994.

Walker, J.A., Estimating velocities and accelerations of animal locomotion: A simulation experiment comparing numerically different algorithms, *J. Exp. Biol.*, 201, 981–995, 1998.

Walker, J.A., and Westneat, M.W., Mechanical performance of aquatic rowing and flying, *Proc. R.S. Lond. B*, 267, 1875–1881, 2000.

Walker, W.F. Jr., A structural and functional analysis of walking in the turtle, *Chrysemys picta marginata*, *J. Morphol.*, 134, 195–214, 1971.

Walker, W.F. Jr., The locomotor apparatus of Testudines, in *Biology of the Reptilia, Vol. 4: Morphology D*, C. Gans and T.S. Parsons (eds.), London: Academic Press, 1973, 1–100.

Webb, R.G., North American recent soft-shelled turtles (Family Trionychidae), *Univ. Ks. Publ. Mus. Nat. Hist.*, 13, 431–611, 1962.

Westerga, J., and Gramsbergen, A., Changes in the electromyogram of two major hindlimb muscles during locomotor development in the rat, *Exp. Brain Res.*, 92, 479–488, 1993.

Westneat, M.W., Feeding, function, and phylogeny: analysis of historical biomechanics in labrid fishes using comparative methods, *Syst. Biol.*, 44, 361–383, 1995.

Westneat, M.W., and Walker, J.A., Motor patterns of labriform locomotion: Kinematic and electromyographic analysis of pectoral fin swimming in the labrid fish *Gomphosus varius*, *J. Exp. Biol.*, 200, 1881–1893, 1997.

Willey, J.S., and Blob, R.W., Tail kinematics of juvenile common snapping turtles during aquatic walking, *J. Herpetol.*, 38, 360–369, 2004.

Wren, K., Claussen, D.L., and Kurz, M., The effects of body size and extrinsic mass on the locomotion of the ornate box turtle, *Terrapene ornata*, *J. Herpetol.*, 32, 144–150, 1998.

Wyneken, J., Sea turtle locomotion: Mechanisms, behavior, and energetics, in *The Biology of Sea Turtles*, P.L. Lutz and J.A. Musick (eds.), Boca Raton, FL: CRC Press, 1997, 165–198.

Zangerl, R., The turtle shell, in *Biology of the Reptilia, Vol. 1, Morphology A*, C. Gans, A.d'A. Bellairs, and T.S. Parsons, (eds.), London: Academic Press, 1969, 311–339.

Zani, P.A., and Claussen, D.L., Voluntary and forced locomotion in juvenile and adult painted turtles, *Chrysemys picta*, *Copeia*, 1994, 466–471, 1994.

Zani, P.A., Gottschall, J.S., and Kram, R., Giant Galapagos tortoises walk without inverted pendulum mechanical-energy exchange, *J. Exp. Biol.*, 208, 1489–1494, 2005.

Zernicke, R.F., and Smith, J.L., Biomechanical insights into neural control of movement, in *Handbook of Physiology Section 12: Exercise: Regulation and Integration of Multiple Systems*, L.B. Rowling and J.T. Shepherd (eds.), New York: American Physiological Society, 1996, 293–330.

Zug, G.R., Buoyancy, locomotion, morphology of the pelvic girdle and hind limb, and systematics of cryptodiran turtles, *Misc. Publ. Mus. Zool. Univ. Mich.*, 142, 1–98, 1971.

7 Cervical Anatomy and Function in Turtles

Anthony Herrel, Johan Van Damme, and Peter Aerts

CONTENTS

7.1 INTRODUCTION

Being mobile is an essential requirement for any animal. Not only do animals need to move about to find food or partners, they also need to be able to escape potential predators (Irschick & Garland, 2001). Interestingly, some vertebrate groups appear to have sacrificed part of their mobility in response to predation pressure by the development of a robust armored body (e.g., pangolins, glyptodonts, turtles, and so on). Although body armor can provide an animal with an adequate protection against predators, it also dramatically reduces its locomotor ability and overall agility (Wren et al., 1998; Zani et al., 2005). Thus, many armored vertebrates have specialized on eating non-mobile food items like plants, or clumped food sources such as ants or termites (King, 1996). However, some groups have developed an alternative strategy for capturing elusive prey by developing long, mobile appendages such as projectile tongues (Deban et al, 1997; Herrel et al., 2000) or a long neck (Gans, 1992). For instance, many semi-aquatic and aquatic turtles have developed remarkably long necks that are used to capture elusive prey under water (Pritchard, 1984).

Although the turtle carapace provides an excellent defense against predators, it is imperative that the long neck and head can be protected as well. To do so, the head-neck system needs to be withdrawn within the margins of the bony shell. This can be done in one of two ways: in the mid-sagittal plane (vertically), which involves the retraction of the head and neck within the bony shell, or laterally (in the horizontal plane), where the neck is folded between the dorsal and ventral rim of the bony shell. In this case, the head and neck do remain partially exposed in the outer carapacial chamber in front of the pectoral girdle (Van Damme et al., 1995). This difference in the mode of neck retraction in turtles has often been used as an important character for the subdivision of the class Testudines into the subclasses Cryptodira and Pleurodira. Whereas cryptodires retract their head-neck system in the vertical plane, pleurodires do so in the horizontal plane (Figure 7.1).

Accurate control of the neck during rapid movements associated with escape head retraction and prey capture appears crucial for turtles. However, the turtle neck is a highly complex multi-jointed system, consisting of eight cervical vertebrae, the head, and the body and a large number of muscles that span anywhere from one to over eight joints. The control of such a multi-joint system with a large number of degrees of freedom appears inherently complex. Although some ways to facilitate the control of the system have been identified previously (Aerts et al., 2001), a better understanding of the detailed structure and function of the musculo-skeletal elements of the cervical system is essential to gain better insight into the control of the cervical system. Moreover, large differences in vertebral structure, in the morphology of associated musculature, and in the control of the cranio-cervical system can be expected for turtles that retract their necks predominantly in either the horizontal or vertical plane. Unfortunately, previous authors have described differences in vertebral structure between cryptrodiran and pleurodiran turtles predominantly from a taxonomic standpoint (Vaillant, 1881; Williams, 1950; but see Weisgram & Splechtna 1990, 1992). Thus, it remains presently unclear whether different functional capacities (i.e., ranges of mobility) are associated with either morphology. Without this type of information, our understanding of the control of the cranio-cervical system in cryptodiran and pleurodiran turtles must, unfortunately, remain limited.

The aim of the present chapter is to give a detailed morphological description of the cervical system in pleurodiran and cryptodiran turtles. In doing so, our emphasis will be on the functional consequences of differences in morphology in the two groups. Additionally, some previously unpublished data on the actual kinematics of neck movement will be presented to highlight the consequences of morphological differences in the two groups. As our type representatives for cryptodiran and pleurodiran turtles we have chosen the genera *Chelodina* and *Apalone*. The Australian pleurodiran turtles of the genus *Chelodina* are renowned for their extreme elongated neck. Because the neck is longer than the carapace, these turtles are also known as snake-necked turtles. These animals use their elongated neck both for quick strikes at prey and for snorkeling, thus aspirating air from the surface without exposing more than the tip of the snout. Species from this genus will be used as a typical representative of the pleurodiran condition. Turtles of the genus *Apalone*, also known as the soft-shelled turtles, represent the cryptodiran counterpart of *Chelodina*. These animals are also characterized by an extremely elongated neck, often longer than the carapace itself (Ernst & Barbour, 1989). Soft-shell turtles typically live in shallow water and will use their elongate neck to breathe at the water surface with minimal movement. Moreover, just like *Chelodina*,

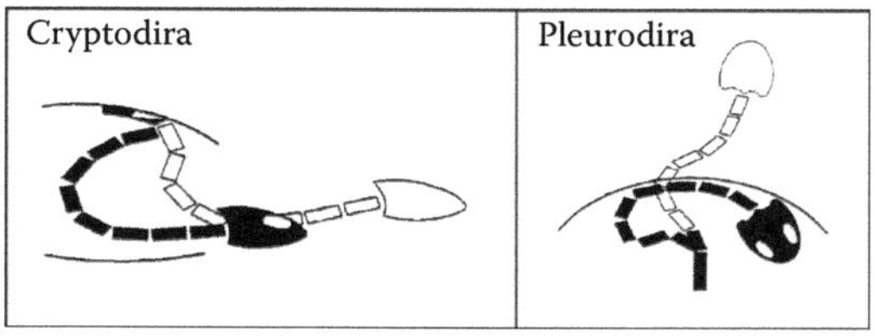

FIGURE 7.1 Schematic representation of the two major modes of head retraction in turtles. Left, lateral view on the cervical system in a cryptodire. In cryptodires, the head is retracted in the vertical plane. Right, dorsal view on the cervical system in a pleurodire. Here the head is retracted in the horizontal plane.

members of the genus *Apalone* are voracious predators that will actively strike at elusive prey under water (Dalrymple, 1977; Ernst & Barbour, 1989).

7.2 MATERIALS AND METHODS

7.2.1 Anatomical Studies

The anatomy of the cervical vertebrae, joint structures, and cervical musculature of *Chelodina longicollis* was studied by means of dissection of a preserved specimen (K.B.I.N., R.G.nr.4566). Additionally, an adult *C. reimani* (cadaver obtained through the commercial pet trade) was used to study the cervical anatomy. One individual of the species *Apalone ferox* and one *A. spiniferus* were dissected to investigate the anatomy of the cervical system in a typical cryptodire. Animals were obtained through the commercial pet trade, sacrificed by means of an overdose of Nembutal and preserved in a 10% aqueous formaldehyde solution for 24 hours. Next, animals were rinsed extensively and transferred to a 70% aqueous ethanol solution.

Cervical morphology was studied by means of dissections. In the morphological descriptions, vertebrae are indicated by capital C or D (cervical and dorsal vertebrae, respectively) followed by their serial number. C1 is closest to the head. Joints are labeled by the number of the adjacent vertebrae: C(n)-($n - 1$), with n the number of the more caudal vertebra and ($n - 1$) the number of the more cranial one as suggested by Heidweiller (1991). For example, the joint between vertebra 5 and 6 is defined as "joint C6-5."

7.2.2 Kinematic Analyses

To study the kinematics of snorkeling, two live adult specimens of *Chelodina longicollis* were used for the experiments (one female of 730 g, 18 cm carapace length and one male of 520 g, 15 cm carapace length). The animals were obtained with the help of the Antwerp Zoo and were housed in a glass aqua-terrarium on a 12-hour light/dark cycle. One live *Apalone ferox* and one live *A. spiniferus*, obtained through the commercial pet trade, were used to study the mobility of the cervical vertebrae in these long-necked cryptodires. The water temperature was kept at 28°C for all species. Twice a week, the turtles were fed with meat, mice, and small invertebrates (crickets, grasshoppers).

Snorkeling movements (neck movements in the vertical plane) of *C. longicollis* were recorded by means of cineradiography in lateral view using a Polydoros 80S generator equipped with a Siemens Siregraph D40 x-ray flash apparatus at 66 kV. The digital cineradiographic recordings (depending on the sequence 4 or 6 frames per second, Fluorospot H) were printed on Scopix laserfilm (35 × 43 cm) by means of an Agfa laser printer. During the recordings, the animals were restrained by means of a body-shaped corselet. This corselet was mounted under the water surface on a fixed frame.

The sequences were projected frame by frame and digitized. Digitization of the position of the joints allowed the calculation of several kinematical parameters (joint angles, elevation of the head, head position) in a turtle-bound frame. The same terminology of joint rotations is used as described for the neck movements in the horizontal plane, i.e., clockwise rotations are defined as positive.

Mobility of the cervical vertebrae in *A. ferox* and *A. spiniferus* was studied by means of CT scanning. The animal was anesthetized by means of intramuscular injection of Ketamine (150 mg/kg body mass). CT scans were recorded using a CT-highlight Advantage scanner at the University of Antwerp Hospital. Of 10 different static neck positions, ranging from fully extended to fully retracted, 3-s long recordings resulting in 1.5 mm thick slices through the vertebral column were made at 140 kV, 140 mA. Scans were printed on Scopix LT2B-100 NIF x-ray film.

7.3 RESULTS

7.3.1 Osteology

The cervical column in all recent turtles consists of eight elongated vertebrae (C1 to C8) and nine joints. The most important element of the vertebrae is the vertebral centrum, which is positioned right below the spinal cord. From the centrum, the neural arch rises and covers the spinal cord (Figure 7.2). The neural spines are positioned midventrally on the neural roof and are reduced in most turtles (Figure 7.2). The articulation of the different vertebrae occurs by means of the vertebral centra and the zygapophyses (Figure 7.2). The prezygapophyses of the vertebra articulate with the postzygapophyses of the more cranially situated vertebra. Whereas the articular facets of the prezygapophyses are positioned dorsally, those of the postzygapophyses are oriented ventrally. Vertebral centra can be biconvex (C5 and C8 in *Chelodina*), biconcave or amphicoelous (C1 in both species; C7 in *Chelodina*), procoelous (C6 in *Chelodina*) or opisthocoelous (C2, C3, and C4 in *Chelodina*; C2 to C8 in *Apalone*).

In a transverse section, the vertebral centrum of *Chelodina* is saddle-shaped (Figure 7.3 and Figure 7.4). The vertebral centrum is well developed only at the level of the anterior and posterior articulations. The neural spine is reduced to an inconspicuous longitudinal rim. The anterior zygapophyses are clearly separated from each other (Figure 7.3 and Figure 7.4). Their articular facets are mainly oriented in the medio-dorsal direction. The posterior zygapophyses are fused to each other except those of C1 (Figure 7.3), but the articular facets remain separated and face mainly ventrally and slightly laterally (Figure 7.3 and Figure 7.4). The elevation of the posterior zygapophysis gradually increases from distal to proximal (Figure 7.3 and Figure 7.4). The transverse processes are strongly developed, especially those on the more proximal vertebrae (Figure 7.3 and Figure 7.4).

The cervical vertebrae in *Apalone* are relatively gracile and elongated (Figure 7.5 and Figure 7.6). They bear well-developed and well-separated pre- and post-zygapophyses (Figure 7.5 and

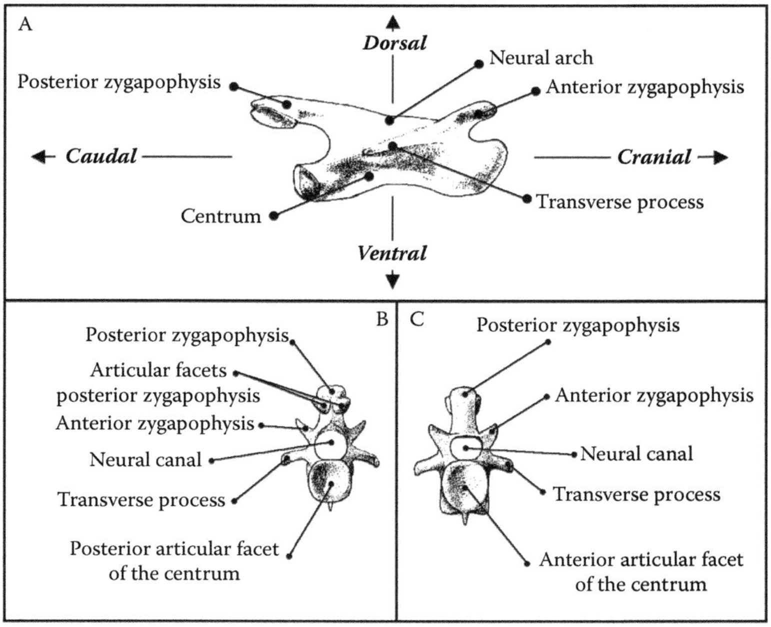

FIGURE 7.2 *Chelodina longicollis*. Terminology of the most important structural elements of a cervical vertebra (C7). (A) Lateral view, (B) posterior view, and (C) anterior view (terminology after Romer & Parsons, 1977).

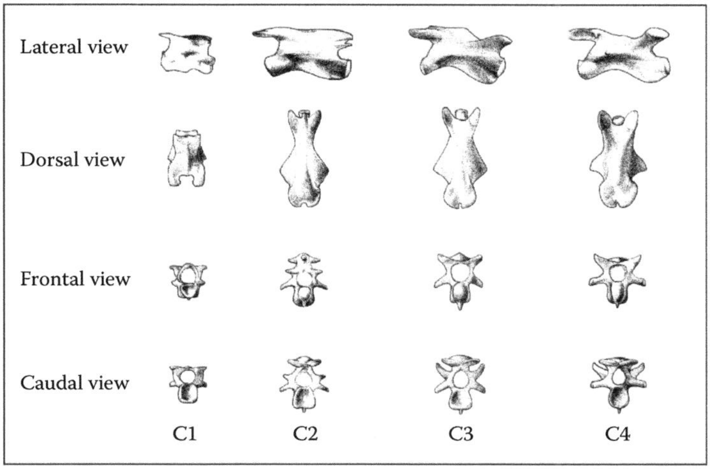

FIGURE 7.3 Lateral, dorsal, frontal, and caudal views of the first four cervical (C1-C4) vertebrae in *C. longicollis*.

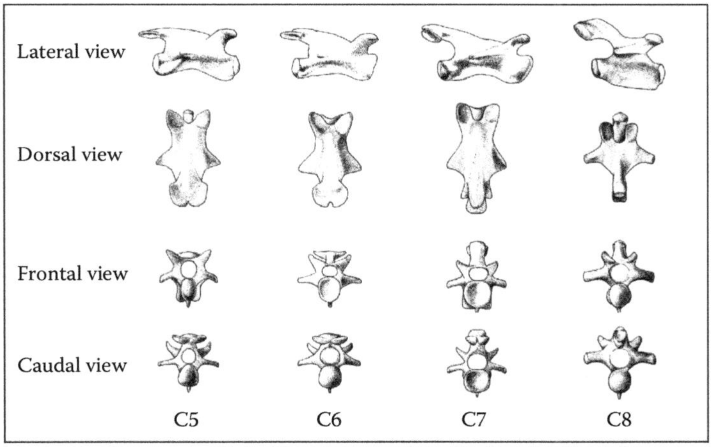

FIGURE 7.4 Lateral, dorsal, frontal, and caudal views of the last four cervical (C5-C8) vertebrae in *C. longicollis*.

Figure 7.6). The orientation of the zygapophyses is variable throughout the neck. Because of the absence of the iliocostalis and logissimus groups (discussed later), the transverse processes are only poorly developed (Figure 7.5 and Figure 7.6). Because of the strong divergence of the post-zygapophyses and the absence of a well-developed neural spine, a hiatus intervertebralis (Bojanus, 1821) is created. In *Apalone*, this hiatus is covered by a thick membrane (Ogushi, 1913).

7.3.2 Cervical Joints

Joint S(kull)-C1 (occipito-cervical joint). This procoelous ball-and-socket type joint allows the skull to move independently from the neck. The occipital is spherical. The ridges of the congruent cotyle on C1 are well developed. C1 lacks anterior zygapophyses. The connection between the

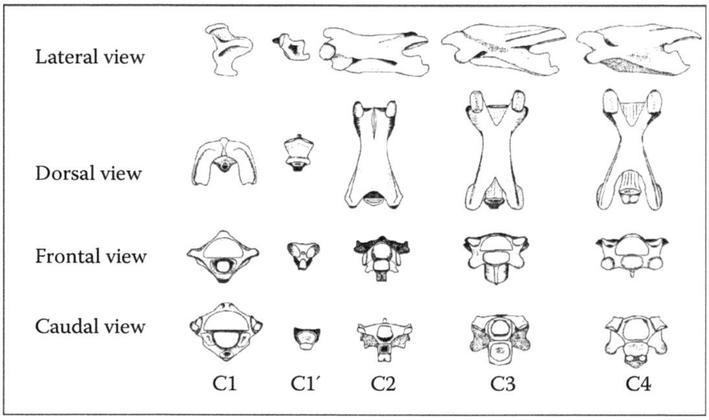

FIGURE 7.5 Lateral, dorsal, frontal, and caudal views of the first four cervical (C1-C4) vertebrae in *A. spinifera*. C1′ is the intercentrum of the atlas.

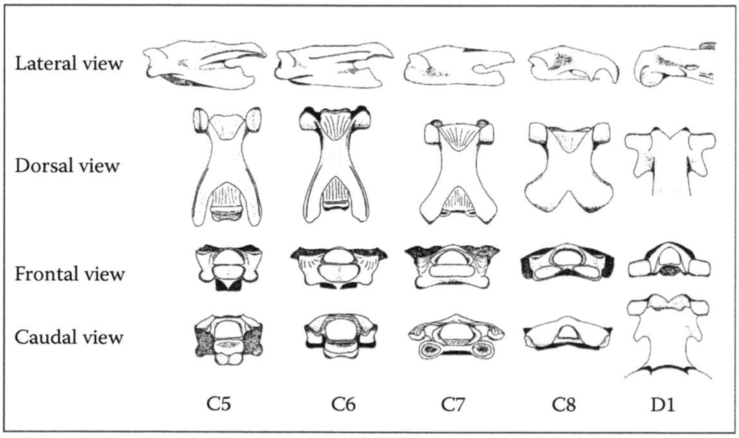

FIGURE 7.6 Lateral, dorsal, frontal, and caudal views of the last four cervical (C5-C8) and the first dorsal vertebra (D1) in *A. spinifera*.

neural arch of C1 and the base of the skull is ligamentous. The joint is similar in both *Apalone* and *Chelodina*.

Joints C2-1, C3-2, C4-3, C5-4 (anterior part of the neck). The articulation facets of vertebrae composing these joints are very similar. The vertebral centra are all opisthocoelous and have an anterior elliptical cotyle, of which the long axis is oriented vertically in *Chelodina* and horizontally in *Apalone*, and a posterior condyle with enlarged lateral ridges. The articular facets of the zygapophyses of the most cranial joint of this group (C2-1) are less strongly developed. The elongated neural spine of C2 extends anteriorly beyond the articular facets of the posterior zygapophyses of C1 in both groups. The posterior zygapophyses of C1 are clearly separated. In the more caudal joints, the zygapophyses are elongated and slightly elevated (Figure 7.3). The articular facets of the anterior zygapophyses are slightly concave and their outline is kidney-shaped. They are oriented in a dorso-medial direction. The posterior zygapophyseal facets are ventrolaterally inclined. Lateral mobility is strongly reduced in *Apalone* compared to *Chelodina* because of the paired post-zygapophyses.

Joints C6-5 and C7-6. The central joints C6-5 and C7-6 are procoelous in *Chelodina* but opisthocoelous in *Apalone*. In posterior view, the posterior condyle of C5 in *Chelodina* has the shape of a raindrop. Alternatively, in *Chelodina* the condyle of C6 is almost spherical. The congruent cotyle on the anterior side of C6 and C7 has pronounced lateral and ventral ridges. The articular facets of the posterior zygapophysis are well developed and also have a kidney-shaped outline. The bilateral posterior zygapophyseal facets contact each other medially, forming a semicircular or horseshoe-shaped articular surface. The articular facets of the anterior zygapophyses are more oval (almost circular) as compared to the more distal joints. In *Apalone*, the articulatory facets of the post-zygapophyses are positioned horizontally. The post-zygapophyses themselves are nearly in line with the long axis of the vertebrae (less than 6°).

Joint C8-7. C8-7 is opisthocoelous and the shape of its condyle and cotyle resemble previously described ophisthocoelous joints in both groups. The cotyle has enlarged lateral ridges in *Chelodina* and is fully split in *Apalone*. The posterior zygapophysis of C7 is elongated, inclined in dorsal direction in both groups (Figure 7.4). The articular facets are clearly distinguishable from those of the remainder of the zygapophyses and are laterally oriented in *Chelodina*. The rather short but strongly elevated anterior zygapophyses of C8 tend to incline dorsomedially in *Chelodina* but more laterally in *Apalone*.

Joint D1-C8 (cervico-dorsal joint). D1-C8 is procoelous in *Chelodina*. In *Apalone*, this joint lacks a central articulation. In *Chelodina*, the rather flattened condyle on the posterior part of C8 is almost circular. The articular facet faces caudo-ventrally. The circular congruent cavity on the anterior part of D1 is shallower than those on the more anterior joints. The elongate posterior zygapophysis in *Chelodina* is more elevated in the dorsal direction than that of joint C8-7 (Figure 7.4). The articular facets of the zygapophyses are more inclined than those in the previously described joints. On C8, the articular facets of both sides fuse and are clearly distinguishable from the posterior zygapophysis. Their appearance is that of a vertically positioned cylinder (approximating a vertical hinge joint). The anterior zygapophyses on D1 are rather short and present an increased dorsomedial inclination of the articular facets. In *Apalone*, the post-zygapophyses of C8 are extremely well developed and wider than those of the more anteriorly situated vertebrae (Figure 7.6). The articulatory facets are strongly concave. The pre-zygapophyses of D1 have been rearranged to horizontally oriented cylinders. Only ventroflexion is possible at this joint.

7.3.3 Musculature

The following short description of the cervical musculature is based on the terminology of Shah (1963) for *Chelodina* and that of Ogushi (1913) for *Apalone*. Other morphological accounts on the cervical musculature in turtles (Bojanus, 1819; George & Shah, 1954, 1955; Scanlon, 1982) were consulted to clarify the position of certain muscles where needed.

7.3.3.1 *Chelodina*

m. constrictor colli. This is a thin sheet-like muscle which covers the first two thirds of the neck. The muscle has a dual origin: some fibers originate on the squamosal and the dorsal connective tissue associated with the occipital spine and others originate on the neural spines of C2-C6. The fibers run ventrally and insert midventrally on a central connective tissue sheet (median raphe).

m. retrahens capitis et collique (Figure 7.7). This muscle consists of a number of well-developed and individualized bundles that connect the carapace to the neck and head. The first bundle is the longest and originates at the lateral aspect of D8 and the adjacent part of the carapace. The muscle runs cranially and inserts by means of a long tendon at the basioccipital. A number of fibers from the first bundle insert on the lateral aspect of C1-C5. The second bundle originates at the level of D5. The muscle runs cranially and inserts tendinously at the lateral aspect of C5. The third bundle

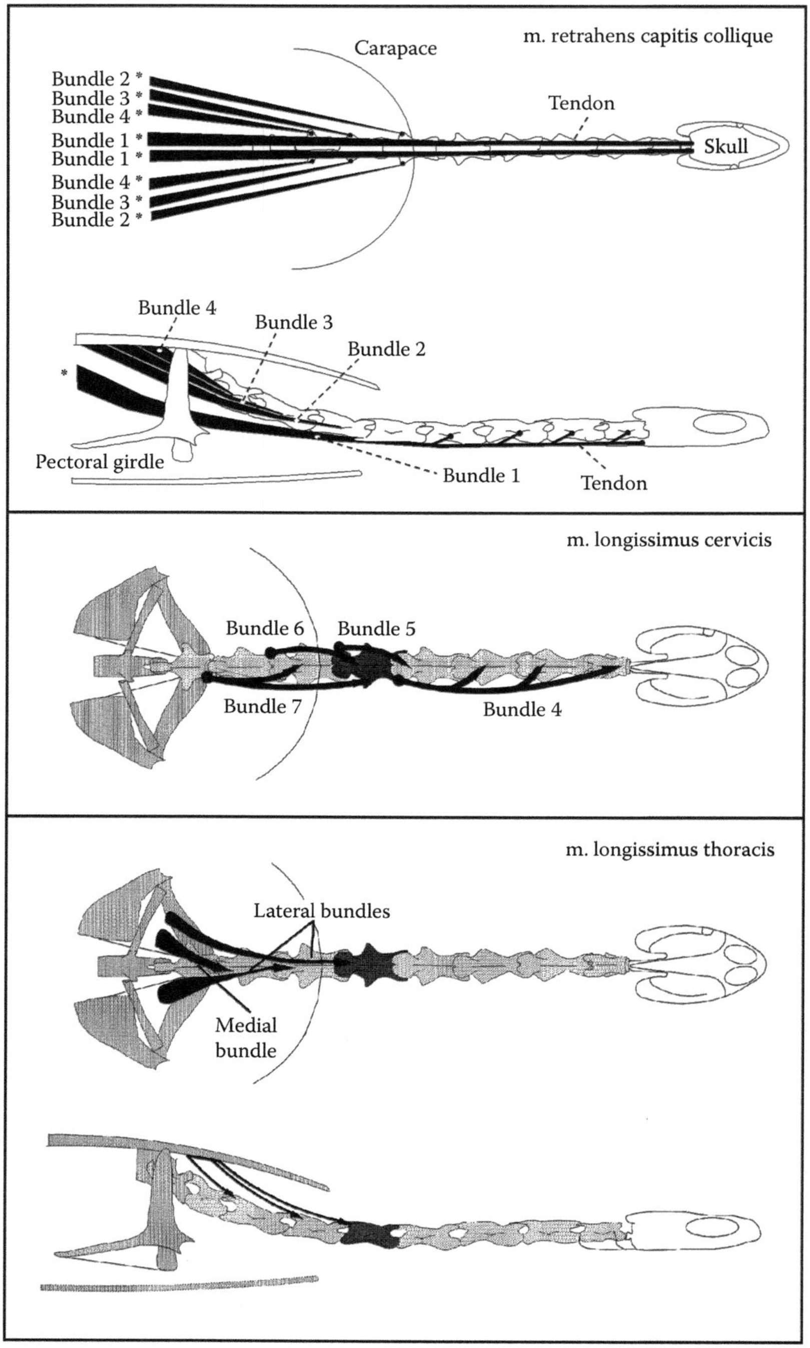

FIGURE 7.7 Schematic representation of the m. retrahens capitis collique, the m. longissimus cervicis, and the m. longissimus thoracis in *C. longicollis*.

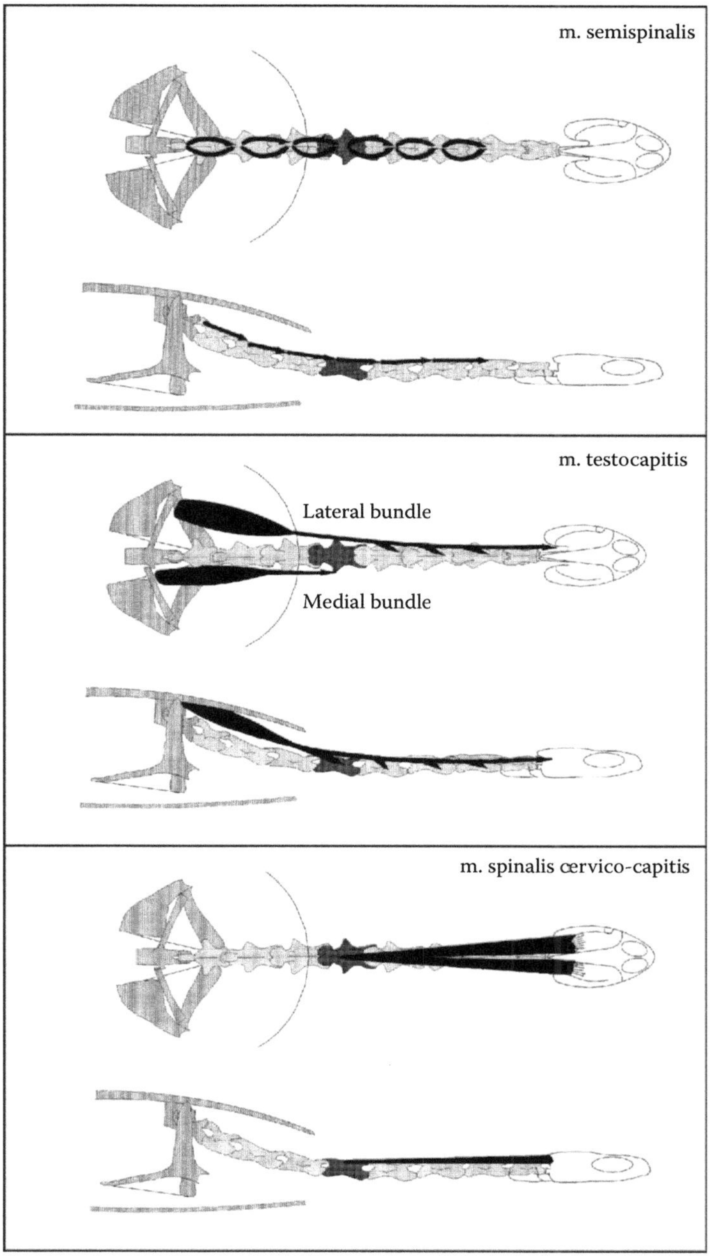

FIGURE 7.8 Schematic representation of the m. semispinalis, the m. testocapitis, and the m. spinalis cervico-capitis in *C. longicollis*.

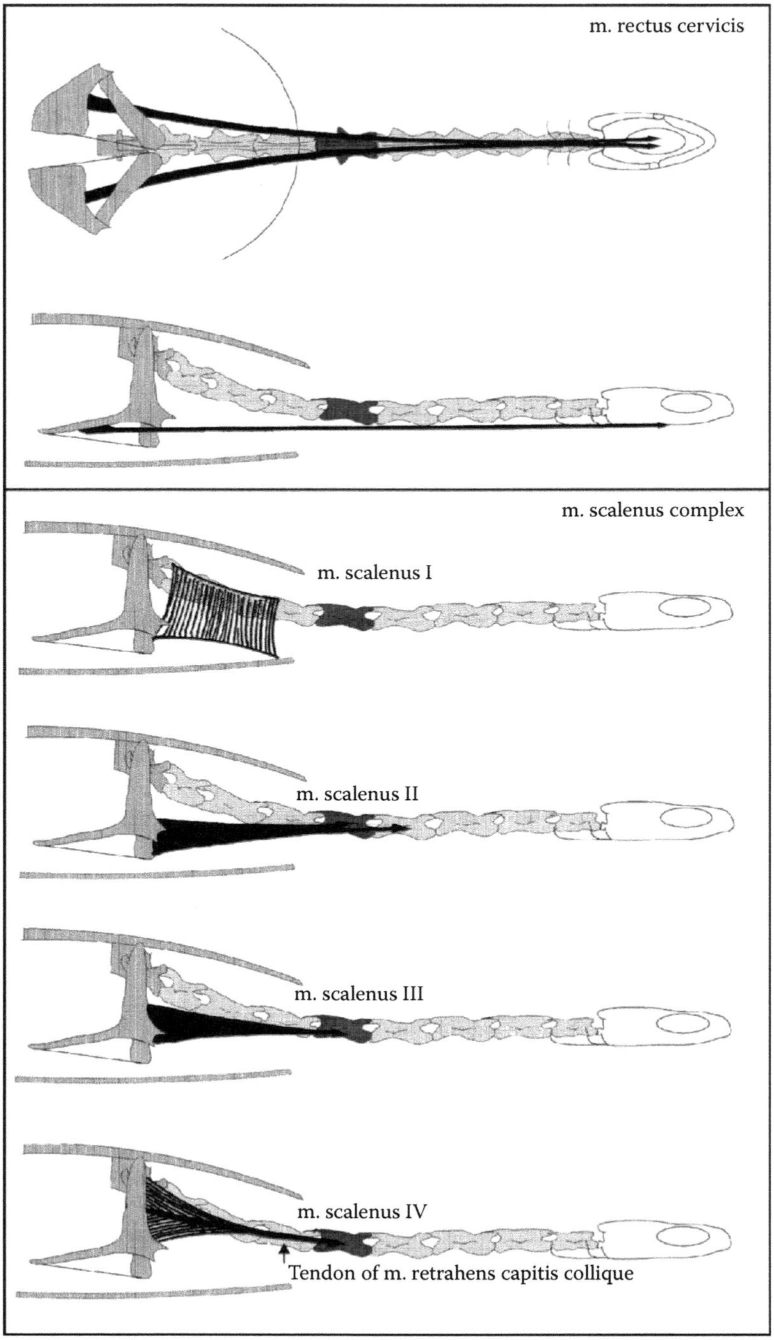

FIGURE 7.9 Schematic representation of the m. rectus cervicis and the m. scalenus complex in *C. longicollis*.

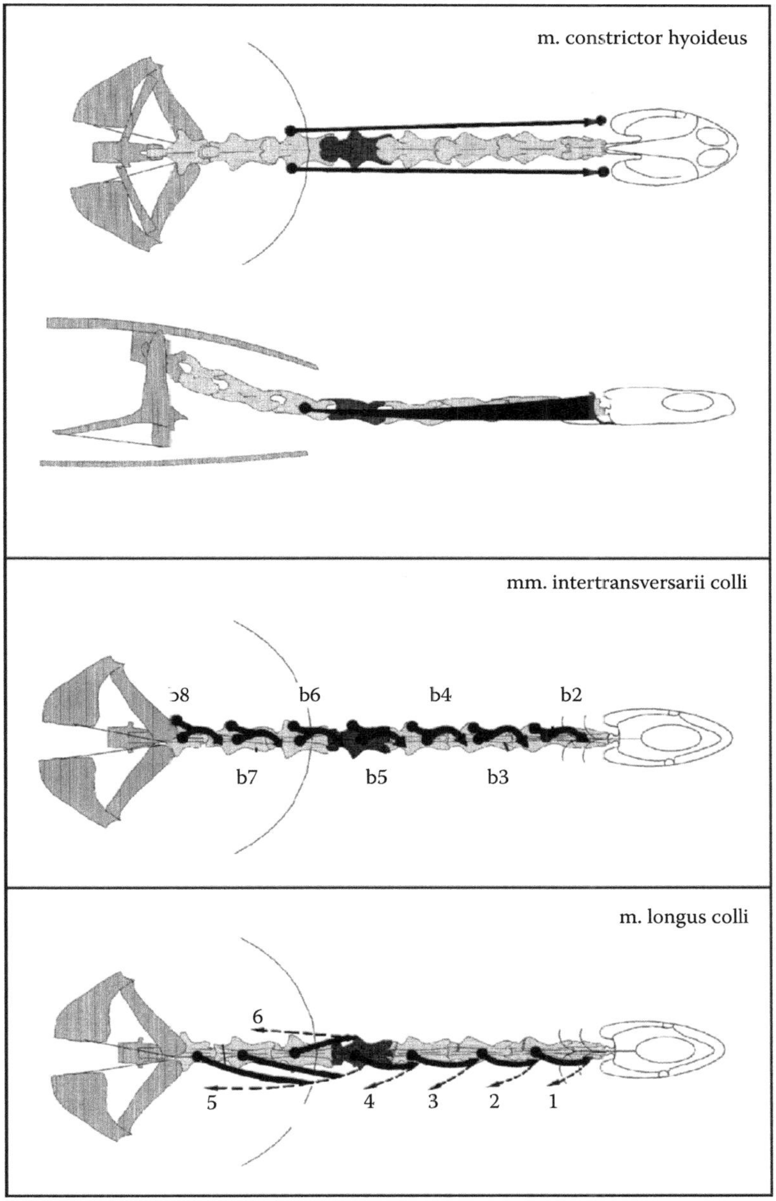

FIGURE 7.10 Schematic representation of the m. constrictor hyoideus, the mm. intertransversarii colli, and the m. longus colli in *C. longicollis*.

originates at D4 and inserts at the lateral aspect of C7. The fourth bundle originates at D3 and inserts at the transverse process of C8.

m. longissimus cervicis (LC, Figure 7.7). This is a segmentally arranged muscle with a more complex arrangement. The LC of C7 originates on the pre-zygapophysis of C8. The muscle belly runs anteriorly and divides into a more medial and a lateral bundle. The lateral bundle inserts on the post-zygapophysis of C6 and the medial on the lateral aspect of C5. The LC 6 originates on the pre-zygapophysis of C7 and inserts on the post-zygapophysis of C5. The LC 5 originates on the pre-zygapophysis of C6 and inserts on the post-zygapophysis of C4. The LC 4 originates on the pre-zygapophysis of C5 and inserts mainly on the neural arch of C1. During the course of the muscle, fibers diverge and insert on the post-zygapophyses of C3 and C2 as well.

m. longissimus thoracis (Figure 7.7). This muscle is responsible for lateral flexion of the posterior part of the neck and consists of one medial and two lateral bundles. All bundles originate at the ventral aspect of the carapace, cranial to the origin of the m. testocapitis. The medial bundle inserts at the lateral aspect of C8. The first lateral bundle inserts by means of a tendon on C6 at the level of the post-zygapophysis. The fibers of the lateral most bundle run initially ventral to the previous bundle but run more laterally at the level of C7. The muscle bundle inserts tendinously at the level of the post-zygapophysis of C5.

mm. semispinalis (Figure 7.8). This is a series of short serially arranged muscles that span adjacent vertebrae and run from C2 to C8. The muscles originate at the neural spine of the more posterior vertebra and run anterolaterally to insert on the post-zygapophysis of the anterior vertebra.

m. testocapitis (Figure 7.8). This is a very long muscle that consists of two distinct muscle bundles that connect the carapace to the head and neck. Both bundles originate at the ventral aspect of the carapace, lateral to the scapula. The fibers of the medial bundle converge onto a strong tendon that inserts at the lateral aspect of C5. The lateral bundle inserts through a complex arrangement of tendons in the anterior part of the neck at the lateral aspect of C4-C2.

m. spinalis cervico-capitis (Figure 7.8). This is superficial dorsal cervical muscle bundle. The muscle originates at the neural spines of C3-C5 and runs obliquely anteriorly. The more medially positioned fibers insert on the lateral aspect of the occipital spine and the more lateral fibers on the dorsal aponeurosis covering the m. adductor mandibulae.

m. rectus cervicis (Figure 7.9). This is a very long and ventrally positioned muscle that connects the pectoral girdle with the ventral aspect of the hyobranchium. The muscle originates at the mid-dorsal aspect of the coracoid. The fibers run ventrally and join the fibers of the m. constrictor hyoideus of the same side. At the level of the hyobranchium, the two muscles separate again and the fibers of the m. rectus cervicis insert partially on the second ceratobranchial and partially on the basibranchium.

m. scalenus complex (Figure 7.9). This is a series of four muscles that connect the pectoral girdle with the lateral aspect of the posterior neck. The most superficial bundle is very thin and runs between the acromion, the dorsal aspect of the plastron, and the posterior most cervical vertebrae (C6-C8). The second and ventral most bundle of the scalenus complex originates on the acromion but distally to the first bundle. The fibers insert at the ventrolateral aspect of C4 and C5. The third bundle originates partly on the scapula and partly on the acromion. The fibers converge on a flat tendon that inserts at the ventrolateral aspect of C5. The fourth part is the shortest and originates across the entire width of the scapula. The muscle inserts on the proximal aspect of the tendinous insertion of the second bundle of the m. retrahens capitis et collique.

m. constrictor hyoideus (Figure 7.10). This muscle originates at the lateral aspect of C6, runs cranially and inserts at the posterior aspect of the first ceratobranchial. Although its primary function is presumably related to the abduction and retraction of the ceratobranchial, it presumably also plays a role during lateral bending of the neck.

mm. intertransversarii colli (Figure 7.10). This is another segmentally arranged muscle that has a dual origin. The lateral bundle originates at the ventral aspect and the medial bundle at the diapophysis of the cervical vertebra. Both bundles join and insert at the posterior aspect of the vertebral centrum of the more cranially positioned vertebra.

m. longus colli (Figure 7.10). This muscle consists of a series of segmentally arranged small muscle bundles that run in close association to the m. retrahens capitis et collique. The longus colli bundles of the first five vertebrae originate at the ventral aspect of the respective vertebrae. Insertion takes place by means of the tendinous insertion of the second bundle of the m. retrahens capitis et collique on the lateral aspect of C4-C1. The muscles in the more posterior part of the neck are somewhat more complex. The m. longus colli 8 originates on the ventral side of C8 and inserts on the tendon of the second bundle of the m. retrahens capitis et collique. The longus colli 7 originates at the ventral aspect of C7 and inserts by means of the insertion of the tendon of the second bundle of the m. retrahens capitis et collique at the lateral aspect of C5. The longus colli 6 originates at the ventral aspect of C6 and inserts at the lateral aspect of C5.

7.3.3.2 *Apalone*

Notable in *Apalone* is the strong reduction of the epaxial musculature (Hofstetter & Gasc, 1969), the absence of the longissimus system, and the absence of the iliocostalis system that is characteristic for all turtles.

m. constrictor colli (m. sphincter colli, Ogushi, 1913). This muscle is poorly developed in *Apalone* and consists of two discrete parts. The anterior part covers the lateral and ventral aspect of the anterior neck region. It originates on the dorsal connective tissue associated with the occipital spine and runs ventrally to insert on the median raphe. The posterior part is often difficult to discern and originates more posteriorly in the neck on the nuchal connective tissue associated with the neural spines and inserts ventrally on the median raphe.

m. rectus cervicis (m. coraco-hyoideus, Ogushi, 1913). This is the ventral-most muscle in the cervical region. The muscle has a dual origin—one part originates on the dorsal aspect of the proximal part of the procoracoid, and the other part originates at the dorsal aspect of the epicoracoid. The muscle runs anteriorly and inserts at the posterior aspect of the basibranchium.

m. carapaco-basioccipitis (Figure 7.11). This is a very long muscle that runs from the pelvic girdle to the cranium. Three distinct origins on the crista mediana ventralis of the first, second, and third caudal vertebrae are present. In addition, three accessory sites of origin can be discerned: on the proximal aspect of the fifth rib, across the entire length of the sixth costal plate, and along the entire seventh rib and associated costal plate. The fibers of this last bundle initially run at an angle of at most 90° to the long axis of the remainder of the muscle.

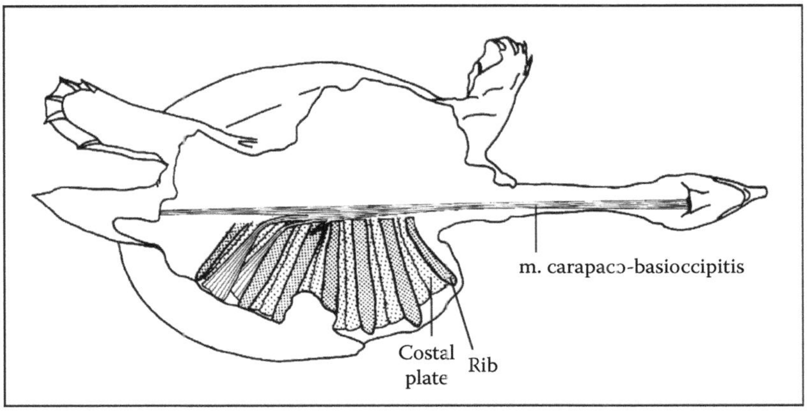

FIGURE 7.11 Schematic representation of the m. caracapo-basioccipitis in *A. spinifera*.

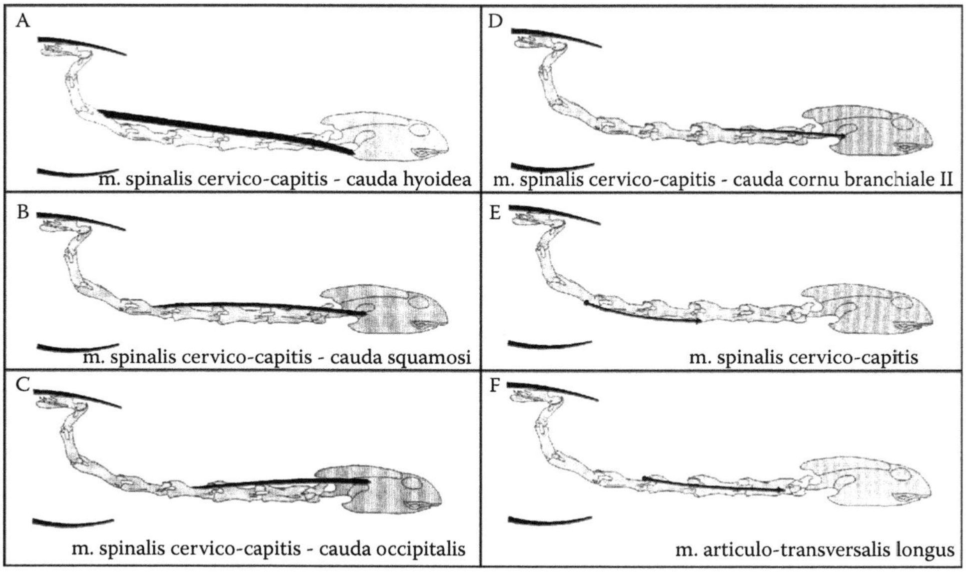

FIGURE 7.12 Schematic representation of the m. spinalis cervico-capitis complex (A–D), the m. cervico-spinalis (E), and the m. articulo-transversalis longus (F) in *A. spinifera.*

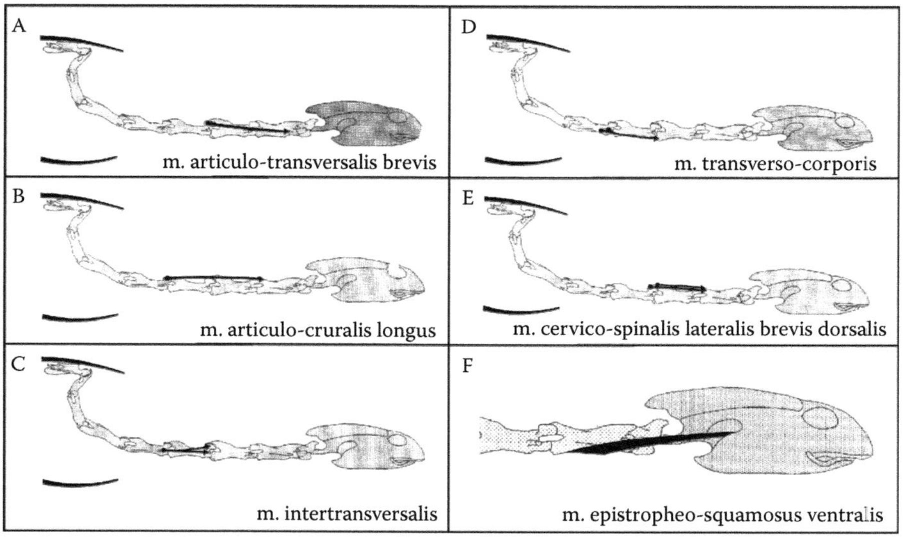

FIGURE 7.13 Schematic representation of the m. articulo-transversalis brevis, the m. articulo-cruralis longus, the m. intertransversalis, the m. transverso-corporis, the m. cervico-spinalis lateralis brevis dorsalis, and the m. epistropheo-squamosus ventralis in *A. spinifera.*

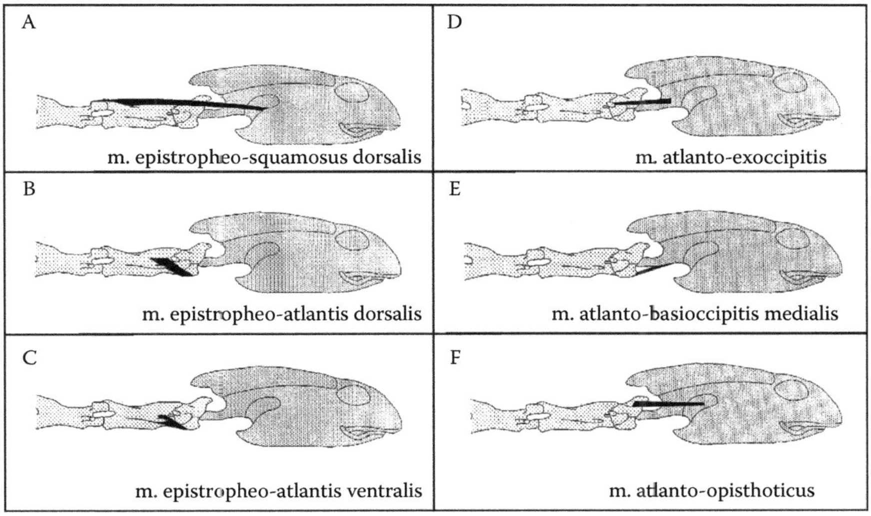

FIGURE 7.14 Schematic representation of the m. epistropheo-squamosus dorsalis, the m. epistropheo-atlantis dorsalis and ventralis, the m. atlanto-exoccipitis, the m. atlanto-basioccipitis medialis, and the m. atlanto-opisthoticus in *A. spinifera.*

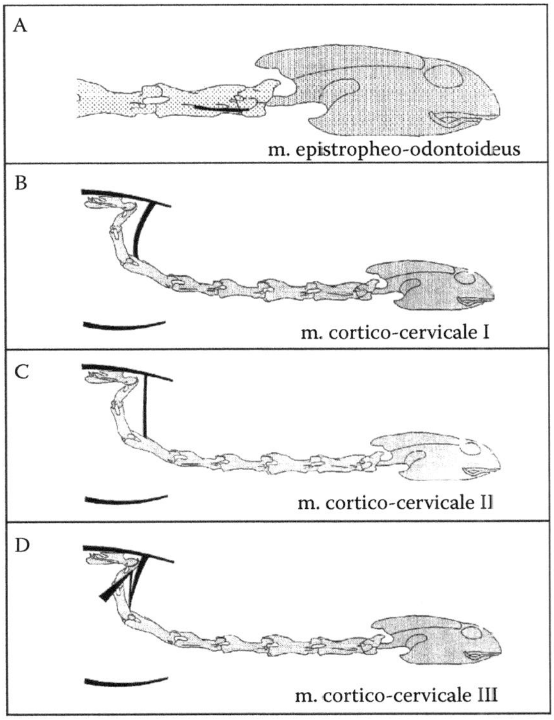

FIGURE 7.15 Schematic representation of the m. epistropheo-odontoideus and the m. cortico-cervicale group in *A. spinifera.*

m. spinalis cervico-capitis (cervico-hyo-capitis, Ogushi, 1913; Figure 7.12). This is a large muscle that consists of four distinct muscle bellies—the cauda hyoidea, the cauda squamosi, the cauda occipitalis, and the cauda cornu branchiale II. The cauda hyoidea (Figure 7.12A) originates at the neural roof of C6, runs anteriorly and merges with the fibers of the m. rectus cervicis. The cauda squamosi (Figure 7.12B) part originates at the neural roof of C5, runs cranially, and inserts by means of a broad aponeurosis at the posterior aspect of the squamosum. The cauda occipitalis (Figure 7.12C) originates at the neural roof of C4, runs cranially, and inserts on the external fascia covering the external adductor. The cauda cornu branchiale II (Figure 7.12D) originates at the neural roof of C3, runs cranially, and inserts at the connective tissue associated with the tip of the second ceratobranchial.

m. cervico-spinalis medialis (Figure 7.12E). This is a segmentally arranged muscle. The muscles originate tendinously at the anterior aspect of the crista mediana ventralis, run anteriorly, and insert tendinously at the posterior aspect of the crista mediana ventralis of the third more cranially positioned vertebra. The muscles originating on the C2-C4 merge and insert jointly at the ventral aspect of the basioccipital by means of a well-developed tendon.

m. cervico-spinalis lateralis longi (Figure 7.12 & Figure 7.13). This consists of three long, superficial muscles that originate at the lateral aspect of the zygapophyses of C5 through C8:

- The m. articulo-transversalis longus (Figure 7.12F) originates on the lateral aspect of the pre-zygapophyses of the last five cranial vertebrae. The muscle runs cranially across the ventral aspect of the two more cranially situated vertebrae and inserts on the transverse process of the following vertebra.
- The m. articulo-transversalis brevis (Figure 7.13A) runs across three vertebrae. It originates at the dorsal aspect of the pre-zygapophyses and inserts jointly with the m. articulo-transversalis longus.
- The m. articulo-cruralis longus (Figure 7.13B) originates tendinously at the dorsal aspect of the pre-zygapophyses, runs cranially, and inserts directly at the dorsal aspect of the post-zygapophyses of the third more cranially positioned vertebra.

mm. cervico-spinalis lateralis breves ventrales (Figure 7.13). This consists of two distinct muscle groups, the more laterally positioned mm. intertransversales (Figure 7.13C) and the more medially positioned mm. transverso-corporis (Figure 7.13D). The former originates tendinously at the transverse process, runs cranially, and inserts directly on the posterolateral aspect of the pre-zygapophysis and the posterior aspect of the transverse process of the more cranially situated vertebra. The m. transverso-corporis originates on the ventral aspect of the transverse process and the vertebral body, runs cranially, and inserts on the condyle of the more cranially situated vertebra.

mm. cervico-spinalis lateralis brevis dorsales (Figure 7.13E). This is a segmentally arranged muscle running from C7 to C3. Each segment consists of two distinct parts. The medial part originates at the lateral aspect of the crista lateralis of the post-zygapophysis and inserts at the dorsal aspect of the base of the post-zygapophyses on the more cranially positioned vertebra. The lateral part originates tendinously on the connective tissue surrounding the zygapophyseal articulation and inserts at the dorsolateral aspect of the post-zygapophysis of the more cranially situated vertebra.

m. epistropheo-squamosus ventralis (Figure 7.13F). The muscle originates aponeurotically at the posterior aspect of the crista mediana ventralis of the axis. The muscle runs anterodorsally to insert at the processus mastoideus (Ogushi, 1913) of the squamosal.

m. epistropheo-squamosus dorsalis (Figure 7.14A). This originates fleshy at the neural arch and the posterior aspect of the neural roof of the axis. The muscle runs anteriorly and inserts at the processus mastoideus of the squamosal and the posterior edge of the opisthoticum.

m. epistropheo-atlantis dorsalis (Figure 7.14B). This originates at the anterior aspect of the neural arch of the axis. The muscle runs anteriorly and inserts tendinously at the processus articularis transversalis ventralis of the atlas.

m. epistropheo-atlantis ventralis (Figure 7.14C). This lies ventral to the previous muscle. It originates fleshy at the lateral aspect of the axis, both dorsal and ventral to the transverse process, and inserts by means of a short aponeurosis at the processus articularis transversalis of the atlas.

m. atlanto-exoccipitis (Figure 7.14D). This originates at the tendinous insertion of the m. rectus lateralis on the atlas and inserts fleshy at the exoccipital.

m. atlanto-basioccipitis medialis (Figure 7.14E). This originates by means of a short tendon at the processus articularis ventralis of the atlas. The muscle runs cranially and diverges toward its origin on the basioccipital.

m. atlanto-opisthoticus (Figure 7.14F). This originates at the dorsolateral aspect of the atlas and inserts at the dorsal surface of the opisthoticum.

m. epistropheo-odontoideus (Figure 7.15A). This originates at the lateral aspect of the atlas, just dorsal to the crista mediana ventralis, and inserts by means of a narrow tendon at the processus odontoideus.

m. cortico-cervicale I (Figure 7.15B). This originates fleshy at the anteromedial aspect of the nuchal plate and inserts musculously at the lateral aspect of the post-zygapophysis of C6.

m. cortico-cervicale II (Figure 7.15C). This originates at the nuchal plate, just posterior to the origin of the m. cortico-cervicale I. The muscle inserts fleshy at the neural roof of C6, just posterior to the insertion of the cauda hyoidea of the m. cervico-hyo-capitis.

m. cortico-cervicale III (Figure 7.15D). This has a dual origin. The lateral part of the muscle originates at the lateral aspect of the ventral side of the nuchal plate by means of a narrow and long muscular head. The medial head originates posterior to the origin of the m. cortico-cervicale II. Both heads merge toward the origin on the neural roof of C7, lateral to the insertion of the m. cervico-spinalis lateralis brevis dorsalis.

m. spinalis dorso-lumbalis. This segmentally arranged muscle is positioned in the canalis collateralis vertebralis (Vallois, 1922). The origin is partly on the inner aspect of the canal and partly on the dorsal aspect of the capitulum of the rib. The muscle leaves the canalis collateralis vertebralis right posterior to the rib associated with D1 and inserts directly at the post-zygapophysis of C8.

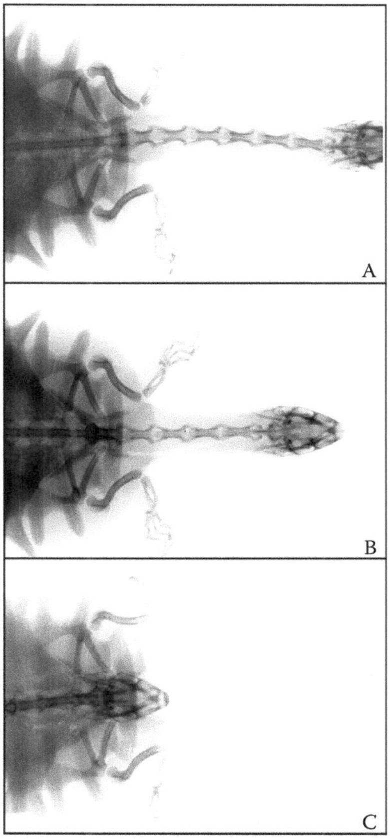

FIGURE 7.16 Static cineradiographs recorded in dorso-ventral view, showing the configuration of the cervical vertebrae in (A) fully extended, (B) relaxed, and (C) fully retracted positions of the neck in *A. spinifera*.

7.3.4 Neck Movements

Neck movements in *Chelodina* are described in detail by Van Damme et al. (1995, 2002), Aerts et al. (2001), Weisgram et al. (1992), and Van Damme and Aerts (1997). Previously unpublished information on neck movements in *Apalone* and *Chelodina* are presented here.

7.3.4.1 Neck Retraction and Cervical Mobility in *Apalone ferox*

In the extended configuration, joints C9-8, C8-7, and C7-6 show the largest initial angles. The more cranially positioned joints are all in the extended configuration. The retraction of the neck in *Apalone* is characterized by relatively small angular changes in the first three (C3-2, C2-1, and S-C1) and the last cervical joint (C9-8). The largest

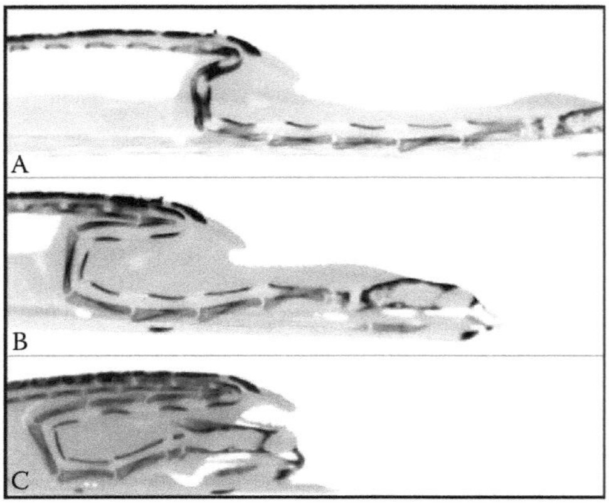

FIGURE 7.17 CT-scan showing the configuration of the neck and the cervical vertebrae of *A. spinifera* in fully extended, relaxed, and fully retracted positions.

angular changes during retraction occur in the middle region of the neck. Joints C8-7 and C7-6 extend during the course of the retraction. The situation for joints C6-5, C5-4, and C4-3 is more complex. C6-5 initially flexes during the retraction but extends again toward the latter part of neck retraction. The two more cranially situated joints start with an initially extended configuration. During retraction, the angle of these joints gradually increases.

The potential range of movement in the dorso-ventral direction is relatively large in most vertebral joints but largest in joints C8-7, C5-4, C4-3, and C3-2. However, joint angles during a simulated snorkeling movement are most pronounced at the most posterior cervical joints (D1-C8, C8-7, and C7-6). The anterior joints are kept relatively constant throughout the movements, as is observed during neck retraction (Figure 7.16 and Figure 7.17). Lateral mobility of the neck in *A. spinifera* is also surprisingly high, with the more proximal joints being the ones allowing most movement in the horizontal plane. In contrast, the last three cervical joints allow almost no lateral movements.

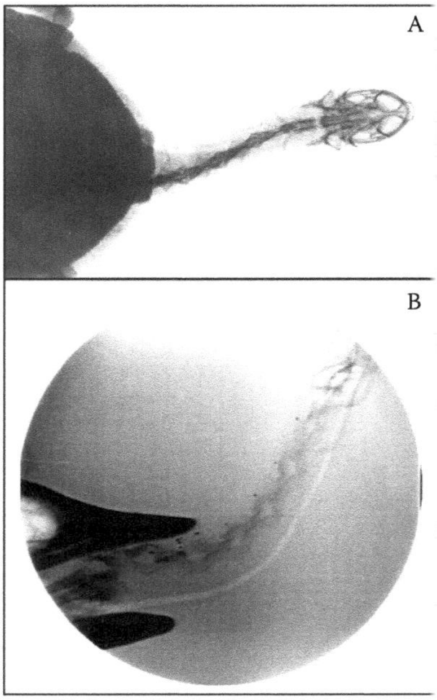

FIGURE 7.18 (A) Static cineradiographs recorded in dorso-ventral view illustrating the position and shape of the cervical vertebrae in *C. longicollis*. (B) Lateral view of the cervical system in *C. longicollis* during snorkeling. Note the radio-opaque markers inserted onto the cervical vertebrae to facilitate the analyses of vertebral movements.

7.3.4.2 Kinematics of Snorkeling in *C. longicollis*

Neck movements during snorkeling in *C. longicollis* are much slower in comparison with those involved during feeding behavior (Aerts et al., 2001; Van Damme et al., 2002) and escape retraction (Van Damme et al., 1995). The total duration of a typical ventroflexion followed by a dorsiflexion is about 6 s. Each component of the movement takes about 3 s. The representative example of a snorkeling movement in *C. longicollis* is represented by successive stick diagrams in Figure 7.19. Starting from an extended configuration in which head and neck are slightly depressed, the animal initially balances its neck around this configuration. This configuration is characterized by very small starting angles in all joints (with exception of D1-C8). Minor changes in joint angles occur during the first phase of the movement mainly in the posterior part of the neck (Figure 7.20). During this phase, no conspicuous changes of the elevation of the head and head position are observed. After approximately 3 s, a conspicuous, rather steady increase in the elevation of the head is observed, mainly as a result of changes in the angle of joints C4-3 to C7-6. In patterns observed in C8-7 and D1-C8 in the most posterior part of the neck and S-C1 and C2-1 in the most anterior part of the neck, changes in vertebral joint angle appear less conspicuously related to the observed movements, show less overall change, and display a generally more irregular movement pattern. At very high degrees of elevation—at which also head position is also starting to change—changes in joint angle are mainly occurring in joints C5-4, C6-7, and especially C4-3 (more than 30° of dorsiflexion). This is not the case for C6-5. This joint reaches up to 20° of dorsiflexion and then levels off. Figure 7.21 illustrates the absolute range of movement (expressed in degrees of flexion) of each joint during the whole sequence. Notice the relatively minor joint rotations in the anterior (S-C1, C2-1, and C3-2) and posterior parts (C8-7, D1-C8) of the neck. As mentioned previously, the most conspicuous changes in joint angle are observed in the middle part of the neck, especially in joints C4-3, C5-4, and C7-6.

In summary, the range of mobility in the vertical plane of the cervical joints is limited, especially in comparison with the changes of joint angle in the horizontal plane. Dorsiflexion of the joints is more conspicuous than ventroflexion. Ventroflexion in the joints is nearly non-existent. The dominant changes in joint angle are observed in the middle part of the neck.

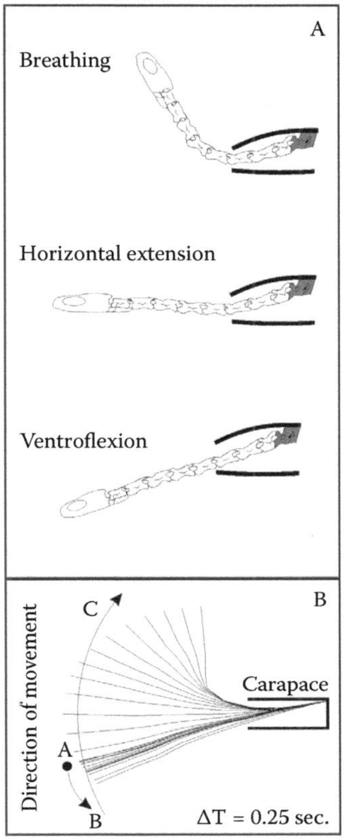

FIGURE 7.19 (A) Schematic illustration of the neck configuration in *C. longicollis*. (B) Stick diagram representing an actual snorkeling movement in *C. longicollis*.

7.4 DISCUSSION

7.4.1 VERTEBRAL STRUCTURE

The nature of retraction of the head-neck system in turtles (in the vertical plane as in *Apalone* or in the horizontal plan as in *Chelodina*) has profound implications on the morphology of the cervical system. Although in both species the cervical vertebrae are markedly elongated, in *Chelodina* the vertebrae are rather narrow and tall. However, in *Apalone* the situation is reversed and the vertebrae are wide and rather shallow. Another marked difference is the presence of a pronounced ventral vertebral crest in *Chelodina*. In *Apalone*, these crests are only present in the most anterior vertebrae. The transverse processes that serve as insertion sites for the muscles of the longissimus are well developed in *Chelodina* and are positioned laterally at the border between the neural roof and the vertebral centrum. Because the longissimus system is absent in *Apalone*, the transverse processes are poorly developed and positioned more cranially. Also, the structure of the posterior zygapophyses is markedly different between the two species. Whereas in *Chelodina* there is only a single basal element carrying the two horizontally oriented articular facets, in *Apalone* two separate post-zygapophyses are present. Additionally, the articular facets are much more curved in *Apalone* compared to those in *Chelodina*. However, in both species the structure of the zygapophyses limits ventroflexion considerably.

7.4.2 CERVICAL MUSCULATURE

The differences in retraction mode in the two groups are also reflected in the structure and differentiation of the cervical musculature. As mentioned previously, for example, the longissimus system appears to be completely absent in *Apalone*. As head-neck movements typically occur in the vertical plane in these animals, this is not unexpected. Alternatively, in *Chelodina* the longissimus system is strongly developed. Notable in the cryptodires is the cortico-cervical muscular system that

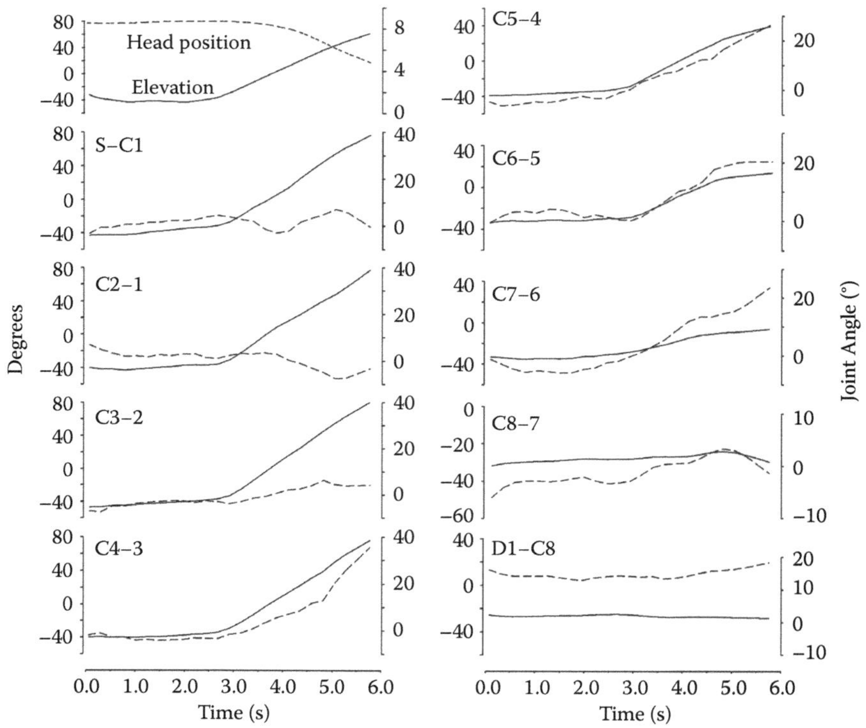

FIGURE 7.20 Kinematic plots illustrating time profiles of joint angle (dashed curves, right vertical axis) and elevation angle with respect to the horizontal of the more distal of the segments constituting the joint (solid curve, left vertical axis) during a snorkeling movement in *C. longicollis*. The first panel represents elevation of the head segment and the rectilinear distance (right axis in cm) between head and carapace.

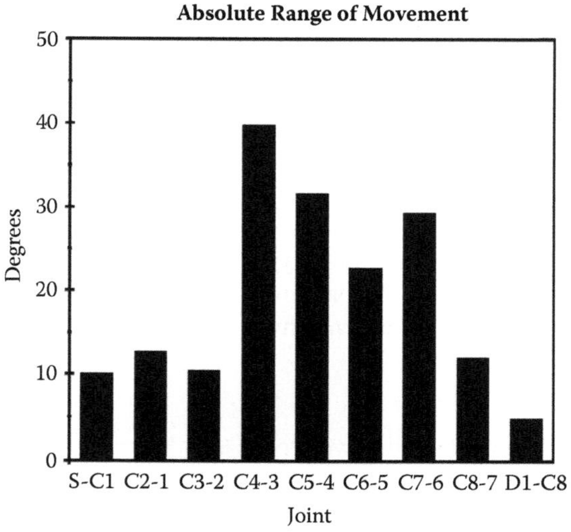

FIGURE 7.21 Movement range of the cervical joints in *C. longicollis* in the vertical plane.

presumably functions to protract and retract the neck. Preliminary electromyographic data for the cryptodire *T. scripta elegans* (Van Damme, unpublished) support this hypothesis. In pleurodires like *Chelodina,* this system is absent. Rather unexpectedly, the m. sphincter colli is poorly developed in *Apalone.* In *Chelodina* and most other pleurodires, this muscle appears much better developed and may actually play a role in aligning the cervical muscle bundles along the cervical column during neck movements. However, this remains speculative and must be tested by electromyography. Also, the dominant head retractor muscle, the m. retrahens capitis et collique in *Chelodina* (Shah, 1963) and the m. carapace-basioccipitis in *Apalone* (Ogushi, 1913), is notably different in the two species. In *Apalone,* this muscle consists of a paired massive muscle bundle running from the first caudal vertebra to the back of the skull. In *Chelodina,* this muscle is more complex and consists of several discrete bundles that insert on specific sites along the cervical column.

However, a striking similarity among the two groups is the extreme length of the head retractor muscle. Presumably, elongation of the retractor muscle allows for more sarcomeres to be placed in series, which may considerably increase the contraction speed of the muscle (Josephson, 1975). Moreover, by increasing the number of sarcomeres in series, each individual sarcomere has to contract less and may thus potentially operate continuously on the plateau of its length-tension relationship. However, for an extremely elongated muscle to contract efficiently, all parts must contract simultaneously. The polyneural, polysegmental innervation of the m. carapaco-basioccipitis in turtles (Guthe, 1981) may allow for this. Unexpectedly, the m. carapace-basioccipitis consists not only of fast twitch fibers but also has a considerable population of tonic fibers (Guthe, 1981), which may help control the position of the head and neck during slow movements or near-stationary behaviors such as snorkeling.

7.4.3 Movement Patterns

The cranio-cervical system in turtles can be characterized as an open kinematic chain of eight vertebrae and nine joints. Because each joint theoretically has three rotational degrees of freedom (and to a limited extent also three translational degrees of freedom), and because the positioning of the vertebrae must be achieved by the combined actions of roughly 50 symmetrical muscle bundles, the control of the position of the head in space obviously presents a challenging control problem (Aerts et al., 2001). However, the control of the system is largely simplified by a number of morphological constraints limiting the mobility at each joint. For example, in *Chelodina* the horizontal nature of the zygapophyseal articular facets and the vertical orientation of the articular facets of the condyle and the vertebral centrum will facilitate movements in the horizontal plane. Additional simplification of the control of the neck in *Chelodina* is achieved by the presence of distinct areas of rotation along the neck (Van Damme et al., 1995). For instance, the joint between cervical vertebrae 7 and 8 and the joints between cervical vertebrae 6-5 and 5-4 typically show the greatest range of angular change. Moreover, the resting position of the neck is characterized by a distinct bend at these locations, facilitating the correct retraction of the neck using only a simple activation of the m. retrahens capitis et collique (Aerts et al., 2001). Dorso-ventral mobility is greatest at the joints C3-2 through C6-5. The joints surrounding the bi-convex articular centrum of C5 show the greatest angular change during dorsiflexion.

The cervical system in *Apalone* is notably divergent and the kinematic patterns are also markedly different from those observed in *Chelodina.* Upon retraction of the neck, the more caudally positioned vertebrae are the first to start their rotation around a transverse axis. The angular changes in all vertebrae are negative during retraction, suggesting that the neck is retracted as a safety-linked bicycle chain. Cervical vertebrae 5, 6, and 7 undergo an angular change of nearly 180°, which implies that these vertebrae end up with their ventral aspect alongside the ventral aspect of the immobile dorsal vertebrae. This is made possible by the nearly complete reduction of the ventral crest of these vertebrae. It is also striking that C8 remains in a near-vertical position at the moment of full protraction. The more anterior vertebrae (C1 to C4) do not contribute to neck retraction at all. The bicycle-chain-like retraction pattern observed for *Apalone* is in contrast to the retraction patterns suggested in other cryptodires such as *Testudo* or *Trachemys* (Scanlon, 1982; Weisgram & Splechtna, 1990). In the latter species, as well as in *Chelodina*, distinct rotation centers are present where the majority of the angular rotation takes place. Thus, the extremely elongated neck and associated morphology of *Apalone* may not be representative for the cryptodiran condition in general but provide an excellent example of a highly specialized cervical system that can be compared to the condition in *Chelodina*. Clearly, further investigation into the morphology and function of the cervical system in turtles is needed to increase our understanding of the evolution of the cervical system and its control in turtles. Especially insightful would be studies exploring cervical structure and function in a broader sample of turtles including short-necked representatives of both cryptodires and pleurodires. Functional approaches including electromyography of the cervical muscles, albeit challenging, are essential to further our understanding of the evolution of the cervical system in turtles.

ACKNOWLEDGMENTS

We thank A. De Schepper (University Hospital Antwerp) for allowing us to use the cineradiography machine and CT scanner; the Antwerp Zoo for providing us with the specimens of *Chelodina longicollis*; G.A. Wood (Wood, 1982) for providing us with a software package for data smoothing and differentiation; and A. Cuylits for making the drawings of the cervical vertebrae in *C. longicollis*. This study was supported by IWONL-grant 910091 (JV) and FKFO-grant 2.9005.90 (FDV). AH is a postdoctoral fellow of the fund for scientific research, Flanders, Belgium (FWO-Vl).

REFERENCES

Aerts, P., Van Damme, J., and Herrel, A., Intrinsic mechanics and control of fast craniocervical movements in aquatic feeding turtles, *Am. Zool.*, 41(6), 129–1310. 2001.

Bojanus, L.H., *Anatome Testudinis Europaeae, Vol. 1.*, Vilnae, Luthuania, republished in 1970, Facsimile reprints in Herpetology, no. 26, Society for the Study of Amphibians and Reptiles, Ohio, 1819.

Bojanus, L.H., *Anatome Testudinis Europaeae, Vol. 2.*, Vilnae, Luthuania, republished in 1970, Facsimile reprints in Herpetology, no. 26, Society for the Study of Amphibians and Reptiles, Ohio, 1821.

Dalrymple, G.H., Intraspecific variation in the cranial feeding mechanism of turtles of the genus *Trionyx*, *J. Herpetol.*, 11, 255–285, 1977.

Deban, S.M., Wake, D.B., and Roth, G., Salamander with a ballistic tongue, *Nature*, 389, 27–28. 1997.

Ernst, C.H., and Barbour, R.W., *Turtles of the World*, Washington, DC: Smithsonian Institution Press, 1989.

Gans, C., Why develop a neck?, in *The Head-Neck Sensory Motor System*, A. Berthoz, P. Vidal, and W. Graf (eds.), New York: Oxford University Press, 1992, 17–21.

George, J.C., and Shah, R.V., The myology of the head and neck of the common Indian pond turtle, *Lissemys punctata granosa* Schoepff, *J. Anim. Morph.*, 1, 1–12, 1954.

George, J.C., and Shah, R.V., The myology of the head and neck of the Indian tortoise, *Testudo elegans*, *J. Anim. Morph. Physiol.*, 2, 1–13, 1955.

Guthe, K.F., Reptilian muscle: Fine structure and physiological parameters, in *Biology of the Reptilia, Vol 1*, C. Gans and T. Parsons (eds.), New York: Academic Press, 1981, 265–354.

Heidweiller, J., Van Der Leeuw, A.H.J., and Zweers, G.A., Cervical kinematics during drinking in developing chickens, *J. Exp. Zool.*, 262, 135–153, 1991.

Herrel, A., Meyers, J.J., Nishikawa, K.C., and Aerts, P., The mechanics of prey prehension in chameleons, *J. Exp. Biol.*, 203, 3255–3263, 2000.

Hofstetter, R., and Gasc, J.P., Vertebrae and ribs of modern reptiles, in *Biology of Reptilia, Vol. 1*, C. Gans and T. Parsons (ed.), New York: Academic Press, 1969, 201–231.

Irschick, D.J., and Garland, T. Jr., Integrating function and ecology in studies of adaptation: Investigations of locomotor capacity as a model system, *Ann. Rev. Ecol. System.*, 32, 367–396, 2001.

Josephson, R.K., Extensive and intensive factors determining the performance of striated muscle, *J. Exp. Biol.*, 194, 135–154, 1975.

King, G., *Reptiles and Herbivory*, London: Chapman & Hall, 1996, 160.

Ogushi, K., Anatomische Studien an der japanischen dreikralligen Lippenschildkröte (*Trionyx japonicus*): Muskel und peripheres nervensystem, *Morph. Jb.*, 46, 299–562, 1913.

Pritchard, P.C.H., Piscivory in turtles, and evolution of the long necked Chelidae, *Symp. Zool. Soc. Lond.*, 52, 87–110, 1984.

Romer, A.S., and Parsons, T.S., *The Vertebrate Body*, Philadelphia: W.B. Saunders, 1977, 161–167.

Scanlon, T.C., Anatomy of the neck of the Western Painted turtle (*Chrysemys picta belli* Gray; Reptilia, Testudinata), unpublished diss., University of Michigan, University Microfilms International, 1982.

Shah, R.V., The neck musculature of a Cryptodire (*Deirochelys*) and a Pleurodire (*Chelodina*) compared, *Bull. Mus. Comp. Zool.*, 129, 343–368, 1963.

Vaillant, M.L., Mémoire sur la disposition des vertebres cervicales chez les chéloniens, *Ann. Sci. Nat. Zool. Paleont.*, Ser. 6, 10, 1–106, 1881.

Vallois, H.V., Les transformations de la musculature de l'épisome chez les vertébrés, *Arch. Morph. Gen. Exp.*, 13, 180–217, 1922.

Van Damme, J., and Aerts, P., Kinematics and functional morphology in aquatic feeding in Australian snake-necked turtles (Pleurodira; *Chelodina*), *J. Morph.*, 233 , 113–125, 1997.

Van Damme, J., Aerts, P., and De Vree, F., Kinematics of the escape head retraction in the common snake-necked turtle, *Chelodina longicollis* (Testudines: Pleurodira: Chelidae), *Belg. J. Zool.*, 125, 215–235, 1995.

Van Damme, J., and Aerts, P., Cervical movements during prey capture in Australian snake necked turtles, genus *Chelodina* (Pleurodira, Chelidae), in *Topics in Functional and Ecological Vertebrate Morphology, A Tribute to Frits De Vree*, P. Aerts, K. D'Août, A. Herrel, and R. Van Damme (eds.), Shaker Publishing BV, 2002, 77–94.

Weisgram, J., and Splechtna, H., Intervertebral mobility in the neck of two turtle species (*Testudo hermanni hermanni*, *Pelomedusa subrufa*), *Zool. Jb. Anat.*, 120, 425–431, 1990.

Weisgram, J., and Splechtna, H., Cervical movements during feeding in *Chelodina novaeguinaeae* (Chelonia, Pleurodira). *Zool. Jb. Anat.*, 122, 331–337, 1992.

Williams, E.E., Variation and selection in the cervical centre articulations of living turtles, *Bull. Amer. Mus. Nat. Hist.*, 94(9), 505–562, 1950.

Wood, G.A., Data smoothing and differentiation procedures in biomechanics, in *Exercise and Sports Sciences Reviews*, 10, 308–362, 1982.

Wren, K., Claussen, D.L., and Kurz, M., The effects of body size and extrinsic mass on the locomotion of the ornate box turtle, *Terrapene ornata*, *J. Herpetol.*, 32, 144–150, 1998.

Zani, P.A., Gottschall, J.S., and Kram, R., Giant tortoises walk without inverted pendulum mechanical-energy exchange, *J. Exp. Biol.*, 208, 1489–1494, 2005.

8 Functional Evolution of Feeding Behavior in Turtles

Vincent Bels, Sabine Baussart, John Davenport, Marc Shorten, Ruth M. O'Riordan, Sabine Renous, and Julia L. Davenport

CONTENTS

8.1 INTRODUCTION

Turtles have one of the most unusual body plans among the amniotes, with a carapace that constrains their behavioral activities differently from all other tetrapods (Gaffney, 1979; Reisz & Laurin, 1991; Laurin & Reisz, 1995; Lee, 1997; Schaffer et al., 1997; Pough et al., 2001; Cebras-Thomas, 2005; Kear & Lee, 2006; Shedlock et al., 2007). Although the origin and evolution of this clade of vertebrates is still unclear despite thorough discussion (Joyce & Gauthier, 2004; Hill, 2005; Nagashima et al., 2005), such a body plan appears to be conservative. However, reduction of the shell has been reported in relation to chelonian life strategies, especially in aquatic turtles (Pritchard, 1979). Evolutionary transformations of the locomotor system in aquatic turtles enable them to display highly adapted mobility and maneuvering, presumably for optimizing searching for food, sexual partners, and areas for egg laying (Davenport & Clough, 1986; Renous & Bels, 1989; Wyneken, 1997; Chapter 5). Although some sea turtle populations nest and feed in the same general areas, others (species or populations within a species) can migrate over great distances. Leatherbacks have probably the longest migration of all sea turtles because they can be found more than 7000 km from their nesting beaches (Hughes et al., 1998).

Despite their "armored tank" shape, turtles have adopted specialized foraging and feeding strategies derived from those of the stem reptiles that have provided enduring advantages while simultaneously imposing basic restrictions. The living species of turtles, distributed among 99 genera and 14 families (Ernst & Barbour, 1989), inhabit a large number of various ecological niches in marine, freshwater, and terrestrial habitats from temperate and tropical regions of all continents except Antarctica, and in all oceans. It is well known that turtles are either specialized or generalized in their diet, with possible ontogenic dietary shifts being common (Pritchard, 1979). Foraging and feeding behaviors have been generally described for almost all families of turtles (Pritchard, 1979). Dietary preferences and ontogenic changes during the life cycles of turtles have been reported in numerous species, particularly in marine turtles. Most chelonians show dietary fluctuations during their lives and can shift from carnivorous to omnivorous and herbivorous diets as they age (Harless & Morloch, 1979; McCauley & Bjorndal, 1999; Bouchard & Bjorndal, 2006). Briefly, true terrestrial turtles (Testudinidae) are the only species able to ingest and swallow food on land, so are subject to the same constraints as all other truly terrestrial vertebrates. Apart from these turtles, a large number of species make forays into terrestrial habitats for feeding and drinking (e.g., Emydidae). Some other species are wholly water-feeders and chase their food by using various strategies from active foraging to sit-and-wait predator behavior (e.g., softshells, Tryonichidae). Some species can capture live food on land but swallow it in water. Several herbivorous species (e.g., Bataguridae) can feed on plant material at the surface of the water or can drag terrestrial vegetation into the water (Davenport et al., 1992). Some species can feed on land or in water, completing the process from food capture to swallowing in either medium (e.g., *Terrapene carolina*, Emydidae). Although basking on land has been reported in a few cases, marine turtles are generally found on land only in two situations: when adult females return to the land for egg laying and when hatchlings must travel from the nest down the beach to the sea. Marine turtles always forage and feed in the water.

A series of phylogenetic studies have demonstrated significant relationships between various characteristics of the trophic system and corresponding ecological specializations in some groups of turtles (Gaffney, 1975, 1979; Claude et al., 2004). From these data, there have been several hypotheses generated to demonstrate a relationship between feeding performances and the phylogenetic hypotheses. For example, the morphological variation of the superfamily Testudinoidea relates to both diet and habitat (Claude et al., 2004). These authors suggest that aspects of the feeding mode (e.g., diet) can be a key factor in determining morphological evolution and diversification of turtle skulls. Phylogeny appears to constrain only localized features of the skull and remains of minor influence.

8.2 OBJECTIVE

There are substantial problems in providing a complete overview of the feeding behavior of turtles using aquatic and terrestrial habitats based only on previously published data: kinematic and functional analyses of complete feeding mechanisms are relatively rare, and feeding behavior has been studied using non-standardized methods and techniques. In his comprehensive volume of evolution of feeding behavior in tetrapods, Schwenk (2000) provides an exhaustive list of literature describing feeding behavior and mechanisms in turtles. The present chapter provides the first comparative analysis of feeding behavior in turtles. Though aquatic feeding has been studied in several very different representative species drawn from fresh water families, few data are available concerning the feeding behavior of marine turtles or of terrestrial turtles. Consequently, this chapter will not only provide information from the literature but will also present data collected by the chapter's authors, together with their analyses of feeding behavior in terrestrial and aquatic species. All of these data are considered from a comparative viewpoint. When necessary, a rapid survey of the methods used for collecting data is provided for each of the feeding modes, terrestrial and aquatic. The mechanism of neck movements that play a key role in feeding in turtles is discussed in Chapter 7.

8.3 FEEDING STRUCTURES

Detailed analyses have been made of the skull, hyobranchium, and tongue in turtles (Lindeman, 2000; Schwenk, 2000; Claude et al., 2004). When needed, we provide morphological data that are necessary to explain the functional and biomechanical aspects of feeding behavior. Some functional analyses demonstrate how structures are used in various phases of feeding in aquatic turtles (Schwenk, 2000; Lemell et al., 2002).

8.4 OVERVIEW OF FEEDING AND DRINKING BEHAVIORS

A rapid survey of the evolution of turtles and their diet show that turtles use two main feeding strategies to gain water, energy, and nutrients: feeding/drinking on land and feeding in water. Drinking water is a particular problem for terrestrial turtles and for some turtles living in saline/marine habitats. All of these turtles need to drink water (when available) to maintain the hydric balance that is essential to welfare and survival. Water intake rates and frequencies are highly variable among turtles. For example, some desert turtles are able to consume a large amount of water in one intake, thereby increasing body mass by up to 40%. Drinking in aquatic turtles has been rarely described. A recent study demonstrated that hatchling sea turtles can osmoregulate effectively and gain mass by drinking sea water (Reina et al., 2002). Although well reported in the literature, drinking behavior has been quantified only in *Malaclemys terrapin* (Davenport & Macedo, 1990; Bels et al., 1995). This species is able to modulate its drinking posture in relation to the water source; it can drink from very thin films of fresh water either on the sea surface or on the surface of intertidal mud, and can even drink directly from falling rain or from water droplets on vegetation (Davenport & Macedo, 1990). *Malaclemys* can also forcibly expel water out of the bucco-pharyngeal cavity when disturbed during the drinking sequence, or when the volume of water within this cavity is too great. *Malaclemys* can expel water very quickly (less than 1 s) by a sudden gape increase accompanied by throat elevation and neck retraction (Bels et al., 1995).

As recognized for all vertebrates, feeding behavior in turtles is a modal-action-pattern (MAD), involving movements of the trophic elements integrated with neck displacements, sometimes associated with complex limb movements, sometimes not. All of these coordinated sensori-motor movements define the feeding strategy of turtles. Any food item consumed by turtles (from plant material to fast-moving mobile prey) must be located and ingested by a rhythmic series of head, jaw, and hyolingual movements. There are three major successive phases in terrestrial turtles (Schwenk, 2000): ingestion (capture), transport and intrabuccal manipulation, and swallowing. A fourth phase called pharyngeal packing occurs at the end of several feeding sequences but does not involve any jaw opening-closing cyclic movement (Figure 8.1). Since the earliest functional studies of vertebrates, a clear difference has been demonstrated between feeding strategies in water and on land (Liem, 1990). Incompressible water can be used to drive a mobile food item (living prey) into the buccal cavity by producing negative pressure inside that cavity. When food is caught, it must be transported and swallowed.

All feeding phases are based on rhythmic coordinated series of movements of the trophic elements (e.g., jaws, tongue, hybranchium) associated with neck movements producing whole displacement of the head (Figure 8.1). These movements are probably the result of rhythmic jaw movements generated by a central neuronal population called the central pattern (or rhythm) generator (CPG) (Schwenk, 2000). All rhythmic jaw opening-closing activities are modulated by sensory feedback generated by various structures within the buccal cavity. For example, taste buds have been described in the buccal and tongue epithelium of several species of turtles (Iwasaki, 1992, 2002; Iwasaki et al., 1996a, 1996b, 1996c, 1996d; Beisser et al., 2001). Although a large number of

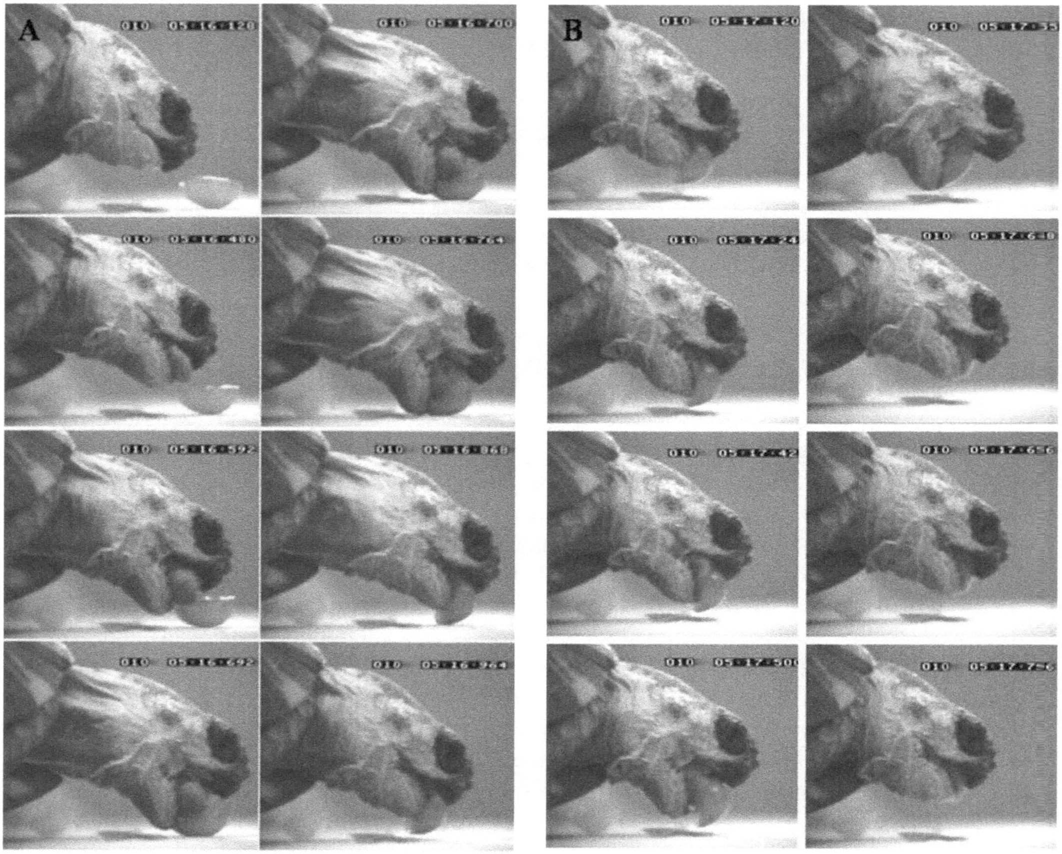

FIGURE 8.1 Series of frames depicting the successive events of a typical feeding sequence in *Geochelone radiata* eating a piece of grape. (A) Ingestion cycle, (B) transport cycle, (C) swallowing or pharyngeal packing, and (D) pharyngeal compression. Food ingestion always involves contact by the tongue prior to surrounding by the jaws for grasping. Food ingestion and transport involve rhythmic cyclic jaw and tongue movements. In transport, the forward movement of the tongue is modulated by the position of the food in the buccal cavity. In swallowing, the mouth is opened and the food moved toward the pharynx. In pharyngeal compression, the jaws always remain closed. The timing of each frame is provided in ms (last three numbers recorded on the frames).

studies provide information on the foraging behavior of turtles, no quantitative analysis has actually demonstrated the effect of food properties (i.e., texture, volume, size) on the modulation of the movements of the trophic system (e.g., jaws, tongue, and hyobranchium). In the majority of studies, turtles have been fed in the laboratory on standardized prey items (e.g., live fish as elusive prey) or plant material (e.g., fruits and lettuce items for terrestrial species, available plants for aquatic species) for herbivorous species.

For aquatic turtles mainly living on the substratum, hunting strategies usually involve immobility while awaiting prey proximity before striking, or moving slowly toward the prey. For example, a very slow (0.4 cms^{-1}) stalking motion is reported in the highly camouflaged pleurodiran *Chelus fimbriatus* that moves toward its prey until the tip of the snout is close to the prey itself (Lemell et al., 2002). Lemell and Weisgram (1997) described the feeding sequence in the aquatic *Pelusios castaneus* while catching prey on the substratum by associating movements of the turtle and the use of the trophic system. Lauder and Prendergast (1992) described a preparatory phase in

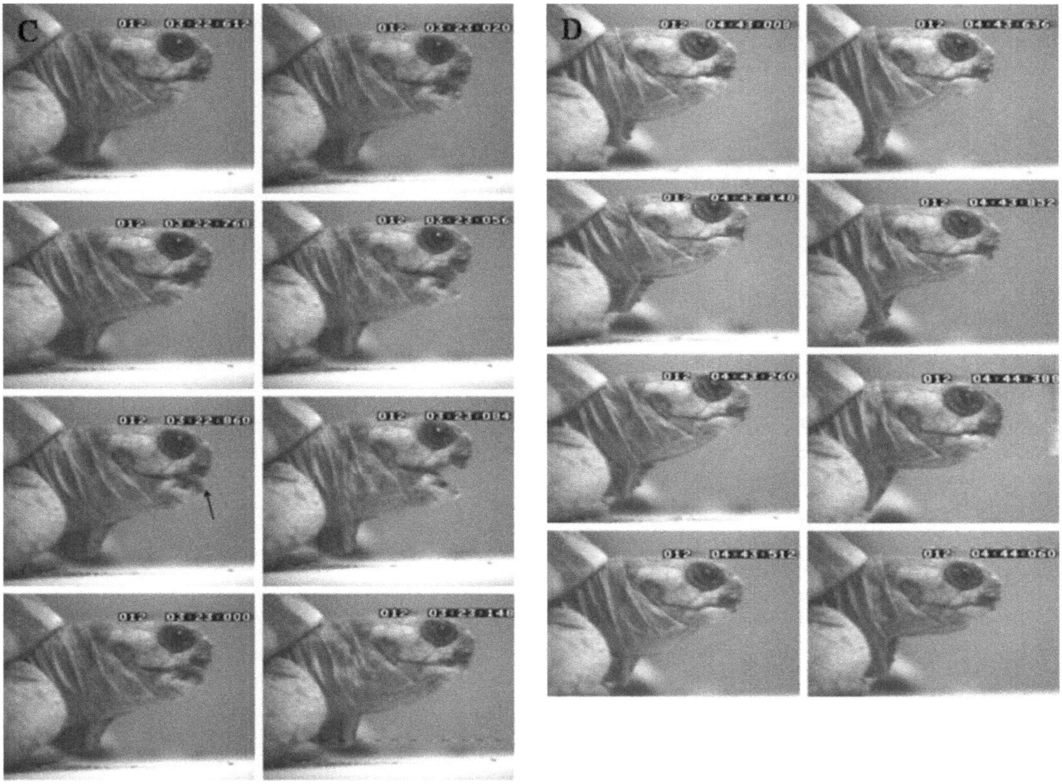

FIGURE 8.1 (continued)

Chelydra serpentina feeding on various prey types that consists of a slow, voluntary stalking movement with little horizontal body and neck movement before a fast opening of the mouth associated with a sudden neck extension (see Chapter 7). Slow stalking movement prior to mouth opening was also reported in *M. terrapin* eating food on the substratum (Bels et al., 1998).

From a functional point of view, it is rather difficult to determine the integration between locomotor and trophic systems for these turtles because in the laboratory, food is often presented at too short a distance from the head to measure the kinematic and hydrodynamic characteristics of the feeding behavior. However, it is clear that the neuromotor integration between both systems plays a particularly key role when turtles are hunting in the water column, or when the prey items show specialized behavior or physical texture (e.g., shells in mollusks and crustaceans). Many aquatic turtles must associate complex limb movements with neck movements and rhythmic movements of the trophic system to support the complex maneuvering ability that improves food capture success (Figure 8.2). Davenport and Clough (1985) report an apparently unique feeding behavior that involved the use of the fore flippers to gain food in the marine turtle *Caretta caretta*. These authors observed the role of "pseudoclaws" (sharp-pointed scales) on the proximal portions of the forelimbs of young specimens to allow the combined use of foreflippers and beak to tear food items that are otherwise too large to be swallowed whole or readily bitten into chunks (thus mimicking the feeding behavior of freshwater turtles such as emydids that use true claws and the beak to achieve a similar goal) and allowing food morsels adhering to the pseudoclaws to be ingested by the turtle.

Feeding behavior in the brackish-water emydid *M. terrapin* is the best example currently available in the literature to demonstrate plasticity in neuromotor integration of limb, neck, and trophic systems during prey capture (Bels et al., 1998). Capturing immobile prey (e.g., mussels) on

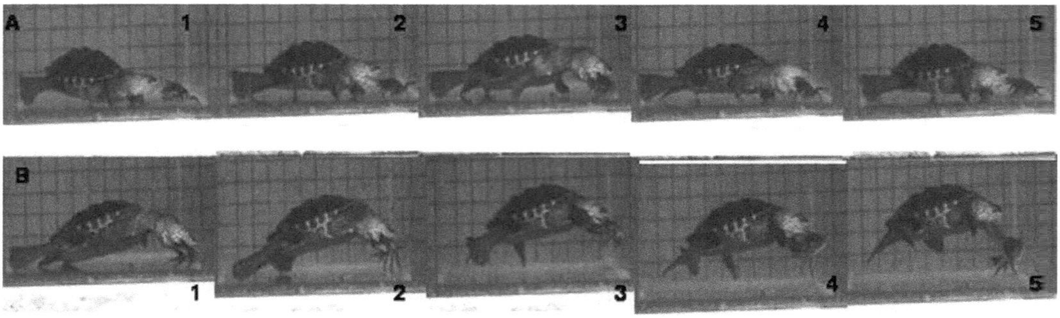

FIGURE 8.2 Two typical feeding sequences of *Malaclemys* attacking a defending crab. These sequences show the complex interaction between the turtle and the living prey. However, images show the stereotyped pattern of neck extension, gape cycle, and throat depression for both attempts to ingest parts of the crab. The arrow shows the jumping behavior of the crab attacking the turtle that defends by biting (corresponding to ingestion cycle).

the substratum is mainly achieved by coordinated neck extension and jaw opening with relatively small inertial suction (Van Damme & Aerts, 1997; see following for discussion). Capturing mobile and aggressively defensive prey in the water column (e.g., living crabs) requires coordination of neck extension and jaw opening with the limb cycles used for swimming. In all cases, the neck is extended to place the opening jaws close to the prey during limb retraction, to increase the thrust of the turtle toward the prey (Figure 8.2). Another sort of coordination of limb, neck, and jaws has also been described. Bels and Renous (1992) and Bels et al. (1998) demonstrated that feeding on soft, slowly swimming prey (jellyfish) by young leatherback turtles always involves placement of the long foreflippers alongside the body during the catching cycles. They suggest that this strategy strongly decreases the possibility of limb movements displacing jellyfish away from the buccal cavity. In terrestrial habitats, turtles often approach food slowly and stop all movements before capturing the prey or biting the plant material.

The use of the limbs in all other feeding phases remains relatively poorly understood. For turtles feeding in the water column (e.g., Emydidae), the limbs play a major rule in optimizing food transport (Bels et al., 1998); typically, the forelimb claws are used to tear large food items into smaller pieces. However, for turtles that feed in terrestrial habitats, the limbs only play an occasional role by stabilizing the position of a food item on the substratum or aiding the seizure of "difficult" hard food items between the jaws (e.g., Testudinidae eating living prey such as snails). All of these examples show that functional studies must be ecologically relevant to yield a full understanding of turtle feeding strategies and their plasticity that plays a key role in the overall fitness of the organism.

8.5 KINEMATICS

8.5.1 Terrestrial Feeding

8.5.1.1 Ingestion

Two modes of food prehension have been reported for terrestrial turtles: lingual prehension and jaw prehension (Figure 8.1 & Figure 8.3). A clear difference in ingestion cycles has been reported for *Terrapene* sp. feeding on insects and *Geochelone* sp. feeding on plant material. *T. carolina* and *T. ornata* never use lingual prehension but only jaw prehension when catching insects (e.g., Bels et al., 1997). *Geochelone* sp. and *Testudo* sp. always use lingual prehension for feeding on plant material. For mobile prey (mealworms, as used in *Terrapene* studies), *Kinixys* also uses lingual prehension. However, no quantitative comparative analyses of the modulatory effects of food items

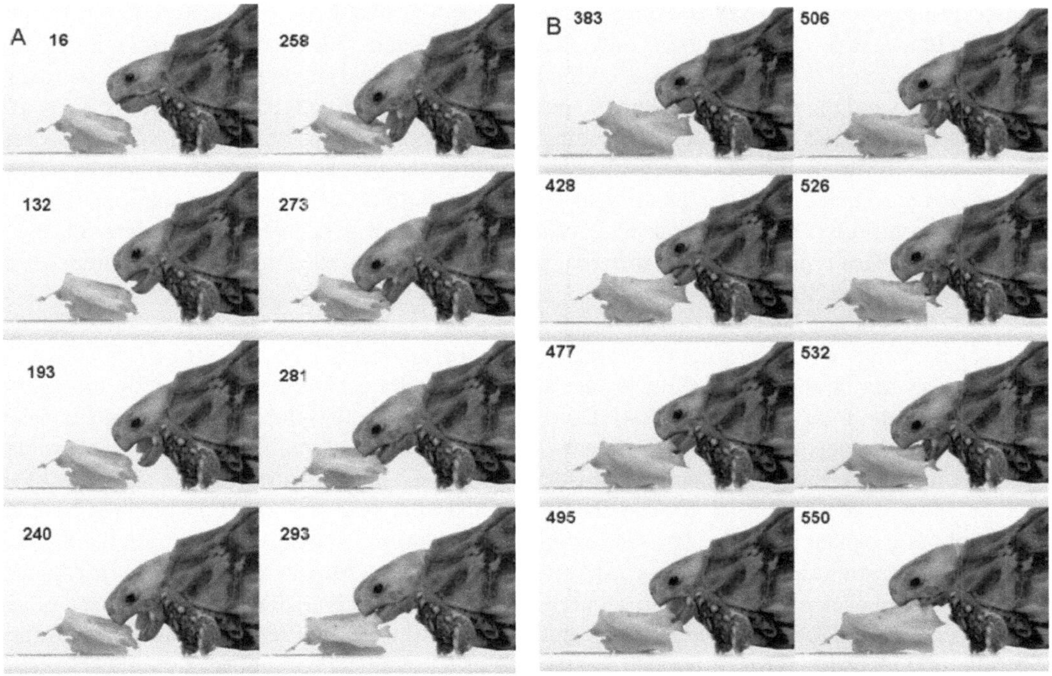

FIGURE 8.3 Series of frames depicting the successive events of a typical feeding sequence in *Kinixys belliana* eating a piece of lettuce: (A) ingestion cycle and (B) transport cycle. Food prehension does not involve contact by the tongue outside the buccal cavity. Food transport involves rhythmic cyclic jaw and tongue movement. Contact between tongue and food occurs within the buccal cavity. In transport, the forward movement of the tongue that plays the key role is modulated by the position of the food in the buccal cavity. The timing of each frame is provided in ms (last three numbers recorded on the frames).

on kinematic and motor variables have yet been reported in these representative tortoises of Testudinidae. In cases of living prey (e.g., insects), only one ingestion cycle is used to catch the food and drive the food item into the front of the buccal cavity. In *Terrapene* sp. studied with this food item, the jaw moves quickly around the prey and the tongue is never used for prehension of insects (Bels et al., 1997). For plant material, the ingestion (capture) cycle is more complex and depends on the size of the food item because the feeding bout is a continuous process that does not terminate at one particular phase. Two examples demonstrate this complexity. For relatively large food items such as grapes, *G. radiata* contacts the food with the tongue and retracts the food item within the buccal cavity while the jaws are moving around the food. The turtle usually uses one to three ingestion cycles to bring the food material successfully into its buccal cavity. Similar behavior has been reported for the herbivorous lizard *Uromastix* fed on endive (Herrel et al., 2002). The *G. radiata* head is then placed in an approximately horizontal position during the immediately subsequent cycles, allowing transport and intrabuccal manipulation; for "long" food items such as lettuce leaves that are drawn into the mouth by rhythmic jaw cycles, the food items are caught by the tongue and the jaws move the piece of food around several times before the food partly or completely enters the buccal cavity. It is rather difficult to separate food ingestion and transport because at each tongue cycle (Figure 8.1 and Figure 8.3), the food within the buccal cavity is moved by the mid and posterior portions of the tongue while the foretongue moves to adhere to the parts of the food item that are still outside of the buccal cavity. When a piece of food is cut off, the turtle can stop to ingest the rest of the food item, which continues to be acted upon only by transport cycles. Such a continuous process of ingestion-transport is strongly influenced by the size of the food material. When a short piece of lettuce is provided to turtles, only one or two ingestion cycles are used to deal with the food. The effect of the size of the food item on the continuous process involving the foretongue to catch the food item and the mid- and hindtongue portions for transporting the food has been clearly demonstrated in all herbivorous mammals feeding on long pieces of hay (Bels, 2006), so this process is not unusual in vertebrates. As soon as the food is within the buccal cavity of *G. radiata*, a series of integrated head (neck), jaw, and hyolingual cycles termed transport-intrabuccal manipulation move the food posteriorly toward the pharynx. At the end of a complete sequence, cycles determining the swallowing phase typically show a different gape profile as has been noted in herbivorous lizards (Herrel & De Vree, 1999).

In both modes of food prehension, turtles always use combined movement of the jaws, the head, and the neck that may or may not be associated with displacements of the hyolingual apparatus. All ingestion cycles in all terrestrial turtles are accompanied by neck extension, moving the opening jaws toward the food item (Figure 8.4). During the next extension, the head is positioned relative to the food item depending on the properties of the food and its proximal cues (i.e., size, volume, texture, mobility), showing a complex sensory-motor mechanism controlling the initiation of this feeding phase. At each ingestion cycle, the turtle rotates the head in various positions to be able to take hold of the plant material. The sensory-motor mechanisms underlying such functional plasticity during ingestion remain understudied. In terrestrial turtles, the question of positioning the head together with a neck extension remains an open question. Probably differences in the speed of head movements toward the food are key parameters that permit these reptiles to modulate head rotation during food ingestion. This difference can be seen as a functional consequence of diet selection in terrestrial turtles.

Two main types of ingestion cycles have been reported: jaw prehension and lingual prehension. In both modes of food prehension, turtles always use combined movements of the jaws, the head, and the neck, associated or not with displacements of the hyolingual apparatus. In all prehension, the mouth is opened in slow (SO) and fast (FO) stages (Figure 8.4). However, the duration of SO stage is variable. In *T. carolina* and *T. ornata*, a clear distinction between SO and FO stage was not always easy to determine because the gape increases rather regularly, and often the mean duration of FO stage is longer than that of the fast closing (FC) stage. The tongue always moves forward within the buccal cavity during gape increases and retracts during the beginning of the FO stage. Upon protrusion, the tongue bulges and the tongue tip curves ventrally, so that the foretongue makes contact with the food item. During the SO phase, the tongue is protruded from the mouth and contacts the food. Next, the tongue continues to be extended and pushed against the food item. The food is then maintained in contact with the tongue because of strong mucous adhesion between the tongue surface and the food. Once the tongue has been retracted within the jaw margins, the jaws are closed quickly, and the slow closing/power stroke phase started. The mean duration of the FO stage is longer than that of the FC stage. During the decrease in gape size, the lower jaw arrives under the beak formed by the upper jaws as in transport cycles. The tongue moves forward while the gape increases and contacts the food before maximal gape is achieved. Retraction of the tongue occurs during the beginning of the FO stage. After closing the jaws on the prey, the head retracts and rotates in reverse. Timing of kinematic events of *G. elephantopus* (lingual prehension) and *T. carolina* (jaw prehension) is similar in these different species (Figure 8.4). For example, peak hyoid retraction occurs just after peak gape angle and maximum throat depression occurs 30 to 100 ms after maximum gape angle. The successive six events in both turtles can be summarized as follows:

- Neck extension to drive the head toward the food
- Opening of the mouth (SO and FO stages)
- Retraction of the tongue during the beginning of the FO stage
- Depression of the throat immediately following the tongue retraction

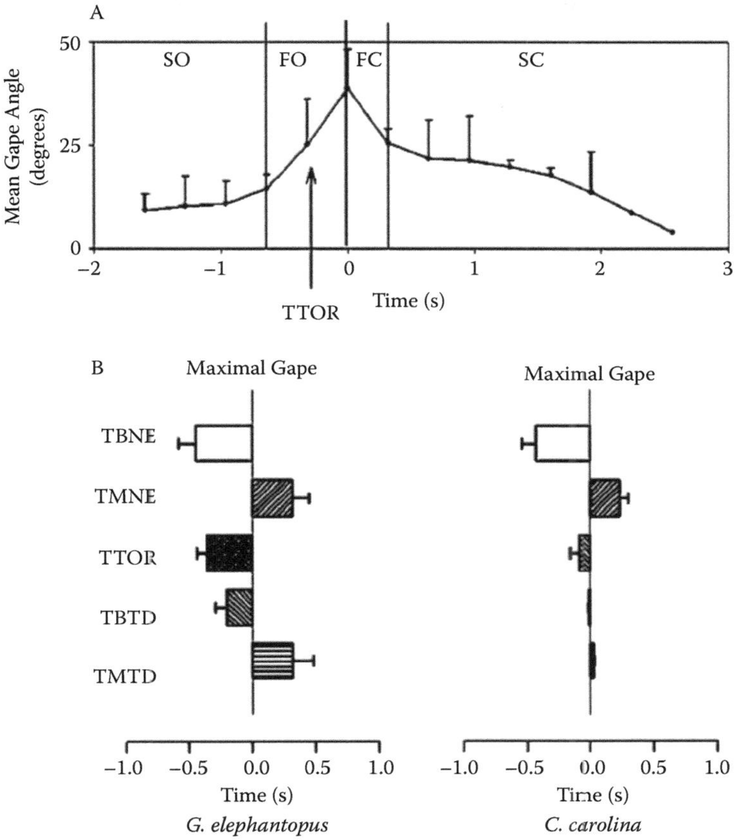

FIGURE 8.4 (A) Mean gape profile (± standard deviation) from 10 lateral ingestion cycles in one specimen of *Geochelone elephantopus*. The arrow indicates the mean time of tongue retraction. (B) Comparison between timing events in *G. elephantopus* and *T. carolina* (data from Bels et al., 1997). Timings were calculated from the time to peak gape angle (time 0). Gape is divided into slow opening (SO), fast opening (FO), fast closing (FC), and slow closing (SC). TBNE: time to beginning of the neck extension, TMNE: time to maximal neck extension during ingestion cycle, TTOR: time to tongue retraction, TBTD: time to beginning of throat depression, TMTD: time to maximum throat depression.

- Closing of the mouth on the food
- Retraction of the head toward the carapace.

8.5.1.2 Intra-oral Transport Cycle

As soon as the food (e.g., living prey, plant material) enters the buccal cavity, transport of the food toward the pharynx begins. For example, after ingestion of a mealworm, the prey is not completely held within the buccal cavity and is reduced (i.e., partially crushed) by the jaw margins at each transport cycle (Figure 8.3 and Figure 8.5). Because the mealworm enters the buccal cavity progressively, successive portions of the mealworm are reduced at the closing stage of each transport cycle.

The slow opening of the mouth during transport is composed of the SO I (opening) and often SO II (stationary stage) stages. The FO stage corresponds to a rapid increase in the vertical displacement of the jaws (Figure 8.5). The FO stage begins after a slight decrease in gape angle at the end

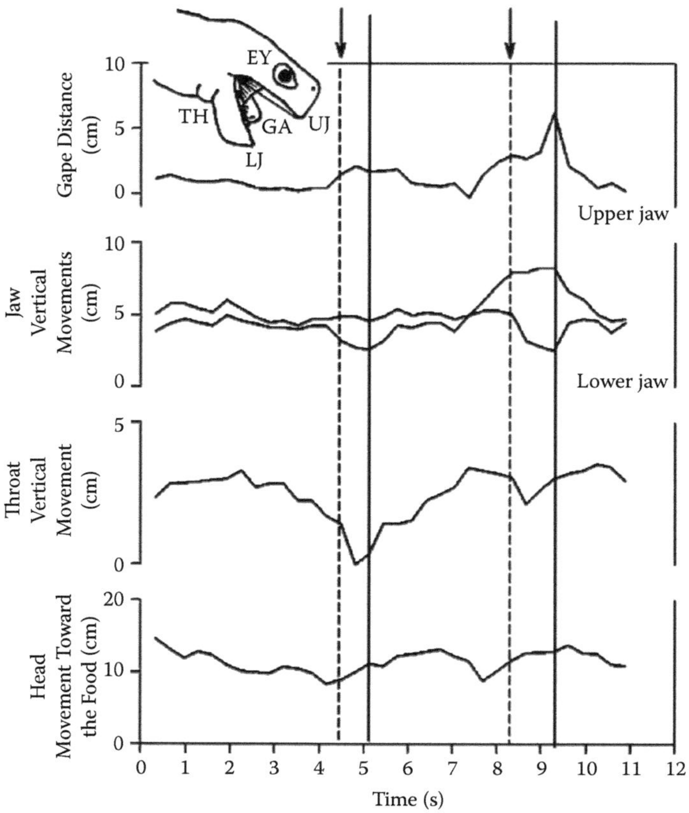

FIGURE 8.5 Kinematic profiles of two successive intraoral food transport cycles in *G. elephantopus.* Throat vertical movement (TH) associated with tongue retraction shows the cyclic depression and elevation of the throat. Head movement toward the food was measured by the horizontal displacement of the eye (EY). This movement shows the cyclic displacement of the jaws toward the food associated with gape angle (GA) produced by combination of vertical movement of the upper (UJ) and lower (LJ) jaws. The arrows indicate the first frame of tongue retraction with the food adhering to the tongue surface and the line indicates peak gape angle for each of the gape cycles.

of SO stage. The closure of the mouth is divided into fast closing (FC) and slow closing (SC) stages. The FC stage corresponds to a rapid decrease in the vertical displacement of the jaws. The SC stage corresponds to the elevation of the lower jaw under the beak of the upper jaw. The horizontal distance between the tip of the upper jaw and lower jaw decreases during the FC stage because vertical displacement of the lower jaw produces an arc, and the upper jaw moves little. The SC stage is not reflected in recordable gape angle and gape vertical distance changes because the lower jaw moves under the edges of the upper jaw. However, during this stage the horizontal distance between the tip of the upper jaw and the most anterior point of the lower jaw increases just after FC stage in the gape cycle, indicating that the lower jaw continues to close against the upper jaw. When gape increases during the SO stage of the next cycle, this horizontal distance decreases again.

In the transport cycle, during the SO stage the tongue is protracted as illustrated by a decrease of the horizontal distance between the tips of the tongue and the lower jaw (Figure 8.3B and Figure 8.5). At the same time, the tongue is slightly elevated. In the transport cycle, the forward displacement of the tongue into the buccal cavity is strongly correlated with the vertical displacement of the lower jaw during the SO stage. When the tongue moves forward, the lower jaw depresses

(SO I), and when the tongue is stationary prior to moving backward during the FO stage, the lower jaw does not move and the gape angle remains stationary (SO II). The head extends during the FO stage and the beginning of the FC stage, and depresses during the SO, FO, and the beginning of FC stages. During half of the duration of the SO stage, the head rotates downward as showed by the vertical displacement of the head. Preliminary analysis of food transport by using x-ray video (250 Hz) in a typical plant feeding turtle, *K. belliana*, shows that the food is accumulated within the pharynx and the rear part of the buccal cavity (Figure 8.6). Subsequently, swallowing of the alimentary bolus begins.

8.5.1.3 Swallowing

According to Schwenk (2000), we can divide swallowing in terrestrial turtles into pharyngeal packing (opening of mouth) and pharyngeal compression (no gape cycle). Figure 8.3 and Figure 8.6 show typical swallowing cycle in two *T. carolina* (Figure 8.3) and *K. belliana* (Figure 8.6) feeding on different food items (mealworms in *T. carolina* and lettuce leaves in *K. belliana*). Probably based on the sensory-motor feedbacks from properties of the alimentary bolus (i.e., volume, size, toughness), the turtle either enters into a new feeding sequence that drives more food into the pharynx,

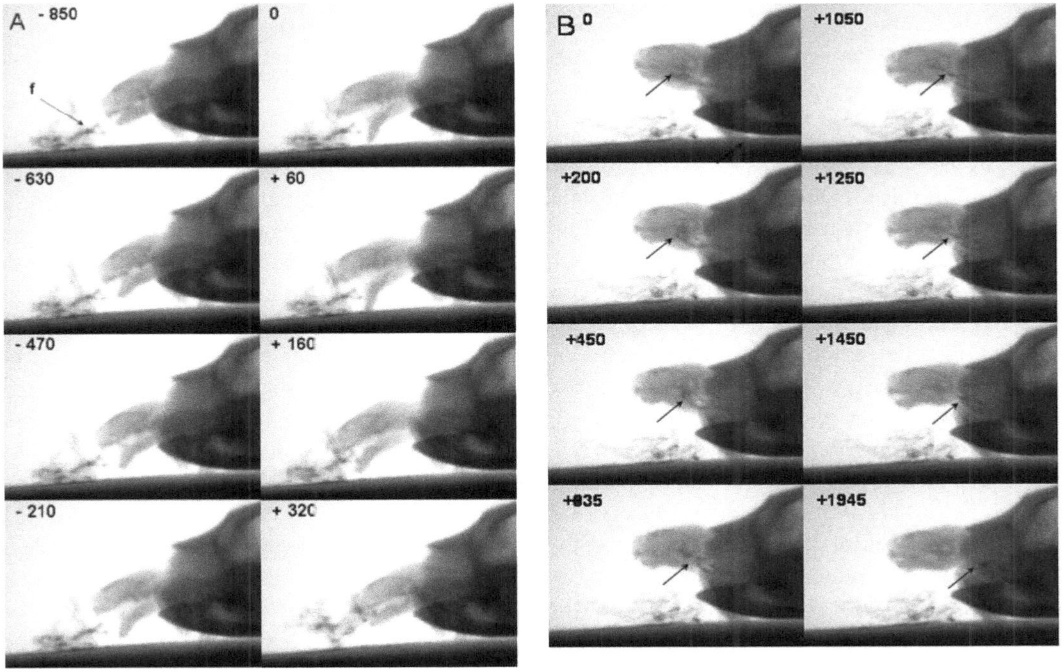

FIGURE 8.6 (A) Typical ingestion cycle of a piece of lettuce in *K. belliana* filmed by x-ray videofluoroscopy (100 Hz), showing the tongue contacting the food whereas the jaws moved to surround the food material (leaf of lettuce). The contact between the tongue and the food occurs within the buccal cavity at time 0. (B) Typical swallowing cycle filmed by x-ray fluoroscopy (200 Hz) involving mouth opening to show the combined movement of the food associated with the swallowing cycle, demonstrating the key role of the tongue and throat in the posterior displacement of the food toward the esophagus. Peristaltic activity of the esophagus helping movement of the food within the esophagus is also shown. The displacement of the food along pharyngeal packing is shown by the arrow on each frame. Time 0 corresponds for this cycle to the first frame of mouth opening. f: markers (the lettuce was covered with a thin film of water charged with barium powder).

or moves the bolus into the esophagus, which structure drives the bolus directly to the stomach by peristaltic movement. For emptying the pharynx, the turtles produce either jaw cycles or pharyngeal compression (Figure 8.6).

8.5.2 Aquatic Feeding

Feeding in aquatic turtles has received more attention than that of terrestrial turtles, although relatively few species have been studied. However, robust experimental conditions with more-or-less similar environmental constraints have been employed. Since the first quantitative analysis of feeding mechanism in *Chelydra* (Lauder & Prendergast, 1992), turtles have been used for testing hypotheses and generalizations regarding morphological and functional patterns associated with aquatic feeding in lower vertebrates.

The large number of species and foraging strategies in aquatic-feeding turtles probably reflects a rather unique group of vertebrates that has evolved aquatic feeding convergently with anamniote feeding systems and with amniotes that have returned secondarily to aquatic environment such as several mammals. Many food resources can be exploited in aquatic habitats. Plant material constitutes a first source of food. This material can be firmly attached to the substrate and must be bitten and extracted or pulled out before being transported and swallowed. Other plant material can be floating and must be bitten and caught from beneath (Davenport et al., 1992). Living prey can be exploited by rapid strikes if they have elusive abilities (e.g., fishes) or must be approached slowly to avoid their movement away of the buccal cavity (e.g., jellyfish). Since the earliest analyses of feeding mechanisms in water, the question of importance of suction has remained paramount (Aerts et al., 2001; Figure 8.7).

8.5.2.1 Ingestion Cycle

Aquatic turtles all show much the same pattern of head, jaws, and hyobranchium movements. To our knowledge, the tongue is not used for capturing aquatic prey. Typical capture cycles for some aquatic turtles (*C. longicollis*, *C. fimbriatus*, *C. serpentina*) involve a sudden forward thrust of the head by neck extension (see Chapter 7). In contrast, other species strike less rapidly (mean duration between 250 and 300 ms in *M. terrapin*, Bels et al., 1998; 400 ms in *T. carolina*, Summers et al., 1998; and up to 800 ms in *D. coriacea*, this chapter), opening the mouth either by using SO and FO stages or regularly without the SO stage. All kinematic profiles available in the literature provide classical examples of neck extension producing a sudden thrust of the opening mouth associated with large depression of the hyobranchium. Depression of the hyobranchium occurs 30 to 50 ms after maximum gape (Lauder & Prendergast, 1992; Bels & Renous, 1992; Van Damme & Aerts, 1997; Bels et al., 1998; Summers et al., 1998; Lemell et al., 2002).

The problem of approaching the food in aquatic environment always remains the same: any flow of water produced by the turtle must be compensated for in efficient capture. Lauder and Prendergast (1992) were the first to stress the hydrodynamic constraints on aquatic prey capture resulting in kinematic similarities among predators such as fish, amphibians, and turtles. Their findings were supported by other analyses of ingestion in turtles from freshwater and marine habitats. All of the aquatic turtles can generate a backward water flow relative to the buccal cavity (called suction) to acquire the food (Van Damme & Aerts, 1997). Van Damme and Aerts (1997) discussed in depth the question of compensatory and inertial suction in prey capture by aquatic turtles. The term "compensatory suction" indicates suction used to maintain food immobile during the strike, and permits engulfing of the food as the jaw apparatus surrounds it. The term "inertial suction" is used to describe suction in which food and water move toward the buccal cavity whereas the head of the predator remains essentially immobile (Van Damme & Aerts, 1997; Aerts et al., 2001). Aerts et al. (2001) state that probably food capture in various aquatic turtles is related to a combination of both modes of suction, as demonstrated by displacements of various food items recorded in different aquatic turtles. For example, inertial suction plays a major role in *Chelodina longicollis*

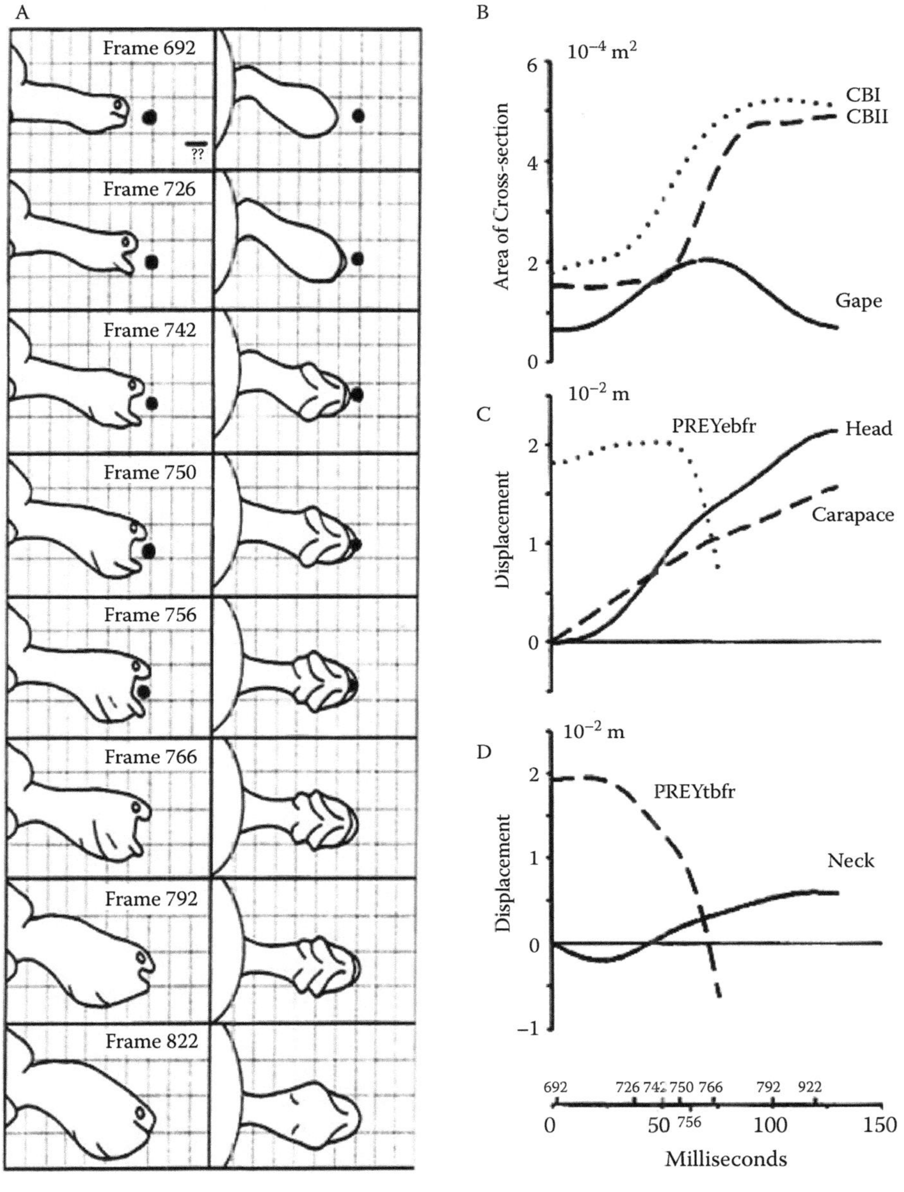

FIGURE 8.7 (A) Drawings of lateral and ventral views of the strike used for illustrating the kinematic profiles. (B–D) Kinematic profiles of one representative strike in *Chelodina longicollis* to show (B) the modification of areas of cross sections through three successive levels on the head of the turtle (Van Damme & Aerts, 1997): the mouth (gape) and at the level of ceratobranchial I (CB I) and II (CB II), the displacement, of the prey, the head, and the carapace of the turtle in the earth bound frame of reference (ebfr = fixed frame of reference). Time scale is given in ms and numbers of the upper time scale correspond to the number of the sequences presented at the left of the figure. (Modified from Van Damme & Aerts, 1997.)

but compensatory suction is dominant in *C. serpentina* (Van Damme & Aerts, 1997). Van Damme and Aerts (1997) and Summers et al. (1998) do not agree with the use of the ram/suction index (RSI) proposed by Norton and Brainerd (1993) to describe the feeding mechanisms of fishes. In their recent study, Lemell et al. (2002) applied this index to capture of by *C. fimbriatus*. The equation for RSI is

$$RSI = \frac{D_{predator} - D_{prey}}{D_{predator} + D_{prey}}$$

where $D_{predator}$ and D_{prey} are the net distances moved by the predator and the prey between the moment the mouth first begins to open and the moment the prey disappears or is seized by the jaws). RSI ranges from +1, indicating a pure ram strike in which only the predator moves, to −1, indicating a pure suction strike in which only the prey moves. The calculated RSI during prey (fish) capture in *P. castaneus* varies between 0.36 and 0.55, and in *C. fimbriatus* was always positive (mean = 0.36 ± 0.23, range = 0.071 to 0.664; prey = fish; $N = 20$). However, these authors think that the equation used for RSI tends to overestimate the ram component.

Whatever the terms used for describing food capture (e.g., suction, RSI), and whatever the hydrodynamic calculations or other considerations involved, all aquatic turtles must expand the oropharyngeal cavity for successful food capture. Summers et al. (1998) demonstrate clearly that prey capture in water involves greater hyoid depression in *T. carolina* feeding in water compared with feeding on land. These authors and Bels et al. (1998) show that hyoid depression is modulated not only from land to water but also in water feeding in relation to food type and behavior (e.g., crabs capable of defensive behavior). It is evident that this expansion plays a key role in the successful capture of elusive prey by any mode of prey displacement toward the buccal cavity (inertial suction) or of maintaining the prey immobile in the water column (compensatory suction). Aerts et al. (2001) suggest that functional demands related to feeding in water always relate to expansion of the area posterior to the jaw apparatus that opens to allow prey and water to enter partly or completely into the oropharyngeal cavity. However, the role of this expansion becomes less important in turtles feeding upon plant material floating at the surface, pulling material out of the substratum, or attacking a vigorously defending prey. In the latter case, biting performance by the jaws probably plays the key role because it allows the turtle to obtain food and also defend against the attack of the self-protecting potential prey (Herrel et al., 2002). Despite large structural variations, the expansion of the oropharyngeal cavity is mainly produced by movement of the hyobranchium as nicely demonstrated by x-ray films (Aerts et al., 2001; Lemell et al., 2002). For illustrating the motor control of throat expansion in *Chelonida*, few data are available. Aerts et al. (2001) provided the first description of motor sequences that show depression and retraction of the hyoid body (Figure 8.8). Lemell et al. (2002) report that the esophagus is filled with the large amount of water sucked in during the gape cycle until the mouth is closed. These authors provide a complete x-ray analysis of prey movement within the bucco-pharyngeal cavity (Figure 8.9). In all turtles, the prey is either sucked inward (inertial suction) or bitten by closing jaws (compensatory suction). In the meantime, the throat is laterally expanded by rotation of lateral elements (ceratobranchials I) of the hyobranchium. The displacement of the elements of the hyoid apparatus has always been assumed to produce throat expansion in all studied aquatic turtles (Lemell & Weisgram, 1997; Van Damme & Aerts, 1997; Lemell et al., 2002), and now there is clear confirmatory evidence. The posterior part of the throat is expanded by rotation of the posterior lateral elements of the hyobranchium (cetarobranchials II; Van Damme & Aerts, 1997; Lemell et al., 2002). The mouth is then again opened slightly to expel the excess water by returning the hyoid apparatus to its starting position, and the prey is retained on the floor of the buccal cavity.

All major data on food ingestion in aquatic turtles have been collected in environmental conditions (e.g., temperature, salinity) close to the classical relevant ecological conditions determined for the studied species. For example, Bels et al. (1998) recorded feeding behavior in *M. terrapin* at 25°C in water with salinity of 33 psu. Leatherback turtles were filmed at 26°C in classical artificial seawater (Bels & Renous, 1992). However, performances in turtles (as in all reptiles) can be modulated in relationship to temperature. It would be expected in ectothermic turtles that speeds of jaw action would increase with higher temperatures within the physiological range of a species, and vice versa.

Given the different thermal conditions of their natural ranges, we compared gape performances of three aquatic species (Figure 8.10). In this study, it was necessary to study jaw action at a range of realistic temperatures for each species. Data were obtained for *Trachemys* (Emydidae) at 18, 24 and 30°C. *Cuora* (Emydidae) and *Siebenrockiella* (Bataguridae) would not feed at 18°C; they were filmed at 24, 28, and 30°C (three specimens of each species were held overnight for more than 14 hours at each study temperature before being filmed). Figure 8.3 and Figure 8.4 respectively show the effects of temperature on bite time (i.e., time to contact food) and jaw opening time (i.e., time to maximum gape) for adults of all three species. It is evident that the two emydid species were both less affected by temperature changes than *Siebenrockiella*, which showed a steep decline in bite time and jaw opening time between 24 and 30°C (at 24and 30°C, one-way ANOVA comparisons among the three species revealed highly significant differences, $p < 0.005$). *Trachemys* showed no change in bite time and jaw opening between 24 and 30°C; even at 18°C there was relatively little slowing of jaw action in comparison with 30°C (bite time by 24%, jaw opening time by 55%). Tukey post hoc tests showed that *Trachemys* was significantly faster (shorter bite/jaw opening times) than *Cuora* at common temperatures (24 and 30°C; $p < 0.05$ in all cases) and also significantly faster ($p < 0.05$) than *Siebenrockiella* at 24 but not at 30°C. Data for bite time were more variable than for jaw opening time; this possibly reflects variation in food morsel size and the positioning of the jaws relative to food during jaw closure. Jaw opening time appears to be more useful for comparisons and is solely presented for the rest of the results. Differences between the two emydid species were particularly marked, with jaw opening time being 227% longer in *Cuora* than in *Trachemys* at 24°C, and 200% longer at 30°C; clearly, *Trachemys* bites far faster than its Asian relative.

These results provide the first relevant functional data to explain the role of feeding performances in the success of invasion of *Trachymys* turtles throughout the aquatic habitats of the world.

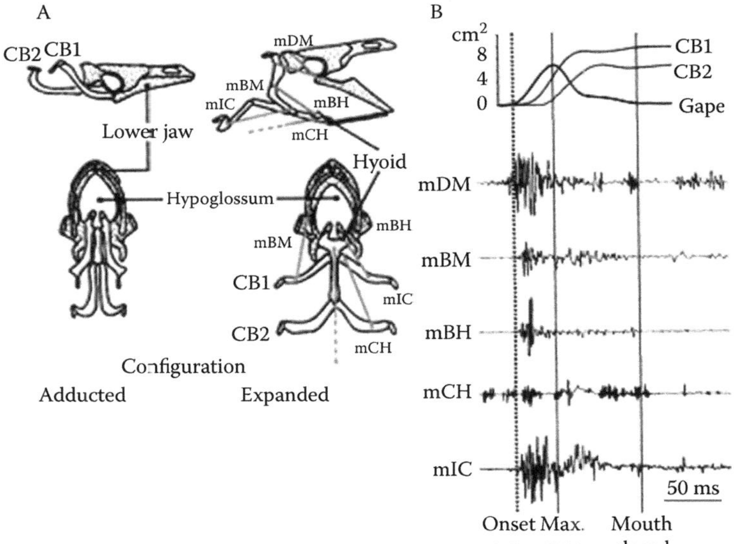

FIGURE 8.8 (A) Schematic representation of the head skeleton of *C. longicollis* in the adducted and expanded configurations. (B) Results of electromyographical analysis during expansion of the hyobranchium in *C. longicollis* (from Aerts et al., 2001). The graph represents the change of the cross sectional surface of the oropharyngeal cavity at the level of the gape and through the throat at the level of ceratobranchials I and II. Arrows indicate position and orientation of the major expansion muscles. CB1: ceratobranchial I; CB2: ceratobranchial II; mBH: m. branchiohyoideus; mBM: m. branchiomandibularis; mCH: m. coracohyoideus; mDM: m. depressor mandibulae; mIC: m. intercornuatus., exp.: throat expansion (Aerts et al., 2001).

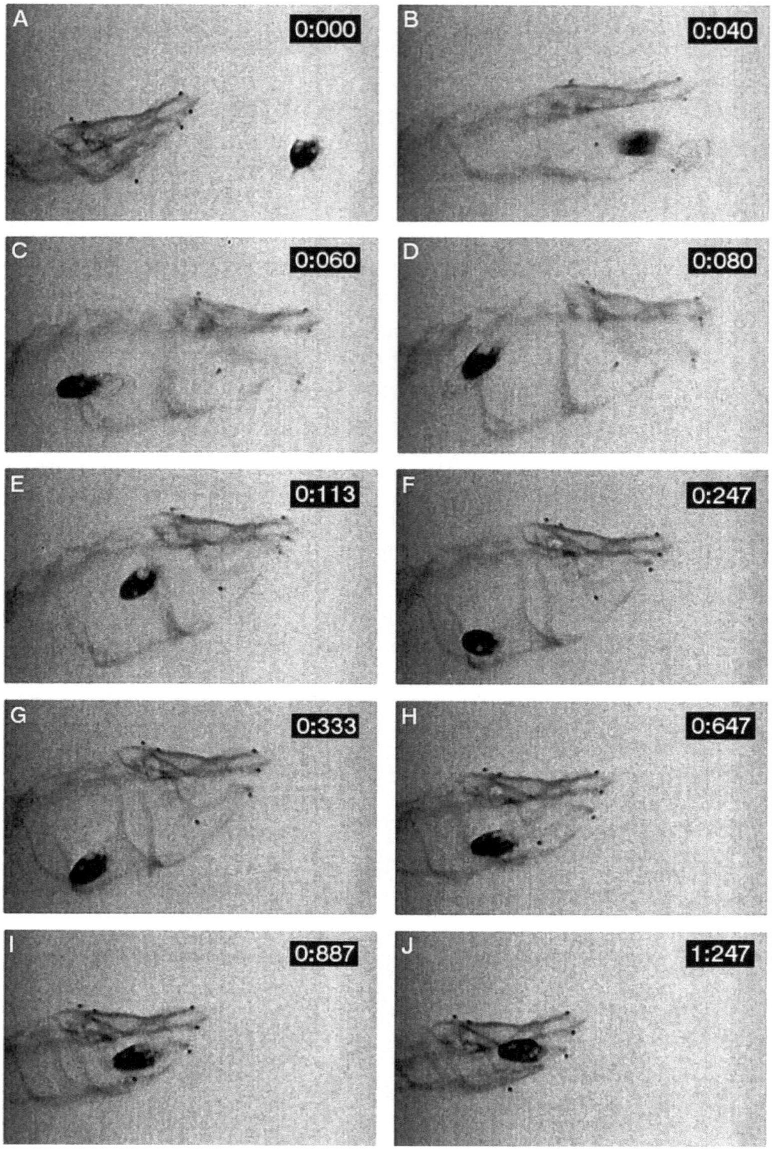

FIGURE 8.9 Successive frames of a high-speed x-ray film sequence (150 frames/s) showing a lateral view of *C. fimbriatus* during food capture. The time (s:ms) is provided for each frame. The prey item appears dark because of the x-ray contrast medium. Lead markers were also positioned on the skull of the turtle. (From Lemell et al., 2002.)

We demonstrate too that the invasive *Trachemys* bites faster over a wide range of temperatures relative to the two Asian species (*Cuora*, *Siebenrockiella*) whose habitat it now shares. This suggests that it can deal more effectively with elusive or dangerous prey as well as feeding on carrion or vegetation more quickly. The thermal data also indicate that *Trachemys* is capable of effective feeding over a wider temperature range than the two native species, allowing it to continue feeding under cooler conditions. The introduction of an omnivorous, highly effective competitor to an ecosystem is clearly undesirable, but the situation is made worse by the unusual characteristics of

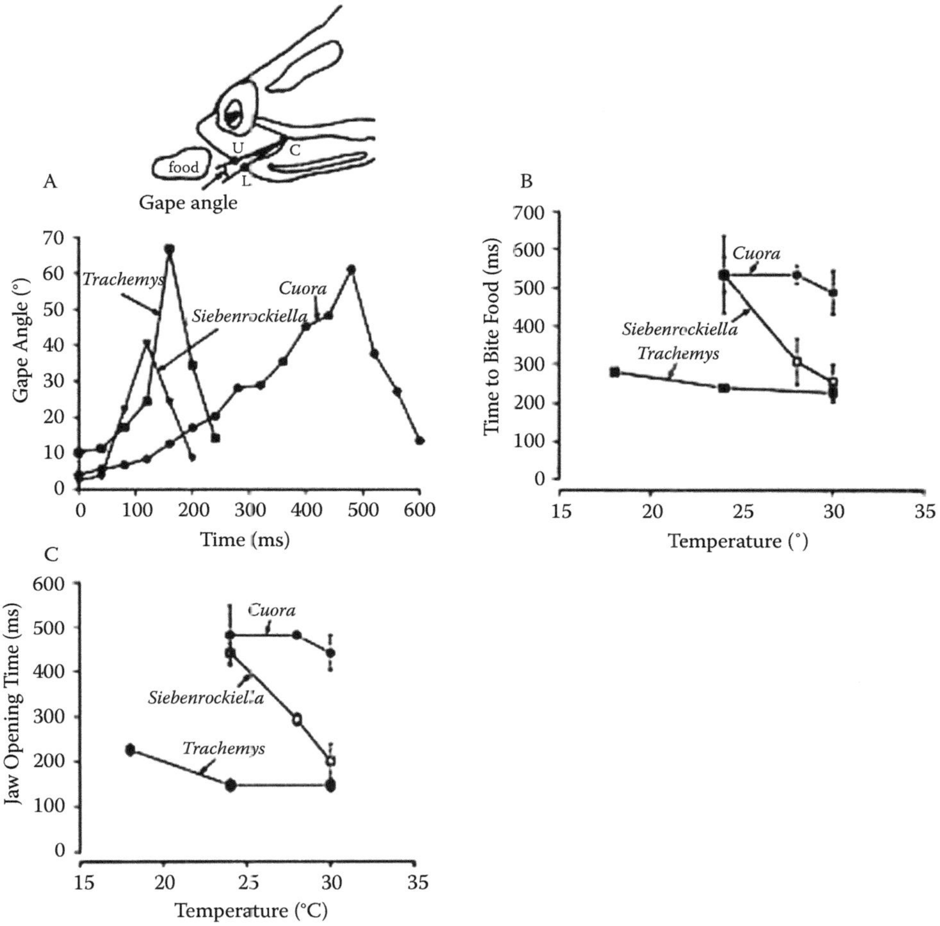

FIGURE 8.10 (A) Examples of gape angle sequences in the three study species at 30°C. The body masses of the three specimens filmed were as follows: *Trachemys* (492 g), *Siebenrockiella* (701 g), and *Cuora* (1041 g). (B) Effect of temperature on bite times (time to contact with food) in adult *Trachemys*, *Cuora*, and *Sieberockiella*. Symbols represent mean values ($n = 3$) and standard deviations. (C) Effect of temperature on jaw opening times (time to maximum gape) in adult *Trachemys*, *Cuora*, and *Sieberockiella*. Symbols represent mean values ($n = 3$) and standard deviations.

Trachemys introductions. Most invasive species do not become established in new habitats because they fail to reproduce, or do not have immunity to pathogens present in their new habitat. Whereas feral *Trachemys* have been shown to breed in Japan, Israel, Germany, and France (Ernst et al., 1994; Cadi et al., 2004), the great bulk of their numbers in Asia (as elsewhere) stem from repeated and continuing large-scale introductions. To some extent, their ability to breed/not breed is irrelevant, particularly as they are very long lived, often surviving 30+ years in the wild (Kuhrt & Dewey, 2002). According to the Turtle Conservation Fund (2002), Asia is "the geographic region that warrants the highest priority if we are to avoid losing [chelonian] species in the near future." The results of our comparative biomechanical study provide further indication of the necessity for trapping and removing this highly competitive introduced turtle from non-native ecosystems.

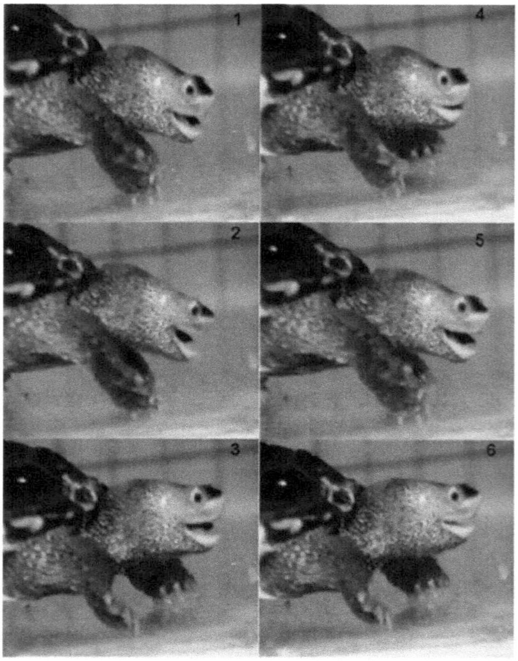

FIGURE 8.11 Typical stages of food (entire mussel) manipulation in *M. terrapin*. The tongue plays a major role in moving food from one side of the jaws to the other. The food is partially crushed at each closing of the jaws. Time between frames: 0.08 s.

8.5.2.2 Manipulation and Transport Cycle

Although poorly analyzed, transport of the food is more complex because it involves two main types of use of both elements of the hyobranchium: the tongue and the hyoid apparatus. In *M. terrapin*, the tongue plays a key role in manipulating food in association with the depression of the hyobranchium (Figure 8.11). For transporting the food, the tongue is also clearly associated with classical depression of the hyobranchium in helping backward displacement of the food within the oropharyngeal cavity. In other species, the food can be sucked in without any action of the tongue. Based on differences in movement of the hyobranchium in *C. fimbriatus*, Lemell et al. (2002) described two modes of transport cycles involving slow suction effects on the prey within the buccal cavity (Figure 8.12). In the first mode, the hyoid is depressed slightly and the mouth is open enough to facilitate the release of the prey from the jaws, with the fish being held at the end of the ceratobranchials. In the second mode, hyoid depression is of the same extent as during prey capture. The volume of the anterior part of the esophagus increases slowly and the prey is held between the ceratobranchials. The turtle expels the water very slowly and the prey remains at the end of the hyoid apparatus.

8.5.3 Feeding in *Dermochelys coriacea*: A Typical Example of a Marine Turtle with a Highly Specialized Diet

8.5.3.1 Materials and Methods

Four young leatherback turtles (80 to 500 g) have been filmed at 200 to 300 Hz using 16 mm films. The turtles from French Guiana were incubated at 30.5°C and kept in an aquarium of 2.3 to 5.0 m^3 at 25°C in sea water (pH = 8.0 to 8.1; salinity = 32 g/l). The animals were filmed under 500 W when fed with crude mussels (Bels et al., 1988). The food was presented gently in front of the turtles. Only true lateral sequences were kept for the analysis using the typical method detailed previously (Bels et al., 1998) and points were digitized to compare movements of the jaws, the hyobranchium, and the limbs during the successive phases of the feeding behavior (Figure 8.12 and Figure 8.13).

8.5.3.2 Results

Feeding bouts in *D. coriacea* consist of successive jaw and hyolingual cycles from prey capture to its transport into the esophagus. During the feeding bouts, the turtles stabilize the body by slow displacements of the forelimbs that are not related to the successive gape openings. Two types of jaw cycles were observed: ingestion (including biting of the smooth material that is cut at each cycle) and transport cycles. We did not record any specific swallowing cycles with the type of soft

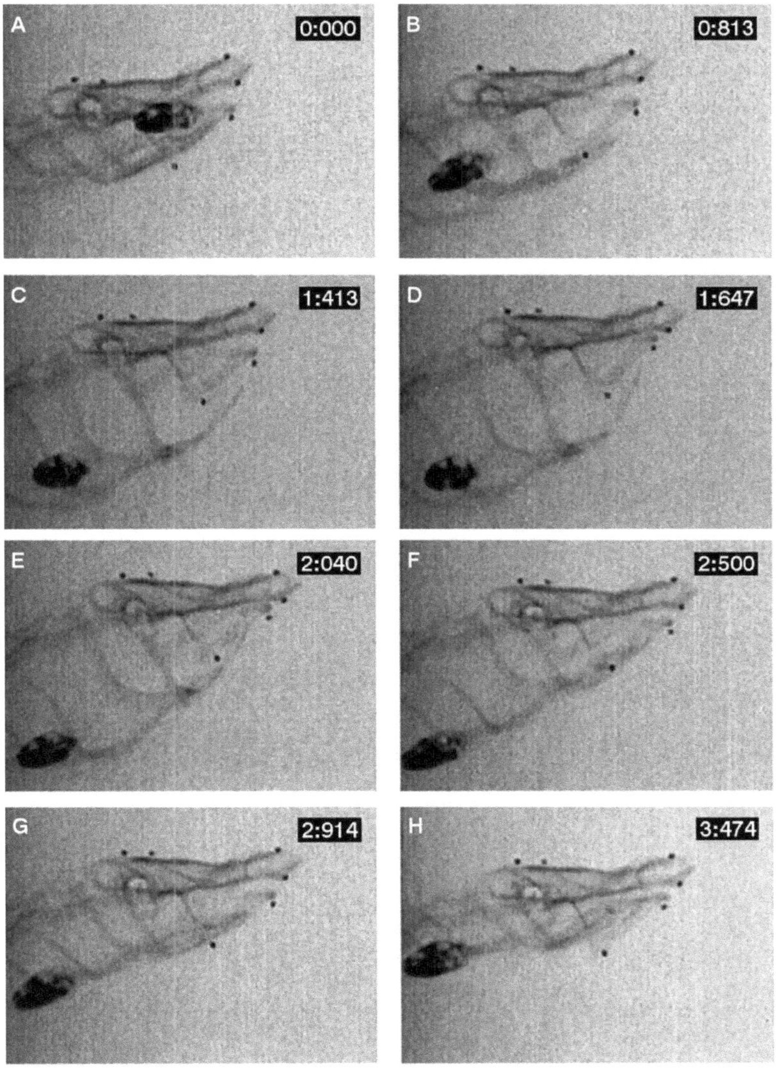

FIGURE 8.12 Successive frames of a high-speed x-ray film sequence (150 frames/s) showing a lateral view of *C. fimbriatus* during food transport. The time (s:ms) is provided for each frame. The prey item appears dark because of the x-ray contrast medium. Lead markers were also positioned on the skull of the turtle (figure from Lemell et al., 2002).

food used in this study (pieces of mussel flesh). In successive ingestion cycles, the food enters the buccal cavity and is reduced between the closing jaws. However, in some of these ingestion cycles, transport also occurs when the food is maintained in front of the turtles. In true transport cycles (Schwenk, 2000), the food is taken completely into the buccal cavity and then moves posteriorly to the esophagus without any reduction between the jaws. However, the food may be reduced between the hard palate and the tongue. Such reduction and biting cycles produce small particulate matter that is ejected from the mouth during the slow opening of the next jaw cycle. Throat elevation during the time between the two FO stages of successive cycles produced movement of water (and particulate matter) out of the buccal cavity (Figure 8.13).

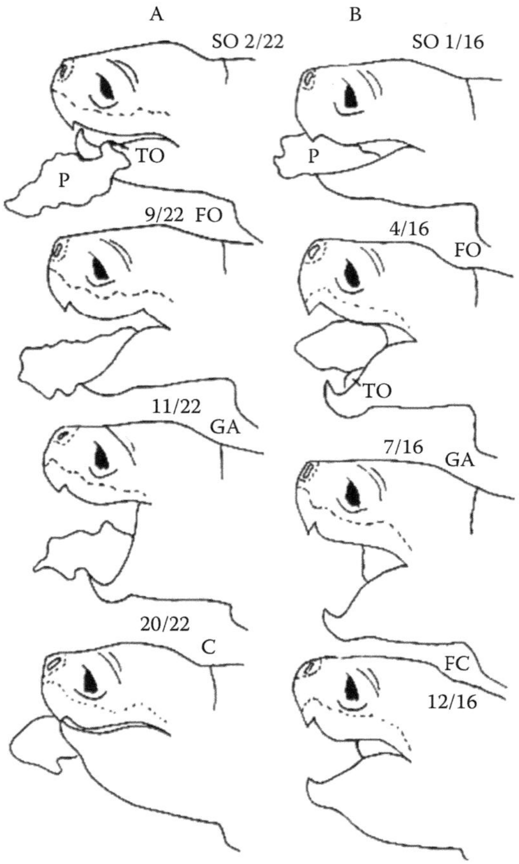

FIGURE 8.13 Successive frames showing (A) ingestion and (B) transport cycles in *D. coriacea* (high-speed 16-mm films at 100 Hz). The gape is divided in slow opening (SO), fast opening (FO), and closing (C) stages—the division in fast closing (FC) and slow closing (SC) stages was not always clear during the ingestion cycle. Inertial suction plays a key role in both feeding phases. The duration of the ingestion cycle is 0.22 s (22 frames) and the duration of the transport cycle is 0.16 s (16 frames). The numbers indicated at each drawing correspond to the number of frames relative to the total number of the cycle. Time to peak gape is indicated by GA. Contact between the food and the tongue was always observed and tongue retracts with the food as the throat is depressed. TO: tongue.

Kinematically, each gape cycle of the biting and transport phases involves a slow opening (SO), fast opening (FO), and closing stage (C). The SO stage of both cycle types is highly variable in duration and amplitude. In some cycles, it involves a "stationary stage" (SO II) prior to the sudden increase of the FO stage; in others, it does not. The SO stage may be also completely absent in some cycles. From high-speed cinematographic films, it was not possible to record clear relationships between the presence and duration of this SO stage, with the cycle function or the tongue-food position. The mean duration of SO stage in biting cycle was significantly longer than of SO stage in the transport cycle (Kruskall-Wallis ANOVA, T = 4.7; df = 18; $P < 0.05$), whereas FO and C stages were not significantly different.

The mouth is opened by a combination of ventral depression of the lower jaw and dorsal elevation of the upper jaw. Typically, the hyoid-tongue displacements are not different in ingestion and transport cycles. During the SO stage and the first half of the FO stage, the hyoid-tongue complex

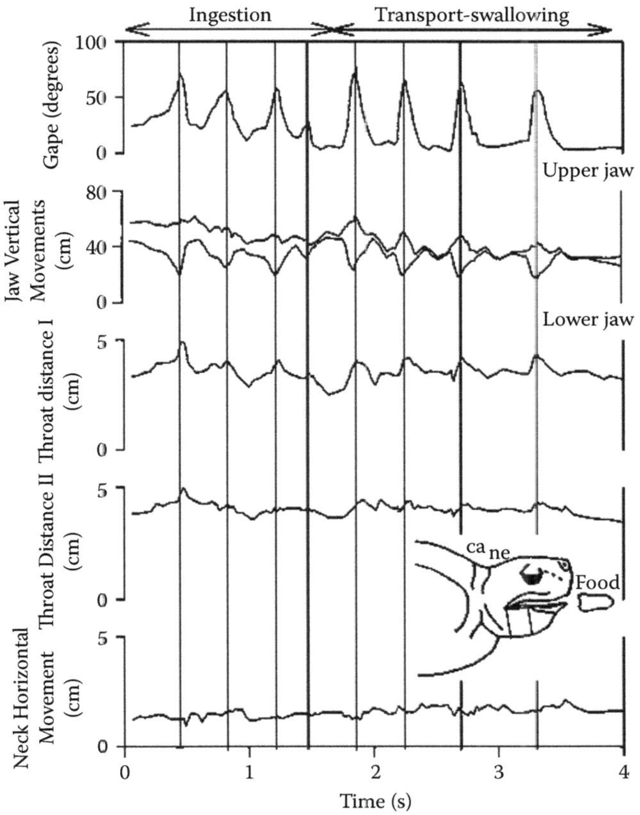

FIGURE 8.14 Kinematic profiles of ingestion and transport cycles in *D. coriacea* feeding on soft material (pieces of mussels) with properties close to the material caught in the natural environment (jellyfish). We categorize ingestion cycles as all cycles used for moving food into the buccal cavity, and transport cycles as all cycles that take place with fcod inside the buccal cavity. Gape cycles in both phases are divided into slow opening (SO), fast opening (FO), and fast closing (FC). Division between closing in fast closing (FC) and slow closing (SC) is not yet clear. However, at the end of the closing stage the angular acceleration of the gape angle decreases slightly. The kinematic profiles illustrating the throat displacement follow a similar pattern during biting and transport cycles. The throat distances I and II were calculated as the distance corresponding to the line traced from one point on the throat to the horizontal line passing through the edge of the mandible at each frame at an angle of 90°. The throat distance I corresponds to the hyoid body and the throat distance II to the region of the ceratobranchials I. During the SO and first half of the FO stages of each cycle, the throat distance I is elevated 0.01 s before the throat distance II (posterior). Retraction of the hyoid-tongue system begins during the FO stage and occurs during the closing stage. Peak hyoid retraction begins with or just before (0.01 s) fast closing. The hyoid apparatus is maximally retracted until the fast closing (FC) is achieved. The neck distance was calculated as the distance between point NE (end of the skull) and CA (carapace).

is elevated and protracted as illustrated by decreasing of the throat angle. The distance between the mandible and the hyoid body visible through the throat skin decreases strongly during SO stage and is 0.07 ± 0.01 s prior to maximal gape amplitude (end of FO stage). The throat expansion then occurs during 0.14 ± 0.05 s at the end of FO and FC stages. In cases of a stationary stage prior to a SO stage or very slow opening stage, the hyoid-tongue distance does not change greatly prior decreasing. There was no significant difference between displacements of the hyoid point on the throat during biting and transport cycles (Figure 8.14).

During SO and FO stages, water moves into the buccal cavity simultaneously as the prey that moves either to the buccal cavity (biting) or posteriorly into the buccal cavity (transport). An additional inertial suction mechanism helps displace prey toward the pharyngeal cavity; this is mainly produced by tongue retraction in biting and transport cycles. Strong contacts between prey surface and tongue dorsal surface were observed in both cycles.

8.6 EVOLUTION OF FEEDING BEHAVIOR IN TURTLES

Discussion of evolution of feeding behavior in turtles remains controversial. Based on their study of prey capture in *T. carolina*, Summers et al. (1998) state that "the currently accepted hypothesis is that the ancestor to recent turtles was terrestrial (Gaffney et al., 1987), but that an aquatic lifestyle evolved early in the history of the clade and is probably primitive for all extant turtles (Gaffney et al., 1987, 1991; Carroll, 1988)." These authors concluded that terrestrial feeding in turtles is a derived behavior (Bels et al., 1997; Summers et al., 1998). Here we provide some discussion that highlights the need for future quantitative kinematic and motor analyses in terrestrial and aquatic turtles.

8.6.1 Ingestion

Terrestrial feeding modes include lingual and jaw prehension. Both modes are used for various food items. Except in *Terrapene* (Bels et al., 1997; Summers et al., 1998; Bels, 2003), lingual prehension is the main mode for all studied turtles of the Testudinidae. The food adheres to the tongue either in (e.g., *Kinixys*) or out (e.g., *Geochelone*) of the buccal cavity. Summers et al. (1998) suggest that *T. carolina* uses a terrestrial feeding mode in water because the velocities of all skull elements during food capture by generalists are slower than those of specialists. From available data, three hypotheses can be suggested. First, the ingestion cycle has been derived in the "true" terrestrial chelonians (i.e., Testudinidae) showing a tongue-based intraoral feeding behavior from food transport and reduction to swallowing. Ingestion can be viewed as an evolutionary transformation of a tongue-based feeding mechanism for efficient procurement of food. The coordination of the tongue movement has not changed greatly, with the tongue moving slightly toward the food during jaw opening and retracting the food after contact; alternatively, the jaw apparatus can simply surround the food item and the tongue makes contact with the food within the buccal cavity prior to being retracted. Second, both modes of prehension (lingual and jaw prehensions) were present in ancestors of turtles and can be used in various ways determined by the properties of the food. Such modulation in food prehension has been reported for squamates (Schwenk, 2000). Third, terrestrial turtles derived from aquatic ancestors have "reinvented" the key role of the tongue for improving efficiency of food procurement together with their conquest of terrestrial habitats. In this case, we must admit that lingual prehension is a derived pattern constrained by the properties of the food resources becoming available during the evolution of terrestrial turtles. In *Terrapene*, which belongs to a primitively aquatic clade, terrestrial feeding is a secondarily derived mode in which the jaw prehension that plays a key role in aquatic feeding has been conserved. Summers et al. (1998) suggest that *T. carolina* would use tongue prehension on prey that is larger than mealworms. We have not yet confirmed this suggestion by examining prehension of various types of food in *Terrapene* (Bels, personal observation).

8.6.2 Transport (and Other Feeding Phases)

The evolution of feeding behavior and motor control of turtles still remains difficult to fully understand. First, the origin of turtles continues to be largely problematic. Second, only a few species have been extensively studied, mainly in terms of feeding mechanisms and not from an evolutionary perspective. However, available data can provide some insight, particularly with respect to the likelihood of some possible hypotheses concerning the evolution of feeding behavior in chelonians.

A comparative analysis of all feeding phases—except food capture—appears interesting. Lauder and Gillis (1997) compared transport kinematics and mechanisms of truly aquatic vertebrates (e.g., fishes and amphibians), first with those of amphibians that feed in both terrestrial and aquatic habitats, and second with those of truly terrestrial Amniota (e.g., squamates). They suggest that several traits recorded for squamates are novel features of the feeding mechanism that appeared with the conquest of terrestrial habitats and can therefore be considered as plesiomorphic for tetrapods. These traits include diversity of intraoral processing, the presence of a SO stage prior to sudden gape increase (FO stage), and hyoid and tongue protraction during SO produced by a unique pattern of hyoid muscle contraction (Lauder & Gillis, 1997). These suggestions agree with the evolutionary approach to feeding behavior proposed by Reilly and Lauder (1990), who advocate a single generalized model of food transport in tetrapods. More recently, McBrayer and Reilly (2002) do not support all the features of Bramble and Wake's (1985) general model (i.e., presence of SO I and SO II in slow opening). These two models have been proposed to describe the generalized gape pattern and tongue movement during food transport in terrestrial vertebrates. Based on available data in 1985, Bramble and Wake (1985) presented a hypothetical generalized model for ancestral tetrapods with gape being divided into slow opening (SO), fast opening (FO), fast closing (FC), and slow closing (SC-PS) stages. Before FO, the gape increase is divided into slow opening I (SO I) involving "a comparatively low gape angle" (Bramble & Wake, 1985) while the tongue slides beneath the food and a slow opening II (SO II), that "is recognized by a distinct decline in the rate of change of gape" (Bramble & Wake, 1985). This latter stage is represented in the model by a plateau in the gape cycle. Then during FO, the food is moved posteriorly by action of the tongue. After FO, the mouth closes rapidly (FC) and then more slowly as the jaws contact the prey during SC-PS stages.

Reilly and Lauder (1990) proposed another generalized model for amniotes. These authors divide the increasing stage of the gape cycle (opening of the mouth) only into slow opening (SO) and fast opening (FO), with no SO II stage. This difference in gape increase is highly significant because it emphasizes a difference in tongue movement during transporting of the food—see McBrayer & Reilly (2002) for a discussion based on quantitative data in a large set of squamates. According to Lee (1997), the development of the shell and rather flattened body of turtles could be linked to an ancestral herbivorous diet. It may be assumed that ancestral forms were terrestrial and fed on plant material (and perhaps small, slowly moving prey or carrion), as is the case for a large number of extant species studied to date (e.g., *Geochelone* sp., *Kinixys* sp.). Based on this hypothesis, we may conclude that available data support the conclusion that the characteristics of transport cycles are plesiomorphic for terrestrial feeding in turtles, although there is a large variety in diet and specialization of the hyobranchium, yielding the presence of a SO stage before FO of the mouth, a movement of the tongue under the food during the SO, and retraction of the food by the tongue at SO-FO transition or at the beginning of SO stage.

In their quantitative analysis, Bels et al. (1997) compared their data in terrestrial feeding by *T. carolina* with the two previously described generalized models for tetrapods. Available data do not permit categorical conclusions. However, a rapid survey of kinematics facilitates the comparison of aquatic and terrestrial transport phases. In all turtles, with all types of food, the jaw cycle is often divided into SO, FO, and closing stages. The SO stage is always present in all terrestrial turtles. This stage has also been reported in a large number of aquatic turtles (*P. castaneus*, Lemell & Weisgram, 1997; *T. carolina*, Bels et al., 1997; Summers et al., 1998; *M. terrapin*, Bels et al., 1998; *D. coriacea*, Bels et al., 1998; *C. frimbriatus*, Lemell et al., 2002). We suggest that there is an intra-oral transport cycle that is similar for all terrestrial turtles. It is also evident that the tongue is used in a large number of aquatic turtles to transport food items, just as in terrestrial turtles. Its protraction occurs during the SO stage and it retracts with the food at the boundary between SO and FO stages or early during FO stage. We may call this mechanism *intra-oral aquatic lingual transport*. In contrast, suction plays the dominant role for transporting food in some species such as *C. fimbriatus* (Lemell et al., 2002). We may call this mechanism *intra-oral aquatic hyoid transport*. Probably, the relative size of the tongue is the limiting morphological factor in determining use of tongue-based intra-oral transport or hyoid-based intra-oral transport. The slowly opening mouth can generate low pressure within the mouth cavity because of protraction of the hyoid during this stage in both cases. However, this is rapidly compensated for by suction that occurs during throat expansion after maximum gape. During the FO stages, rapid mouth opening is immediately followed by a posteroventral movement of the hyoid apparatus; high negative pressure occurs within the oropharyngeal cavity, which is produced by peak hyoid depression following peak gape in all studied turtles. During the FC phase, the hyoid apparatus remains depressed until the mouth is closed.

ACKNOWLEDGMENTS

The authors thank all those who helped to gather the data used in this paper. They are very grateful to S. Attere for her help in writing this paper, and to persons for discussion and help in various parts of this work (G. Daghfous, S. Montuelle). We thank two anonymous reviewers for their helpful comments.

REFERENCES

Aerts, P., Van Damme, J., and Herrel, A., Intrinsic mechanics and control of fast craniocervical movements in aquatic feeding turtles, *Am. Zool.*, 41, 1299, 2001.

Beisser, C.J., Lemell, P., and Weisgram, J., Light and electron microscopy of the dorsal lingual epithelium of *Pelusios castaneus* (Pleurodira, Chelidae) with special respect to its feeding mechanics, *Tiss. Cell*, 33, 63, 2001.

Bels, V.L., Evaluating the complexity of the trophic system in Reptilia, in *Biomechanics and Evolution of Vertebrates*, V.L. Bels, J.-P. Gasc, and A. Casinos (eds.), Oxford: Bios Publishers, 2003, 185.

Bels, V.L., *Feeding in Domestic Vertebrates: From Structure to Behaviour*, Wallingford, CT: Cabi Publishing, 2006.

Bels, V.L., Davenport, J., and Delheusy, V., Kinematic analysis of the feeding behavior in the box turtle *Terrapene carolina* (L.), (Reptilia: Emydidae), *J. Exp. Zool.*, 277, 198, 1997.

Bels, V.L., Davenport, J., and Renous, S., Food ingestion in the estuarine turtle *Malaclemys terrapin*: Comparison with marine leatherback turtle *Dermochelys coriacea*, *J. Mar. Biol. Assoc. UK*, 78, 953, 1998.

Bels, V.L., Davenport, J., and Renous, S., Drinking and water expulsion in the diamondback terrapin *Malaclemys terrapin* (Latreille). J. Zool., 236:483–497, 1995.

Bels, V.L., and Renous, S., Kinematics of feeding in two marine turtles (*Chelonia mydas* and *Dermochelys coriacea*), in *Proceedings of the Sixth Ordinary General Meeting S.E.H.*, Z. Korsos and I. Kiss (eds.), Budapest: Hungarian Natural History Museum Press, 1992, 73.

Bels, V.L., Rimblot-Baly, F., and Lescure, J., Croissance et maintien en captivité de la tortue Luth, *Dermochelys coriacea* (Vandelli, 1761), *Rev. Fr. Aquariol.*, 15, 59, 1988.

Bouchard, S.S., and Bjorndal, K.A., Nonadditive interactions between animal and plant diet items in an omnivorous freshwater turtle *Trachemys scripta*, *Comp. Biochem. Phys. B*, 144, 77, 2006.

Bramble, D.M., and Wake, D.B., Feeding mechanisms of lower tetrapods, in *Functional Vertebrate Morphology*, M. Hildebrand, D.M. Bramble, K.F. Liem, and D.B. Wake (eds.), Cambridge, MA: Harvard University Press, 1985, 230.

Cadi, A., Delmas, V., Prévot-Julliard, A.-C., Joly, P., Pieau, C., and Girondot, M., Successful reproduction of the introduced slider turtle (*Trachemys scripta elegans*) in the South of France, *Aquat. Conserv.*, 14, 237–246, 2004.

Carroll, R.L., *Vertebrate Paleontology and Evolution*, New York: W.H. Freeman & Co, 1988.

Cebra-Thomas, J., Tan, F., Sistla, S., Estes, E., Bender, G., Kim, C., Riccio, P., and Gilbert, S.F., How the turtle forms its shell: A paracrine hypothesis of carapace formation, *J. Exp. Zool. B*, 304, 558, 2005.

Claude, J., Pritchard, P., Tong, H., Paradis, E., and Auffray, J.-C., Ecological correlates and evolutionary divergence in the skull of turtles: A geometric morphometric assessment, *Syst. Biol.*, 53, 937, 2004.

Cleuren, J., and De Vree, F., Kinematics of the jaw and hyolingual apparatus during feeding in *Caiman crocodilus*, *J. Morphol.*, 212, 141, 1992.

Davenport, J., and Clough, W., The use of limbscales or "pseudoclaws" in food handling by young loggerhead turtles, *Copeia*, 1985, 786, 1985.

Davenport, J., and Clough, W., Swimming and diving in young loggerhead sea turtles (*Caretta caretta* L.), *Copeia*, 1986, 53, 1986.

Davenport, J., and Macedo, E., Behavioural osmotic control in the euryhaline diamondback terrapin *Malaclemys terrapin* Latreille: Responses to low salinity and rainfall, *J. Zool.*, 220, 487,1990.

Davenport, J., Spikes, M., Thornton, S.M., and Kelly, O.B., Crab-eating in the diamondback terrapin *Malaclemys terrapin*: Dealing with dangerous prey, *J. Mar. Biol. Assoc. UK*, 72, 835, 1992.

Davenport, J., Wong, T.M., and East, J., Feeding and digestion in the omnivorous estuarine turtle *Batagur baska* (Gray), *Herpetol. J.*, 2, 133, 1992.

Ernst, C.H., and Barbour, R.W., *Turtles of the World*, Washington, DC: Smithsonian Institution Press, 1989.

Ernst, C.H., Lovich, J.E., and Barbour, R.W., *Turtles of the United States and Canada*, Washington, DC: Smithsonian Institution Press, 1994.

Gaffney, E.S., A phylogeny and classification of the higher categories of turtles, *Bull. Am. Mus. Nat. Hist.*, 155, 387, 1975.

Gaffney, E.S., Comparative cranial morphology of recent and fossil turtles, *Bull. Am. Mus. Natur. Hist.*, 164, 67, 1979.

Gaffney, E.S., Hutchinson, J.H., Jenkins, F.A., and Meeker, L.J., Modern turtle origins: The oldest known cryptodire, *Science*, 237, 289, 1987.

Gaffney, E.S., Meylan, P.A., and Wyss, A.R., A computer assisted analysis of the relationships of the higher categories of turtles, *Cladistics*, 7, 313, 1991.

Harless, M., and Morlock, H., *Turtles: Perspectives and Research*, New York: John Wiley & Sons, 1979.

Herrel, A., and De Vree, F., Kinematics of intraoral transport and swallowing in the herbivorous lizard *Uromastix acanthinurus*, *J. Exp. Biol.*, 202, 1127, 1999.

Herrel, A., O'Reilly, J.C., and Richmond, A.M., Evolution of bite performance in turtles, *J. Evol. Biol.*, 15, 1083, 2002.

Hill, R.V., Integration of morphological data sets for phylogenetic analysis of Amniota: The importance of integumentary characters and increased taxonomic sampling, *Syst. Biol.*, 54, 530, 2005.

Hughes, G.R., Luschi, P., Mencacci, R., and Papi, F., The 7000-km oceanic journey of a leatherback turtle tracked by satellite, *J. Exp. Mar. Biol. Ecol.*, 229, 209, 1998.

Iwasaki, S., Fine structure of the dorsal epithelium of the tongue of the freshwater turtle, *Geoclemys reevesii* (Chelonia, Emydinae), *J. Morph.*, 211, 125, 1992.

Iwasaki, S., Evolution of the structure and function of the vertebrate tongue, *J. Anat.*, 201, 1, 2002.

Iwasaki, S., Asami, T., and Wanichanon, C., Fine structure of the dorsal lingual epithelium of the juvenile hawksbill turtle, *Eretmochelys imbricata bissa*, *Anat. Rec.*, 244, 437, 1996a.

Iwasaki, S., Asami, T., and Wanichanon, C., Ultrastructural study of the dorsal lingual epithelium of the soft-shell turtle, *Trionix cartilaginous* (Chelonia, Trionychidae), *Anat. Rec.*, 246, 305, 1996b.

Iwasaki, S., Wanichanon, C., and Asami, T., Histological and ultrastructural study of the lingual epithelium of the juvenile Pacific ridley turtle, *Lepidochelys olivacea* (Chelonia, Chelonidae), *Ann. Anat.*, 178, 243, 1996c.

Iwasaki, S., Wanichanon, C., and Asami, T., Ultrastructural study of the dorsal lingual epithelium of the Asian snail-eating turtle, *Malayemys subtrijuga*, *Ann. Anat.*, 178, 145, 1996d.

Joyce, W.G., and Gauthier, J.A., Palaeoecology of Triassic stem turtles sheds new light on turtle origins, *Proc. R.S. B.-Biol. Sci.*, 271, 1, 2004.

Kear, B.P., and Lee, M.S., A primitive protostegid from Australia and early sea turtle evolution, *Biol. Lett.*, 2, 116, 2006.

Kuhrt, T., and Dewey, T., "Trachemys scripta", Animal Diversity Web, http://animaldiversity.ummz.umich.edu/site/accounts/information/Trachemys_scripta.html, 2002.

Lauder, G.V., and Gillis, G.B., Origin of the amniote feeding mechanism: Experimental analyses of outgroup clades, in *Amniote Origins: Completing the Transition to Land*, S. Sumida and K. Martin (eds.), San Diego, CA: Academic Press, 1997, 152.

Lauder, G.V., and Prendergast, T., Kinematics of aquatic prey capture in the snapping turtle *Chelydra serpentina*, *J. Exp. Biol.*, 164, 55, 1992.

Laurin, M., and Reisz, R.R., A reevaluation of early amniote phylogeny, *Zool. J. Linn. Soc.*, 113, 165, 1995.

Lee, M.S.Y., Correlated progression and the origin of turtles, *Nature*, 379, 812, 1997.

Lemell, P., Beisser, C.J., and Weisgram, J., Feeding patterns of *Pelusios castaneus* (Chelonia: Pleurodira) with comments on the morphology of its tongue, *J. Morphol.*, 232, 285, 1997.

Lemell P., Lemell, C., Snelderwaard, P., Gumpenberger, M., Wocheslander, R., and Weisgram, J., Feeding patterns of *Chelus fimbriatus* (Pleurodira: Chelidae), *J. Exp. Zool.*, 205, 1495, 2002.

Lemell, P., and Weisgram, J., The feeding patterns of *Pelusios castaneus* (Chelonia: Pleurodira), *Neth. J. Zool.*, 47, 429, 1997.

Liem, K.F., Aquatic versus terrestrial feeding modes: Possible impacts on the trophic ecology of vertebrates, *Am. Zool.*, 30, 209, 1990.

Lindeman, P.V., Evolution of the relative width of the head and alveolar surfaces in map turtles (Testudines: Emydidae Graptemys), *Biol. J. Linn. Soc.*, 69, 549, 2000.

McBrayer, L.D., and Reilly, S.M., Testing amniote models of prey transport kinematics: A quantitative analysis of mouth opening patterns in lizards, *Zoology*, 105, 71, 2002.

McCauley, S.J., and Bjorndal, K.A., Response to dietary dilution in an omnivorous freshwater turtle: Implications for ontogenetic dietary shifts, *Physiol. Biochem. Zool.*, 72, 101, 1999.

Nagashima, H., Uchida, K., Yamamoto, K., Kuraka, S., Usuda, R., and Kuratani, S., Turtle-chicken chimera: An experimental approach to understanding evolutionary innovation in the turtle, *Dev. Dyn.*, 232, 149, 2005.

Norton, S.F., and Brainerd, E.L., Convergence in the feeding mechanics of ecomorphologically similar species in the Centrarchidae and Cichlidae, *J. Exp. Biol.*, 176, 11, 1993.

Pough, F.H., Andrews, R.M., Cadle, J.E., Crump, M.L., Savitzky, A.H., and Wells, K.D., *Herpetology*, Upper Saddle River, NJ: Prentice-Hall, 2001.

Pritchard, P.C.H., *Encyclopedia of Turtles*, Neptune, NJ: T.F.H. Publications, 1979.

Reilly, S.M., and Lauder, G.V., The evolution of tetrapod feeding behavior: Kinematic homologies in prey transport, *Evolution*, 44, 1542, 1990.

Reina, R.D., Jones, T.T., and Spotila, J.R., Salt and water regulation by the leatherback sea turtle *Dermochelys coriacea*, *J. Exp. Biol.*, 205, 1853, 2002.

Reisz, R.R., and Laurin, M., Owenetta and the origin of turtles, *Nature*, 349, 324, 1991.

Renous, S., and Bels, V., Etude cinématique de la palette natatoire antérieure de la tortue luth *Dermochelys coriacea* (Vandelli, 1761) au cours de sa locomotion terrestre, *Can. J. Zool.*, 69, 495, 1989.

Schwenk, K., *Feeding: Form, Function and Evolution in Tetrapod Vertebrates*, San Diego, CA: Academic Press, 2000.

Shaffer, B.H., Meylan, P., and McKnight, M.L., Tests of turtle phylogeny: Molecular, morphological and paleontological approaches, *Syst. Biol.*, 46, 235, 1997.

Shedlock, A.M., Botka, C.W., Zhao, S., Shetty, J., Zhang, T., Liu, J.S., Deschavanne, P.J., and Edwards, S.V., Phylogenomics of nonavian reptiles and the structure of the ancestral amniote genome, *Proc. Natl. Acad. Sci. USA*, 104, 2767, 2007.

Smith, T., Kardong, K., and Bels, V., Prey capture behavior in the blue-tongued skink, *Tiliqua scincoides*, *J. Herpetol.*, 33, 362, 2000.

Summers, A.P., Darouian, K.F., Richmond, A.M., and Brainerd, E.L., Kinematics of aquatic and terrestrial prey capture in *Terrapene carolina*, with implications for the evolution of feeding in cryptodire turtles, *J. Exp. Zool.*, 281, 280, 1998.

Van Damme, J., and Aerts, P., Kinematics and functional morphology of aquatic feeding in Australian side necked turtles (Pleurodira: Chelodina), *J. Morphol.*, 233, 113, 1997.

Wyneken, J., Sea turtle locomotion: mechanisms, behavior and energetics, in *The Biology of Sea Turtles*, P.L. Lutz and J.A. Musick (eds.), Boca Raton, FL: CRC Press, 1997, 165.

9 The Structure of Cardiopulmonary Systems of Turtles: *Implications for Behavior and Function*

Jeanette Wyneken

CONTENTS

9.1 TURTLE CARDIOPULMONARY SYSTEM

The circulatory and pulmonary systems are structurally and functionally related in all vertebrates and are often discussed together as the cardiopulmonary system. The lungs of turtles are suspended from the visceral side of the vertebrae and ribs that form much of the carapace. Suspended between them is the heart, which externally shows just two atria and one ventricle. This has led to the common but over-simplified description of the turtle heart as three chambered heart. This description ignores both structural and functional specializations that correlate with physiological, ecological, and behavioral adaptations specific to different testudine taxa. Turtles and tortoises are episodic breathers and can shunt blood to or away from the lungs via vascular and intracardiac shunting routes and mechanisms (Glass et al., 1978).

Additionally, turtles are particularly adept at breath-holding compared to other reptiles. The adaptations that are associated with this trait include hypoxia tolerance, large lung volume, structures that allow rapid and extensive air exchange during ventilation, physiological buffering of the blood and pericardial fluid, and behavioral changes such as aestivation (or brumation).

9.2 HEART

The heart is located within the pericardium and is bordered cranially by the acromion processes and laterally or anterolaterally by procoracoid processes (Figure 9.1). Ventrally, it is covered by the pectoral muscles. Dorsally, it is sandwiched between the lungs and laterally, it is adjacent to the left and right lobes of the liver. The pericardial sac typically contains clear pericardial fluid. The posterior aspect of the pericardium and ventricle are attached to the peritoneum by a collagenous cord, the gubernaculum cordis. This cord anchors the caudal apex of the heart and allows the muscular walls to come into opposition during contraction rather than to retract the heart cranially. The gross myocardial structure tends to be low in outer compact muscle, internally very spongy and rich in trabeculae.

The heart is multichambered and serves as the main circulatory pump. All turtles have a large thin-walled sinus venosus, two large atria, and a dorsoventrally flattened ventricle (Figure 9.2) that houses three communicating compartments. One might argue that the turtle heart functionally has five or even six chambers—the sinus venosus, two atria, and the internally subdivided ventricle comprising the cavum venosum, cavum arteriosum, and cavum pulmonale. Blood flow within and out of the ventricle may be complex (Figure 9.3), moving from the atria to the ventricle and out from two of the three ventricular compartments (the cavum venosum and cavum pulmonale) to the great vessels.

9.3 GREAT VESSELS AND MAJOR TRIBUTARIES

The great vessels are the three major arteries that emerge from the cranial and ventral parts of the ventricle: the left aorta, the right aorta, and a pulmonary trunk (Figure 9.2). These three vessels and their tributaries have thick, muscular and elastic walls and are typically considered to be "high pressure" vessels. The left aorta arches dorsally and to the left of the peritoneal cavity, traveling caudally before giving off several smaller arteries to the abdominal viscera before joining the right aorta to form the dorsal aorta. The right aorta arches dorsally then travels caudally near the ventromedial surface of the right lung; it gives off major branches that supply blood to the head, stomach, pancreas, spleen, and duodenum before joining the left aorta. The pulmonary arteries usually arise as a short pulmonary trunk that quickly forms the right and left pulmonary arteries to the right and left lungs, respectively. The pulmonary arteries enter the lung with each bronchus and tend to remain in the more dorsal aspects of the lung while giving off multiple branches, some of which connect directly to the more ventrally positioned pulmonary veins.

The right aorta gives off a single short branch right away, the brachiocephalic trunk that bifurcates to form the subclavian arteries. Small thyroid arteries, carotid arteries, then the ventral cervical arteries arise from either the brachiocephalic trunk or the subclavian arteries, lateral to the thyroid arteries (Figure 9.3). The carotids supply blood to the head. They bifurcate near the skull to form the small external and large internal carotid arteries (Bojanus, 1817–1821; Jamniczky and Russell, 2007). The subclavian arteries continue laterally toward the forelimbs; they become the axillary arteries, the junction of the scapula and procoracoid. The axillary arteries give off branches to the scapular musculature, lateral aspect of the carapace, and ventral locomotor muscles before becoming the brachial artery of the forelimb. In at least sea turtles, the brachial artery gives off multiple branches as part of a rete system (a series of arteries and veins that parallel one another associated with countercurrent heat exchange). The more distal vasculature includes peripheral shunts, such as the circumflex of manus; details of peripheral shunts are beyond the scope of this chapter and may be found elsewhere. The circulatory anatomy of one species, *Emys orbicularis*, by Bojanus (1819–1821), includes such structural detail. This work still serves as the template for understanding other species.

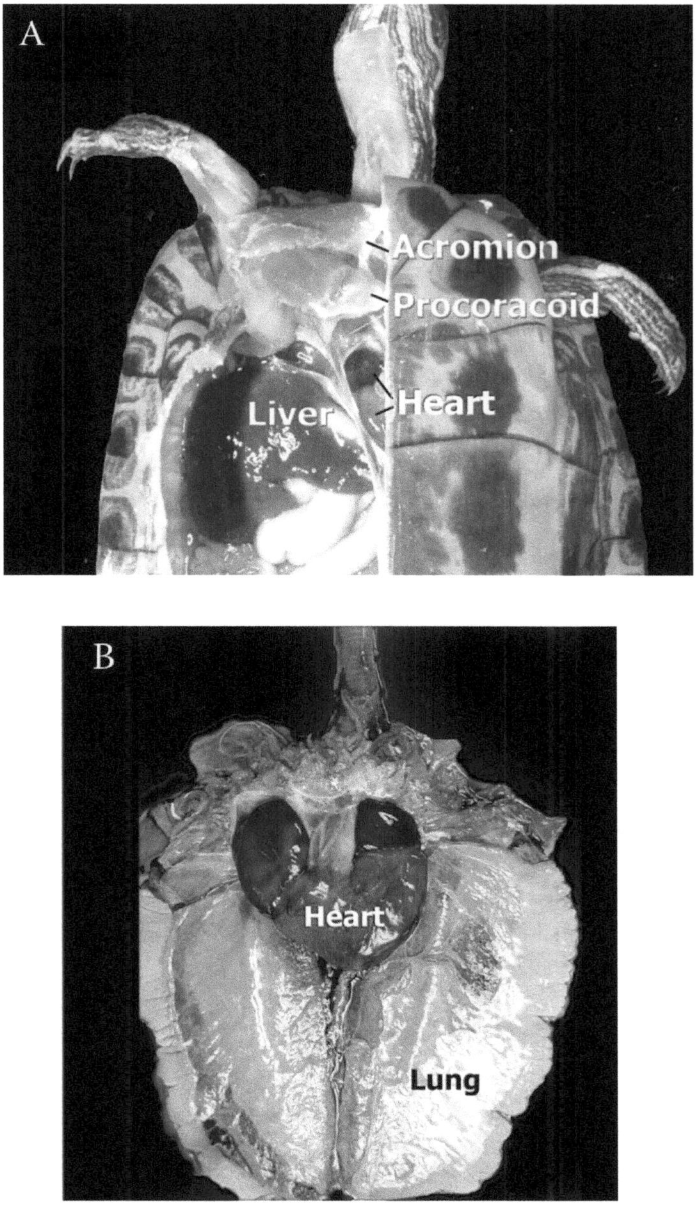

FIGURE 9.1 (A) The location of an emydid turtle heart *in situ* (*Trachemys scripta*) with the right forelimb protracted and the procoracoid and acromion rotated cranially. The right half of the heart can be seen relative to the landmarks provided by the scutes. (B) The ventral position of the heart relative to the lungs is clear from dissection of the cardiopulmonary system of the marine turtle *Eretmochelys imbricata*. The ridge from the long bronchus can be seen in this ventral view running most of the length of each lung. (From Wyneken, 2001. With permission.)

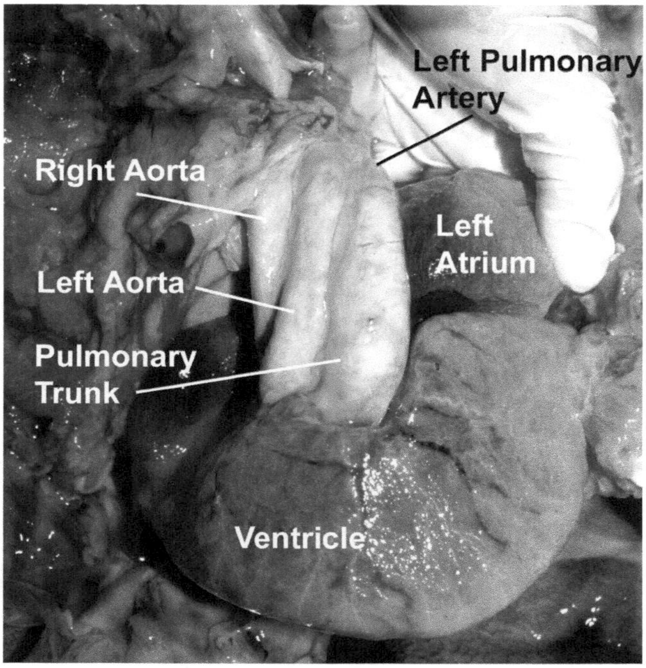

FIGURE 9.2 Adult turtle heart after removal of the pericardium and gubernaculum cordis. The atria are large and relatively thin-walled. The muscular ventricle shows no external indication of the three internal compartments. The great vessels are large and prominent as they emerge from the anteroventral aspect. (From Wyneken, 2001. With permission.)

Physiological studies of other reptiles link intracardiac shunting and digestion. The drainage of the upper gastrointestinal (GI) system suggests a functional link in turtles. Much of the blood supply to the turtle GI system arises from the left aorta. The great vessels and the GI system receive vagal innervation. The left aorta, the middle of the three great vessels, turns dorsolaterally and passes the level of the stomach before producing three branches: the gastric artery, the celiac artery, and the superior mesenteric artery draining to the stomach, pancreas, duodenum stomach, liver, and gallbladder. The superior (or anterior) mesenteric artery gives off many branches that fan out through the intestinal mesenteries and supply the small intestines. After giving off the superior mesenteric artery, the left aorta continues posteriorly where it joins the right aorta (typically) to form a single dorsal aorta. The position where the two join is variable but generally occurs within the middle third of the body.

Within the lung, capillaries join venules that flow into veins that drain into the pulmonary veins. The pulmonary veins travel along the ventral surface of each bronchus, then exit the lung anteriorly and arch medially. They enter the left atrium dorsolaterally. Venous blood from the body drains into the sinus venosus from four major veins: the left precava (left superior vena cava), the right precava (superior vena precava), the left hepatic vein, and the postcava (posterior vena cava).

9.4 LUNGS

Turtles have multichambered (multicameral) lungs located in the dorsal carapace (Perry, 1998; Wyneken, 2001). They extend caudally from the nuchal region for approximately 50 to 78% of the carapace length (for example, *Caretta caretta, Gopherus agasszii, Testudo graeca,* and *T. scripta* span this range; Figure 9.4). The lungs are located dorsally and are

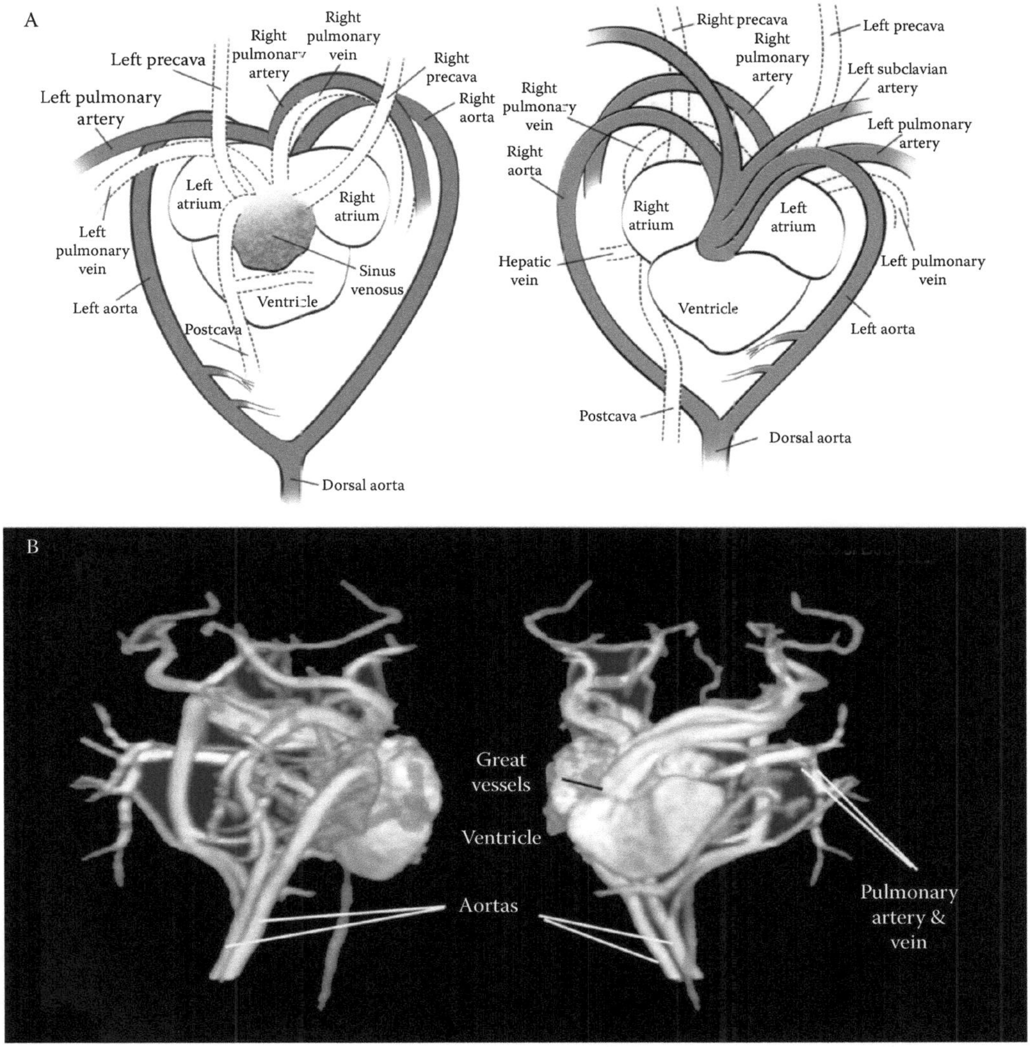

FIGURE 9.3 (A) Diagram of the dorsal (left) and ventral (right) turtle cardiopulmonary system vasculature. (B) The same views of the vasculature and heart as an MRI taken *in vivo*. The turtle's neck was retracted during imaging, so the carotids and their branches are not fully extended.

attached to the carapace and vertebral column via a pulmonary ligament. In some species (e.g., *Lepidochelys kempii* and *C. caretta*), the lungs are more closely attached to the vertebral column than in other species (*T. scripta* and *Geochelone carbonaria*). Turtles have an intrapulmonary bronchus that extends most or all of the length of each lung and structurally separates the lungs of most species into medial and lateral chambers. Freshwater turtles and some tortoises show torsion of the lobes so that the structure appears more complex in the species examined (Perry, 1998).

Turtle lungs are structured differently than mammalian lungs (Figure 9.5). The bronchus is unbranched, so there is no bronchial tree. There are no alveoli. The parenchyma is organized into thin walls (trabeculae) that surround niches (Figure 9.5). The niches contain shallow open box-like ediculi (or ascini) or faveolar spaces that are deeper than they are wide (Duncker, 2004; Perry, 1998). These air exchange structures are associated with suites of other morphological, behavioral,

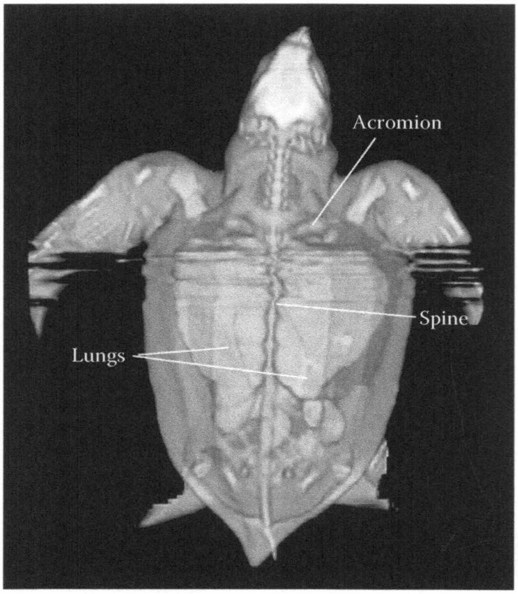

FIGURE 9.4 A three-dimensional CT reconstruction of the compact leatherback (*D. coricea*) turtle's lungs. The lungs in this species are relatively short but dense. The horizontal lines are due to a movement artifact when the turtle ventilated.

and physiological adaptations. The lungs of very active species—particularly sea turtles—have dense edicular parenchyma that is associated with extensive elastic tissue; this morphology differs greatly from that of less migratory, shallow-diving, or terrestrial taxa. For example, the lungs of a "sit-and-wait" predator, such as the common snapping turtle, *Chelydra serpentina* have few niches and these relatively simple lungs have edicular and homogeneous parenchyma. Perry (1998) described and classified reptilian lung form in detail and summarizes the specifics of lung structure in turtles.

9.4.1 Form of the Airways

The airways begin at the glottis, which is located in the anterior portion of the tongue. The glottis and its muscles are supported by the hyoid apparatus (the skeletal elements located between the lower jaws). The chelonian glottis is located at the base of the muscular, fleshy tongue in the anterior to middle portion of the oropharynx. Within the shell, the trachea passes ventral to the esophagus as a single structure. Cranial and dorsal to the heart, the carina forms and bifurcates the airways well cranial of the formation of separate bronchi. The two bronchi enter the anterior part of the lungs along with the pulmonary arteries and extend far into the lungs. They have

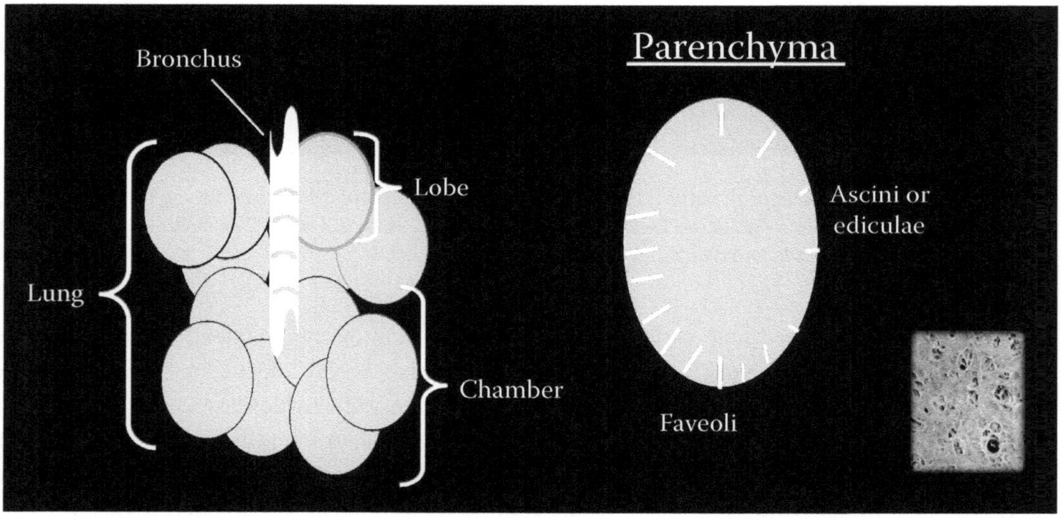

FIGURE 9.5 Diagram of the terminology that defines the lung air exchange surfaces. Each lung has a single unbranched bronchus. Chambers are formed of lobes. The parenchyma of each lobe is partitioned into niches that contain gas exchange units (faveoli, ediculae, or ascini) that differ in shape, shown together for comparison. A photo of the niche trabecular walls, faveoli, and educular surfaces is in the lower right.

many openings into the complex internal lobes of the lungs. These openings are supported airways that lead to cartilage-free airways.

9.5 CARDIOPULMONARY CIRCULATION

The extent of separation of the pulmonary and systemic circuits of flow is thought to be directed in part by a muscular ridge (Van Mierop & Kutsche, 1985) within the heart. The details of the intraventricular structure differ with taxon. Vertical septa in the ventricle are mostly absent. The muscular ridge of the ventricle, a feature that is important in directing flow within the ventricle, is present. The degree of development of the ventricle's muscular ridge is relatively low in freshwater species, such as *Trachemys*, but much more robust in animals that are active divers (sea turtles) and in large tortoises. Although a number of species have been studied generally (summarized by Farrell et al., 1998), relative few studies experimentally address both the turtle heart's structural and functional relationships (summarized by Hicks, 1998) and fewer yet are comparative (White, 1968, 1978; Burggren, 1975, 1977; Shelton & Burggren, 1976; Van Mierop & Kutsche, 1985).

Circulation through the heart differs depending upon whether blood is shunted toward the lungs and the body or primarily toward the body. Unlike mammalian cardiopulmonary systems, the pulmonary and systemic blood flows are not completely separate. The extent of separation between the pulmonary and systemic circuits of flow differs somewhat within reptiles and even within turtles with the extent of development of the ventricle's muscular ridge. For example, in the leatherback turtle, *Dermochelys coriacea*, an active and somewhat endothermic species, the muscular ridge is exceptionally well developed when compared with other marine turtles. The examination of dissected dermochelyid hearts suggests that there may be a nearly complete separation of systemic (body) and pulmonary (lung) circulation during ventricular contraction as opposing walls meet, but the physical separation is less well developed in the cheloniid species. In some marine turtles and the Galapagos tortoise (*Geochelone elephantopus*), pulmonary artery walls are thickened near the lungs, similar to a muscular sphincter (Sapsford, 1978), presumably acting to reroute blood to the body. The aortae and pulmonary trunk receive vagal innervation and can locally increase resistance to flow. Vagal tone is correlated with development and regulation of left- to-right (L-R) shunts (Comeau & Hicks, 1994; Hicks 1998; Wang et al., 2001). Experimental studies of turtles generally show that whether blood is shunted to or away from the lungs is a function of blood arterial gas levels (Shelton & Burggren, 1976; Burggren 1977).

Venous blood returning from the head, limbs, and body enters the sinus venous and then flows to the right atrium. From the right atrium, blood enters the ventricle, where it flows along at least two possible routes: (1) into the cavum venosum and out the systemic aortas, or (2) some blood may flow into the other two compartments and is ejected through the aortas and the pulmonary artery.

Computed tomography (CT) and functional magnetic resonance imaging (fMRI) can add insight into the function of the cardiopulmonary system in unanesthetized animals in a noninvasive manner. CT and fMRI images, plus dissections of fresh material, suggest the following: first, the pulmonary arteries intermittently perfuse the lung, whereas venous flow appears to be more constant; second, when shunting from the pulmonary to systemic systems occurs, several anatomical sites may be involved (within the ventricular compartments, from the pulmonary circuit to the great vessels via selective constriction of the pulmonary arteries and, more peripherally, between the pulmonary arteries and veins), suggesting control of perfusion at several levels. Experimental study of the roles each plays is warranted.

9.6 VENTILATION

Air passes through a glottis that is closed except during ventilation. Air is supplied by a long, large-bore trachea and paired bronchi (Figure 9.6). Turtles accomplish the movement of respiratory gases across the gas exchange surface by changing intracoelomic pressure using shoulder and inguinal

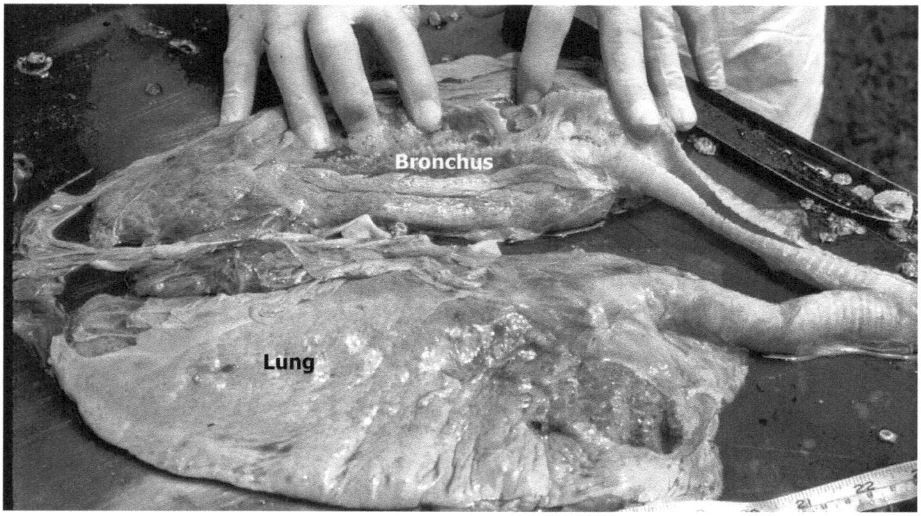

FIGURE 9.6 Dissection of the ventral surface of the lungs of *C. caretta* showing the long intrapulmonary bronchus. (From Wyneken, 2001. With permission.)

muscles (Gans & Hughes, 1967; Gaunt & Gans, 1969). Rates and flows vary depending on whether the animal is aquatic or terrestrial. Inspiration and expiration are active processes and are decoupled from locomotor activity (Landberg et al., 2003). Turtles have no diaphragm muscle separating pulmonary and coelomic compartments, but some species possess a membranous post-pulmonary ligament, which partially separates the pulmonary and coelomic compartments.

9.7 CARDIOVASCULAR SHUNTS

Shunts are responsible for rerouting blood. Intracardiac shunting mechanisms have anatomical and hydrodynamic basis; they route blood between systemic and pulmonary circuits (Figure 9.7). They are involved in changing myocardial oxygenation and coordinating the movement of blood among the ventricular compartments. Within reptiles, there are at least three other groups of peripheral shunts that can reroute blood: the pulmonary artery–systemic aorta shunt, the intrapulmonary shunt, and peripheral shunts such as the vascular circumflex of the distal limbs.

9.7.1 Cardiac Shunts

Cardiac shunts are associated with intraventricular flow, rerouting of blood between pulmonary and systemic circuits (Figure 9.7), and require an incomplete ventricular septum; these are found in noncrocodilian reptiles, including turtles. Shunts regulate blood flow and respond, in part, to blood gases; they are often described as "right to left" (R-L) referring to the shift in blood from the pulmonary circulation to the body (the system), and "left to right"(L-R), referring to the shift of some blood back to the lungs and the body when physiological conditions permit (Wang et al., 1997; Hicks, 2002).

Wang et al. (2001) demonstrated that the low systemic blood oxygen levels that result from the large R-L shunt may serve to regulate metabolism. Yet, when any of several physiological or environmental events (temperature, exercise, or digestion) increase metabolism, the R-L cardiac shunt becomes less strong and oxygen delivery to the body increases. Hicks (2002) explored this and other "adaptive hypotheses" for the functional significance of cardiac shunts. He concluded that comparative and experimental studies document the function of shunts but do not provide supporting data for the adaptive value of cardiac shunts.

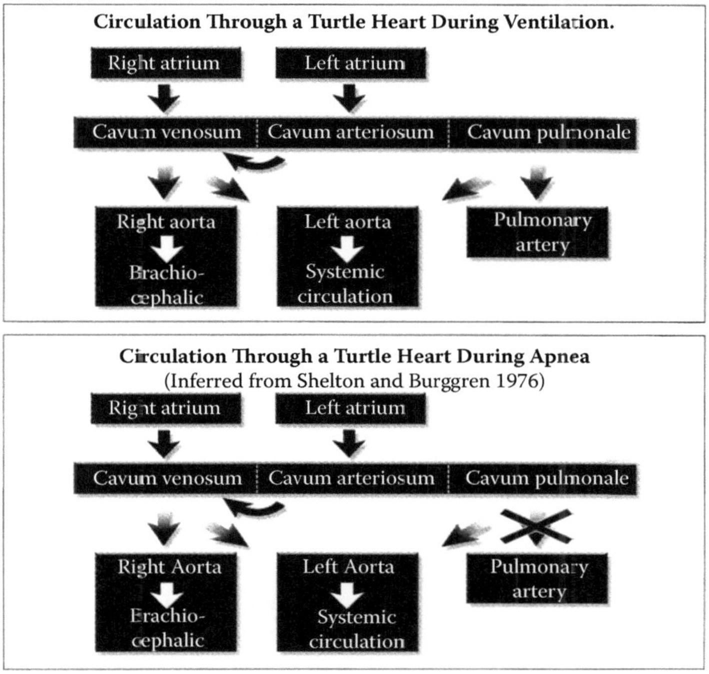

FIGURE 9.7 A diagrammatic summary of the presumed blood flow pattern in turtles during ventilation compared with apnea. Pulmonary arterial resistance often increases during apnea, resulting in a left-to-right shunting of blood. This shift routes the blood in the ventricle away from the cavum pulmonale. (From Wyncken, 2001. With permission.)

Sphincters associated with a shunting mechanism were described in the pulmonary arteries of *C. caretta*, *Chelonia mydas*, and *D. coriacea* in the 1930s (Koch, 1934). Dunlap (1955) and Sapsford (1978) redescribed the occurrence of the sphincter in *D. coriacea*, several cheloniid sea turtle species, some nonmarine turtles, and in some other reptiles. In each case, the presence of the sphincter was functionally linked to compensation for the incomplete division of the ventricle and to R-L shunting of blood to bypass the pulmonary circuits. Clearly, turtles and tortoises shunt blood between pulmonary and systemic paths. Yet, to date, the pulmonary sphincters and some degree of development of an intraventricular muscular ridge remain as the only morphological structures identified that may functionally control L-R and R-L shunting in testudines. Interestingly, these structures are not well developed in some species that develop clear L-R shunts.

9.7.2 Other Shunts

It is worth noting that there are several other types of shunts, including microcirculation that bypasses capillary beds, peripheral shunts such as the vascular circumflex in the limb, intrapulmonary shunts between the pulmonary artery and vein, at least in sea turtles (Wyneken, 2004). Any or all of these shunts may function antagonistically with the heart to redirect blood flow appropriately.

9.8 RESPIRATORY PHYSIOLOGY

Reptiles have unique systems for the assimilation and elimination of respiratory gases (Figure 9.5 and Figure 9.6). Whereas the morphology for these processes is quite varied within the reptile class, the remainder of the respiratory process is rather consistent. Of particular significance is the turtle's ability to use anaerobic metabolism for relatively long periods. The buffering systems within the body (blood, pericardial fluid, and shell) are able to compensate for the buildup of lactic acid and hydrogen ions during anaerobiosis (Stecyk et al., 2004).

Turtles are not limited to gas exchange through the pulmonary tissues. Some aquatic species have an ability to exchange respiratory gases through the integument, pharynx, or cloaca. Soft-shelled turtles (Trionychidae), may obtain up to 70% of their oxygen during submergence through the leathery shell (Stone et al., 1992). Highly vascularized pharyngeal papillae are also able to extract dissolved oxygen from the water. In some Australian sideneck turtles, cloacal bursae function in gas exchange (King and Heatwole, 1994).

9.9 OVERVIEW

The functional relationships of the cardiopulmonary system components (the pump, pipes, and ventilation) remain enigmatic. Physiologically invasive studies show that cardiac shunting occurs, yet morphological correlations are loose at best. At present, separation of systemic and pulmonary blood flows has been compared *in vivo* in few species. Even the limited survey of intracardiac and great vessel structures greatly exceeds the measurement of function. Correlations between structure and function are more closely associated with behaviors that include diving habits, migratory activity, and prolonged apnea.

Major morphological differences in cardiopulmonary structure among turtle taxa include the degree of separation of the ventricular compartments through muscular ridge development, the presence and form of pulmonary artery sphincters, and, in the lungs, size and density of air exchange structures. Differences tend to correlate with ecological specialization; however, the adaptive value of some aspects has not been experimentally demonstrated. The structural–functional features of the cardiopulmonary systems generally are correlated to the ecology and behavior; however, this is not always the case. Comparisons among species in the morphology of the structures directing L-R and R-L shunt development indicate moderate development in the loggerhead sea turtle (*C. caretta*), a migratory, intermediate depth, diving species, greater development in Galapagos tortoises (*G. elephantopus*). For example, anatomical evidence of mechanisms for the development of strong L-R and R-L shunts is weak (small muscular ridge development or no clear vascular sphincters in pulmonary circuit) in shallow diving freshwater turtles, *T. scripta*, the most studied species. Yet, the physiological evidence for L-R and R-L shunting in these turtles and some small tortoises is robust. Wang et al. (2001) reason that species that can sustain the highest metabolic rates also should possess the highest degree of anatomical ventricular separation and therefore, weaker intracardiac shunting. Turtles generally lack high metabolic rates. Arguably the most endothermic species (*D. coriacea*) may maintain high metabolic rates, but the cardiac and great vessel anatomies imply mechanisms for developing strong L-R and R-L shunting. Theoretically, this species may retain the heart morphology of its smaller, primarily ectothermic youth. Juvenile stages are ectothermic at smaller body sizes (Eckert, 2002). Unfortunately, outflow from the heart has yet to be measured in this species. The ontogeny of cardiopulmonary structure and function has not been studied in any species.

9.10 COMPARISONS WITH OTHER REPTILES

The respiratory tracts of most reptiles are anatomically and physiologically different from those of mammals. Reptiles lack a bronchial tree, and air exchange surfaces (ascini and faveoli) differ in structure from the alveoli of mammals.

In addition to airway structure, there are significant differences in lung structure and function among the four major reptilian groups: turtles, snakes, lizards, and crocodilians. Turtles all have multichambered lungs with an unbranched bronchus. Lizards cover the range from simple lungs in geckos, to multichambered lungs of iguanas, to those with long diverticula or lobes (chameleons), and the multichambered lungs with branched bronchi in varanids and helodermids. In snakes, the left lung is reduced or absent. Snakes lungs are elongate and have a highly vascular anterior part (the respiratory lung) and the caudal portion is an avascular sac (saccular lung). Many turtle and tortoise species have poorly vascularized lobes along their lateral and caudal periphery; however, the distinction is not as great as in snakes. Not surprisingly, species-specific specializations in lung structure and function are related to the ecology and behavior of the species; those lungs with the greatest surface area and complexity are found in the more active, wide-ranging, or predatory reptilian species.

Ventilation of the lungs occurs without the assistance of a diaphragm. Movements of the dorsal and ventral ribs by intercostal muscles drive ventilation in lizards. Lizards may also supplement air movement with gular pumping. Snakes use movements of the dorsal ribs for ventilation. Turtles and tortoises ventilate by movements of inguinal, axial, and shoulder muscles to change the pressure within the pleuroperitoneal cavity. Buccal oscillation is functionally separate from ventilation in turtles (Brainerd and Owerkowicz, 2006). Some aquatic species of turtle (some Australian side-necks and mud turtles) may supplement lung-based gas exchange through highly vascular gas exchange surfaces in the cloaca.

9.11 CONCLUSIONS

The turtle cardiopulmonary systems vary in the extent of lung surface area, architecture of the ventricle, and the form of the pulmonary arteries. These three characteristics are grossly correlated with behavior and ecology. Turtle hearts may be viewed as five- or six-chambered, with the chambers having context-specific roles in routing blood to the body or to the lungs and the body. Unique features of testudine cardiopulmonary systems include a weakly to moderately developed muscular ridge, pulmonary artery sphincters, highly variable cardiac output, intracardiac and intrapulmonary shunts, episodic ventilation, and diversity in lung form, size, and surface area.

ACKNOWLEDGMENTS

I thank S. Epperly, C. Farmer, M. Godfrey, B. Jensen, C. Johnson, K. Kardong, P. Lutz, S. Milton, J.A.G. Rhodin, M.D., A.G.J. Rhodin, M.D., F. Steinberg, M.D., M. Salmon, M. Starck, B. Tuller, T. Wang, and D. Wilke for technical assistance, thoughtful discussions, and/or critical review. W. Teas, J. Weege, D.V.M., D. Mader, D.V.M., S. Milton, Gumbo Limbo Environmental Center, Harbor Branch Oceanographic Institute, Loggerhead Marinelife Center, National Marine Fisheries Service, and the ARAV provided access to specimens.

REFERENCES

Bojanus, L.H. Anatome Testudinis Europaeae: an anatomy of the turtle. 1819. Reprinted Athens, OH: SSAR, 1970.

Brainerd, E.L. and Owerkowicz, T. Functional morphology and evolution of aspiration breathing in tetrapods. *Respir. Physiol. Neurobiol.* 154(1-2), 73–88. 2006.

Burggren, W.W. A quantitative analysis of ventilation tachycardia and its control in two chelonians, *Chrysemys scripta* and *Testudo graeca*. *J. Exp. Biol.* 63, 367–380, 1975.

Burggren, W.W. The pulmonary circulation of the chelonian reptile: morphology, haemodynamics and pharmacology. *J. Comp. Physiol.* 116, 303–323, 1977.

Burggren, W.W. and Shelton, G. Gas exchange and transport during intermittent breathing in chelonian reptiles. *J. Exp. Biol.* 82, 75-92, 1979.

Comeau, S.G. and Hicks, J. W. Regulation of central vascular blood flow in the turtle. *Am. J. Physiol. Regulatory Integrative Comp. Physiol.* 267: 569–578, 1994.

Dunlap, C.E. Notes on the visceral anatomy of the giant leatherback turtle (*Dermochelys coriacea* Linneaus). *Bull. Tulane Univ. Med. Fac.* 14(2): 55–69, 1955.

Duncker, H.R. Vertebrate lungs: structure, topography and mechanics. A comparative perspective of the progressive integration of respiratory system, locomotor apparatus and ontogenetic development. *Respir. Physiol. Neurobiol.* 144: 111–124, 2004.

Eckert S.A. Distribution of juvenile leatherback sea turtle *Dermochelys coriacea* sightings. *MEPS* 230: 289–293, 2002.

Eckert, S.A., Nellis, D.W., Eckert, K.L., Kooyman, G.L. Diving patterns of two leatherback sea turtles (*Dermochelys coriacea*) during internesting intervals at Sandy Point, St. Croix, U.S. Virgin Islands. *Herpetologica*, 42(3). 381–388, 1986.

Farrell, A.P. Gamperl, A.K. and Francis T.B. Comparative aspects of heart morphology. Pp. 375–424. In: Gans, C., Gaunt, A.S. (Eds.), *Biology of the Reptilia,* vol. 19. Society for the Study of Amphibians and Reptiles, Ithaca, NY, 1998.

Gans, C. and Hughes, G.M. The mechanism of lung ventilation in the tortoise *Testudo graeca* Linne. *J. Exp. Biol.* 47, 1–20, 1967.

Gaunt, A.S. and Gans, C. Mechanics of respiration in the snapping turtle, *Chelydra serpentina. J. Morph.* 128: 195–218, 1969.

Glass, M., Burggren, W.W., and Johansen, K. Ventilation in an aquatic and a terrestrial chelonian reptile. *J. Exp. Biol.* 73: 165-179, 1978.

Hicks, J.W. The physiological and evolutionary significance of cardiovascular shunting patterns in reptiles. *News Physiol. Sci.* 17(6): 241–245, 2002.

Hicks J.W. Cardiac shunting in reptiles: mechanisms, regulation and physiological function. p. 425–483. In: Gans, C., Gaunt, A.S. (Eds.), *Biology of the Reptilia*, vol. 19. Society for the Study of Amphibians and Reptilians, Ithaca, NY, 1998.

Hicks, J.W. and Wang, T., Functional role of cardiac shunts in reptiles. *J. Exp. Zool.* 275, pp. 204–216. 1996.

Johansen, K., Lenfant, C. and Hanson, D. Phylogenetic development of pulmonary circulation. *Feder. Proc.* 29(3): 1135–1140, 1970.

King, P. and Heatwole, H. Partitioning of aquatic oxygen uptake among different respiratory surfaces in a freely diving pleurodiran turtle, *Elseya latisternum. Copeia*, 1994(3), 802–806, 1994.

Koch, W. Lungengefässe und Kreislauf der Schildkröten. Biologia Generalis (Vienna).10: 359–382. 1934.

Landberg, T. J, D. Mailhot, E.L. Brainerd, E. L. Lung ventilation during treadmill locomotion in a terrestrial turtle, *Terrapene carolina. J. Exp. Biol.* 206: 3391–3404, 2003

Perry, S.F., Lungs: comparative anatomy, functional morphology, and evolution. pp. 1–92. In: Gans, C., Gaunt, A.S. (Eds.), *Biology of the Reptilia*, vol. 19. Society for the Study of Amphibians and Reptilians, Ithaca, NY, 1998.

Sapsford, C. W. Anatomical evidence for intracardiac blood shunting in marine turtles. *Zool. Afr.,* 13(1), 57–62, 1978.

Shelton G. and Burggren W.W. Cardiovascular dynamics of the Chelonia during apnoea and lung ventilation. *J. Exp. Biol.* 64, 323–343, 1976.

Stecyk, J.A.W., Overgaard, J., Farrell, A.P. and Wang, T. α-Adrenergic regulation of systemic peripheral resistance and blood flow distribution in the turtle *Trachemys scripta* during anoxic submergence at 5°C and 21°C. *J. Exp. Biol.* 207, 269–283, 2004.

Stone, P. A., Dobie, J.L. and Henry, R.P. Cutaneous surface area and bimodal respiration in soft-shelled (*Trionyx spiniferus*), stinkpot (*Sternotherus odoratus*) and mud turtles (*Kinosternon subrubrum*). *Physiol. Zool.* 65, 311–330, 1992.

Van Mierop L.H.S. and Kutsche L.M. Some aspects of comparative anatomy of the heart. pp. 38–53. In: Alfred Benzon Symposium 21; *Cardiovascular Shunts; Phylogenetic, Ontogenetic and Clinical Aspects*, (Eds.) Johansen K and Burggren W. Copenhagen: Munksgaard, 1985.

Wang, T. Krosniunasi E.H. and J.W. Hicks. 1997. The role of cardiac shunts in the regulation of arterial blood gases. *Am. Zool.* 37(1): 12-22, 1997

Wang T., Warburton, S., Abe, A. and Taylor. T. Vagal control of heart rate and cardiac shunts in reptiles: Relation to metabolic state. *Exp. Physiol.* 86(6): 777–784, 2001.

White, F. Functional anatomy of reptile hearts. *Am. Zool.* 8: 211–219, 1968.

White FN. Comparative aspects of vertebrate cardiorespiratory physiology. *Annu. Rev. Physiol.* 40: 471-99, 1978.

Wyneken, J. Guide to the Anatomy of Sea Turtles. NMFS Tech. Publication. NOAA Tech., Memo NMFS-SEFSC-470, 2001.

10 Reproductive Structures and Strategies of Turtles

Jeffrey D. Miller and Stephen A. Dinkelacker

CONTENTS

10.1 INTRODUCTION

The ecology and reproductive biology of chelonians have received a great deal of attention. Many publications focus on the ecology or reproductive biology of a single species (e.g., *Macrochelys temminckii*: Pritchard, 2006; *Chelydra serpentina*: White & Murphy, 1973; *Geochelone radiata*:

Auffenberg, 1978; *Trachemys scripta*: Moll & Legler, 1971; Gibbons, 1990; *Gopherus polyphemus*: Iverson, 1980; *Chrysemys picta*: Zweifel, 1989; Iverson & Smith, 1993; *Apalone ferox*: Iverson & Moler, 1997; Webb, 1956; *Chelodina novaeguineae*: Kennett et al., 1992). Some studies provide comparative information on reproductive behaviors and biology of two or more species (e.g., *Chelydra serpentina*, *Apalone*, *Macrochelys*, *Trachemys*), and some contain drawings or pictures of pre-copulatory and mating behavior of different species (e.g., Legler, 1955; Webb, 1961, 1962; Auffenberg, 1978; Bellairs, 1970; Bustard, 1972; LeBuff, 1990; Obst, 1986; Oliver, 1955; Dodd, 2001; Ewert, 1976; Harrel et al., 1996). Some provide comparative information on the biology and ecology of multiple species grouped by general habitat; for example, river turtles (Moll & Moll, 2004) and sea turtles (Lutz & Musick, 1997; Lutz et al., 2002). Still others summarize the information concerning multiple species over large geographic areas (Australia: Legler, 1985; Cann, 1998; Venezuela: Pritchard & Trebbau, 1984). Some species and groups have received more attention than others as a result of their accessibility (e.g., *Pseudemys concinna*: Seidel & Dreslik, 1996; *Trachemys scripta*: Seidel & Ernst, 2006; *Terrapene carolina*: Ernst & McBreen, 1991), distribution (e.g., *Chelydra serpentina*: Ernst et al., 1988; Gibbons et al., 1988; Lovich, 1993; *Apalone*: Webb, 1990), or conservation status (Swingland & Klemens, 1989; Van Abbema, 1997; Klemens, 2000).

The focus of this chapter is on the morphological diversity of chelonian reproductive structures and the effects these have on the range of reproductive strategies exhibited by turtles. Our goals are to provide a summary of information and ideas and to provide access to the literature in hope of stimulating further research. The literature cited throughout has been selected to demonstrate a point and to provide access to more detailed literature. No attempt has been made to cite every exemplary document that exists in the vast literature concerning chelonian morphology and reproductive ecology.

10.2 GENERAL MORPHOLOGY

The overall the anatomy of chelonians is consertative in that they have a shell encasing the soft parts of the body. However, the turtle shell—which incorporates the ribs and encases the girdles—is not just a two-part, ridged box connected around the body of the turtle by bridging bones. In some groups (e.g., Dermochelyidae, Trionyichidae), the dermally derived (thecal) portion of the shell is greatly reduced; in these, the pliable shell structure is provided by epithecal ossicles and connective tissue covered by epidermis (Zangerl, 1969). Among hard-shelled chelonians, the carapace is formed by thecal dermal elements and is more rigid than the plastron, with the exception of *Kinixys* in which the carapace is movable (Zangerl, 1969). Plastronal kinesis allows some chelonians to close the gap between the carapace and plastron, thereby providing increased protection of the soft parts. Kinesis occurs in both pleurodires (e.g., *Pelusios*) and cryptodires (e.g., *Kinosternon*, *Sternotherus*, *Emydoidea*, *Emys*, *Terrapene*, *Cuora*, *Kinixys*, *Pyxis*, *Testudo*) but is not universal, even among closely related species or between the sexes within a species (e.g., *Testudo graeca*) (Zangerl, 1969). Hinges occur in different places on the plastron and may have weak to strong mobility. For a hinge to work, the sutures between both the epidermal shields and the underlying bones must be aligned. Hinges may occur anterior to the bridge (*Pelusios*, *Terrapene*, *Pyxis*), both anterior and posterior (*Kinosternon*, *Sternotherus*), or on the bridge (emydids) (Zangerl, 1969).

The evolution of the turtle shell occurred in concert with a number of other anatomical changes that had physiological and behavioral consequences (Zangerl, 1969). For example, changes occurred in the shape and articulations of the bones of the neck (Zangerl, 1969); also, muscles of the head and jaw region altered over time, thus allowing some diversity in foraging (Schumacher, 1973; Shah, 1963) while retaining one of two general patterns of neck articulation (pleurodires, cryptodires). Further, alterations of size and shape of limbs and feet that occurred with the evolution of the shell in conjunction with specializations associated with habitat use affected patterns of locomotion (Walker, 1973, 1979). As a result, some chelonians have paddle-shaped flippers well suited to an aquatic life (Cheloniidae, Carettochelyidae, Trionychidae), whereas others have expansible and webbed hind feet and reduced webbing on their forefeet (many Emyididae, Chelidae, Pelomedusidae, Geomydidae) suited for aquatic and semi-aquatic habitats. Others have elephantine feet (Testudinidae, some Emydidae) specialized for terrestrial locomotion.

10.3 REPRODUCTIVE STRUCTURES

The reproductive structures of turtles include the gonads (testes, ovaries) and associated ducts to carry sperm (epididymis, vas deferens) or ova (oviducts), and the cloaca containing the clitoris or penis and associated structures.

10.3.1 Gonads

The development of the urinogenital system in reptiles, including reproductive organs and associated ducts of turtles, was reviewed by Fox (1977) and Raynaud and Pieau (1985). Miller and Limpus (2003) provide a detailed description of the ontogeny of the gonads of marine turtles; a similar pattern occurs in other cryptodires (Fox, 1977). No similar description exists for pleurodires, but it is assumed that the process is similar.

The gonads develop on the germinal ridges on the ventro-medial surfaces of the mesonephros (*Sternotherus*: Risley, 1933a; *Chrysemys picta marginata*: Allen, 1906; Fox, 1977). As differentiation continues, testis and ovary become distinguishable by the histological differences in the relative amounts of the cortex (thin layer in males) or medulla (dense, non-tubular in females) during the last half of embryonic development (Fox, 1977). By hatching, the gonads of most species are distinguishable based on their microscopic structure (e.g., *Caretta caretta*: Mrosovsky & Yntema, 1980; *Chelonia mydas*: Miller & Limpus, 1981, 2003). Vestigial oviducts (Müllerian ducts) occur in many species of male chelonians (e.g., *Emydoidea blandingii*: Nicholson & Risley, 1941; *Mauremys caspica leprosa*: Stefan, 1959, 1963; Fox, 1977), at least for a while after hatching. Conversely, the embryonic expression of the vas deferens (Wolffian duct or mesonephric duct) may persist in the female but tends to degenerate to become part of the peritoneal wall as the turtle ages (Fox, 1977; Wibbels et al., 1999).

10.3.1.1 Testis

Details of the structure of the testis and the spermatogenetic cycle have been described—at least in part—for a number of species in six families (Emydidae, Chelydridae, Cheloniidae, Kinosternidae, Testudinidae, Trionychidae) (Table 10.1; Gist, 1999; Girling, 2002). Testes have been illustrated in varying detail in drawings and photographs (e.g., Agassiz, 1857; Bojanus, 1819; Thomson, 1932; Noble & Noble, 1940; Ashley, 1962; Miller & Limpus, 2003; Owens, 1980, 1997; Lance & Rostal, 2002; Hamann et al., 2003; Rostal, 2005). The testes are located toward the dorsal-midline of the body cavity, attached mediolaterally to the kidneys (*Emydoidea blandingii*, *Trachemys scripta*, *Glyptemys insculpta*, *Chelydra serpentina*: Fox, 1977). The hatchling testis shows clear immature seminiferous tubules within the medulla (e.g., *Caretta caretta*: Mrosovsky & Yntema, 1980; *Chelonia mydas*: Miller & Limpus, 1981, 2003). The testis does not become functional until puberty; however, it does increase in size as the turtle grows. Maturation is a slow process and may require a decade in some species (*Caretta caretta*: Limpus & Limpus, 2003). During puberty, the testis enlarges and the epididymis extends from the peritoneal wall until it is pendulous (Miller & Limpus, 2003).

Testes of mature turtles are oval in cross section but may be cylindrical or flattened; typically, they have a smooth surface with seminiferous tubules visible within the tissue (Moll, 1979; Owens, 1980, 1997; Miller & Limpus, 2003). The histological structure of the adult testis reveals the thin epithelium of the cortex on the tunica albuginea. The seminiferous tubules are contained within the vascularized connective tissue of the medulla. Leydig's cells occur throughout the stroma in

TABLE 10.1
Descriptions of the Spermatogenetic Cycle of Turtles

Species	Aspect	Reference
Emydidae		
Chrysemys picta	Description of testicular cycle	Ernst (1971a); Christiansen & Moll (1973); Moll (1973); Dubois et al. (1988)
Chrysemys picta	Seasonal microscopic changes	Gibbons (1968b); Sprando & Russell (1988)
Mauremys caspica	Seasonal changes in lipids	Lofts & Boswell (1961)
Mauremys leprosa	Description of testicular cycle	Combescot (1955b)
Tarrapene ornata	Description of testicular cycle	Legler (1960a)
Terrapene carolina	Brief description of testicular cycle	Hansen (1938); Atland (1951); Sprando & Russell (1988)
Trachemys scripta	Description of testicular cycle	Moll & Legler (1971); Sprando & Russell (1988)
Kinosternidae		
Kinosternon flavescens	Description of testicular cycle	Mahmoud & Klicka (1972); Christiansen & Dunham (1972)
Kinosternon odoratum	Spermatogenesis	Sprando & Russell (1988)
Kinosternon subrubrum	Description of testicular cycle	Mahmoud & Klicka (1972)
Sternotherus carinatus	Description of testicular cycle	Mahmoud & Klicka (1972)
Sternotherus odoratus	Description of testicular cycle Environmental timing mechanisms	Risley (1933a, 1933b, 1938)
Chelydridae		
Chelydra serpentina	Description of testicular cycle	White & Murphy (1973)
Macrochelys temminckii	Description of testicular cycle	Dobie (1971)
Trionychidae		
Lissemys punctata	Subcellular spermatogenesis	De & Maiti (1989)
Pelodiscus sinensis	Description of testicular cycle	Lofts & Tsui (1977)
Cheloniidae		
Caretta caretta	Description of testicular cycle	Wibbels et al. (1990)
Lepidochelys kempii	Description of testicular cycle	Rostal (2005)
Testudinidae		
Testudo ibera	Hormonal control of testicular cycle	Combescot (1955c)
Testudo hermanni	Description of testicular cycle	Kuchling (1981, 1982a); Kuchling et al. (1981)

See also reviews by Lofts (1968, 1969), Moll (1979), Licht (1984), Owens (1997), Fox (1977), Gist (1999), Kuchling (1999), and Hamann et al. (2003).

the tissue between the testicular tubules. These interstitial cells secrete testosterone and lipoidal substances (Pellegrini, 1925a, 1925b) under pituitary hormonal control (Licht, 1974; Fox, 1977; Crews & Garrick, 1980). Interstitial cells enlarge prior to spermatogenesis then regress as the reproductive cycle ends. During their active period, they stimulate secondary sexual characteristics such as changes in coloration (Cooper & Greenberg, 1992) and courtship behavior (Moore & Lindzey, 1992). Spermatogonia occur inside the basement membrane of the seminiferous tubules; during the breeding season, primary spermatocytes are also visible in this area. Sertoli cells are situated inside the seminiferous tubules next to the basement membrane (*Sternotherus odoratus*: Risley, 1938; *Terrapene carolina*: Altland, 1951). Sertoli cells facilitate spermatogenesis via adenohypophysial stimulation (Lofts, 1968). The cells may fill the tubule lumen at the end of the reproductive period (*Mauremys caspica*: Lofts & Boswell, 1961). Sertoli cells eliminate remaining sperm during regression of the seminiferous tubules (Fox, 1952).

Licht (1984) defined four phases of the spermatogenetic cycle for reptiles: regenerative, progressive, cumulative, and regressive. The spermatogenetic cycle can be further divided into five phases (Moll, 1979) or eight (McPherson & Marion, 1981; Wibbels et al., 1990). The differences in the definitions of phases result from different emphasis being placed on particular characteristics of spermatogenesis. At the cellular level, spermatogenesis appears to follow a general pattern of sperm production that is regulated by pituitary hormones (Licht, 1974; Crews & Garrick, 1980; Owens 1997).

Mature testes change size in reflection of spermatogenesis: size increases as spermatogenesis increases and regression occurs as activity decreases (e.g., *Kinosternon subrubrum hippocrepis*, *K. flavescens*, *Sternotherus odoratus*, *S. carinatus*: Mahmoud & Klicka, 1972; *Clemmys*: Lofts & Boswell, 1961; *Trachemys*: Moll & Legler, 1971; *Chrysemys picta*: Gibbons, 1968a; Ernst, 1971a; *Terrapene ornata*: Legler, 1960a; *Testudo hermanni*: Kuchling, 1982a, 1982b; *Mauremys caspica leprosa*: Combescot & Guyon, 1955). Seminiferous tubules increase in diameter as they fill with sperm (*Mauremys caspica*: Lofts & Boswell, 1961; *Sternotherus odoratus*: Risley, 1938b); seminiferous tubules may regress following the reproductive season (*Trachemys*: Moll & Legler, 1971) or may contain sperm throughout the year (*Macrochelys temminckii*: Dobie, 1971; *Chelydra serpentina*: Conant, 1938). Testis size and pattern of enlargement as well as the timing of sperm in the seminiferous tubules may vary within a species exhibiting temperate and tropical patterns of reproduction (Moll & Legler, 1971; Moll, 1979). The temperate pattern shows "peak Leydig cell activity during gametogeneic quiescence" (Moll, 1979), whereas in the tropical pattern, the period of quiescence is short and the peak Leydig cell activity does not necessarily occur during the quiescent period (Moll, 1979).

The epididymis is attached to the peritoneal wall near the testis and is located between the rete testis and the vas deferens. The epididymis is a low ridge until puberty, but as puberty progresses it becomes more convoluted and extended from the peritoneal wall (Miller & Limpus, 2003; Limpus & Limpus, 2003). During the reproductive cycle, the epididymis becomes enlarged and turgid-white in appearance; it regresses but remains obvious during the non-reproductive period (Miller & Limpus, 2003). The vas deferens provides for sperm transport between the epididymis and the ostia leading into the cloaca near the proximal end of the penile groove; the histology of this has not been described.

Turtles do not have spermatheca; instead, sperm are stored in the epididymis until passed to the female during coitus. Some turtles may have sperm in the epididymis year-round even though spermatogenesis occurs seasonally (Moll, 1979).

10.3.1.2 Ovary

Details of the structure of the ovary and the changes that occur throughout the reproductive cycle have been described in few turtle species from three families (Cheloniidae, Geoemydidae, Kinosternidae) (Table 10.2; Girling, 2002). The hatchling ovary shows a compact (homogeneous) medulla separated by the tunica albuginea from the columnar epithelium of the cortex. Germinal

TABLE 10.2
Descriptions of the Ovarian Cycle and Changes in the Structure of the Oviducts in Turtles

Species	Aspect	Reference
Ovary		
Several species	Microscopic oogenesis	Agassiz (1857); Munson (1904); Altland (1951)
Cheloniidae		
Chelonia mydas	Fine structure	Aitken et al. (1976)
Marine turtles	Ontogeny of gonads	Miller & Limpus (2003); Hamann et al. (2003)
Lepidochelys kempii	General female cycle	Rostal (2005)
Geoemydidae		
Rhinoclemys funerea	General female cycle	Moll & Legler (1971)
Kinosternidae		
Kinosternon leucostomum	General female cycle	Moll & Legler (1971)
Trionychidae		
Lissemys p. punctata	Maturation of gonad	Sarkar et al. (1995); Sen & Maiti (1990)
Oviduct		
Emydidae		
Terrapene carolina	Seasonal changes in fine structure	Hansen & Risley (1941); Legler (1960a)
Mauremys leprosa	Seasonal changes in fine structure	Combescot (1954, 1955a)
Mauremys caspica	Seasonal changes in fine structure	Lofts & Boswell (1961)
Chrysemys picta	Seasonal changes in fine structure	Moll (1973); Christiansen & Moll (1973); Ernst (1971d); Gibbons (1968a)
Trachemys scripta	General female cycle	Moll & Legler (1971)
Chelydridae		
Chelydra serpentina	Seasonal changes in fine structure	White & Murphy (1973)
Macrochelys temminckii	Seasonal changes in fine structure	Dobie (1971)
Kinosternidae		
Kinosternon flavescens	Seasonal changes in fine structure	Christiansen & Dunham (1972); Mahmoud & Klicka (1972)
Kinosternon subrubrum	Seasonal changes in fine structure	Mahmoud & Klicka (1972)
Sternotherus carinatus	Seasonal changes in fine structure	Mahmoud & Klicka (1972)
Sternotherus odoratus	Seasonal changes in fine structure	Mahmoud & Klicka (1972)
Testudinidae		
Gopherus polyphemus	Regions and fine structure	Palmer & Guillette (1988)
Trionychidae		
Lissemys p. punctata	Seasonal changes in fine structure	Sarkar et al. (1995); Sen & Maiti (1990)

oocytes appear in the cortex as large cells typically with an obvious space within the cell membrane (Miller & Limpus, 1981, 2003) until the turtle begins puberty. During puberty, the stroma of the ovary increases in size, providing the support for the first development of the oocytes in late puberty and regular follicular development in adulthood.

The mature ovary is comprised of stroma, pre-vitellogenetic follicles, maturing follicles, atretic follicles, corpus lutea, and corpus atretica; the relative numbers of each depend on the maturity of the individual and its phase in its reproductive cycle. Ovaries have been illustrated in varying detail in drawings and photographs (e.g., Agassiz, 1857; Bojanus, 1819; Thomson, 1932; Noble & Noble, 1940; Ashley, 1962; Moll, 1979; Miller & Limpus, 2003; Rostal, 2005; Owens, 1980, 1997). The morphology of the ovary has not been described for pleurodires; it is presumed to be essentially the same as in cryptodires.

In general, the descriptions of the morphology of the ovary of the adult green turtle (*Chelonia mydas*: Aitken et al., 1976; Owens, 1980; Miller & Limpus, 2003), the soft-shelled turtle (*Lissemys punctata punctata*: Sarkar et al., 1996), and Blanding's turtle (*Emydoidea blandingii*: Nicholson & Risley, 1941) represent the ovaries of all chelonians. The ovary is a "membranous, curtain-like structure with a relatively short attached border" (*Chelonia mydas*: Aitken et al., 1976). In smaller species, the regressed ovaries appear to be more compact but stretch, becoming curtain-like during the reproductive period. Regardless of species, the expression of the curtain-like stroma is greatest in mature, reproducing individuals. Follicles are present on both sides of the ovarian stroma. Various aspects of the structure of follicles have been described for several species (*Testudo*: Loyez, 1906; *Clemmys guttata*, *Graptemys geographica*, *Emydoidea blandingii*, *Sternotherus odoratus*, *Chrysemys picta*: Thing, 1918) and illustrated by Agassiz (1857). The ovarian cycle—including vitellogenesis, ovulation, the retention of corpora lutea, and their eventual regression as corpora albucantia—is controlled by hormones (Licht et al., 1979, 1980; Licht, 1980; Owens, 1980, 1997). Owens (1997) and Hamann et al. (2003) provide integrated reviews of the role of hormones in the reproductive ecology of sea turtles; based on the existing information, the general pattern holds true for other chelonians, especially for those species that produce multiple clutches in a breeding season (Lofts, 1968, 1969; Moll, 1979; Licht, 1984; Owens, 1997; Kuchling, 1999).

The ovarian cycle is divided into four phases (Moll & Legler, 1971; Moll, 1979):

Enlargement of follicles
Ovulation and preoviposition (intrauterine) period
Oviposition
Follicle regression and quiescence (between reproductive seasons)

Toward the end of puberty, the turtle ovary is stimulated to initiate vitellogenesis, wherein multiple follicles enlarge but do not ovulate (Limpus & Limpus 2003). These follicles become atritic; as they regress, they are invaded by blood vessels and eventually reduce in size. As they become atritic, the color changes from the yellow-cream color of a maturing follicle to become orange and eventually red-orange as the invading blood vessels proliferate. They eventually become small reddish lumps (corpora atretica) in the stroma.

In reproductively quiescent turtles, no enlarged follicles occur in the stroma but previtellogenetic follicles are abundant. Previtellogeneic follicles contain an oocyte surrounded by a layer of cuboidal granulosa that, in turn, is surrounded by the vacuolated cells of the gonadal mound. Previtellogenetic follicles appear yellow-cream in color; several different size classes may be evident but all are much smaller than a follicle in vitellogenesis. Small white corpora lutea may be visible in the stroma. The expanded stroma contains many arteries and veins (Aitken et al., 1976).

In reproducing individuals, follicles undergo vitellogenesis and enlarge. In temperate freshwater species, follicular enlargement may occur before or after brumation, depending on geographic location, age, or species (Table 10.3). For *Chelonia mydas*, vitellogenesis requires a year or two, possibly longer, before ovulation (Limpus & Nicholls 1988, 2000). In species that produce multiple

TABLE 10.3
Examples of the Timing of Follicular Development in Relation to Hibernation in Several Species of Turtles

Genus/Species	Follicular Development	Reference
Macrochelys temminckii, Chelydra serpentina	Before hibernation	Dobie (1971); White & Murphy (1973)
Kinosternon sternotherus	After hibernation, just before ovulation	Christiansen & Dunham (1972); Mahmoud & Klicka (1972)
Mauremys caspica	After hibernation, just before ovulation	Lofts & Boswell (1961)
Chrysemys picta	In Northern populations, before hibernation; in southern populations, after hibernation	Moll (1973); Christiansen & Moll (1973); Ernst (1971d); Gibbons (1968a); Powell (1967)

clutches, enlarged follicles will not be the same size (Gibbons & Tinkle, 1969). The largest will be ovulated in the first clutch; follicles for the second clutch will increase in size while the eggs are in the oviduct or immediately following oviposition. In general, the number and diameter of enlarged follicles correlates with the number and the size of the eggs to be deposited in the clutch; a larger number of follicles yields a larger egg count in the clutch and larger diameter follicles yield larger diameter eggs. However, one or more follicles may enlarge but not be ovulated; these will regress to become corpora atretica.

Following ovulation, the stroma and theca surrounding the follicle collapse to form a corpus luteum. This is the primary site of steroidogenesis converting pregnenolone and cholesterol to progesterone (Callard et al., 1976, 1991; Klicka & Mahmoud, 1972, 1973; Licht & Crews, 1976). The corpus luteum is a disc-shaped structure comprised of the theca externa and theca interna plus a thick layer of granulose cells; the granulose cells go through luteinization to become granulose lutein cells (Cyrus et al., 1978). The supporting ovarian tissue swells and at least in some species becomes vesicular (e.g., Cheloniidae: Miller & Limpus, 2003). The theca interna begins to invade the granulose lutein layer with collagenous fibers, fibroblasts, and blood vessels (Cyrus et al., 1978). Corpora lutea have been reported in *Trionyx* (Agassiz, 1857), *Terrapene ornata*, *Emydoidea blandingii* (Legler, 1958), *Chrysemys picta* (Powell, 1967), *Trachemys scripta* (Moll & Legler, 1971), *Chelydra serpentina* (Klicka & Mahmoud, 1973), and *Geochelone gigantea* (Swingland & Coe, 1978). The corpora lutea remain large until the uterine eggs are oviposited, after which they regress to become corpora albicantia. Given the function of the corpora lutea, it is likely that they occur in all chelonians.

As regression progresses, the corpora lutea decrease in size, first becoming semi-white then white as connective tissue replaces gransulosa lutein cells. Eventually, the structure becomes a small scar (corpora albicantia) on the ovarian stroma. The scar remains visible for the current breeding season and may be retained for several years, at least in some species (Cheloniidae: Miller & Limpus, 2003). When corpora albicantia remain as small scars embedded in the stroma, they provide clear evidence that the turtle has reproduced (*Apalone spinifera*: Moll, 1979; *Chelonia*: Miller & Limpus, 2003; *Caretta caretta*: Limpus & Limpus, 2003). However, corpora albicantia may disappear within a short time following ovulation (*Trachemys*: Moll & Legler, 1971; *Chrysemys picta*: Ernst, 1971a). The ovaries of more species should be examined carefully to look for the presence of small scars (approx 1 to 2 mm) that carry over between reproductive episodes.

At the end of the reproductive season, any remaining follicles begin to regress. Following regression, the ovaries enter quiescence and remain essentially quiescent between reproductive seasons. It is assumed that the quiescent period allows the female to regain the energy expended in maintaining her system and required for the production of her eggs. In reproductively quiescent turtles, the ovary undergoes a general restoration via the regression of the corpus luteum, absorption of non-ovulated, enlarged follicles that become corpora atretica, and a general shrinkage of the stretched stroma. Existing previtellogeneic follicles remain scattered throughout the unattached regions of the ovarian stroma.

Occasionally, during ovulation a blood vessel in the stroma ruptures and bleeds onto the surface of the follicle. The blood spot remains even if the embryo dies before obvious development of embryonic blood islands. When the egg is opened at the end of incubation, ovarian blood spots may be confused with embryonic blood islands.

Extra-uterine migration of follicles has been noted to occur in several turtles (*Terrapene, Emydoidea, Chrysemys, Pseudemys*: Legler, 1958; *Sternotherus*: Tinkle, 1959). There is no apparent advantage to extra-uterine migrations of oocytes; the proximity of the infundibulum may facilitate the exchange.

10.3.2 Oviduct

Each female turtle has two oviducts that open independently into the cloaca. The oviducts have multiple functions; in addition to providing for the passage of the egg from the ovary to the nest, the oviduct facilitates sperm storage, fertilization, albuminization, deposition of the shell membrane and shell matrix, and storage of the eggs before oviposition (Girling, 2002). In cross-section, each oviduct shows two general layers: mucosa and muscularus. The mucosa includes the epithelial lining, as well as the glandular and connective tissue of the lamina propria. The glandular cells may be branched saccular or branched acinar, tubo-alveolar, or simple tubular (Girling, 2002). The muscularus contains inner circular and outer longitudinal muscle layers. The oviduct is encased in a serosa and is attached to the peritoneum by the mesotubarium mesentery.

Each oviduct is divided into five functional regions that differ in structure and function: infundibulum, uterine tube, isthmus, uterus, and the vagina (Aitken & Solomon, 1976; Fox, 1977; Palmer & Guillette, 1988; Girling, 2002; Table 10.4). The ostium of the infundibulum is suspended from the peritoneal wall. The infundibulum is comprised of two portions—the anterior portion is expanded whereas the posterior portion is contracted, giving the structure a funnel-like shape (*Gopherus polyphemus*: Palmer & Guillette, 1988) into which the ovulated follicle enters from the coelom. The membranous tissue of the anterior infundibulum is lined with ciliated and non-ciliated cells, some of which appear to be glandular. The posterior portion becomes folded into longitudinal grooves and contains ciliated and non-ciliated cells (Palmer & Guillette, 1988). In addition, the posterior portion of the infundibulum appears to be the aglandular zone (Aitken & Solomon, 1976; Girling, 2002) because it is transitional between the infundibulum and the uterine tube. The uterine tube is

TABLE 10.4
Definition of Recognized Sections of the Chelonian Oviduct*

Section	Function
Infundibulum, including anterior and posterior regions	Entry of follicles into oviduct Epithelium consists of probable site of fertilization, at least in posterior region
Uterine tube	Secretes albumen; contains sperm storage glands
Isthmus	Narrow, non-glandular section between the uterine tube and uterus
Uterus	Secretes egg shell membrane fibers from endometrial glands; contains sperm storage glands
Vagina	Holds eggs from entering cloaca before oviposition

* *Based on Palmer & Guillette (1988), Palmer & Guillette (1991), and Girling (2002).*

also lined with ciliated and non-ciliated cells, as well as glandular epithelial cells (Palmer & Guillette, 1988). The glandular structure of the uterine tube suggests that it produces albumen from the numerous mucosal glands (Palmer & Guillette, 1988; Aitken & Solomon, 1976). This region has internal longitudinal folds; externally, it appears "pleated" transversely (*Gopherus polyphemus*: Palmer & Guillette, 1988). The oviductal isthmus appears to be transitional between the uterine tube and the uterus; this short section may be aglandular (*Gopherus polyphemus*: Palmer & Guillette, 1988) or contain active glandular cells, at least during ovulation (*Lissemys p. punctata*: Sarkar et al., 1995).

The endometrial glands of the uterine section secrete the fibers of the eggshell (*Gopherus polyphemus*: Palmer & Guillette, 1988). The epithelial lining of the shell-forming region contains ciliated and non-ciliated columnar cells (Aitken & Solomon, 1976; Palmer & Guillette, 1988; Motz & Callard, 1991; Sarkar et al., 1995). The shell-forming region in turtles sequentially secretes both the supporting shell membrane and the calcareous matrix of the shell (Aitken & Solomon, 1976). In the shell-secreting region, the crystalline structure of the eggshell is deposited onto the strands of the membrane matrix. Differences in the structure of the eggshells produced may reflect differences in the glandular structure of the uterine region (Girling, 2002). The heavily folded vagina is lined with ciliated epithelial cells, and the muscularus layer is thick. The short, muscular vagina keeps the eggs from entering the cloaca until oviposition (Palmer & Guillette, 1988). The vagina may or may not contain glands; if present, the glands support sperm storage (Sever & Hamlett, 2002).

Details of the impact from changes in hormonal levels on oviductal tissues throughout the reproductive cycle have been described in few turtle species from three families: Emydidae, Testudinidae, Trionychidae (Table 10.5; Girling, 2002). Plasma levels of estradiol seem to control the changes in the oviductal tissues (Girling, 2002). Secreted by the corpora lutea, progesterone helps to maintain the function of the oviduct while the eggs are being coated with albumen and covered by the shell material. The openings of the two oviducts enter the cloaca from the ventral medio-lateral aspect (*Emydoidea blandingii*: Nicholson & Risley, 1941), the dorsal-lateral aspect (*Trionyx euphraticus*: Salih, 1965), or the ventro-lateral aspect (*Emys (Testudo) europaea*: Bojanus, 1819; *Testudo graeca*: Thomson, 1932; *Chelonia mydas, Caretta caretta, Eretmochelys imbricata*: Miller, unpublished).

10.3.3 Cloaca and Associated Structures

The cloaca is an expandable tube that has a multi-ridged, longitudinally folded epithelium surrounded by longitudinal and transverse muscles, and connects between the sphincter of the colon and the sphincter at the anus. The cloaca of the female is relatively short, whereas the cloaca of the male is much longer. The increased length allows for the penis to be accommodated inside the cloaca. The bladder, any ancillary bladders, ureters, and the oviducts or vas deferens enter the cloaca near the proximal end, usually from the ventrolateral aspect (Kuchling, 1999).

Kuchling (1999) described the cloaca as being partially divided into "an upper and lower chamber" by septae and stated that "the vasa deferentia and ureters of each side open into the lower chamber" called the sinus urogenitalis. Older descriptions of the cloaca show some disagreement in terminology and observed structures (Bojanus, 1819; Thomson, 1932). However, these descriptions agree that there is a small vestibule formed by the ostia of the oviducts and folds of the cloacal wall in the floor of the cloaca. There has been no modern detailed description of the cloaca of turtles that allows comparison within and among taxa.

All chelonians have urinary bladders (Minnich, 1982) comprised of transitional epithelium that allows expansion and has at least a few muscle fibers that facilitate contraction. The bladder is filled by urine carried through ureters from the kidneys; the ureters connect at the base of the bladder in the short muscular neck area between the bladder and the cloaca (Thomson, 1932; Noble & Noble, 1940; Ashley, 1962). The bladder empties into the cloaca through a short urethra that connects between the two structures. The urinary bladder may be singular, as in sea turtles, or bi-lobed, as in most freshwater species.

TABLE 10.5
Changes in Hormonal Levels Correlated with Changes in the Oviduct during the Reproductive Cycle of Three Chelonians

Species	Change in Hormonal Level	Change in Oviduct Tissue	Source
Chrysemys picta	Increased levels of estradiol and progesterone result in preovulatory increase in uterine tube and uterus thickness and gland activity; progesterone receptors in nuclei of uterine epithelial cells, sub-mucosal glands, smooth muscle	Increase in size of glands in uterine tube and uterus; thickness increased maintained until after oviposition	Motz & Callard (1991); Callard et al. (1978); Salhanick et al. (1979); Giannoukos & Callard (1996)
Gopherus polyphemus	Increase in mucosal thickness during vitellogenesis when estradiol is secreted by the ovary	Epithelia cell height and mucosal thickness increased in uterine tube and uterus during vitellogenesis; epithelium remained enlarged but mucosal thickness declined following ovulation	Palmer & Guillette (1990)
Lissemys p. punctata	Increasing plasma estradiol levels until peak at ovulation then declined; progesterone peaked post ovulation	Oviductal weight increased prior to ovulation and decreased following ovulation in coordination with epithelial cell height, gland size, and activity	Sarkar et al. (1995, 1996)

Not all turtles have cloacal bursae (Minnich, 1982); some turtles have cloacal bursae that are small and easily missed in gross dissection. For example, sea turtles have rudimentary cloacal bursae contained in the connective tissue on either side of the anterior part of the cloaca (Wyneken, 2001); their function, if any, is unknown. The ancillary bladders (cloacal bursae) are also comprised of transitional epithelium, which allows them to expand and store fluid. These empty into the cloaca separately from the urethra.

In general, functional cloacal bursae occur in those species likely to undergo water stress. In these cases, the cloacal bursae are used to store water that may be absorbed to keep the body hydrated during periods of water stress (Munnich, 1982; Peterson, 1996). In addition, water in the bursae or bladder may be used by female terrestrial turtles in xeric environments or ones with compacted substrates to soften soil to aid digging the egg chamber prior to oviposition (Patterson, 1971). The wetting of the surrounding substrate may also serve to buffer the hydric environment for the eggs after oviposition (*Chelodina longicollis*, *Geochelone elephantopus*, *G. pardalis*, *Apalone ferox*, *Gopherus berlandieri*, *Trachemys scripta*, *Chrysemys picta*, *Malaclemys terrapin*, *Terrapene carolina*: Ehrenfeld, 1979). Although some of these species regularly nest in areas where wetting the substrate might assist digging, some do not. In one study, painted turtles (*Chrysemys picta marginata*) transported an average of 44.7 g (range: 24.4 to 79.1 g) of water that was voided during the excavation of their nests (Kinney et al., 1998); these turtles carried an additional 11% of their gravid mass to the nest site. Interestingly, if they voided during handling prior to nesting, they returned to the water before attempting to nest, suggesting that the carriage of the water was purposeful (Kinney et al., 1998). Possible ecologically related explanations for such behavior include "(1) softening of the substrate, (2) providing moisture for the eggs, (3) bacteriostatic effect,

(4) repulsion of egg predators through taste or smell, (5) compaction of dirt after nesting to ward off predators, and (6) camouflage" (Ehrenfeld, 1979, based on Patterson, 1971).

The inner surfaces of the ancillary bladders (and the cloaca) may be vascularized to assist in regulation of electrolyte balance by absorbing sodium (*Chrysemys picta*: Dunson, 1967; Minnich, 1982) and contribute to respiration as aquatic respiratory organs (*Apalone spinifera asper*: Dunson, 1960; *Trionyx triunguis*: Girgis, 1961). Other species may also use aquatic respiration, at least to some degree (*Podocnemis* spp.: Steen, 1971, cited by Seymour, 1982). Respiratory cloacal bursae are found in *Rheodytes leukops,* where they augment respiration (Legler, 1979). The inner lining is modified into vascularized villiform projections that facilitate the uptake of oxygen from the water and the release of carbon dioxide to the water; water is circulated by muscular contractions of the bursae and cloaca (Legler, 1979; Legler & Cann, 1980). The augmentation of respiration allows this species to remain submerged and active for extended periods to time (Gordos et al., 2003a, 2003b); typically, it remains in highly oxygenated water located downstream of riffles (Cann, 1998).

Male chelonians have a single penis situated in the ventral midline of the cloaca. The penis is comprised of erectile tissue (Andersson & Wagner, 1995). In cryptodires, the penis is divided into two unequal portions: the long proximal shaft and the shorter distal glans (Zug, 1966). In pleurodires, the shaft portion is virtually nonextant and the entire structure appears to be the glans (McDowell, 1983).

Two types of erectile tissue occur in the turtle penis: corpus spongiosum and corpus fibrosum (Zug, 1966). McDowell (1983) uses the term corpus cavernosum instead of corpus spongiosum based on the lack of clear homology. In cryptodires, the corpus spongiosum contains a thick layer of vascularized sinusoidal tissue situated on top of the corpus fibrosum, which is denser, less erectile connective tissue. The penis of pleurodires is morphologically quite different from that of cryptodires. The corpus cavernosum exists as a veneer with just enough "to form and close the lips of the seminal groove" (McDowell, 1983). The corpus fibrosum is not cavernous or erectile; instead, it is comprised of dense, cartilage-like material (McDowell, 1983).

The glans of both cryptodires and pleurodires range in structure from simple to elaborate. The glans of sea turtles and snapping turtles (Cryptodira: Cheloniidae, Chelydridae) are relatively simple (Zug, 1966). In contrast, the glans of *Emydura australis* (Pleurodira: Chelidae) is relatively elaborate, looking more like an opening flower (McDowell, 1983). Zug (1966) used the morphology of the glans to reconstruct the basic relationships among families within the cryptodires. For example, Williams (1950) placed the kinosternid genera into Chelydridae; however, Zug (1966) found the similarity of the "plica media suggests the kinosternids and the dermatemydids arose from a common ancestor," as proposed by McDowell (1961). In the context of modern genetically based classifications extending Zug's work to include more species of both cryptodires and pleurodires, their sexual dimorphism and their ecological characteristics (Stephens & Wiens, 2003, 2004) will aid analysis and understanding of the evolution of reproductive structures and strategies found in chelonians.

Although in both groups the glans becomes engorged with blood, the supportive components of the erection are not the same. During erection in cryptodires, the engorged glans and the shaft extrude from the cloaca of the male into the cloaca of the female (Zug, 1966). The length of the shaft ensures the proper placement of the glans in the anterior part of the cloaca of the female for sperm transfer. In contrast, the copulatory organ of pleurodires (e.g., *Emydura*, *Phrynops*, *Platemys*) includes both the glans and the partial eversion of the cloaca (McDowell, 1983), which functionally acts like the shaft of cryptodires and extends the total length of the organ, allowing insertion into the female cloaca.

Although the penis must have some rigidity to be inserted into the female cloaca, it is not fully erect before insertion. Once the distal portion has been inserted, the expansion of the glans, and supporting region (in conjunction with the partial relaxation of the female cloaca), moves the penis further into the female cloaca. The engorged erectile tissue of the seminal ridges presses them together, thereby ensuring the lumen of the seminal groove (now formed into a tube) remains open for the passage of sperm. In pleurodires, because the seminal groove does not extend to the end of the glans the erect penis appears to be "doubled over" (McDowell, 1983), which facilitates the placement of sperm from the seminal groove into the cloacal vestibule of the female. During intromission, the penis fills the cloaca and is held in place by the engorged erectile tissue in the glans (McDowell, 1983). The enlarged glans orients the tip of the seminal groove in the vestibule at the opening of the oviducts. The enlarged glans helps to seal the boundary of the distal penis against the epithelium of the female's cloaca.

While flaccid in the male's cloaca, the more expansible corpus spungiosum is on the dorsal surface and the less expansible corpus fibrosum is on the ventral surface. As both become engorged with blood, the greater expansion of the corpus spongiosum causes the penis to extend from the cloaca and curve forward. As a result, the glans—which is posteriorly directed in the cloaca—becomes anteriorly directed and the dorsal surface of the flaccid penis becomes ventrally directed during the erection. Because the male is mounted above and behind the female, he must bring his cloaca into contact with the opening of her cloaca. This means that the penis enters the female cloaca upside-down (or at least partially on its side) relative to the flaccid position in his cloaca. Penetration may be aided, at least to some degree, in some groups (e.g., *Testudo*, *Pyxis*: Obst, 1986) by the hardened tip of the male's tail; the cap scale on the tail tip may be used to locate the opening leading into the female's cloaca or simply to guide the engorging penis to the area preceding intromission. The role of the tail in copulation needs to be better defined by observations in the field and in captivity.

The duration of intromission is difficult to determine. Harrel et al. (1996) observed alligator snapping turtles (*Macrochelys temminckii*) in coitus for between 5 and 25 minutes. Wood and Wood (1980) reported that the average time spent mounted for green turtles (*Chelonia mydas*) in captivity was 25.5 hours. Their observations on captive sea turtles indicated higher fertility was positively correlated with longer periods of intromission (Wood & Wood, 1980); similar information is not available for other chelonians. The stimulus for ejaculation is unknown for any turtle. Ejaculation does not appear to be spasmodic, but rather sperm transfer seems to occur as a flow over time. However, this may be an illusion associated with the tendency of the male to remain mounted on the female for an extended period; this behavior may facilitate uptake of sperm through the ostia of the oviducts. The time involved may reflect the process of sperm transfer not in an ejaculatory rush but rather a more pulsed flow, the distance the sperm must travel, and the need for sperm to enter the ostia of the oviducts once the seminal fluid has been placed into the anterior of the cloaca. Intromission is an understudied part of chelonian reproduction.

The clitoris arises from the medial-ventral wall of the cloaca as a ridge in the proximal region of the cloaca. Unfortunately the size, shape, and histological structure of the clitoris have not been described for any species of turtle.

10.3.4 Hormonal Changes during Reproduction

The general pattern of endocrine control of the reproductive cycle in male and female sea turtles (species that produce multiple clutches in a single reproductive season) shows increases and decreases in four critical hormones and changes in associated behaviors (Licht et al., 1979; Owens & Morris, 1985; Owens, 1997; Lance & Rostal, 2002; Rostal, 2005; Hamann et al., 2003). For males, sperm formation occurs prior to migration and in conjunction with an increase in testosterone and a slight rise in gonadotropin. Levels of both hormones decrease during migration and mating periods, and by the time the male begins his remigration to his foraging area, the levels are greatly reduced. In females, estrogen reaches its maximum level during preparation of the follicles in the foraging area, then reduces during the migration to the mating and nesting areas, and is lower still during mating receptivity. Levels of testosterone are low in the foraging area but increase preceding and during the migration and mating periods. Mating ceases about two weeks prior to ovulation, which precedes nesting by about two weeks. In this time, the testosterone level dips but recovers prior to ovulation; during each subsequent inter-nesting period, testosterone decreases and recovers but each recovery is less than the previous recovery. Estrogen level recovers just prior to ovulation from a low during the mating period. The dip and recovery occurs during each successive inter-nesting period; however, the recovery and dip levels stay about the same each time.

The level of gonadotropin is steady during follicle preparation and migration; it reduces slightly during mating and prior to the first ovulation. At this time, blood concentration of gonadotropin surges and crashes within a few days. The pattern is followed closely by a surge in progesterone such that the peak concentrations are offset slightly. All hormone levels reduce dramatically following the last oviposition of the nesting season. Sequential changes in the levels of hormones (estradiol, progesterone, AVT, neurophysin, PGE_2, PGF) circulating during the nesting process correlate well with changes in observed behaviors (Guillette et al., 1991; Owens, 1997). It is likely that in species that produce two or three clutches of eggs, the pattern of cyclic fluctuations in hormone concentration levels is similar to that described for sea turtles, except for the number of sequential ovipositional cycles (e.g., *Geochelone nigra*; Schramm et al., 1999). In species that ovulate a single clutch, the cyclic pattern of hormonal concentration increase and decrease is truncated. Variation is to be expected in the actual levels of hormone concentrations when the general cycle is applied to a specific species.

10.3.5 Eggs

All turtles lay amniotic eggs (Ewert, 1979, 1985). By definition, the structure of the cleidoic egg allows the exchange of only gases between the external environment and the embryonic tissue. This condition is met by the hard-shelled eggs of terrestrial chelonians, but more pliable eggs may allow liquid water to penetrate the shell matrix and enter the egg in a fluid state as well as a vapor (Thompson, 1983, 1987; Ackerman et al., 1985). The egg shape is defined by the shell. Inside the shell matrix is a fibrous membrane that surrounds the albumen (Packard & DeMacro, 1991). Within the albumen is the yolk membrane that contains the yolk material and on which the embryonic disc is situated. Regardless of the actual size or shape of the egg, or the type of egg shell, the proportions of egg components are relatively constant among the species. In terms of overall egg mass, the shell typically contributes about 11 to 12% (range: 4.3 to 24.7%), the albumen contributes about 46 to 47% (range: 34.4 to 61.9%), and the yolk contributes about 41 to 42% (range: 32.1 to 55.0%) (Gibbons, 1990; see Table 7 in Ewert, 1979).

The shell is composed of two general layers: an outer crystalline structure and the inner, supporting proteinaceous, fibrous membrane (Packard & DeMacro, 1991; Schleich & Kästle, 1988; Gad, 1994). The flexibility of the egg shell can be categorized into two groups: inflexible (hard-shelled) and flexible (soft-shelled). The difference in the rigidity of the shell is the result of differences in the matrix of the calcified layer (Packard & DeMarco, 1991). In hard-shelled eggs, the calcified matrix is densely packed calcium columns (Packard & DeMarco, 1991). In contrast, in soft-shelled eggs the matrix is comprised of aragonite crystals that form rosettes on the fibrous inner shell membrane (Packard & DeMarco, 1991). In both types, the shell matrix must allow the movement of oxygen and carbon dioxide; more hydric flux occurs through the shells of soft-shelled eggs (*Chelydra serpentina*: Packard et al., 1987, *Chrysemys picta*: Paukstis et al., 1984). In general, the structure of the egg shell correlates with the environmental conditions in which the eggs incubate (Ewert, 1979). Typically, soft-shelled eggs are laid by marine turtles and some river turtles; these species incubate their eggs in a moist substrate and the eggs gain weight during their relatively short incubation period. Some pliable eggs also follow this pattern. Hard-shelled eggs are laid by terrestrial turtles nesting in arid and semi-arid environments and by species nesting in areas that become dry during their longer incubation periods (see Table 5 in Ewert, 1979). Equally as important in considering the size and shape of turtle eggs are the thickness, functional pore area, and structure of the shell because the egg shell provides the critical boundary between the external and internal environments affecting the developing embryo by regulating the movement of gases, including water vapor (Packard & Packard, 1988a; Thompson & Speake, 2004). The shell structure is not continuous; the pores facilitate gas and vapor exchange. The shell also provides calcium for embryonic bone development (*Apalone spinifera*: Packard & Packard, 1991; *Chrysemys picta*: Packard & Packard, 1986; *Chelydra serpentina*: Packard & Packard, 1989; *Dermochelys coriacea*: Simkiss, 1962).

Within the protective boundary of the egg shell, the albumen serves the embryo in three important ways (Palmer & Guillette, 1991; Thompson & Speake, 2004). First, the albumen provides antimicrobial protection through its fibrous structure and its colloid form. Second, its viscosity provides support for the expanding vitelline membrane in early development and to the growing embryo later in development. Third, albumen also provides for at least some water storage and may assist in accumulating water from the substrate, depending on the egg shell structure (Palmer & Guillette, 1991; Thompson & Speake, 2004).

The yolk constitutes the primary source of energy provided by the female (Ar et al., 2004). The actual energy content varies among the species (see Table 7.1 in Ar et al., 2004); dry mass energy content of turtle eggs ranged between 24.18 kJ/g for *Chelydra serpentina* and 28.46 kJ/g for *Caretta caretta*. Congdon et al. (1983c) reported differences in the percentage of yolk lipid used during incubation in seven species of freshwater turtles (*Chrysemys picta*, *Emydoidea blandingii*, *Graptemys ouachitensis*, *Trachemys scripta*, *Sternotherus odoratus*, *Graptemys geographica*, *Chelydra serpentina*) incubated under common hydric conditions. Although their data were not unequivocal, they speculated the emergence tactics of hatchlings were linked in some part to lipid levels in turtle eggs (Congdon et al., 1983c). Certainly, the hydric conditions at the nest site influence yolk utilization and the hatchling size in *Chrysemys picta* (Packard et al., 1991). Chelonians seem to be using a different strategy than lizards because their hatchlings contain a relatively larger energy reserve at hatching (Ar et al., 2004); this reserve supports the hatchling as it moves away from the nesting area to a reasonably safe area before beginning to feed (Packard & Packard, 1988a; Ar et al., 2004). If the embryo remains unhatched in the nest (e.g., *Deirochelys reticularia, Kinosternon baurii*: Ewert & Wilson, 1996), or if the hatchling remains in the nest over winter, its yolk reserves must sustain it through changes in temperature (and hence changes in metabolism) and moisture with enough to provide for it until it can reach food in the spring (e.g., *Chrysemys picta*: Costanzo et al., 1995).

10.4 REPRODUCTIVE STRATEGIES

Reproductive strategies in reptiles tend to be conservative (Shine, 1985); yet within and among groups, there is variation. The general reproductive cycle of reptiles is under environmental control (Duvall et al., 1982). The reproductive ecology of chelonians has received more emphasis than morphology in recent years (Fox, 1977), with the primary focus on sexual dimorphism, mating system/pattern, fecundity, seasonal timing of reproduction, ovipositional site, incubation period, mode of sex determination, embryology, and hatchling size and survivorship (Shine, 1985; Ewert, 1979; Moll, 1979; Bury, 1979; Gibbons, 1990; Lutz & Musick, 1997; Lutz et al., 2003). Reproductive cycles have been described in varying detail for several turtle species (e.g., *Sternotherus odoratus*: Risley, 1938; *Terrapene carolina*: Atland, 1951; *Terrapene carolina*, *T. ornata*, *Sternotherus*: Legler, 1960a; *Trachemys scripta elegans*: Burger, 1937; Moll & Legler, 1971; Gibbons, 1990).

10.4.1 Sexual Dimorphism

Chelonians exhibit a variety of sizes and shapes. Although some fossil forms are much larger (e.g., *Archelon ischyros*: Carr, 1964; Obst, 1986), the largest living chelonian is the leatherback sea turtle, *Dermochelys coriacea*—the largest individual on record was 256 cm curved carapace length and 915 kg (Eckert & Luginbuhl, 1988). The smallest chelonian is the speckled cape or padloper tortoise, *Homopus signatus*, at less than 10 cm, less than 140 g (Boycott & Bourquin, 1988). Between these extremes are large Galapagos tortoises and smaller freshwater turtles such as the Kinosternids; some river turtles are quite big (more than 100 cm carapace length) and some terrestrial turtles are rather small (less than 15 cm) (for dimensions, see Carr, 1952; Ernst & Barbour, 1972, 1989; Ewert, 1979; Pritchard, 1979; Pritchard & Trebbau, 1984; Cann, 1998; Bonin et al., 2006).

Often, the adult females are larger than adult males but the generalization is not universally true, even within a single family (Carr, 1952; Ernst & Barbour, 1972, 1989; Pritchard, 1967, 1979; Ewert, 1979; Bonin et al., 2006). For example, in freshwater turtles (Emydidae), females are larger than males in some genera (e.g., *Chrysemys*, *Graptemys*, *Kachuga*), whereas in other genera, the difference in size is slight (e.g., *Emys*, *Clemmys*, *Mauremys*). In sea turtles (Cheloniidae, Dermochelyidae), the adult females are larger than the males (Wibbels et al., 1991); in contrast, the sexes are about equal in size in several species of mud turtles (Kinosternidae) and land tortoises (Testudinidae). In a few chelonians, the males are noticeably larger than females (e.g., *Geochelone*, *Melanochelys*, *Macrochelys*).

Regardless of their size, mature males can be easily identified by their longer tail, which houses the penis. Other distinguishing sexual characteristics found in turtles include eye color (e.g., orange-red in *Terrapene carolina*: Ernst & Barbour, 1972; *Emys orbicularis*: Pritchard, 1966), color of head skin marking (e.g., orange spots in *Clemmys guttata*: Ernst & Barbour, 1972), extended gular spurs in Gopher tortoises, plastral concavity in some sea turtles (Wibbels et al., 1991), and extra-long claws on the front feet of male slider turtles (*Trachemys scripta*: Viosca, 1933). Sexual dichromatism is widespread in both pleurodires and cryptodires (Cooper & Greenberg, 1992); seasonal dichromatism, related to reproduction, is less well represented but definitely occurs (Cooper & Greenberg, 1992). More fieldwork must be conducted to elucidate the roles of these secondary sexual characteristics in the reproductive behavior of chelonians.

10.4.2 Mating Behavior

Although turtle reproductive behavior is complex, descriptions of pre-copulatory and mating behavior of turtles are relatively few (e.g., *Pseudemys concinna*, *P. floridana*: Cagle, 1944; *Gopherus*: Auffenberg, 1978; *Chelydra serpentina*, *Apalone*: Legler, 1955; *Chelonia mydas*: Bustard, 1972; *Macrochelys*: Harrel et al., 1996; *Macrochelys*, *Chelydra*: Ewert, 1976; *Caretta caretta*: LeBuff, 1990; *Terrapene*: Dodd, 2001; Table 10.6) and often repeated (see illustrations in Oliver, 1955; Bellairs, 1970; Obst, 1986). Other reproductive behaviors include nest construction and concealment, as well as other actions (e.g., thermoregulation) that increase the fitness of an individual adult and, potentially, its eggs. Turtles exhibit stereotyped nesting behaviors that can be compared among groups because they contain several easily described phases (Ehrenfeld, 1979; Carpenter & Ferguson, 1977). The sequential behaviors include finding the site, selecting the specific location for the nest, digging the nest cavity, oviposition, covering the nest, and leaving the nest site (Ehrenfeld, 1979; Carpenter & Ferguson, 1977).

Chelonian courtship involves a variety of behaviors including following, head bobbing or swaying, nudging or biting, cloacal touching or sniffing, titillation via long foreclaws (Carpenter, 1980; Carpenter & Ferguson, 1977; Harless, 1979; Bels & Crama, 1994; Mason, 1992). Three general types of courtship patterns have been proposed based primarily on male behavior: premounting, mounting, and intermediate (Bels & Crama, 1994). Premounting courtship behaviors are found among aquatic species (e.g., Cheloniidae, Dermochelyidae, Emydidae, Chelidae) swimming in the three dimensional environment; mounting courtship behaviors occur while the male is mounted and preparing for intromission. The intermediate classification allows for variation in some precopulatory behaviors but does not contain species-specific behaviors (Bels & Crama, 1994). All three types are preceded by approaching, following/chasing or sniffing behaviors (Bels & Crama, 1994). Premounting and mounting courtship occur in both pleurodires and cryptodires. Premounting courtship seems to be widespread in cryptodires (Bels & Crama, 1994). The courtship type seems to be correlated with habitat type (i.e., open water, bottom dwelling, terrestrial), which may limit the range of movement of the turtle. The most elaborate courtship rituals occur in the three-dimensional environment of water.

TABLE 10.6
Examples of Sensory Reproductive Communication in Turtles*

Sense	Genus or Group	Observation or Evidence	Source
Olfactory	*Gopherus*	Enlargement of sub-dentary glands; secretions of chin glands and cloaca used in sex and species identification	Weaver (1970); Auffenberg (1977); Rose et al. (1969)
	Geochelone *Homopus*	Smelling and sniffing behavior	Auffenberg (1965); Weaver (1970)
	Kinosternids	Nudging and sniffing of musk glands	Mahmoud (1967)
	Trionyx	Probing skin areas	Plummer (1977)
	Chrysemys	Cloacal sniffing	Jackson & Davis (1972a); Cagle (1950)
Visual	*Geochelone*	Specific head movements; color change	Eglis (1962); Auffenberg (1965)
	Chrysemys *Graptemys*	Posturing, orientation of male relative to female visual area (e.g., backward swimming, face to face)	Jackson & Davis (1972b); Cagle (1950) Ernst (1974)
Tactile	Kinosternids Emydids	Snout nudging and rubbing, nipping, and biting	Mahmoud (1967) Evans (1953); Ernst (1970); Davis & Jackson (1970, 1973); Baker & Gillingham (1983)
	Testudinids	Shell knocking, pushing, ramming	Auffenberg (1964); Evans (1953, 1961)
	Emydids	Titillating of female cheeks by male's long front claws	Jackson & Davis (1972b); Ernst (1974); Cagle (1950)
	Chelydrids Emydids Cheloniids	Mounted position with male clasping onto female's carapace with two or four feet	Legler (1955); Mahmoud (1967); Evans (1953); Booth & Peters (1972)
Auditory	Testudinids *Gopherus*	Vocalizations before and during coitis	Ernst & Barbour (1989); Auffenberg (1977)

* Based on Carpenter (1980); see also Carpenter & Ferguson (1977) and Harless (1979).

Courtship involves both male-female interactions and male-male interactions (Carpenter & Ferguson, 1982; see illustrations in Obst, 1986). The courtship includes olfactory recognition, gaping, and biting (Table 10.6). Multiple males may be attracted to a female in estrus and compete among themselves for the privilege to court her. Interactions among male terrestrial turtles (e.g., *Gopherus*) may be combative. The rival males may circle each other, head butt, bite, and ram each other in the flanks; this "jousting" may result in the loser being turned over, possibly to die. Male-female courtship is reminiscent of male-male interactions without attempting to turn her over, and involves olfactory recognition, ramming, and biting before mounting.

Among aquatic turtles, males interact by gaping at each other, biting, and shoving with their claws. The sequence of premating behaviors includes circling, sniffing, biting, or tickling. Painted turtles (*Chrysemys, Trachemys*) have evolved elaborate courtship rituals (Cagle, 1944). For example, both *Pseudemys concinna* and *P. floridana* position themselves above their chosen female and extend their front legs and long claws down to tickle the sides of her neck. In both *Trachemys scripta* and *T. terrapen*, the male approaches the front of the female, almost touching nose to nose, and reaches forward to tickle the female along the sides of her neck before attempting to mount. Among some groups of freshwater turtles (e.g., Kinosternidae, Trionychidae), courtship rituals are simple and almost nonexistent; often, receptive females are mounted following olfactory inspection and a minimum of other interaction.

Among sea turtles, two or more males may attempt to court the same female; they bite and push each other while trying to attract her attention (Booth & Peters, 1972). As the male quickly moves to attempt to secure his position attached to her carapace, the female attempts to twist away. If successful in mounting, the male holds on to the lateral front and lateral rear parts of the female's carapace with his longer, curved (first digit) claws. In this position, he is carried by the female; if she surfaces to breathe, he can breathe. He is also vulnerable to biting by attendant males along the exposed edges of his flippers and the dorsal part of his tail. It is not uncommon to find males with the posterior margins of their flippers raw and swollen—sometimes bleeding—during the mating season (Limpus, 1993). Flippers and tails may be gnawed to expose bone. These interactions may not dislodge the mounted male but serve to reduce his ability (fitness) to court and mount a female in subsequent mating activity, at least during the current breeding season.

10.4.3 Seasonal Timing of Reproduction

Although turtles lay their eggs on land, they do not show a uniform pattern of nesting. Some species show highly seasonal patterns (e.g., *Chrysemys picta*), whereas others may nest almost all year (e.g., *Pseudemys floridana*) (Figure 10.1). The pattern displayed by a species often correlates well with the temperatures required for adult activity, embryonic development, and hatchling survival (Moll, 1979; Georges et al., 1993). Males and females may share the same habitat but do not necessarily exhibit the same patterns of sexual selection or selection for fecundity. For example, male *Chrysemys picta* mature (become sexually active) at a smaller size than females (Wilbur, 1975b); this may result from the different responses to a reduced risk of predation with increasing size and the reproductive constraint imposed by egg production.

Chelonians breed when they have accumulated adequate stores of energy (Congdon & Tinkle, 1982; Congdon et al., 1983b; Gibbons, 1982). This depends on food quality and quantity, which are impacted—if not regulated—by environmental conditions. Some species residing at the same location may be affected more than others by environmental conditions, depending on their food requirements (Gibbons et al., 1983). The correlation between the El Nino-Southern Oscillation and the breeding biology of green turtles (*Chelonia mydas*) suggests that a two-year time lag is required between favorable foraging and oviposition (Limpus & Nichols, 1988, 2000); the proportion of females breeding each year varies as well (Limpus & Reed, 1985a, 1985b). Variation in environmental conditions experienced by *Geochelone gigantea* on Aldabra resulted in differences in the number of females breeding, clutch size, egg mass, and size at maturity in three subpopulations when comparisons were made over a long period of time (Swingland, 1977; Swingland & Coe, 1978, 1979).

The time of oviposition varies considerably among turtles and is often related to latitudinal distribution and the seasonality of rainfall. Licht (1984) described three patterns of reproduction for reptiles. Reproduction can either be continuous with comparable breeding levels all year, continuous with variable reproductive intensities, or discontinuous. Turtles fit into the second or third patterns. For instance, green turtles in the tropics nest all year round at some sites, although there is a substantial change in numbers between the summer and winter periods (Miller, 1997). An example of the discontinuous pattern occurs with the painted turtle in North America, where nesting occurs in the late spring or early summer regardless of latitude (Moll, 1973).

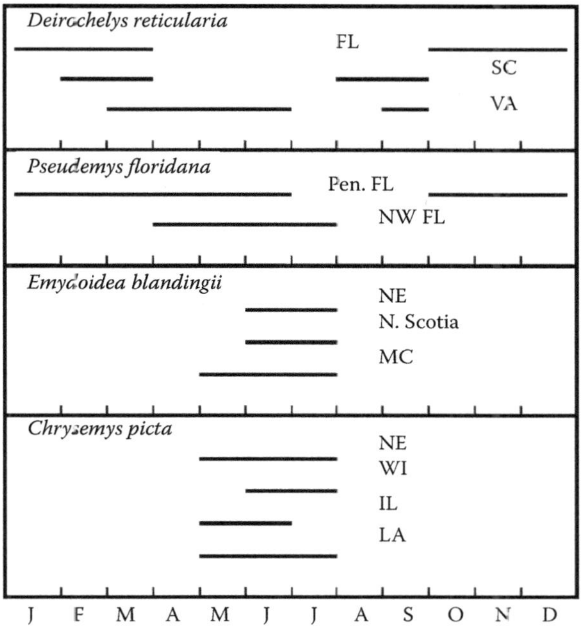

FIGURE 10.1 Patterns of nesting seasons of North American freshwater turtles, showing changes in season associated with different latitudes. *Emydoidea blandingii* and *Chrysemys picta* show a northern nesting pattern with the nesting season restricted to the late spring and summer months. *Deirochelys reticularia* and *Pseudemys floridana* show a southern nesting pattern. FL = Florida, SC = South Carolina, VA = Virginia, Pen. FL= Peninsular Florida, NW FL = Northwestern Florida, MC = Michigan, WI = Wisconsin, IL = Illinois, LA = Louisiana, NE = North East, N. Scotia = Nova Scotia. Data: Aresco (2004); Gibbons & Greene (1978); Buhlmann (1995); Jackson (1988); Congdon et al. (2000); Iverson & Smith (1993); Moll (1973); Dinkelacker (unpublished).

Temperate species typically have discontinuous nesting seasons that are primarily influenced by seasonal temperatures (Legler, 1985; Kuchling, 1999). Kuchling (1999) describes the basal asynchronous reproductive cycles of temperate turtles. Briefly, most freshwater turtle species exhibit a seasonal nesting pattern in which eggs are typically oviposited in the late spring and early summer, development occurs during the summer, and hatching occurs in the early fall. There are no examples of continuous reproducing species in North America. However, several species approach a nearly continuous cycle in Florida (e.g., *Deirochelys reticularia*, *Pseudemys floridana*, and *Kinosternon baurii*), although in more northerly populations the nesting season becomes progressively more discontinuous and similar to other temperate species (Figure 10.1) (Gibbons & Greene, 1978; Jackson, 1988; Buhlmann, 1995; Ewert & Wilson, 1996; Aresco, 2004). Jackson (1988) provides an interesting discussion of "winter nesting" patterns in these species, as well as the effects of radiation into northern habitats. Legler (1985) described a similar temperate pattern for several species of Australian turtles in which females produced small flexible-shelled eggs with short incubation periods and the emergence of hatchlings before winter. The "temperate" pattern was exemplified by *Emydura* and *Elseya*, which nested during the early southern spring. All species in this study were widely distributed across the temperate and tropical zones, yet retained the cycle (i.e., temperate or tropical) of their origin. In other words, *Emydura* and *Elseya* retained a temperate nesting pattern even when found in tropical zones.

The nesting patterns of temperate species are closely related to their gonadal cycles (Kuchling, 1999). Most temperate freshwater turtles show post-nuptial gamete maturation (*Terrapene carolina*:

Atland, 1951; *Terrapene ornata*: Legler, 1960a; *Chrysemys picta*: Ernst, 1971a, 1971b; *Chelydra serpentina*: White & Murphy, 1973; *Sternotherus odoratus*: McPherson et al., 1982; *Chelodina longicollis*: Parmenter, 1985). In males, spermatogenesis routinely occurs during the hottest months of the year or immediately following the nesting season. The sperm are then stored in the epididymus during the fall and winter for use during spring mating, or are used during fall mating (Kuchling, 1999). In females, vitellogenesis may take up to a year or longer to complete (Ernst & Zug, 1994). Typically, vitellogenesis begins following oviposition in the summer or fall (Moll, 1979). At latitudes with sufficiently cold winters, there may be a pause in vitellogenesis during the coldest months. Vitellogenesis is completed either before or after hibernation, and in some cases just prior to ovulation (Moll, 1979). Thus, the reproductive cycles of males and females may be asynchronous, with the possible exception of turtles living in high latitudes (Kuchling, 1999). An explanation for the asynchrony may stem from the relative costs of the gonadal cycles. Spermatogenesis is relatively cheap and can be accomplished quickly at a time of year when more vigorous reproductive events are absent. Alternatively, vitellogenesis is energetically expensive and may require a prolonged period for the female to accumulate or convert a substantial quantity of her fat reserves into yolk. As such, vitellogenesis may begin when resources (i.e., food) are high and other reproductive activities are reduced (i.e., during the summer and fall). In addition, adaptations such as sperm storage and multiple paternity have permitted the female to mate at a time when vitellogenesis is not completed (Moll, 1979; Kuchling, 1999). Because the female can store the sperm from a single mating event in the fall, she may fertilize her eggs in the absence of a spring mating event. In addition, she may use stored sperm as well as "fresh" sperm to fertilize her eggs (Roques et al., 2006). The advantages and disadvantages of multiple paternity have received considerable attention in the literature (Galbraith, 1993; Valenzuela, 2000; Pearse et al., 2001, 2002; Lee & Hays, 2004; Roques et al., 2004, 2006; Jensen et al., 2006).

Tropically distributed species often have discontinuous nesting patterns as well. Pre-nuptial gamete maturation occurs in tropical and subtropical species (*Lissemys punctata*: Singh, 1977; *Chelonia mydas*: Licht et al., 1985; *Caretta caretta*: Wibbels et al., 1990; *Geocholone nigra*: Schramm et al., 1999). Rainfall seems to be the main factor dictating nesting in tropical species (Kuchling, 1999). Legler (1985) reported that in Australia, tropical species nested during the first six months of the year, which correlates to the dry season. Vogt (1997) described five patterns of nesting in South America: rainy season, end of rainy season, initiation of lowering rivers, lowest river depth, and spring nesters. Those species that nest during the dry season probably rely on exposed beaches, sand banks, or other nesting habitats that are not available during the rainy seasons. Data regarding gonadal cycles are scarce for freshwater turtles (Kuchling, 1999), and as a result the proximate and ultimate factors regulating reproduction and nesting in tropical turtles are still poorly understood.

10.4.4 Insemination and Sperm Storage

Fertilization is achieved internally via the penis being inserted into the cloaca of the female, sperm being transferred to the cloacal vestibule of the female, and sperm being accepted into the ostia of the oviducts. Once in the oviducts, sperm are carried via cilia to glands (*Chrysemys picta*: Parker, 1928, 1931; *Chrysemys picta*, *Trachemys scripta elegans*: Crowell, 1932; *Mauremys japonica*: Yamada, 1952) where, presumably, they mix with other sperm from previous copulations. The ciliated bands that move the sperm may be single or double. As an ovulated follicle enters the infundibulum, the folds are expanded; if any folds contain sperm, they are spread onto the surface of the follicle. As with birds, when one sperm penetrates the follicular layers, the coating of the follicle changes to prevent further penetrations. Polyspermy probably occurs at about the same rate as in birds and other animals, but there are no data available for polyspermy in turtles.

Female turtles from at least six families possess tubules in the oviduct in which sperm can be stored (Hattan & Gist, 1975; Palmer & Guillette, 1988; Gist & Jones, 1989; Sarkar et al., 2003). Sperm storage tubules have been reported from the anterior half of the oviduct, including albumen-secreting glands and in the "uterine region of the oviduct near the uterovaginal junction" (Palmer & Guillette, 1988; Gist & Congdon, 1998). Sperm were not found in the center of the glandular region (Gist & Congdon, 1998).

An ultrastructural study of sperm storage in *T. carolina* (Gist & Fischer, 1993) reported that the albumen glands containing sperm were cytologically the same as those that lacked sperm. Because sperm were observed in the lumen and not in contact with the epithelium, the sperm were considered to be resident in the tubules, rather than tubules being "specialized for the maintenance of stored sperm" (Gist & Congdon, 1998). The oviducts of *Trachemys scripta* and *Sternotherus odoratus* store sperm in "undifferentiated tubules and ducts of submucosal glands at the periphery of large glandular segments of the albumen and uterine regions of the oviduct" (Gist & Congdon, 1998). Sarkar et al. (2003) reported sperm storage tubules in the wall of the isthmus of the oviduct of the Indian freshwater soft-shelled turtle (*Lissemys punctata punctata*). The tubules were lined by both ciliated and nonciliated epithelial cells. The depth and activity of the tubule epithelia were highest during the breeding season and lowest in the non-breeding season. The tubules of the isthmus also regress in the non-breeding season. Scattered, short tubules appeared in the infundibulum only during the breeding phase. Sperm were retained in the tubules of the isthmus from the time of mating until just prior to ovulation. Sperm then moved into the tubules of the infundibulum via material secreted from the tubule epithelium. They speculated that the release of sperm "might have been stimulated by estrogen secreted from the ovarian follicles of pre-ovulatory turtles" (Sarkar et al., 2003).

Gist and Congdon (1998) found that sperm transfer began in October for *Sternotherus odoratus* and in January for *Trachemys scripta*. They speculated that stored sperm were most likely used for fertilization of eggs in the second and subsequent egg clutches. In *Chrysemys picta*, testis weight changes as a result of annual sperm production cycle but the epididymis does not, although the epididymis contains sperm throughout the entire year in *Chrysemys picta* and *Trachemys scripta* (Gist et al., 2002). Sperm of both in *Chrysemys picta* and *Trachemys scripta* exhibit relatively low motility and velocity (Gist et al., 2002).

Sperm may be stored in the oviduct for extended periods. Sperm were found in seminal receptacles in the oviducts of *Terrapene carolina* 14 months after mating (Hattan & Gist, 1975). Sperm were identified in tubules of female turtles isolated from males for as long as 423 days (Gist & Jones, 1989). Viable eggs were produced by isolated females after 2 and 4 years in *Malaclemys* (Hildebrand, 1929; Barney, 1922) and after 4 years in *Terrapene carolina* (Ewing, 1943). These may be extreme cases but they provide evidence that sperm can be stored beyond the current mating season. Storage of viable sperm in females is facilitated by low motility and longevity; this in turn allows time for sperm movement within the oviduct (Gist et al., 2002). Pearse and Avise (2001) reported that one female *Chrysemys picta* used the sperm from the same male to fertilize her eggs in three consecutive years; given the lack of pair bonding, sperm storage seems the mostly likely explanation. Actual fertilization may be offset from the time of insemination (e.g., *Chrysemys picta*: Gist et al., 1990). The storage of sperm in males ensures that sperm will be available for transfer to the female, and sperm storage in the female ensures that sperm will be available for fertilization, even though the seasonal transfer may be restricted.

Pearse and Avise (2001) reported multiple-paternity in nine species (Pleurodira: *Podocnemis expansa*; Cryptodira: *Chelydra serpentina*, *Glyptemys insculpta*, *Chrysemys picta*, *Dermochelys coriacea*, *Caretta*, *Chelonia mydas*, *Lepidochelys kempi*, *Gopherus agassizii*) with 33% of 321 clutches analyzed having multiple paternity (range per species studied: 0 to 100% in 14 studies) and concluded that multiple paternity is a "common phenomenon" (albeit not universal). Many temperate species produce two or more clutches per reproductive season, depending on endogenous and exogenous factors (Ewert, 1979; Kuchling, 1999). Short-term sperm storage alleviates the female's need to re-mate between oviposition of sequential clutches, whereas long-term sperm storage "puts strong selective pressure on males to produce sperm that can survive long-term storage" (Pearse & Avise, 2001).

10.4.5 Fecundity

The shape and number of chelonian eggs vary among species (Ewert, 1979, 1985; Iverson & Ewert, 1991), and may vary within species (Finkler & Claussen, 1997). Further, the number of clutches oviposited per annum varies among species (Ewert, 1979; Moll, 1979). Chelonian eggs range in size from small (23.6 × 13.5 mm; weight = 2.5 g) in *Sternotherus odoratus* to large (59.7 × 55.2 mm; weight = 106.9 g) in *Geochelone elephantopus* (Ewert, 1979, 1985; Iverson & Ewert, 1991).

10.4.5.1 Shape of Eggs

The shape of turtle eggs is the result of the interplay between the needs of the embryo and the constraints of the female's morphology (Iverson & Ewert, 1991). Three general categories (oblong, elliptical, spherical) can be defined based on divisions of the length/width ratios. In oblong eggs, the length is approximately twice (or more) the diameter; the ratio of length to width in elliptical eggs varies between 1.99 and 1.10. Spherical eggs have ratios between 1.10 and 1.00. The vast majority of chelonians produce ellipsoidal eggs (Ewert, 1979; Wilbur & Morin, 1988; Bonin et al., 2006). The relationship between the ratio of egg length divided by egg width and female carapace length among cryptodiran families shows an obvious trend in egg shape from oblong eggs (greatest ratio) being produced by small turtles to spherical eggs (least ratio) being produced by large turtles (Figure 10.2). Within families, members of the Cheloniidae, Dermochelydidae, Carettochelydae, and Chelydidae produce essentially spherical eggs and range in size from about 250 to more than 1500 mm in carapace length. Among the Geomydidae, Emydidae, Kinosternidae, and Platysternidae, females range in size from about 100 to 700 mm and produce ellipsoidal eggs (with a few exceptions). Several Geoemydidae, a few Kinosternidae, and a single Emydidae produce oblong eggs. In contrast, most Testudinidae produce ellipsoidal eggs but some larger species produce spherical eggs. Among the Trionychidae, the largest species and several of the smaller species produce spherical eggs, whereas many mid-sized species produce ellipsoidal eggs. Among the pleurodires, the two families show a similar pattern of the majority of species producing ellipsoidal eggs and a few producing oblong or spherical eggs (Figure 10.2).

In both oblong and elliptical eggs, the diameter of the yolk nearly equals the inside diameter of the shell. Only a thin layer of albumen separates the yolk from the inner shell membrane; the majority of the albumen is displaced toward the ends of the egg. In contrast, the albumen in spherical eggs is equally distributed between the yolk and the inner shell membrane.

10.4.5.2 Number of Eggs

The number of eggs oviposited varies (Pritchard, 1967, 1979; Cann, 1998; Bonin et al., 2006; Ernst & Barbour, 1972, 1989; Ernst et al., 1994). A few species produce only one egg per clutch (e.g., *Chersina angulata*, *Homopus boulengeri*, *Malacochersus tornieri*, *Psammobates geometrica*, *Kinosternon angustipons*); others lay 2 to 4 eggs per clutch (e.g., *Sternotherus*, *Platysternon*, *Indotestudo*) (Ewert, 1979). All *Gopherus* and many species of *Terrapene* and *Graptemys* produce up to 10 eggs per clutch (Ewert, 1979). Most medium- to large-sized turtles, including *Testudo*, *Geochelone*, *Chelodina*, *Asterochelys*, and *Macrochelys* (among others), produce clutches containing 10 to 30 eggs. *Natator depressus* and larger species of *Trionyx*, *Chitra*, and *Pelochelys* lay about 50 eggs per clutch (Ewert, 1979). Interestingly, *Chelydra serpentina* produce an extraordinary range of 6 to 109 eggs per clutch, depending on female body size and geographic locality (Iverson et al., 1997). The other sea turtles (i.e., *Chelonia*: Hirth, 1971; *Eretmochelys*: Witzell, 1983; *Caretta*: Dodd, 1988; *Dermochelys*: Pritchard, 1971; *Lepidochelys*: Marquez et al., 1976) and some species of *Chitra* (Bonin et al., 2006) regularly produce clutches of 100 eggs or more; clutches containing over 150 eggs have been recorded (Pritchard, 1979; Ewert, 1979; Bonin et al., 2006).

In general, chelonians smaller than 200 mm in carapace length lay 2 to 7 eggs per clutch; whereas medium-sized turtles (200 to 300 mm carapace length) produce 2 to 20 eggs (Moll, 1979).

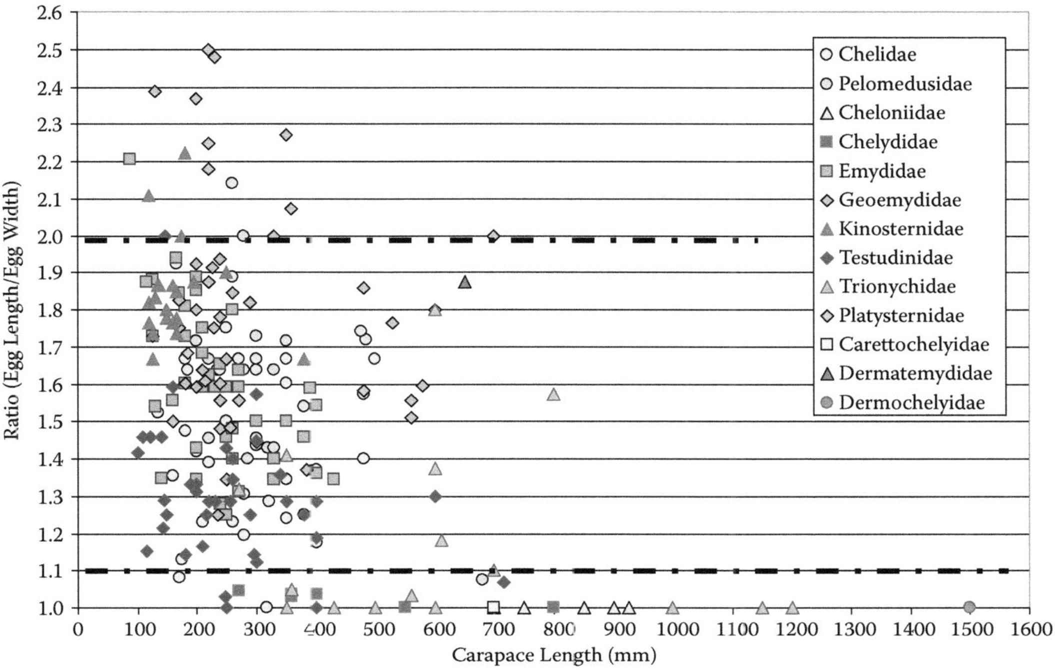

FIGURE 10.2 Relationship between the ratio of egg length/width and female carapace length for 238 species of chelonians grouped by family, including 14 genera of Pleurodires (circles) and 73 genera of Cryptodires (all other shapes). Dotted lines indicate boundaries between oblong, ellipsoidal, and spherical eggs. Data from Bonin et al. (2006); Ewert (1979); Moll (1979); Pritchard (1979); Ernst & Barbour, (1989).

Chelonians greater than 300 mm in carapace length typically lay 15 to 30 eggs, but there are exceptions (Moll, 1979). Moll (1979) plotted the number of eggs per clutch for 109 species and commented that "the majority (over 60%) lay clutches averaging less than 10 eggs and less than 10% regularly lay clutches exceeding 30 eggs." When he considered the same data arranged by geography (temperate verses tropical), Moll (1979) found no "absolute trend." Temperate species typically produced 2 to 10 eggs per clutch, with exceptions ranging up to about 30 eggs per clutch, whereas a large proportion of tropical species produce 2 to 4 eggs with a large number of species producing from 1 to over 100 eggs per clutch (Moll, 1979).

10.4.5.3 Number of Clutches

Some species typically produce a single clutch (i.e. *Chelydra*, many *Kinosternon*) whereas others produce two clutches (e.g., *Sternotherus*) or three clutches (e.g., *Batagur, Malacochersus, Deirochelys, Graptemys*) in a reproductive season (see Figure 5 in Moll, 1979). Other turtles typically produce four (e.g., *Terrapene carolina, Natator depressus*) or five (e.g., *Malaclemys, Caretta, Chrysemys picta, Apalone spinifera*) clutches per season. An individual *Geochelone paradalis* may produce six clutches of eggs per season (Bonin et al., 2006) and *Chelonia mydas* may lay eight or nine clutches per season, although the mean for their nesting populations are typically less (Ewert, 1979). There is variation in the number of clutches produced within a nesting season by turtles of the same species (e.g., *Sternotherus odoratus*: McPherson & Marion, 1983), in the same genus (e.g., *Kinosternon* species lay one, two, or three clutches per season), or even the same family (Ewert, 1979). Problems in determining the number of clutches produced by a nesting female can be compounded by individual identification, e.g., problems of tag loss (Limpus, 1992), movement of nesting turtles among different nesting sites, monitoring efforts, and potential differences in the number of clutches laid by turtles of different ages. Ewert (1979) suggested that "two or three clutches per season are probably the mode for most species, and 10 or more are a rare exception."

Several approaches have been used to identify relationships among the size and shape of eggs, the number of clutches produced, and the size of the reproducing female. Ewert (1979) catalogued information about the size and shape of chelonian eggs in a series of tables. He observed that:

> Spherical eggs are usually laid by species that produce more than 10 eggs per clutch, and elongate eggs are produced by species that lay less than 10 eggs per clutch.
> Eggs in the Kinosternidae and Emydidae are elongate.
> Eggs from the neotropical pleurodira and Testudinidae vary in shape.
> Most members of the Trionychidae produce spherical eggs.

Comparing female carapace length to egg size, Ewert (1979) concluded that, in general, larger species produce larger eggs but among medium-sized species, "considerable variation in eggs size remains." He also noted that when species are grouped by broad climatic regions, those from rain forests and those from wet tropics "produce proportionally larger eggs" (Ewert, 1979). That larger turtles tend to lay larger numbers of eggs per clutch than smaller species was noted by Moll (1979), who stated that "within a species, clutch size increases with body size and age, although this trend is less evident and may be absent in species laying small clutches."

Wilbur and Morin (1988) used a multivariate approach to assess the life history diversity in 80 chelonian species from 11 families (2 pleurodires, 9 cryptodires). They examined egg volume, clutch size, and total clutch volume in relationship to habitat (freshwater, marine, terrestrial), food (herbivore, omnivore, carnivore), and latitude (tropical, temperate), with carapace length as the covariate. They concluded that habitat was the most important variable determining life history pattern. Then they used a pair-wise process to compare egg volume, clutch size, and total clutch volume between habitats. When the data were adjusted for the size of the female, they found no difference in these variables between freshwater and marine species. In contrast, terrestrial species increased egg size but reduced clutch size. Although they argued that the larger size and number of sea turtle eggs compared to freshwater species is to be expected given the larger size of the females, they acknowledge that the data used were limited, especially concerning the proportion of breeding females per year (Wilbur & Morin, 1988).

Testudinids produce fewer but larger eggs per clutch when compared to Emydids, but the volumes of the clutches are not significantly different when adjusted for differences in body size (Wilbur & Morin, 1988). They argue that diet as well as growth rate may offset the risks associated with producing small hatchlings for the freshwater turtles. The primary trade-off that turtles face is between the production of more but smaller eggs and the production of fewer but larger eggs in conjunction with the number of clutches produced during each reproductive season. An increase in the number of clutches produced in a reproductive season requires an increase in the number of follicles prepared in the ovary. This requires the female to have gathered and stored more energy (Ar et al., 2004). Another consequence of increasing the number of clutches is that the female will have to commence nesting earlier or nest later in the season; this could alter the sex ratio among all surviving hatchlings. Equally as important as the impact on the incubation of the eggs and survivorship of the hatchlings is the increased risk to the female, who must go ashore to nest. It also spreads the risk of loss spatially, thus potentially reducing offspring mortality from flooding, erosion, other natural events, and from depredation. However, spreading oviposition temporally may increase the risk of loss, especially when the reproductive period is limited by environmental change (e.g., lowering of temperature, flooding).

Moll (1979) described two "diametrically opposed" reproductive patterns in chelonians. In Pattern I, turtles produce "large clutches of relatively small eggs; multiple clutches during a well defined nesting season; communal nesting in well defined ancestral nesting areas; careful construction of covered nests." In Pattern II, turtles produce "small clutches of relatively large eggs; acyclic or continuous nesting; solitary nesting with no special nest area; nests poorly constructed or not even attempted." He speculated on the advantages and disadvantages of each pattern while acknowledging between these extreme patterns lies a wide range of intermediates that utilize certain advantages of each pattern" (Moll, 1979).

Intraspecific variation in egg size and female size occurs as a result of both local and regional geography (Gibbons & Tinkle, 1969; Gibbons, 1970b; Moll, 1973; Iverson et al., 1993, 1997) and there are several morphological considerations that limit clutch and egg size. The coelomic cavity of a turtle is limited by the shell and the volume of the internal organs, including any food in the gut, which in turn limits the space available for follicular development. A second controlling aspect is the strength of the ovarian stroma that must support the enlarged follicles. Third, the turtle must be able to pass the egg through the pelvic canal once it has been produced.

For a given population, larger hatchlings emerge from larger eggs (Congdon & Tinkle, 1982; Congdon et al., 1983a; Cox & Marion, 1978; Ewert, 1979; Iverson, 1979) and egg size varies within and among populations (e.g., *Chrysemys picta bellii*: Rowe, 1994). Although some data provide a positive correlation between hatchling survival rate and egg size/hatchling size within a species (Janzen, 1993; Janzen et al., 2000a; Tucker, 2000; Kolbe & Janzen, 2001; *Chelydra serpentina*: Congdon et al., 1999; *Chrysemys picta*: Janzen & Morjan, 2002; *Dipsochelys elephantina*: Swingland & Coe, 1979), more data are needed to examine the relationship, as well as the impact of habitat use on hatchling survivorship (Janzen & Morjan, 2001). Among sea turtles that use the same beach at the same time but produce eggs and hence hatchlings of different sizes, the morphology of the predators as well as their numbers influence hatchling survival on shore. Land-based differential predation by silver gulls (*Larus novaehollandiae*) and ghost crabs (*Ocypoda ceratophthalma*) on hatchling *Chelonia mydas* in preference to hatchling *Natator depressus* (Bustard, 1979), and between *Chelonia mydas* and *Caretta caretta* (Miller, personal observation) may be mitigated by near-shore predation by reef sharks (*Carcharhinus spallanzani*). Embryonic growth and hatchling size may be important to the survival of other species as well (Janzen et al., 2000a, 2000b). The contributions of hatchling behavior, cryptic coloration, and temporal and spatial habitat use should be considered more thoroughly in future studies.

10.4.6 Nesting Patterns

Female and male turtles do not necessarily follow the same reproductive cycles; males may reproduce more frequently than females. For example, female green turtles nesting on Heron Island, Great Barrier Reef, Australia, do not nest in sequential summers and have a mean remigration interval of 4.65 ± 1.5 years (range: 2 to 7 years) (Limpus, 1993). Alternatively, males reproduce on a biannual cycle (mean interval: 2.08 ± 1.14 years, range: 1 to 5 years) (Limpus, 1993). The difference is thought to reflect the time required to recover from the reproductive investment: males do not need the recovery period required by females. After reproducing, the female must recoup the energy reserves used for making the follicles, migration, multiple nesting attempts, and other reproductive activities, whereas males need only to recoup energy expended for migration and courtship.

In freshwater species, only a portion of the adult females may be involved in annual reproduction; mature females are known to reproduce irregularly in several species (*Chrysemys picta*: Congdon & Tinkle, 1982; Tinkle et al., 1981; *Emydoidea blandingii*: Congdon et al., 1983b; *Caretta caretta*: Frazer, 1984; *Kinosternon subrubrum*: Gibbons, 1983; *Geochelone gigantea*: Swingland, 1977; Swingland & Coe, 1978; Coe et al., 1979). Estimates of the reproductive portion of a population vary among species from 23 to 80% (see references in Frazer et al., 1990) and vary within a particular population from year to year—e.g., 23 to 48% of female *Emydoidea blandingii* (Congdon et al., 1983b). Yearly variation may be influenced by available energy reserves and previous breeding efforts. In addition, some species appear not to produce multiple clutches (e.g., *Sternotherus odoratus*: Tinkle, 1961; *Terrapene nelsoni*: Milstead & Tinkle, 1967; *Chelydra serpentina osceola*: Iverson, 1977; *Gopherus polyphemus*: Iverson, 1980) but some individuals may, at least under some circumstances.

Turtles usually nest successfully unless they are disturbed during the process or encounter unfavorable conditions in the potential nest site; in addition, predators take a high toll on eggs and hatchlings at many nesting sites (Auffenberg & Iverson, 1979; Bury, 1979; Wilbur & Morin, 1988). Multiple nesting, such as exhibited by sea turtles and many other species, varies in both spatial and temporal distribution. This certainly reduces the impact of storm and erosion events, and may reduce predation on the eggs and hatchlings.

10.4.6.1 Nesting Density

Some turtles nest in large numbers at restricted nesting locations (e.g., marine turtles: Miller, 1997; river turtles: Moll & Moll, 2004). It can be difficult to separate individual behaviors of multiple individuals nesting at a restricted number of nesting sites (such as turtles nesting on isolated sand banks along freshwater streams) and actual groups of turtles nesting together (e.g., *Lepidochelys olivacea*: Plotkin et al., 1995).

The density of nesting turtles at a particular site can have an impact on the success of nesting and the number of clutches hatching, as well as nest site selection. Green turtles (*Chelonia mydas*) nesting year-round at Ras Al-Hadd (Oman) show both high-density and low-density patterns (Al-Kindi et al., 2003). During the high-density period, the area available is restricted by the number of turtles attempting to nest; about 20% of the turtles attempting to nest were prevented from doing so. Some turtles moved elsewhere on the beach to continue their nesting process, whereas others returned to the ocean. Also, more turtles nested in suboptimal areas (at or near the tide zone) in the high-density period than during low-density period. Interestingly, winds during the low-density period dried the sand sufficiently to cause more egg chambers to collapse and, hence, "significantly higher abandonment" of the nesting effort (Al-Kindi et al., 2003).

10.4.7 Ovipositional Site Selection

The external environmental requirements (temperature, gas exchange, moisture) of the embryo must be met for development to be successful. Because chelonians exhibit limited parental care, nest site selection has a profound impact on successful reproduction, including embryo survivorship and female survivorship (Schwarzkopf & Brooks, 1987; Spencer & Thompson, 2003). As such, the selection of a nest site represents a synthesis of benefits and costs among multiple criteria (Spencer, 2002; Spencer & Thompson, 2003; Restrepo et al., 2006). An examination of the literature suggests that three groups of ideas have emerged regarding nest site-selection criteria: females select the nest site to manipulate the sex ratios among the hatchlings (e.g., Roosenburg, 1996; Spencer, 2002), to protect the eggs from predators, parasites, or environmental extremes, and to reduce their own risk of predation (Spencer, 2002; Spencer & Thompson, 2003).

In fact, nest site selection appears to be nonrandom in several species (e.g., *Kinosternon baurii*: Wilson, 1988; *Chrysemys picta*: Janzen & Morjan, 2001; *Chelydra serpentina*: Kolbe & Janzen, 2002; *Malaclemys terrapin*: Burger & Montevecchi, 1975; *Terrapene carolina carolina*: Flitz & Mullin, 2006). Turtles use a variety of environmental clues to select a nest site, including but not limited to slope, temperature, distance from water, substrate composition, moisture content, compactness, pH, salinity, and vegetation cover (*Chelonia mydas*: Mortimer, 1990; *Caretta caretta*: Miller et al., 2003; *Chrysemys picta*: Spencer & Thompson, 2003). The importance of each seems to vary among species (Ehrenfeld, 1979). For example, Ehrenfeld (1979) noted that some species are not particularly selective of substrate (*Chrysemys picta picta*, *Malaclemys terrapin macrospilota*), whereas others are more selective (sandy or friable, well drained: *Apalone ferox*, *A. mutica*, *Trachemys scripta*, *Graptemys geographica*, *Clemmys guttata*, *Gopherus polyphemus*). Some select nest sites are exposed to direct sunlight (*Geochelone elephantopus porteri, Apalone ferox, A. mutica, A. spinifera, Graptemys geographica, G. pseudogeographica*). Some species travel greater distances to nest away from water than others (closer: *Pseudemys concinna, Apalone muticus, Macrochelys temminckii*; further: *Malaclemys terrapin macrospilota, Platemys platycephala, Emydoidea blandingii*). Colombian slider turtles (*Trachemys callirostris callirostris*) nested under vegetation about 5 m from the water line, but whether the turtle selected the sites for the cover type or the underling substrate or moisture was not determined (Restrepo et al., 2006). Sliders (*Trachemys scripta venusta*: Moll, 1994), the painted terrapin (*Callagur borneoensis*: Pritchard, 1979; Bonin et al., 2006), the river terrapin (*Batagur baska*: Pritchard, 1979; Bonin et al., 2006), the Nile softshell turtle (*Trionyx triunguis*: Carr & Carr, 1985) the giant Asian softshell (*Pelochelys bibroni*: Rhodin et al., 1993), as well as the pignosed turtle (*Carettochelys insculpta*: Georges & Rose, 1993) live in coastal freshwater swamps and rivers but enter the ocean and nest on oceanic beaches. Currently, we have little understanding of the importance and interrelationships of these environmental variables in determining nesting success. Additional field data are needed that characterize nest sites selected for oviposition as well as available sites that are not selected.

With only one known exception, turtles oviposit by digging a hole in soil on land. The land may be an oceanic beach, a sandy river bank, sandy loam beside a freshwater creek, or clay soil away from the water's edge. The "hole" may be a deep chamber (as in the Cheloniidae), a scrape on the surface of the soil, or even among debris on the surface of the substrate (e.g., *Geochelone denticulate, Sternotherus* spp., *Kinosternon* spp.: Ehrenfeld, 1979). One Australian pleurodire species (*Chelodina rugosa*) lays its eggs underwater during flood inundation in areas where the nest will be exposed when the flood recedes (Kennett et al., 1993a). Embryonic development is arrested until the eggs are no longer submerged. This suggests that the egg shell, and possibly the oviductal fluid that fills the inter crystal spaces, act to prevent the movement of water vapor into or out of the egg; in conjunction with an embryo in diapause whose oxygen requirements are extremely low, this strategy allows oviposition, development, and emergence at times that facilitate survival. In addition, Polisar (1996) provided evidence that *Dermatemys mawii* may use a similar strategy in that nesting occurs during the rainy season and the eggs have similar inundation capabilities as *Chelodina rugosa*. However, no direct observations of nesting were made.

10.4.8 Oviposition

The depth and shape of the egg chamber is determined by the size and shape of the hind foot and to some degree by the substrate in which the nest is constructed. Some species dig nests as deep as 45 to 75 cm (sea turtles), whereas others dig shallow nests 10 to 30 cm deep (most other chelonians), and some make virtually no nest at all (Ehrenfeld, 1979). The depth and shape of the nest is a consequence of the length of the hind limb and the ability of the turtle to manipulate it in the substrate (e.g., Tiwari & Bjorndal, 2000). During oviposition, terrestrial and most aquatic chelonians use their hind feet to dig the egg chamber. Some species (e.g., all sea turtles: Ehrenfeld, 1979; Miller, 1997) use their front limbs to prepare the nesting site. The front flippers are used to clear debris and dry surface sand from the nest site as the turtle "swims" forward, throwing sand backward and slightly to the side. This behavior exposes more moist sand that is less likely to collapse while the egg chamber is being dug.

Considering the total reproductive output, several aspects of female morphology and egg production are involved. First, the number of eggs produced reflects, in part, the structure of the ovary from which the follicles (eventual egg yolks) are ovulated, as well as the nutritional basis from which the energy for reproduction is gleaned. Second, the oviduct must be long enough to contain the shelled eggs before oviposition, and the eggs must be of a size to fit into the oviduct until oviposition. The oviduct may provide some constraint but the expansion of the oviduct accommodates a variety of egg sizes and shapes (Ewert, 1985; Iverson & Ewert, 1991; Tucker & Janzen, 1998; Clark et al., 2001). The role of the oviduct in the shaping of eggs has not been defined, nor has the length of the oviduct been considered in evaluations of clutch number in concert with egg size. Third, the size of the pelvic arch is determined by the gap within the pelvic bones, coupled with the tissue surrounding the cloaca within the pelvic bones (pelvic gap). The size of the pelvic gap and egg width increase with female in size (*Deirochelys reticularia*: Congdon et al., 1983a); however, in *Trachemys scripta* the pelvic aperture does not limit egg diameter (Congdon & Gibbons, 1987). Long and Rose (1989) reported that the pelvic aperture was larger in females that produced ellipsoidal eggs (e.g., *Terrapene ornata*, *Kinosternon flavescens*) but not for females that produced spherical eggs (e.g., *Gopherus berlandieri*).

In assessing the constraints imposed on egg size and shape by the morphology of female *Sternotherus odoratus*, Clark et al. (2001) found that the pelvic aperture was typically wider than the width of the female's largest egg, suggesting that the pelvic aperture is of minor importance in limiting egg diameter compared to the caudal gap (distance between the carapace and the plastron), which was too narrow to pass the widest egg in a quarter of the turtles. Additionally, the caudal gap may constrain the size of the egg, particularly in smaller turtles (*Sternotherus odoratus*: Clark et al., 2001). This constraint is partially ameliorated by the notching of the plastron, the expansion of the posterior carapace in a supra-pygal hump, and by plastronal kinesis. However, a downward extension of the carapace may require the egg to alter direction (at least slightly) during oviposition. Kinesis of the plastron alleviates the problem by reducing the angle through which the egg must pass; the hinge flexes to allow the passage of the egg that would otherwise be too big to fit through the gap or make the required angular change. Several families (e.g., Testudinidae, Kinosternidae, Emydidae: Zangerl, 1969) exhibit a plastronal hinge or a dome-shape that facilitates oviposition of oblong and ellipsoidal eggs. In *Testudo marginata*, the posterior marginal scutes flair to form a "skirt" that may assist in oviposition but may also impede copulation (Bonin et al., 2006).

The pelvic aperture and the shell gap may exert pressure that affects the egg diameter, but the hormonal system that controls vitellogenesis (Ho, 1987; Ho et al., 1982, 1985) and, ultimately, the yolk diameter may play an important role, also. Yolks of eggs produced by younger (smaller) female painted turtles contained more testosterone, and the eggs were "nearly 20% smaller" than eggs from older (larger) females (*Chrysemys picta*: Bowden et al., 2004). Hormone levels declined in yolks of *Trachemys scripta elegans* during incubation (Bowden et al., 2002) and may be involved in determining the sex of the turtle (Bowden et al., 2000, 2001). These results suggest that egg size may be constrained by hormonal levels as well as physical characteristics of the female. This possibility must be examined in a wide selection of species to establish the relative importance of hormonal contribution to egg size.

When oviposition is finished, the female may sit for a few minutes before starting to cover the eggs. Using her hind feet, she scrapes the nearby material into the hole, and at intervals typically packs the material into the neck of the egg chamber. When the chamber is full, the female will scrape other material over the nest site using her front and hind feet. This sequence is particularly noticeable in marine turtles but occurs with variations in all species (Ehrenfeld, 1979). These behaviors re-establish the insulative qualities of the thermal and hydic environments of the substrate in which the eggs will incubate and help to buffer the eggs against rapid changes.

In considering the strict oviparity among chelonians, Kuchling (1999) argued, using sea turtles as the focus but including all chelonians, that "most threats to progeny occur immediately after eggs are laid and when hatchlings disperse," "egg incubation itself [was] a minor hazard," and "a reduction of its duration [was] without strong selective advantage." Seemingly, the risks involved with terrestrial nesting do not "compensate for the disadvantage to the female in carrying the embryos for a longer time" (Kuchling, 1999). Obviously, chelonians have retained oviparity; the reasons may remain obscure for some time.

10.4.9 Embryonic Development and Incubation

The embryological development of turtles occurs within the egg shell and is supported by the nutrients contained in the yolk. In the nest, environmental conditions (i.e., temperature, moisture, gas) affect the rate of development and, in some cases, the sex of the hatchlings (Ewert, 1979, 1985; Miller, 1985b; Packard & Packard, 1988a). The embryo develops to middle gastrulation while within the oviduct and then enters diapause until after oviposition (Ewert, 1979, 1985; Miller, 1985b). Development may resume within a matter of hours, as in sea turtles (Miller, 1985b), or it may not resume until floodwaters have receded (Kennett et al., 1993b).

Moll (1979) reported that the intrauterine period between ovulation and oviposition is "highly variable among species." The internesting period reported for marine turtles is typically about 14 days (range: 9 days in *Dermochelys* to 28 days in *Lepidochelys*), which represents the minimum intrauterine period (Miller, 1997); freshwater and terrestrial species also fit this general pattern (Moll, 1979). During this period, ova are fertilized and the embryos develop to middle gastrulation simultaneously with the processes of albuminization and shell production to complete the egg. The process of albuminization and formation of the inner shell membrane are completed relatively quickly, whereas the completion of the shell matrix requires several days (Cheloniidae: Owens, 1980; Owens & Morris, 1985; Miller, 1985b).

Several sets of developmental stages have been described and most deal with the period between oviposition and hatching. All schema use sequential morphological changes visible on the whole embryo as a basis for defining stages of development. All turtles pass through a series of early developmental stages that are virtually the same, including the formation of the head, somites, pharyngeal clefts, eye spots, and limb-buds (Agassiz, 1857; Pasteels, 1957, 1970; Yntema, 1968; Ewert, 1979, 1985; Miller, 1985b). As the embryos continue to develop, the specific characteristics become more definite: the type of foot, the carapacial scale pattern, and the coloration become obvious (Ewert, 1979, 1985; Miller, 1985b). Most authors place the stages along a timeline from oviposition until hatching and define the conditions of incubation in terms of temperature and moisture (Miller, 1985a; Greenbaum & Carr, 2002). Several older studies provide drawings and descriptions of partial developmental series (Agassiz, 1857).

Stages of development have been described for species in five families: Emydidae (*Testudo hermanni*: Guyot et al., 1994; *Chelydra serpentina*: Yntema, 1968; *Chrysemys picta belli*: Mahmoud et al., 1973; *Trachemys scripta*: Greenbaum, 2002), Cheloniidae (*Chelonia*: Miller, 1985b; Kaska & Downie, 1999; *Natator*: Miller, 1985b; *Eretmochelys*: Miller, 1985b; *Caretta*: Miller, 1985b; Billett et al., 1992; Kaska & Downie, 1999; *Lepidochelys olivacea*: Crastz, 1982), Dermochelyidae (*Dermochelys coriacea*: Miller, 1985b; Renous et al., 1989), Carettochelyidae (*Carettochelys insculpta*: Webb et al., 1986; Beggs et al., 2000), and Trionychidae (*Pelodiscus sinensis*: Tokita & Kuratani, 2001; *Trionyx sinensis*: Cherepanov, 1995).

The conditions under which the eggs incubate have a profound impact on embryonic physiology and development, hatchling and emergence success, and subsequent neonatal survival (Harry & Williams, 1991; O'Steen & Janzen, 1999; Thompson, 1983; Packard & Packard, 1988a; Ackerman, 1997; Ackerman & Lott, 2004; Deeming, 2004). Incubation of the egg is important to the survival of the embryo and the hatchling because the thermal, hydric, and gaseous conditions within the nest interact synergistically with the developing embryo, altering its metabolism.

Birchard (2004) described the thermal environment of reptile embryos as being built upon two larger scale factors: the behavior of the female and the thermal environment around the embryos during development. Female nest site selection and the thermal environment of the nest play predominant roles in influencing the rate of embryonic development and characteristics of the hatchlings. Most chelonian embryos develop successfully in the range of 24 to 32°C (Köhler, 2005); temperature variation above this range may result in the death of the embryo. Daily fluctuations in nest temperatures may be small (±0.5°C, *Dermochelys coriacea*: Godfrey et al., 1997) or large (±9°C, *Emydura macquarii*: Packard & Packard, 1988a; Thompson, 1988; *Emydoidea blandingii*,

Macrochelys temminckii: Figure 10.3). Further, the temperature within the nest may not be uniform top to bottom (±6°C: Thompson, 1988). Embryos incubated at different temperatures use their yolk energy differently to "produce hatchlings with different probabilities of surviving the neonatal period" (Packard & Phillips, 1994).

10.4.9.1 Sex Determination

Temperature-dependent sex determination has been found in several cryptodirian families but in no families of pleurodires (Paukstis & Janzen, 1990; Ewert & Nelson, 1991; Valenzuela, 2004). In those chelonians that exhibit environmentally determined sex (Cheloniidae, Carettochelyidae, Chelydridae, Dermatemydidae, Dermochelyidae, Emydidae, Kinosternidae, Pelomedusidae, Testudinidae), females are produced from eggs incubated at warmer temperatures (typically above 29°C) and males produced from eggs incubated at cooler temperatures (Alho et al., 1985; Bull, 1985; Bull & Vogt, 1979; Miller & Limpus, 1981; Morreale et al., 1982; Mrosovsky et al., 1984a, 1984b; Pieau, 1982; Schwarzkopf & Brooks, 1985, 1987; Vogt & Bull, 1984; Wilhoft et al., 1983; Rostal et al., 2002). Although hatchlings of the same sex are produced toward the upper and lower ends of the embryonic-tolerance temperature range, different ratios of male to female hatchlings may be produced in nests wherein the temperature is closer to the pivotal temperature or if the nest

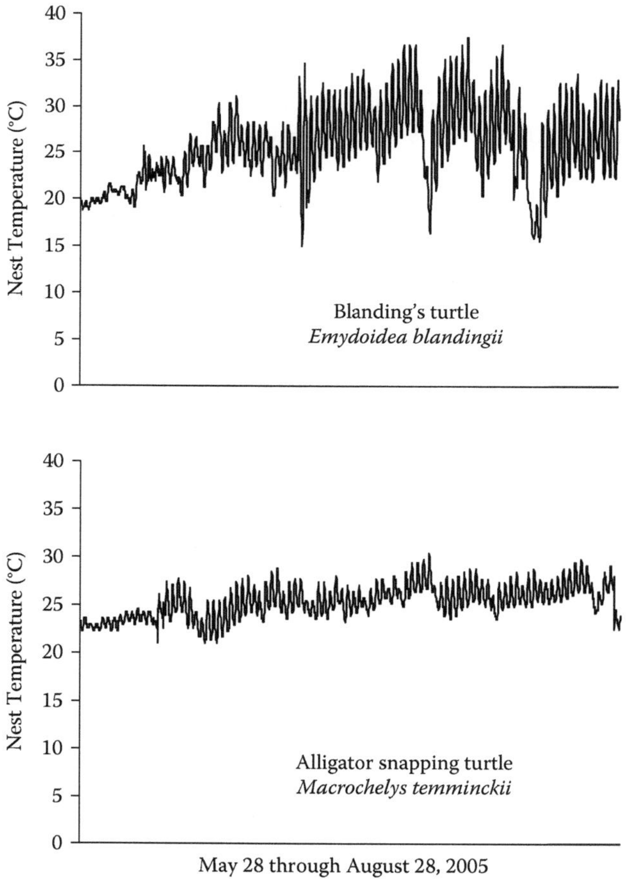

FIGURE 10.3 Daily fluctuations in nest temperature for a shallow nest (12 cm, *Emydoidea blandingii*) and a deep nest (20 cm, *Macrochelys temminckii*) recorded for the same period, albeit at different locations (Dinkelacker, unpublished data).

experiences thermal layering (*Chelydra serpentina*: Wilhoft et al., 1983). At least one species (*Chelydra serpentina*) produced females at high and low temperatures, with males being produced at cooler temperatures in-between (Yntema, 1976, 1979). Additionally, Ewert et al. (1994) report that females are produced at all viable incubation temperatures, whereas males are only produced between 25 and 27°C for *Macrochelys temminckii*. The sex of hatchling *Caretta caretta* is determined by the proportion of developmental time spent in the female (or male) determining temperature rather than the duration of daily exposure (Georges et al., 1994). Nest site selection among several reproductive seasons can have an impact on the sex ratios among the age classes in the population (Valenzuela & Janzen, 2001; Ackermann & Lott, 2004). Hulin and Cuillon (2007) modeled female philopatry in heterogenous habitats using data from sea turtles with a view to assessing the evolution of habitat-dependent sex ratios. Based on their model, they argue that the evolutionary stable strategy is to overproduce females in good habitat and to produce males in poor habitats. They noted that the predicted sex ratios were very dependant on the type of density-dependant regulation assumed to be functioning in the population. They suggest that including more biologically realistic variables (i.e., multiple reproductive episodes, overlapping generations, temporally variable environments) may change the predictions.

10.4.9.2 Incubation

The duration of incubation varies inversely with temperature (Yntema, 1978; Ewert, 1979, 1985; Miller & Limpus, 1981; Miller, 1985b; Choo & Chou, 1987; Gutzke & Packard, 1985; Packard et al., 1987; Mrosovsky, 1988; Birchard, 2004). Over the term of lower temperature incubation, development occurs more slowly (Packard & Packard, 1988a) and hatchlings tend to have larger bodies and smaller residual yolk mass (Packard & Packard, 1988a; Packard & Phillips, 1994). However, at least in some widespread species there appears to be a genetic component to the duration of incubation; eggs from northern populations of snapping turtles (*Chelydra serpentina*) required less time to hatch than eggs from southern populations incubated at the same temperature (Ewert, 1985), although this may be confounded by differences in egg size.

The dynamics of moisture and gas movement in nests are complex (Ackerman, 1977, 1981, 1991; Ackerman et al., 1985; Ackerman & Lott, 2004). Water exchange of chelonian eggs is influenced by temperature: higher temperatures increase the rate of water vapor diffusion (Packard & Packard, 1988a; Ackerman & Lott, 2004). Water exchange is also affected by the amount of egg surface area in contact with the substrate and the water potential gradient between the egg and the substrate (Ackerman et al., 1985; Packard & Packard, 1988a; Ackerman & Lott, 2004). In small clutches, all eggs are in contact with the substrate, to some degree. However, in large clutches the outer eggs are in direct contact with the substrate and the inner eggs only touch other eggs, which creates air spaces (Ackerman, 1977; Packard et al., 1981; Packard & Packard, 1988a).

The structure of the egg shell and the structure of the surrounding material affect the water relations of turtle eggs (Packard & Packard, 1984). In general, hard-shelled eggs are independent from the surrounding hydric conditions compared to flexible-shelled eggs (*Mauremys caspica*: Ackerman, 1991; *Chrysemys picta*: Ratterman & Ackerman, 1989; *Chelydra serpentina*: Rimkus et al., 2002; *Apalone spinifera*: Packard & Packard, 1983, 1990). However, pliable-shelled eggs are subject to hydric flux (Packard, 1991; Packard & Packard, 1984, 1988a; Packard & Phillips, 1994). The amount of available environmental water depends on the structure of the substrate and the moisture gradient (Packard & Packard, 1988b; Packard et al., 1980, 1987). The hydric environment surrounding incubating pliable-shelled turtle eggs affects embryonic uptake and loss of water (Packard & Packard, 1986, 1988b; Finkler, 2001). Over the term of incubation, if conditions are too dry, the eggs will lose mass to the environment and may die if the loss is large enough. Conversely, if too much water is present, oxygen may be excluded from the developing embryo and it might die. Among surviving embryos, those from wetter environments have higher metabolic rates, higher growth rates, remain in the egg longer, and have less residual yolk available for neonate (Packard & Phillips,

1994). Embryonic metabolism of yolk lipids and production of nitrogenous waste (Packard & Packard, 1983, 1988a; Packard et al., 1984; Janzen et al., 1990) and embryonic gas exchange (Ackerman & Lott, 2004) are impacted by the surrounding hydric conditions. The temperature coupled with moisture available in the substrate influences the duration of incubation (Birchard, 2004; Belinsky et al., 2004; Godfrey & Mrosovsky, 2001), calcium metabolism (Packard & Packard, 1986, 199 ; Packard & Clark, 1996), and hatchling mass (Birchard, 2004). Energy reserves (Gutzke et al., 198 ; Packard et al., 1988; Janzen et al., 1990), growth (Brooks et al., 1991; O'Steen, 1998; Spotila et al., 1994), locomotor performance, and other behaviors (Miller et al., 1987; Janzen, 1995) are also affected by the incubation conditions.

Lower than atmospheric oxygen concentrations reduce the survivorship to hatching, and extremely low levels of oxygen can increase the length of incubation (*Trachemys scripta*: Etchberger et al., 1991). Because most chelonian nests are relatively shallow (with the exception of sea turtles) and typically contain a small number of eggs, the oxygen demand of the developing embryos is usually not a constraint on successful development (Packard & Phillips, 1994; Packard & Packard, 1988a). In sea turtle nests, the partial pressure of oxygen declines between 25 to 50% of the levels at the start of incubation (*Chelonia mydas, Caretta caretta*: Ackerman, 1977; *Dermochelys coriacea*: Wallace et al., 2004). Interestingly, in the case of developing leatherback sea turtle eggs the rate of hatching was not coupled to the oxygen partial pressure in the clutch (Wallace et al., 2004).

Changes in other abiotic conditions also increase mortality. Eggs of turtles that nest on sand bars (e.g., *Podocnemis expansa*: Roze, 1964; *Apalone muticus*: Plummer, 1976) and experience flooding from excessive rain (e.g., *Caretta caretta*: Ragotzkie, 1959) or tidal inundation may drown (Foley et al., 2006). Experimental studies on eggs of *Trachemys scripta elegans* indicate that simulated inundation for more than 12 hours "dramatically reduced survivorship" (Tucker et al., 1997 . On ocean beaches, marine turtle eggs may be exposed to the atmosphere by beach erosion and dehydrate to death or become overheated by the sun or both; the hydric conditions may change during incubation by wave wash-over and salt water percolation into the beach. Temporal and spatial variation in the choice of nesting site may increase or decrease exposure to these risks and may alter the total hatchling production of the female's reproductive effort.

10.4.9.3 Developmental Arrest

Two categories of developmental arrest are used by different species of turtles: pre- and post-ovipositional. In the oviduct, turtle eggs remain in arrested development until oviposition. During this time, the eggs are typically in a late gastrulae stage (Ewert, 1985). This pattern is considered universal, as all turtles studied to date exhibit this pattern. The benefits of this arrest include allowing the entire clutch to form in synchrony following oviposition, but also to allow the female to postpone oviposition until a suitable nest site is found (Ewert, 1985).

In some circumstances, chelonians can retain eggs in their oviducts for extended periods of time beyond the normal intrauterine period when conditions/circumstances for nesting are not appropriate (Cagle & Tihen, 1948). This idea was echoed by Pritchard (1979) in reference to arribada (mass) nesting in *Lepidochelys*; he argued, in part, that the turtles could retain their eggs until suitable cues/conditions occurred. Prolonged egg retention may also occur as a consequence of not being able to construct a nest chamber; Limpus (1985) reported two *Caretta caretta* being unable to nest for 41 and 42 days, respectively. When dissected, the eggs in the oviducts were found to be from the most recent ovulation sequence; eggs from one clutch were placed in a hatchery and produced an emergence success of 76.2%. Additionally, some turtles may carry shelled eggs in their oviducts over winter, at least in some years (e.g., *Deirochelys reticularia*: Gibbons & Greene, 1978).

Following oviposition, embryos of some species begin development within the first few hours, whereas others remain in a post-ovipositional arrest. Ewert and Wilson (1996) describe four types of post-ovipositional arrest. In the first type, post-ovipositional arrest can be an extension of the pre-ovipositional arrest where the yolk is partitioned and the embryo is still pre-somite. *Chelodina rugosa* and *Elseya dentata* are examples of this post-ovipositional extension (Kennett, 1999; Kennett et al., 1993b; Andrews, 2004). The eggs of *Chelodina rugosa* are deposited at the bottom of flooded wetlands, where they remain in an arrested state until the wetland dries and the eggs are exposed to atmospheric oxygen (Kennett et al., 1993a, 1993b). The second type of post-ovipositional arrest is embryonic diapause. Some embryos may actually begin to develop and then enter a period of developmental arrest. *Deirochelys reticularia* and *Kinosternon baurii* are both examples of the second type. The turtles deposit their eggs either in the fall or the spring of the year (Ewert & Wilson, 1996). The fall eggs often require chilling (such as winter temperatures) before diapause can be broken and development can resume (e.g., *Deirochelys reticularia*: Ewert, 1985). The other types of post-ovipositional arrest are "delayed hatching" and "embryonic aestivation" (Ewert & Wilson, 1996). These strategies are used by full-term embryos that have internalized their yolks while in a dry nest environment; they are stimulated to hatch by a change to a wet/anoxic environment (e.g., *Carettochelys insculpta*: Webb et al., 1986; *Kinonsternon scorpioides*: Ewert, 1991).

For some species, post-ovipositional arrest is obligatory, whereas in others it is facultative or there is a seasonal component. *Chelodina rugosa* exhibits an obligatory diapause because eggs rarely hatch unless they are deposited underwater for a duration of time; once the egg dries, development will resume (Kennett et al., 1993a, 1993b). If the eggs are subjected to incubation conditions without submergences, hatching rarely ensues (Kennett et al., 1998). At least one species exhibits both a continuation of pre-ovipositional arrest and embryonic diapause. *Chelodina expansa* eggs are deposited during the autumn, yet the hatchlings do not emerge for up to a year later (Ewert, 1979). In *Chelodina expansa*, post-ovipositional arrest can last for up to six weeks after oviposition and can the lead into an embryonic diapause (Booth, 2000). For *Kinosternon baurii*, eggs collected in the fall require a chilling period to break diapause (Ewert & Wilson, 1996). However, eggs collected in the spring do not typically show a diapause.

The cues for breaking diapause range from temperature and oxygen diffusion to moisture availability. Most subtropical to temperate distributed turtles use temperature as their cue. In comparison, most tropical species use rainfall (i.e., moisture) as their cue. The adaptive value of diapause is subject to debate (Spencer et al., 2001). A prevailing thought in the literature suggests that diapause allows for the avoidance of winter mortality or positions hatching at a time more suitable for hatchling survival.

Most research on embryonic diapause has focused on semi-aquatic and aquatic freshwater turtles. However, terrestrial turtles have extremely long incubation periods in comparison, and may express developmental arrest (Ewert, 1985). Little research has been conducted to confirm this possibility, much less its commonality, relevance, or the cues for onset and exit from arrest.

10.4.10 Hatching and Hatchling Survivorship

Embryos of most species complete development within a few months and the hatchlings emerge to fend for themselves (Ewert, 1979, 1985). Examples of this general pattern occur in all sea turtles and many freshwater turtles. The process of hatching is similar in all chelonians, regardless of their egg shape, size, or structure. Once stimulated to hatch, the embryos struggle to split open the eggshell by using their caruncle. During the process, circulation in the extra-embryonic membranes shuts down. The head of the hatchling is typically the first part to extend through the eggshell; this facilitates access to the atmosphere in the nest chamber. Further struggling using the front legs creates other cuts in the shell.

Unfortunately, the stimulus for hatching is poorly defined for most species. In *Carettochelys*, eggs hatch as the water rises at the initiation of the rainy season (Webb et al., 1986), which allows them to emerge at a time of greater resource availability and reduced predation risk. Carr and Hirth (1961) described social facilitation of emergence in hatchling sea turtles. The process of digging out of the nest chamber occurs as hatchlings struggle against the upper surface, freeing sand that trickles down among the lower hatchlings that also struggle, thereby moving the sand to the bottom of the chamber. The group moves upward in bursts of activity that appear to be regulated by the availability of oxygen and by ambient temperature. As oxygen levels are reduced and carbon dioxide levels increase, the hatchlings become quiescent; final emergence onto the beach requires a reversal of temperature (Mrosovsky, 1968; Gyuris, 1993). Sea turtle nests are deep in sand and usually contain many hatchlings; other species of turtles may not experience the same pattern of events during hatching or emergence. Successful incubation is the result of a combination of events, including the risk of predation and variations in the surrounding environment.

Once the hatchlings emerge from the egg, they do not necessarily emerge from the nest (e.g., *Chrysemys picta*: Woolverton, 1961; Bleakney, 1963). Gibbons and Nelson (1978) suggested that a "delayed emergence" may permit the hatchlings to enter the environment at a time when resources are increasing, not decreasing, and growth potential is maximized (Carr & Ogren, 1959; Wilbur, 1975a). As a strategy, delayed emergence is widespread; it occurs in five families and is practiced by at least 19 species (Gibbons & Nelson, 1978). For some species such as *Chelodina rugosa*, this occurs with the onset of the rainy season. Some hatchlings may overwinter as a group in the nest (Gibbons & Nelson, 1978). In other species, the individual hatchlings may dig deeper into the substrate below the nest to overwinter, whereas others from shallower nests emerge (*Terrapene o. ornata*: Converse et al., 2002). Although the sample sizes were small, these observations raise the question of what set of environmental or biological cues stimulate emergence.

Emergence time in turtles is typically synchronous. For example, sea turtles typically show a synchronous emergence (Houghton & Hays, 2001); that is, all hatchlings break free of the nest and make a run for the water at the same time. The advantages of each system are subject to debate and probably dependant upon local environmental conditions and biotic factors, such as the presence of predators.

10.4.10.1 Overwintering

Environmental stresses associated with the impending winter in temperate environments can dictate the strategies used by hatchlings. Costanzo et al. (1995) reported that hatchlings in Nebraska might emerge from the nest in the fall and presumably overwinter in aquatic habitats, or remain in the natal nest, or dig below the nest. Recent research has suggested an alternative strategy exists. Diamondback terrapins (*Malaclemys terrapin*) emerge from the nest in the fall but rebury themselves in suitable terrestrial environments (Draud et al., 2004). Blanding's turtles (*Emydoidea blandingii*) are also suspected of using the same strategy (Dinkelacker et al., 2004). Some species (e.g., *Chrysemys picta*) use both strategies, even at the same locality (DePari, 1996). Regardless of the emergence strategy used by hatchlings, the onset of winter brings a host of environmental stresses with which the hatchling must cope. Should hatchlings remain in the nest or in a terrestrial environment, they must cope with desiccating conditions as well as variable and often extreme temperatures (Costanzo et al., 1995). Although overwintering in aquatic habitats provides thermal stability, hatchlings must cope with osmotic perturbations and potentially hypoxic or anoxic conditions (Ultsch, 1989).

To overwinter in a terrestrial environment, a hatchling must be able to cope with desiccating conditions and potential dehydration. This is especially problematic for hatchlings considering their high surface area to mass ratio. Costanzo and colleagues have provided a wealth of information regarding the dehydration resistance of hatchling turtles (Costanzo et al., 2001b; Baker et al., 2003; Dinkelacker et al., 2004). Their results suggest species that overwinter in terrestrial habitats have much lower rates of evaporative water loss (EWL). However, species such as *Chelydra serpentina* and *Apalone spinifera* have high EWL rates and, therefore, may not be able to survive in terrestrial environments. This may be the force that drives fall emergence in these species. Related to the ability to resist dehydration is the ability to tolerate dehydration. Currently, there is limited research on dehydration tolerance and this avenue of research would certainly be beneficial.

In the higher latitudes, hatchlings overwintering terrestrially may experience temperatures below the freezing point of body fluids (Costanzo et al., 1995; Packard, 1997; Packard et al., 1997a;

Costanzo et al., 2004). Currently, there are two strategies used to survive subzero temperatures: freeze tolerance and supercooling. Freeze tolerance is a strategy in which ice forms within the hatchlings but is typically limited to extra-cellular spaces (Storey et al., 1988; Churchill & Storey, 1992a, 1992b; Costanzo & Lee, 1994). Hatchlings depend upon ice inoculation at high subzero temperatures and a relatively slow cooling rate to limit ice to the extra-cellular spaces. During a freezing event, respiration and cardiovascular function stops (Storey & Storey 1988; Costanzo et al., 2001b). Presumably, membrane potentials are maintained at very low levels. Numerous North American species are capable of surviving at least seven days at –2.5°C (e.g., *Emydoidea blandingii, Crysemys picta, Malaclemys terrapin, Terrapene ornata*: Dinkelacker et al., 2005b); however, no species tested has survived lower than –4°C (Storey et al., 1988; Churchill & Storey, 1992b; Costanzo et al., 1995; Packard et al., 1999; Baker et al., 2004; Dinkelacker et al., 2005b).

Supercooling is a strategy in which body fluids remain liquid below their freezing point. Again, cardiovascular function and respiration ultimately cease (below –8°C), but hatchlings have been reported to survive exposure to –20°C (Birchard & Packard, 1997; Hartley et al., 2000). The danger of supercooling is that the body fluids are held in a metastable state and inoculation by ice causes rapid lethal freezing (Costanzo et al., 2001a; Packard & Packard, 2003b). The ability to resist freezing, termed "inoculation resistance," is dependant upon morphological adaptations as well as environmental variation. Endogenous nucleators inside the hatchling are purged during the fall, and failure to purge these nucleators results in a poor supercooling capacity (Costanzo et al., 2000b; Packard et al., 2001; Costanzo et al., 2003). Exogenous nucleators are found within the nest itself and can include bacteria, minerals, and ice (Packard & Packard, 1993a). The effectiveness of these nucleators is often determined by soil moisture, temperature, and morphological considerations (Costanzo et al., 1998; Costanzo et al., 2000a; Costanzo et al., 2001c). Some species possess subdermal deposits of lipids, which have been suggested to help prevent inoculation of the turtle (Willard et al., 2000; Packard & Packard, 2003a). Species lacking these deposits are thought to be susceptible to freezing.

Whether turtles use freeze tolerance or supercooling to survive subzero temperatures is a topic that has received considerable attention and has spawned much debate. Ultsch (1989) suggested the ability to tolerate freezing is widespread among members of the family Emydidae. The painted turtle (*Chrysemys picta*) has been the focus of much of the attention, and the debate centers around its winter strategy. Packard and colleagues have studied the species in Nebraska where nest temperatures may reach –20°C. These researchers originally endorsed the notion of freeze tolerance as a viable strategy (Packard et al., 1989) soon after its discovery in turtles (Storey et al., 1988); however, they later changed their view to believing that freeze tolerance is widespread among all hatchling turtles (Packard et al., 1999), and that supercooling is the only strategy that permits survival because freeze tolerance is limited to –4°C (Packard & Packard, 1993a, 1993b; Packard, 1997; Packard et al., 1997b; Hartley et al., 2000; Packard & Packard, 1997, 2001, 2003b, 2004). The counterargument is presented by Costanzo and colleagues, who have studied the same species in Nebraska. They believe that the environment dictates the strategy used by this species (Costanzo et al., 1998; Costanzo et al., 2001c; Baker et al., 2003; Costanzo et al., 2004). In cold dry years, inoculation resistance is sufficient to prevent freezing and allow survival by supercooling. However, during warm wet years, inoculation resistance is overcome by the high soil moisture content and freezing ensues (Costanzo et al., 2004). Because wet winters are often mild and not all sites are as extreme as Nebraska (e.g., Indiana), hatchling painted turtles may survive by using freeze tolerance (Costanzo et al., 2004).

Some species forgo the challenges of terrestrial overwintering and instead overwinter in the unfrozen depths of aquatic habitats. Although these habitats are thermally stable, ice cover may force hatchlings to remain submerged for extended periods of time (Ultsch, 1989). In addition, if snow cover accumulates over the ice, bacterial respiration may predominate and the water could become anoxic. Adult turtles are well equipped to deal with prolonged submergence. In anoxic water, painted turtles and snapping turtles rely on a depressed metabolic rate and a highly efficient buffering system to survive for up to five months (Jackson & Heisler, 1982, 1983; Ultsch & Jackson, 1982a, 1982b, 1995; Ultsch & Wasser, 1990; Jackson, 1997, 2000; Ultsch et al., 1999; Jackson et al., 2000; Reese et al., 2000, 2002b; Ultsch & Reese, in press). Softshell turtles are limited to normoxic habitats simply because they lack a pronounced anoxia tolerance. This has been attributed to a poorly ossified skeleton (Reese et al., 2003). Hatchlings are similar in that their shell is incompletely ossified, which prohibits an effective buffering system and results in submergence times for hatchlings in anoxic water that are short in comparison to adults (Reese et al., 2002a; Dinkelacker et al., 2005a). In normoxic water, hatchlings are capable of surviving for up to two months, although osmotic perturbations may ultimately limit survival (Reese et al., 2002a; Dinkelacker et al., 2005a).

10.4.10.2 Growth

Bobyn and Brooks (1994) successfully demonstrated that post-hatchling growth is primarily determined by incubation at intermediate temperatures for *Chelydra serpentina*. Although increased soil moisture influenced the size of hatchlings, no advantage in growth rate was gained. Janzen and Morjan (2002) reported that female painted turtle hatchlings grow faster than males during the first year of life. In addition, egg mass and incubation temperature positively affected juvenile mass at one year of age.

Juvenile chelonian growth is typically supported by a high protein diet, although as they grow toward maturity, many species switch from high- to low-protein diets by switching from insects, fish, and other small animals to vegetation. The duration of the juvenile period varies among species: some species may breed at 2 years of age (*Sternotherus odoratus*: Tinkle, 1961), whereas others may not reproduce for 25 years, at least in some populations (*Dipsochelys elephantina*: Swingland & Coe, 1978). Sea turtles may require decades to reach sexual maturity (Limpus, 1990; Limpus & Reed, 1985b; Heppell et al., 2003).

Local environmental conditions influence growth rates and female size at maturity (*Trachemys scripta elegans*: Tucker, 2001) and, as a consequence, different clutch counts. Chelonians respond to local conditions with variation in some life history traits (Wilbur & Morin, 1988). For instance, *Chrysemys picta* from three different habitats (lake, marsh, river) in Michigan laid significantly different numbers of eggs per clutch that may result from differences in female size related to diet (Gibbons, 1967; Gibbons & Tinkle, 1969). Female *Trachemys scripta* matured at a larger size and produced larger clutches in heated pools than females in natural ponds (Gibbons, 1970b; Gibbons et al., 1981); female *Trachemys scripta* may also respond by commencing nesting earlier and producing more clutches in heated pools compared to those from natural pools (Thornhill, 1982). Research into the effects of natural environmental variation as well as anthropogenic modifications on the plasticity of chelonian life history traits will undoubtedly produce important evolutionary insights.

Recently, growth, age, and population dynamics of sea turtles have been reconsidered (Chaloupka & Musick, 1997) with the conclusions that "it is premature for a full-scale comparison of sea turtle growth dynamics" and "sea turtle population modeling is very much in its infancy." While acknowledging the value of data collected in the past, issues of statistical methodology, sample size, and duration of the study play major roles in limiting the understanding of turtle populations. Field data suggest that growth may be quite different in different populations (Heppell et al., 2003). The same is probably true for freshwater and terrestrial species.

ACKNOWLEDGMENTS

We would like to thank P. Pritchard for allowing JDM to examine specimens in the collection at the Chelonian Research Institute, Florida. Renn Tumilson of Henderson University, Arkansas, loaned specimens of Terrapene for examination. We also wish to acknowledge the helpful comments of the reviewers and the editors.

REFERENCES

Ackerman, R.A., The respiratory gas exchange of sea turtle nests (Chelonia, Caretta), *Respir. Physiol.*, 31, 19, 1977.

Ackerman, R.A., Growth and gas exchange of embryonic sea turtles (Chelonia, Caretta), *Copeia*, 1981, 757, 1981.

Ackerman, R.A. Physical factors affecting the water exchange of buried reptile eggs, in *Egg Incubation: Its Effects on Embryonic Development in Birds and Reptiles*, D.C. Deeming and M.W.J. Ferguson (eds.), Cambridge, UK: Cambridge University Press, 1991, 193.

Ackerman, R.A., The nest environment and the embryonic development of sea turtles, in *The Biology of Sea Turtles*, P.L. Lutz and J.A. Musick (eds.), Boca Raton, FL: CRC Press, 1997, 83.

Ackerman, R.A., and Lott, D.B., Thermal, hydric, and respiratory climate of nests, in *Reptilian Incubation: Environment, Evolution and Behaviour*, D.C. Deeming (ed.), Nottingham, UK: Nottingham University Press, 2004, 15.

Ackerman, R.A., Seagrave, R.C., Dmi'el, R., and Ar, A., Water and heat exchange between parchment-shelled reptile eggs and their surroundings, *Copeia*, 703, 1985.

Agassiz, L., *Contributions to the Natural History of the United States of America*, Boston: Little, Brown & Co., 1857.

Aitken, R., and Solomon, S., Observations on the ultrastructure of the oviduct of the Costa Rican green turtle (*Chelonia mydas* L.), *J. Exp. Mar. Biol. Ecol.*, 21, 75, 1976.

Aitken, R., Solomon, S., and Amoroso, E., Observations on the histology of the ovary of the Costa Rican green turtle, *Chelonia mydas* L., *J. Exp. Mar. Biol. Ecol.* 24, 189, 1976.

Alho, C.J.R., Danni, T.M.S., and Pádua, L.F.M., Temperature-dependent sex determination in *Podocnemis expansa* (Testudinata: Pelomedusidae), *Biotropica*, 17, 75, 1985.

Al-Kindi, A.Y.A., Mahmoud, I.Y., Al-Gheilani, H.M., Bakheit, C.S., Al-Habsi, A.A., and Al-Kiyumi, A., Comparative study of the nesting behavior of the green turtle, *Chelonia mydas*, during high- and low-density nesting periods at Ras Al-Hadd Reserve, Oman, *Chel. Cons. Biol.*, 4, 603, 2003.

Allan, B.M., The origin of the sex cells, *Anat. Anz.*, 29, 217, 1906.

Altland, P., Observation on the structure of the reproductive organs of the box turtle, *J. Morphol.*, 89, 599, 1951.

Andersson, K.E., and Wagner, G., Physiology of penile erection, *Physiol. Rev.*, 75, 191, 1995.

Andrews, R.D., Patterns of embryonic development, in *Reptilian Incubation: Environment, Evolution and Behavior.* D.C. Deeming (ed.), Nottingham, UK: Nottingham University Press, 2004, 75.

Ar, A., Belinsky, A., Dmi'el, R., and Ackerman, R.A., Energy provision and utilization, in *Reptilian Incubation: Environment, Evolution and Behaviour*, D.C. Deeming (ed.). Nottingham, UK: Nottingham University Press, 2004, 143.

Aresco, M.J., Reproductive ecology of *Pseudemys floridana* and *Trachemys scripta* (Testudines: Emydidae) in Northwestern Florida, *J. Herpetol.*, 38, 249, 2004.

Ashley, L.M., *Laboratory Anatomy of the Turtle*, Dubuque, IA: William C. Brown, 1962.

Atland, P.D., Observations on the structure of the reproductive organs of the box turtle, *J. Morphol.*, 89, 599, 1951.

Auffenberg, W., A first record for breeding colour change in a tortoise, *J. Bomb. Nat. Hist. Soc.*, 61, 191, 1964.

Auffenberg, W., Sex and species discrimination in two sympatric South American tortoises, *Copeia*, 1965, 335, 1965.

Auffenberg, W., Display behavior in tortoises, *Am. Zool.*, 17, 241, 1977.

Auffenberg, W., Courtship and breeding behavior in *Geochelone radiata* (Testudines: Testudinidae), *Herpetologica*, 34, 277, 1978.

Auffenberg, W., and Iverson, J.B., Demography of terrestrial turtles, in *Turtles: Perspectives and Research*, M. Harless and H. Morlock (eds.), New York: John Wiley, 1979, 541.

Baker, P.J., Costanzo, J.P., Herlands, R., Wood, R.C., and Lee, R.E., A tolerance for freezing promotes winter survival of hatchlings of the northern diamondback terrapin, *Malaclemys terrapin*, *Integr. Comp. Biol.*, 43, 964, 2004.

Baker, P.J., Costanzo, J.P., Iverson, J.B., and Lee, R.E., Adaptations to terrestrial overwintering of hatchling northern map turtles, *Graptemys geographica*, *J. Comp. Physiol. B*, 173, 643, 2003.

Baker, R., and Gillingham, J.C., An analysis of the courtship behavior in Blanding's turtle, *Emydoidea blandingi*, *Herpetologica*, 39, 166, 1983.

Barney, R., Further notes on the natural history and artificial propagation of the diamond-backed terrapin, *Bull. US Bur. Fish*, 38, 9, 1922.

Beggs, K., Young, J., Georges, A., and West, P., Ageing the eggs and embryos of the pig-nosed turtle, *Carettochelys insculpta* (Chelonia: Carettochelyidae), from northern Australia, *Can. J. Zool.*, 78, 373, 2000.

Belinsky, A., Ackerman, R.A., Dmi'el, R., and Ar, A., Water in reptilian eggs and hatchlings, in *Reptilian Incubation: Environment, Evolution and Behaviour*, D.C. Deeming (ed.), Nottingham, UK: Nottingham University Press, 2004, 125.

Bellairs, A., *The Life of Reptiles, Vol. 2*, New York: Universe Books, 1970.

Bels, V.L., and Crama, Y.J.M., Quantitative analysis of the courtship and mating behavior in the loggerhead musk turtle *Sternotherus minor* (Reptilia: Kinosternidae) with comments on the courtship behavior in turtles, *Copeia*, 1994, 676, 1994.

Billett, F.S., Collins, P., Goulding, D.A., and Sutherland, J., The development of *Caretta caretta*, at 25 to 34°C, in artificial nests, *J. Morphol.*, 213, 251, 1992.

Birchard, G.F., Effects of incubation temperature, in *Reptile Incubation: Environment, Evolution, and Behavior*, D.C. Deeming (ed.), Nottingham, UK: Nottingham University Press, 2004, 103.

Birchard, G.F., and Packard, G.C., Cardiac activity in supercooled hatchlings of the painted turtle (*Chrysemys picta*), *J. Herpetol.*, 31, 166, 1997.

Bleakney, J., Notes on the distribution and life histories of turtles in Nova Scotia, *Can. Field-Nat.*, 77, 67, 1963.

Bobyn, M.L., and Brooks, R.J., Interclutch and interpopulation variation in the effects of incubation conditions on sex, survival and growth of hatchling turtles (*Chelydra serpentina*), *J. Zool.*, 233, 233, 1994.

Bojanus, L.H., *Anatome Testudinis Europaeae*, An anatomy of the turtle, Vilnae Impensis Auctoris (reprint SSAR, 1970) 1819.

Bonin, F., Devaux, B., and Dupré A., *Turtles of the World*, Baltimore, MD: John Hopkins University Press, 2006.

Booth, D.T., Incubation of eggs of the Australian broad-shelled turtle, *Chelodina expansa* (Testudinata: Chelidae), at different temperatures: Effects on pattern of oxygen consumption and hatchling morphology, *Aust. J. Zool.*, 48, 369, 2000.

Booth, J., and Peters, J., Behavioural studies on the green turtle (*Chelonia mydas*) in the sea, *Anim. Behav.*, 20, 808, 1972.

Bowden, R.M., Ewert, M.A., Lipar, J.L., and Nelson, C.E., Concentrations of steroid hormones in layers and biopsies of chelonian egg yolks, *Gen. Comp. Endocrinol.*, 121, 95, 2001.

Bowden, R.M., Ewert, M.A., and Nelson, C.E., Environmental sex determination in a reptile varies seasonally and with yolk hormones, *Proc. R.S. B-Biol. Sci.*, 267, 1745, 2000.

Bowden, R.M., Ewert, M.A., and Nelson, C.E., Hormone levels in yolk decline throughout development in the red-eared slider turtle (*Trachemys scripta elegans*), *Gen. Comp. Endocrinol.*, 129, 171, 2002.

Bowden, R.M., Harms, H.K., Paitz, R.T., Janzen, F.J., Does optimal egg size vary with demographic stage because of a physiological constraint?, *Funct. Ecol.*, 18, 522, 2004.

Boycott, R.C., and Bourquin, O., *The South African Tortoise Book: A Guide to South African Tortoises, Terrapins and Turtles*, Johannesburg: Russel Friedman Books, 1988.

Brooks, R.J., Bobyn, M.L., Galbraith, D.A., Layfield, J.A., and Nancekivell, E.G., Maternal and environmental influences on growth and survival of embryonic and hatchling snapping turtles (*Chelydra serpentina*), *Can. J. Zool.*, 69, 2667, 1991.

Buhlmann, K.A., Habitat use, terrestrial movements, and conservation of the turtle, *Dierochelys reticularia*, in Virginia, *J. Herpetol.*, 29, 173, 1995.

Bull, J.J., Sex ratio and nest temperature in turtles: Comparing field and laboratory data, *Ecology*, 66, 1115, 1985.

Bull, J.J., and Vogt, R.C., Temperature-dependent sex determination in turtles, *Science*, 206, 1186, 1979.

Burger, J., and Montevecchi, W.A., Nest site selection in the terrapin *Malaclemys terrapin*, *Copeia*, 1975, 113, 1975.

Burger, J.W., Experimental sexual photoperiodicity in the male turtle *Pseudemys elegans* (Wied.), Am. Nat., 71, 481, 1937.

Bury, R.B., Population ecology of freshwater turtles, in *Turtles, Perspectives and Research*, M. Harless and H. Morlock (eds.), New York: John Wiley, 1979, 571.

Bustard, H.R., *Australian Sea Turtles: Their Natural History and Conservation*, Sydney: Collins, 1972.

Bustard, H.R., Population dynamics of sea turtles, in *Turtles: Perspectives and Research*, M. Harless and H. Morlock (eds.), New York: Wiley, 1979, 523.

Cagle, F.R., Sexual maturity in the female of the turtle, *Pseudemys scripta elegans*, *Copeia*, 1944, 149, 1944.

Cagle, F.R., The life history of the slider turtle, *Pseudemys scripta troostii* (Holbrook), *Ecol. Monogr.*, 20, 31, 1950.

Cagle, F.R., and Tihen, J., Retention of eggs by the turtle *Deirochelys reticularia*, *Copeia*, 1948, 66, 1948.

Callard, I.P., Etheridge, K., Giannoukos, G., Lamb, T., and Perez, L., The role of steroids in reproduction in female elasmobranches and reptiles, *J. Steroid Biochem.*, 40, 571, 1991.
Callard, I.P., Lance, V., Salhanick, A.R., and Barad, D., The annual ovarian cycle of *Chrysemys picta*: Correlated changes in plasma steroids and parameters of vitellogenesis, *Gen. Comp. Endocrinol.*, 35, 245, 1978.
Callard, I.P., McChesney, I., Scannes, C., and Callard, G., The influence of mammalian and avian gonadotropins on in vitro ovarian steroid synthesis in the turtle, *Chrysemys picta*, *Gen. Comp. Endocrinol.*, 28, 2, 1976.
Cann, J., *Australian Freshwater Turtles*, Singapore: Beaumont Publishing, 1998.
Carpenter, C., An ethological approach to reproductive success in reptiles, in *Reproductive Biology and Diseases of Captive Reptiles*, J.B. Murphy and J.T. Collins (eds.), SSAR Contributions to Herpetology No. 1., 33, 1980.
Carpenter, C.C., and Ferguson, G.W., Variation and evolution of stereotyped behavior in reptiles, in *Biology of the Reptilia, Vol. 7*, C. Gans and D.W. Tinkle (eds.), London: Academic Press, 1977, 335.
Carr, A.F., *Handbook of Turtles*, New York: Cornell University Press, 1952.
Carr, A.F., *The Reptiles*, The Netherlands: Time-Life International, 1964.
Carr, A.F., and Hirth, H., Social facilitation in green turtle siblings, *Anim. Behav.*, 9, 68, 1961.
Carr, A.F., and Ogren, L., The ecology and migration of sea turtles. 3. *Dermochelys* in Costa Rica, *Am. Mus. Nov.*, No. 1958, 1959.
Carr, T., and Carr, N., *Trionyx triunguis*: Habitat, *Herp. Rev.*, 16, 30, 1985.
Chaloupka, M.Y., and Musick, J.A., Age, growth, and population dynamics, in *The Biology of Sea Turtles*, P. Lutz and J. Musick (eds.), Boca Raton, FL: CRC Press, 1997, 233.
Cherepanov, G.O., Ontogenetic development of the shell in *Trionyx sinensis* (Trionychidae, Testudinata) and some questions on the nomenclature of bony plates, *Russ. J. Herpetol.*, 2, 129, 1995.
Choo, B.L., and Chou, L.M., Effects of temperature on the incubation period and hatchability of *Trionyx sinensis* Wiegmann eggs, *J. Herpetol.*, 21, 230, 1987.
Christiansen, J.L., and Dunham, A.E., Reproduction of the yellow mud turtle (*Kinosternon flavescens flavescens*) in New Mexico, *Herpetologica*, 27, 130, 1972.
Christiansen, J.L., and Moll, E.O., Latitudinal reproductive variation within a single subspecies of painted turtle, *Chrysemys picta bellii*, *Herpetologica*, 29, 152, 1973.
Churchill, T.A., and Storey, K.B., Natural freezing survival by painted turtles *Chrysemys picta marginata* and *C. picta bellii*, *Am. J. Physiol.*, 262, R530, 1992a.
Churchill, T.A., and Storey, K.B., Responses to freezing exposure of hatchling turtles *Trachemys scripta elegans*: Factors influencing the development of freeze tolerance by reptiles, *J. Exp. Biol.*, 167, 221, 1992b.
Clark, P.J., Ewert, M.A., and Nelson, C.E., Physical apertures as constraints on egg size and shape in the common musk turtle, *Sternotherus odoratus*, *Funct. Ecol.*, 15, 70, 2001.
Coe, M.J., Bourn, D., and Swingland, I.R., The biomass, production and carrying capacity of giant tortoises on Aldabra, *Phil. Trans. Roy. Soc. B.*, 286, 163, 1979.
Combescot, C., Sexualité et cycle genital de la tortue d'eau algérienne, *Emys leprosa* Schw, *Bull. Soc. Hist. Nat. Afr. N.*, 45, 366, 1954.
Combescot, C., Connées histophysiologiques sur l'oviducte de la tortue d'eau algérienne: *Emys leprosa* Schw, *C.R. Soc. Biol.*, 149, 93, 1955a.
Combescot, C., Action de l'hypophysectomie sur les elements testiculaires de la tortue d'eau algérienne, *Emys leprosa* Schw, *Bull. Soc. Hist. Nat. Afr. N.*, 46, 30, 1955b.
Combescot, C., Action de l'hypophysectomie sur les elements testiculaires de la tortue terrestre, *Testudo ibera. Pal. Bull. Soc. Hist. Nat. Afr. N.*, 46, 98, 1955c.
Combescot, D., and Guyon, L., Les variations hypophysaires chez la Tortue d'eau algérinne: *Emys leprosa* Schw., C.R. Acad. Sci., Paris, 241, 1079, 1955.
Conant, R.C., The reptiles of Ohio, *Am. Midl. Nat.*, 20, 1, 1938.
Congdon, J.D., and Gibbons, J.W., Morphological constraint on egg size: A challenge to optimal egg size theory?, *Proc. Natl. Acad. Sci. USA*, 84, 4145, 1987.
Congdon, J.D., Gibbons, J.W., and Greene, J.L., Parental investment in the chicken turtle (*Deirochelys reticularia*), *Ecology*, 64, 419, 1983a.
Congdon, J.D., Nagle, R.D., Dunham, A.E., Beck, C.W., Kinney, O.M., and Yeomans, S.R., The relationship of body size to survivorship of hatchling snapping turtles (*Chelydra serpentina*): An evaluation of the 'bigger is better' hypothesis, *Oecologica*, 121, 224, 1999.

Congdon, J.D., Nagle, R.D., Kinney, O.M., Osentoski, M., Avery, H.W., van Loben Sels, R.C., and Tinkle, D.W., Nesting ecology and embryo mortality: Implications for hatchling success and demography of Blanding's turtles (*Emydoidea blandingii*), *Chel. Cons. Biol.*, 3, 569, 2000.

Congdon, J.D., and Tinkle, D.W., Reproductive energetics of the painted turtle (*Chrysemys picta*), *Herpetologica*, 38, 228, 1982.

Congdon, J.D., Tinkle, D.W., Breitenbach, C.L., and Van Loben Sels, R.C., Nesting behavior and hatching success in the turtle *Emydoidea blandingi*, *Herpetologica*, 39, 417, 1983b.

Congdon, J.D., Tinkle, D.W., and Rosen, P.C., Egg component and utilization during development in aquatic turtles, *Copeia*, 1983, 264, 1983c.

Converse, S.J., Iverson, J.B., and Savidge, J.A., Activity, reproduction and overwintering behavior of ornate box turtles (*Terrapene ornata ornata*) in Nebraska sandhills, *Am. Midl. Nat.*, 148, 416, 2002.

Cooper, W.E., and Greenberg, N., Reptilian coloration and behavior, in *Biology of the Reptilia, Vol. 18*, C. Gans and D. Crews (eds.), Chicago: University of Chicago Press, 1992, 298.

Costanzo, J.P., and Lee, R.E., Biophysical and physiological responses promoting freeze tolerance in vertebrates, News Physiol. Sci. 9, 252, 1994.

Costanzo, J.P., Baker, P.J., Dinkelacker S.A., and Lee, R.E., Endogenous and exogenous ice-nucleating agents constrain supercooling in the hatchling painted turtle, *J. Exp. Biol.*, 206, 477, 2003.

Costanzo, J.P., Dinkelacker, S.A., Iverson, J.B., and Lee, R.E., Physiological ecology of overwintering in the hatchling painted turtle: Multiple-scale variation in response to environmental stress, *Physiol. Biochem. Zool.*, 77, 74, 2004.

Costanzo, J.P., Iverson, J.B., Wright, M.F., and Lee, R.E., Cold hardiness and overwintering strategies of hatchlings in an assemblage of northern turtles, *Ecology*, 76, 1772, 1995.

Costanzo, J.P., Jones, E.E., and Lee, R.E., Physiological responses to supercooling and hypoxia in the hatchling painted turtle, *Chrysemys picta*, *J. Comp. Physiol. B*, 171, 335, 2001a.

Costanzo, J.P., Litzgus, J.D., Iverson, J.B., and Lee, R.E., Soil hydric characteristics and environmental ice nuclei influence supercooling capacity of hatchling painted turtles, *Chrysemys picta*, *J. Exp. Biol.*, 201, 3105, 1998.

Costanzo, J.P., Litzgus, J.D., Iverson, J.B., and Lee, R.E., Ice nuclei in soil compromise cold hardiness of hatchling painted turtles, *Chrysemys picta*, *Ecology*, 81, 346, 2000a.

Costanzo, J.P., Litzgus, J.D., Iverson, J.B., and Lee, R.E., Seasonal changes in physiology and development of cold hardiness in the hatchling painted turtle, *Chrysemys picta*, *J. Exp. Biol.*, 203, 3459, 2000b.

Costanzo, J.P., Litzgus, J.D., Iverson, J.B., and Lee, R.E., Cold-hardiness and evaporative water loss in hatchling turtles, *Physiol. Biochem. Zool.*, 74, 510, 2001b.

Costanzo, J.P., Litzgus, J.D., Larson, J.L., Iverson, J.B., and Lee, R.E., Characteristics of nest soil, but not geographic origin, influence cold hardiness of hatchling painted turtles, *J. Therm. Biol.*, 26, 65, 2001c.

Cox, W.A., and Marion, K.R., Observations on the female reproductive cycle and associated phenomena in spring-dwelling populations of *Sternotherus minor* in north Florida (Reptilia Testudines), *Herpetologica*, 32, 20, 1978.

Crastz, F., Embryological stages of the marine turtle *Lepidochelys olivacea* (Eschscholtz), *Rev. Biol. Trop.*, 30, 113, 1982.

Crews, D., and Garrick, L., Methods of inducing reproduction in captive reptiles, in *Reproductive Biology and Diseases of Captive Reptiles*, J. Murphy and J. Collins (eds.), SSAR Contribution to Herpetology 1, Society for the Study of Amphibians and Reptiles 49, 1980.

Crowell, P.C., The ciliation on the oviduct of reptiles, *Proc. Natl. Acad. Sci. USA*, 18, 7, 1932.

Cyrus, R., Mahmoud, I., and Klicka, J., Fine structure of the corpus luteum of the snapping turtle, *Chelydra serpentina*, *Copeia*, 1978, 622, 1978.

Davis, J., and Jackson, C., Copulatory behavior in the red-eared turtle, *Pseudemys scripta elegans*, *Herpetologica*, 26, 238, 1970.

Davis, J., and Jackson, C., Notes on the courtship of a captive male *Chrysemys scripta taylori*, *Herpetologica*, 29, 62, 1973.

De, T.K., and Maiti, B.R., Study of the genital tract during the annual testicular cycle of the soft-shelled turtle, *Lissemys punctata punctata* (Lacépède), *Zool. Anz.*, 223, 116, 1989.

Deeming, D.C., Post hatching phenotypic effects of incubation in reptiles, in *Reptilian Incubation: Environment, Evolution and Behaviour*, D.C. Deeming (ed.), Nottingham, UK: Nottingham University Press, 2004, 229.

DePari, J.A., Overwintering in the nest chamber by hatchling painted turtles, *Chrysemys picta*, in northern New Jersey, *Chel. Cons. Biol.*, 2, 5, 1996.

Dinkelacker, S.A., Costanzo, J.P., Iverson, J.B., and Lee, R.E., Cold-hardiness and dehydration resistance of hatchling Blanding's turtles (*Emydoidea blandingii*): Implications for overwintering in a terrestrial habitat, *Can. J. Zool.*, 82, 594, 2004.

Dinkelacker, S.A., Costanzo, J.P., and Lee, R.E., Anoxia tolerance and freeze tolerance in hatchling turtles, *J. Comp. Physiol. B*, 175, 209, 2005b.

Dinkelacker, S.A., Costanzo, J.P., Iverson, J.B., and Lee, R.E., Survival and physiological responses of hatchling Blanding's turtles (*Emydoidea blandingii*) to submergence in normoxic and hypoxic water under simulated winter conditions, *Physiol. Biochem. Zool.*, 78, 356, 2005a.

Dobie, J.L., Reproduction and growth in the alligator snapping turtle, *Macroclemys temmincki* (Troost), *Copeia*, 1971, 645, 1971.

Dodd, C.K., Synopsis of the biological data on the loggerhead sea turtle *Caretta caretta* (Linneaus 1758), *US Fish Wild. Serv. Biol. Rep.*, 88, 1, 1988.

Dodd, C.K., *North American Box Turtles: A Natural History*, Norman, OK: University of Oklahoma Press, 2001.

Draud, M., Bossert, M., and Zimnavoda, S., Predation on hatchling and juvenile diamondback terrapins (*Malaclemys terrapin*) by the Norway rat (*Rattus norvegicus*), *J. Herpetol.*, 38, 467, 2004.

Dubois, W., Pudney, J., and Callard, I.P., The annual testicular cycle in the turtle, *Chrysemys picta*: A histochemical and electron microscopic study, *Gen. Comp. Endocrinol.*, 71, 191, 1988.

Dunson, W.A., Aquatic respiration in *Trionyx spinifer asper*, *Herpetologica*, 16, 277, 1960.

Dunson, W.A., Sodium fluxes in fresh-water turtles, *J. Exp. Zool.*, 165, 171, 1967.

Duvall, D., Guillette, L.J. Jr., and Jones, R.E., Environmental control of reptilian reproductive cycles, in *Biology of the Reptilia, Vol. 13*, C. Gans and H. Pough (eds.), New York: Academic Press, 1982, 201.

Eckert, K.L., and Luginbuhl, C., Death of a giant, *Mar. Turt. News.*, 43, 2-3, 1988.

Eglis, A., Tortoise behavior: A taxonomic adjunct, *Herpetologica*, 18, 1, 1988, 1962.

Ehrenfeld, D.W., Behavior associated with nesting, in *Turtles: Perspectives and Research*, M. Harless and H. Morlock (eds.), New York: Wiley, 1979, 417.

Ernst, C.H., Sexual cycles and maturity of the turtle *Chrysemys picta*, *Biol. Bull. Mar. Biol. Lab. Woods Hole*, 140, 191, 1971a.

Ernst, C.H., Population dynamics and activity cycles of *Chrysemys picta* in southeastern Pennsylvania, *J. Herpetol.*, 5, 151, 1971b.

Ernst, C.H., Observations on the courtship of male *Graptemys pseudogeographica*, *J. Herpetol.*, 8, 377, 1974.

Ernst, C.H., The effects of hurricane Agnes on a painted turtle population, *J. Herpetol.*, 8, 237, 1974.

Ernst, C.H., and Barbour, R., *Turtles of the United States*, Lexington, KY: University of Kentucky Press, 1972.

Ernst, C.H., and Barbour, R., *Turtles of the World*, Washington, DC: Smithsonian Institution Press, 1989.

Ernst, C.H., Gibbons, W., and Novak, S.S., Chelydra Schweigger, Snapping Turtles, in *Catalogue of American Amphibians and Reptiles*, SSAR, 419.1, 1988.

Ernst, C.H., Lovich, J.E., and Barbour, R.W., *Turtles of the United States and Canada*, Washington, DC: Smithsonian Institution Press, 1994.

Ernst, C.H., and McBreen, J.F., *Terrapene carolina* (Linnaeus) Eastern Box Turtle, in *Catalogue of American Amphibians and Reptiles*, SSAR, 512.1, 1991.

Ernst, C.H., and Zug, G.R., Observations on the reproductive biology of the spotted turtle, *Clemmys guttata*, in southeastern Pennsylvania, *J. Herpetol.*, 28, 99, 1994.

Etchberger, C.R., Phillips, J.B., Ewert, M.A., Nelson, C.E., and Prange, H.D., Effects of oxygen concentration and clutch on sex determination and physiology in red-eared slider turtles (*Trachemys scripta*), *J. Exp. Zool.*, 258, 394, 1991.

Evans, L., The courtship pattern of the box turtle, *Terrapene carolina carolina*, *Herpetologica*, 9, 189, 1953.

Evans, L., Aquatic courtship of the wood turtle, *Clemmys insculpta*, *Am. Zool.*, 1, 210, 1961.

Ewert, M.A., Nests, nesting and aerial basking of *Macroclemys* under natural conditions, and comparison with *Chelydra* (Testudines: Chelydridae), *Herpetologica*, 32, 150, 1976.

Ewert, M.A., The embryo and its egg: Development and natural history, in *Turtles: Perspectives and Research*, M. Harless and H. Morlock (eds.), New York: J. Wiley & Sons, 1979, 333.

Ewert, M.A., Embryology of turtles, in *Biology of the Reptilia, Vol. 14*, C. Gans, F. Billett, and P.F.A. Maderson (eds.), New York: John Wiley & Sons, 1985, 75.

Ewert, M.A., Cold torpor, diapause, delayed hatching and aestivation in reptiles and birds, in *Egg Incubation: Its Effects on Embryonic Development in Birds and Reptiles*, D.C. Deeming and M.W.J. Ferguson (eds.), Cambridge, UK: Cambridge University Press, 1991, 173.

Ewert, M.A., Jackson, D.R., and Nelson, C.E., Patterns of temperature-dependent sex determination in turtles, *J. Exp. Zool.*, 270, 3, 1994.

Ewert, M.A., and Nelson, C.E., Sex determination in turtles: Diverse patterns and some possible adaptive values, *Copeia*, 1991, 50, 1991.

Ewert, M.A., and Wilson, D.S., Seasonal variation of embryonic diapause in the striped mud turtle (*Kinosternon baurii*) and general considerations for conservation planning, *Chel. Cons. Biol.*, 2, 43, 1996.

Ewing, H.E., Continued fertility in female box turtles following mating, *Copeia*, 1943, 112, 1943.

Finkler, M.S., Rates of water loss and estimates of survival time under varying humidity in juvenile snapping turtles (*Chelydra serpentina*), *Copeia*, 2001, 521, 2001.

Finkler, M.S., and Claussen, D.L., Within and among clutch variation in the composition of *Chelydra serpentina* eggs with initial egg mass, *J. Herpetol.*, 31, 620, 1997.

Flitz, B.A., and Mullin, S.J., Nest-site selection in the Eastern Box Turtle, *Terrapene carolina carolina*, in Illinois, *Chel. Cons. Biol.*, 5, 309, 2006.

Foley, A.M., Peck, S.A., and Harman, G.R., Effects of sand characteristics on the hatching success of loggerhead sea turtles (*Caretta caretta*) clutches on low-relief mangrove islands in southwest Florida, *Chel. Cons. Biol.*, 5, 32, 2006.

Fox, H., The urogenital system of reptiles, in *Biology of the Reptilia, Vol. 6*, C. Gans and T.S. Parsons (eds.), London: Academic Press, 1977, 1.

Fox, W., Seasonal variation in the male reproductive system of Pacific coast garter snakes, *J. Morphol.*, 90, 481, 1952.

Frazer, N.B., A model for assessing age-specific fecundity in sea turtle populations, *Herpetologica*, 40, 281, 1984.

Frazer, N.B., Gibbons, J.W., and Greene, J.L., Life tables of a slider turtle population, in *Life History and Ecology of the Slider Turtle*, J.W. Gibbons (ed.), Washington, DC: Smithsonian Institution Press, 1990, 183.

Gad, J., Die Ultrastruktur Der Eischalen Von *Geoemyda spengleri, Indotestudo elongata und Sacalia bealei*, *Salamandra*, 30, 277, 1994.

Galbraith, D.A., Multiple paternity and sperm storage in turtles, *Herpetol. J.*, 3, 117, 1993.

Georges, A., and Rose, M., Conservation biology of the pig-nosed turtle, *Carettochelys insculpta*, *Chel Cons. Biol.*, 1, 3, 1993.

Georges, A., Limpus, C., and Parmenter, J., Natural history of the *Chelonia*, in *Fauna of Australia, Vol. 2A, Amphibia and Reptilia*, C.J. Glasby, G.J. Ross, and P.L. Beesley (eds.), Canberra: Australian Gov. Pub. Serv., 1993, 120.

Georges, A., Limpus, C., and Stoutjesdijk, R., Hatchling sex in the marine turtle *Caretta caretta* is determined by proportion of development at a temperature, not daily duration of exposure, *J. Exp. Zool.*, 270, 432, 1994.

Giannoukos, G., and Callard, I.P., Radioligand and immunocytochemical studies of the turtle oviduct progesterone and estrogen receptors: Correlations with hormone treatment and oviduct contractility, *Gen. Comp. Endocrinol.*, 101, 63, 1996.

Gibbons, J.W., Variation in growth rates in three populations of the painted turtle, *Chrysemys picta*, *Herpetologica*, 23, 269, 1967.

Gibbons, J.W., Reproductive potential, activity, and cycles in the painted turtle, *Chrysemys picta*, *Ecology*, 49, 399, 1968a.

Gibbons, J.W., Growth rates of the common snapping turtle, *Chelydra serpentina* in a polluted river, *Herpetologica*, 24, 266, 1968b.

Gibbons, J.W., Reproductive dynamics of a turtle (*Pseudemys scripta*) population in a reservoir receiving heated effluent from a nuclear reactor, *Can. J. Zool.*, 48, 881, 1970.

Gibbons, J.W., Reproductive patterns in freshwater turtles, *Herpetologica*, 38, 222, 1982.

Gibbons, J.W., Reproductive characteristics and ecology of the mud turtle *Kinosternon subrubrum* (Lacepede), *Herpetologica*, 39, 254, 1983.

Gibbons, J.W., Sex ratios and their significance among turtle populations, in *Life History and Ecology of the Slider Turtle*, J.W. Gibbons (ed.), Washington, DC: Smithsonian Institutional Press, 1990, 171.

Gibbons, J.W., and Greene, J.L., Selected aspects of the ecology of the chicken turtle, *Deirochelys reticularia* (Latrielle) (Reptilia, Testudines, Emydidae), *J. Herpetol.*, 12, 237, 1978.

Gibbons, J.W., Greene, J.L., and Congdon, J.D., Drought-related responses of aquatic turtle populations, *J. Herpetol.*, 17, 242, 1983.

Gibbons, J.W., and Nelson, D.H., The evolutionary significance of delayed emergence from the nest by hatchling turtles, *Evolution*, 32, 297, 1978.

Gibbons, J.W., Novak, S.S., and Ernst, C.H., *Chelydra serpentina* (Linnaeus) Snapping Turtle, in *Catalogue of American Amphibians and Reptiles*, SSAR, 420.1, 1988.

Gibbons, J.W., Semlitsch, R.D., Greene, J.L., and Schubauer, J.P., Variation in age and size at maturity of the slider turtle (*Pseudemys scripta*), *Am. Nat.*, 117, 841, 1981.

Gibbons, J.W., and Tinkle, D.W., Reproductive variation between turtle populations in a single geographic area, *Ecology*, 50, 340, 1969.
Girgis, S., Aquatic respiration in the common Nile turtle, *Trionyx triunguis* (Forskål), *Comp. Biochem. Physiol.*, 3, 206, 1961.
Girling, J.E., The reptilian oviduct: A review of structure and function and directions for future research, *J. Exp. Zool.*, 293, 141, 2002.
Gist, D.H., Reptilian reproductive system, male, in *Encyclopedia of Reproduction*, E. Knobil, and J.D. Neill (eds.), San Diego, CA: Academic Press, 1999, 60.
Gist, D.H., and Congdon, J.D., Oviductal sperm storage as a reproductive tactic of turtles, *J. Exp. Zool.*, 282, 526, 1998.
Gist, D.H., and Fischer, E.N., Fine structure of the sperm storage tubules in the box turtle oviduct, *J. Reprod. Fert.*, 97, 463, 1993.
Gist, D.H., and Jones, J.M., Sperm storage within the oviduct of turtles, *J. Morphol.*, 199, 379, 1989.
Gist, D.H., Dawes, S.M., Turner, T.W., Sheldon, S., and Congdon, J.D., Sperm storage in turtles: A male perspective, *J. Exp. Zool.*, 292, 180, 2002.
Gist, D.H., Jones, J.M., and Michaelson, J.M., Fall mating in the painted turtle *Chrysemys picta*, *Herpetologica*, 46, 331, 1990.
Godfrey, M.H., Barreto, R., and Mrosovsky, N., Metabolically-generated heat of developing eggs and its potential effect on sex ratio of sea turtle hatchlings, *J. Herpetol.*, 31, 616, 1997.
Godfrey, M.H., and Mrosovsky, N., Relative importance of thermal and nonthermal factors on the incubation period of sea turtle eggs, *Chel. Cons. Biol.*, 4, 217, 2001.
Gordos, M.A., Franklin, C.E., and Limpus, C.J., Seasonal changes in the diving performance of the bimodally respiring freshwater turtle *Rheodytes leukops* in a natural setting, *Can. J. Zool.*, 81, 617, 2003a.
Gordos, M.A., Franklin, C.E., and Limpus, C.J., Seasonal changes in the diel surfacing behaviour of the bimodally respiring turtle *Rheodytes leukops*, *Can. J. Zool.*, 81, 1614, 2003b.
Greenbaum, E., A standardized series of embryonic stages for the emydid turtle *Trachemys scripta*, *Can. J. Zool.*, 80, 1350, 2002.
Greenbaum, E., and Carr, J.L., Staging criteria for embryos of the spiny softshell turtle, *Apalone spinifera* (Testudines: Trionychidae), *J. Morphol.*, 272, 2002.
Guillette, J.L. Jr., Bjorndal, K.A., Bolten, A.B., Gross, T.S., Palmer, B.D., Witherington, B.E., and Matter, J.M., Plasma estradiol-17(B), progesterone, prostaglandin F, and prostaglandin E2 concentration during natural oviposition in the loggerhead turtle (*Caretta caretta*), *Gen. Comp. Endocrinol.*, 82, 121, 1991.
Gutzke, W.H.N., and Packard, G.C., Hatching success in relation to egg size in painted turtles (*Chrysemys picta*), *Can. J. Zool.*, 63, 67, 1985.
Gutzke, W.H.N., Packard, G.C., Packard, M.J., and Boardman, T.J., Influence of the hydric and thermal environment on eggs and hatchlings of painted turtles (*Chrysemys picta*), *Herpetologica*, 43, 393, 1987.
Guyot, G., Pieau, C., and Renous, S., Développement embryonnaire d'une tortue terrestre, la tortue d'Hermann, *Testudo hermanni* Gmelin, 1789, *Ann. Sci. Nat. Zool. Paris*, 15, 115, 1994.
Gyuris, E., Factors that control the emergence of green turtle hatchlings from the nest, *Wild. Res.*, 20, 345, 1993.
Hamann, M., Limpus, C.J., and Owens, D.W., Reproductive cycles of males and females, in *The Biology of Sea Turtles, Vol. II*, P. Lutz, J. Musick, and J. Wyneken (eds.), Boca Raton, FL: CRC Press, 2003, 135.
Hansen, I.B., Studies on the reproductive tract of the male box turtle, *Anat. Rec.*, 72, 121, 1938.
Hansen, I.B., and Risley, A.S., Seasonal changes in the oviduct of the turtle *Terrapene carolina*, *Anat. Rec.*, 81, 116, Suppl. 1941.
Harless, M., Social behavior, in *Turtles: Perspectives and Research*, M. Harless and H. Morelock (eds.), New York: John Wiley, 1979, 475.
Harrel, J.B., Douglas, N.H., Haraway, M.M., and Thomas, R.D., Mating behavior in captive alligator snapping turtles (*Macroclemys temminckii*), *Chel. Cons. Biol.*, 2, 101, 1996.
Harry, J.L., and Williams, K.L., Differential growth of male and female urinogenital systems of *Caretta caretta* within the sex-determining period, *J. Exp. Zool.*, 258, 204, 1991.
Hartley, L.M., Packard, M.J., and Packard, G.C., Accumulation of lactate by supercooled hatchlings of the painted turtle (*Chrysemys picta*): Implications for overwinter survival, *J. Comp. Physiol. B*, 170, 45, 2000.
Hattan, L.R., and Gist, D.H., Seminal receptacles in the eastern box turtle, *Terrapene carolina*, *Copeia*, 1975, 505, 1975.
Heppell, S.S., Snover, M.L., and Crowder, L.B., Sea turtle population ecology, in *The Biology of Sea Turtles, Vol. II*, P. Lutz, J.A. Musick, and J. Wyneken (eds.), Boca Raton, FL: CRC Press, 2003, 275.

Hildebrand, S.F., Review of experiments on artificial culture of diamond-backed terrapins, *Bull. US Bur. Fish.*, 45, 25, 1929.
Hirth, H.F., Synopsis of the biological data on the green turtle, *Chelonia mydas* (Linneaus) 1758, FAO Fish, Synop. 1, 1971.
Ho, S., Wangh, L.J., and Callard, I.P., Sexual differences in the in vitro induction of vitellogenesis in the turtle: Role of the pituitary and growth hormone, *Comp. Biochem. Physiol.*, 81B[2], 467, 1985.
Ho, S.M., Endocrinology of vitellogenesis, in Hormones and Reproduction in Fishes, Amphibians, and Reptiles, New York: Plenum Press, 1987, 145.
Ho, S.M., Kleis, S., McPherson, R., Heisermann, G.J., and Callard, I.P., Regulation of vitellogenesis in reptiles, *Herpetologica*, 38, 40, 1982.
Houghton, J.D.R., and Hays, G.C., Asynchronous emergence by loggerhead turtle (*Caretta caretta*) hatchlings, *Naturwissenschaften*, 88, 133, 2001.
Hulin, V., and Cuillon, J., Female philopatry in a heterogeneous environment: Ordinary conditions leading to extraordinary ESS sex ratios, BMC Evolutionary Biology, 7, 13, http://www.biomedcentral.com/1471-2148/7/13, 2007.
Iverson, J.B., Reproduction in freshwater and terrestrial turtles of north Florida, Herpetologica 33, 205, 1977.
Iverson, J. B., The female reproductive cycle in North Florida *Kinosternon baurii* (Testudines: Kinosternidae), Brimleyana 1, 37, 1979.
Iverson, J.B., The reproductive biology of *Gopherus polyphemus* (Chelonia: Testudinidae), *Am. Midi. Nat.*, 103, 353, 1980.
Iverson, J.B., Balgooyen, C.P., Byrd, K.K., and Lyddan, K.K., Latitudinal variation in egg and clutchsize in turtles, *Can. J. Zool.*, 71, 2448, 1993.
Iverson, J.B., and Ewert, M.A., Physical characteristics of reptilian eggs and a comparison with avian eggs, in *Egg Incubation: Its Effects on Embryonic Development in Birds and Reptiles*, D.C. Deeming and M.W.J. Ferguson (eds.), Cambridge, UK: Cambridge University Press, 1991, 87.
Iverson, J.B., Higgins, H., Sirulnick, A., and Griffiths, C., Local and geographic variation in the reproductive biology of the snapping turtle (*Chelydra serpentina*), *Herpetologica*, 53, 96, 1997.
Iverson, J.B., and Moler, P.E., The female reproductive cycle of the Florida soft shell turtle (*Apalone ferox*), *J. Herpetol.*, 31, 399, 1997.
Iverson, J.B., and Smith, G.R., Reproductive ecology of the painted turtle (*Chrysemys picta*) in the Nebraska Sandhills and across its range, *Copeia*, 1993, 1, 1993.
Jackson, C., and Davis, J., A quantitative study of the courtship display of the red-eared turtle *Chrysemys scripta elegans*, *Herpetologica*, 28, 58, 1972a.
Jackson, C., and Davis, J., Courtship display behavior of *Chrysemys concinna suwaniensis*, *Copeia*, 1972, 385, 1972b.
Jackson, D.C., Lactate accumulation in the shell of the turtle *Chrysemys picta bellii* during anoxia at 3°C and 10°C, *J. Exp. Biol.*, 200, 2295, 1997.
Jackson, D.C., How a turtle's shell helps it survive prolonged anoxic acidosis, *News. Physiol. Sci.*, 15, 181, 2000.
Jackson, D.C., Crocker, C.E., and Ultsch, G.R., Bone and shell contribution to lactic acid buffering of submerged turtles *Chrysemys picta bellii* at 3°C, *Am. J. Physiol.*, 278, R1564-R1571, 2000.
Jackson, D.C., and Heisler, N., Plasma ion balance of submerged anoxic turtles at 3°C: The role of calcium lactate formation, *Respir. Physiol.*, 49, 159, 1982.
Jackson, D.C., and Heisler, N., Intracellular and extracellular acid-base and electrolyte status of submerged anoxic turtles at 3°C, *Respir. Physiol.*, 53, 187, 1983.
Jackson, D.R., Reproductive strategies of sympatric freshwater emydid turtles in northern peninsular Florida, *Bull. Flor. St. Mus. Biol. Sci.*, 33, 113, 1988.
Janzen, F.J., An experimental analysis of natural selection on body size of hatchling turtles, *Ecology*, 74, 332, 1993.
Janzen, F.J., Experimental evidence for the evolutionary significance of temperature-dependant sex determination, *Evolution*, 49, 864, 1995.
Janzen, F.J., and Morjan, C.L., Repeatability of microenvironment-specific nesting behaviour in a turtle with environmental sex determination, *Anim. Behav.*, 62, 73, 2001.
Janzen, F.J., and Morjan, C.L., Egg size, incubation temperature, and posthatching growth in painted turtles (*Chrysemys picta*), *J. Herpetol.*, 36, 308, 2002.
Janzen, F.J., Packard, G.C., Packard, M.J., Boardman, T.J., and Zum Brunnen, J.R., Mobilization of lipid and protein by embryonic snapping turtles in wet and dry environments, *J. Exp. Zool.*, 255, 155, 1990.

Janzen, F.J., Tucker, J.K., and Paukstis, G.L., Experimental analysis of an early life-history stage: Selection on size of hatchling turtles, *Ecology*, 81, 2290, 2000a.

Janzen, F.J., Tucker, J.K., and Paukstis, G.L., Experimental analysis of an early life-history stage: Avian predation selects for larger body size of hatchling turtles, *J. Exp. Biol.*, 13, 947, 2000b.

Jensen, M.P., Abreu-Grobois, F.A., Frydenberg, J., and Loeschcke, V., Microsatellites provide insight into contrasting mating patterns in arribada vs. non-arribada olive ridley sea turtle rookeries, *Mol. Ecol.*, 15, 2567, 2006.

Kaska, Y., and Downie, R., Embryological development of sea turtles (*Chelonia mydas*, *Caretta caretta*) in the Mediterranean, *Zool. Mid. East*, 19, 55, 1999.

Kennett, R., Reproduction of two species of freshwater turtle, *Chelodina rugosa* and *Elseya dentata*, from the wet-dry tropics of northern Australia, *J. Zool.*, 247, 457, 1999.

Kennett, R., Christian, K., and Bedford, G., Underwater nesting by the Australian freshwater turtle *Chelodina rugosa*: Effect of prolonged immersion and eggshell thickness on incubation period, egg survivorship, and hatchling size, *Can. J. Zool.*, 76, 1019, 1998.

Kennett, R., Christian, K., and Fritchard, D., Underwater nesting by the tropical freshwater turtle, *Chelodina rugosa* (Testudinata: Chelidae), *Austr. J. Zool.*, 41, 47, 1993a.

Kennett, R., Georges, A., and Palmer-Allen, M., Early developmental arrest during immersion of eggs of a tropical freshwater turtle, *Chelodina rugosa* (Testudinata: Chelidae), from northern Australia, *Austr. J. Zool.*, 41, 37, 1993b.

Kennett, R.M., Georges, A., Thomas, K., and Georges, T.C., Distribution of the long-necked freshwater turtle *Chelodina novaeguineae* and new information on its ecology, *Mem. Queensl. Mus.*, 32, 179, 1992.

Kinney, O.M., Nagle, R.D., and Congdon, J.D., Water transport by nesting painted turtles (*Chrysemys picta marginata*) in Michigan, *Chel. Cons. Biol.*. 3, 71, 1998.

Klemens, M. (ed.), *Turtle Conservation*, Washington, DC: Smithsonian Institution Press, 2000.

Klicka, J., and Mahmoud, I., Conversion of pregnenolone-4-C14 to progesterone-4-C14 by turtle corpus luteum, *Gen. Comp. Endocrinol.*, 19, 367, 1972.

Klicka, J., and Mahmoud, I., Conversion of cholesterol to progesterone by turtle corpus luteum, *Steroids*, 21, 483, 1973.

Köhler, G., *Incubation of Reptile Eggs: Basics, Guidelines, Experiences*, Malbar, FL: Krieger Publishing, 2005.

Kolbe, J.J., and Janzen, F.J., The influence of propagule size and maternal nest-site selection on survival and behaviour of neonate turtles, *Funct. Ecol.*, 15, 772, 2001.

Kuchling, G., Le cycle sexuel male de la tortue *Testudo hermanni hermanni* Gmelin dans une population naturell et en captivité, *Bull. Soc. Herpetol. Fr.*, 19, 29, 1981.

Kuchling, G., Effect of temperature and photoperiod on spermatogenesis in the tortoise, *Testudo hermanni hermanni* Gmeln, *Amph.-Rept*,, 2, 32, 1982a.

Kuchling, G., Environmental temperature, spermatogenesis and plasma testosterone concentration in the tortoise, *Testudo hermanni hermanii* Gmelin, *Acta Endocrinol.*, 99 Suppl. 246, 29, 1982b.

Kuchling G., *The Reproductive Biology of the Chelonia, Vol. 38: Zoophysiology*, S.D. Bradshaw, W. Burggren, H.C. Heller, S. Ishii, H. Langer, G. Neuweiler, and D.J. Randall (eds.), New York: Springer, 1999, 223.

Kuchling, G., Skolek-Winnisch, R., and Bamberg, E., Histochemical and biochemical investigation on the annual cycle of testis, epididymis, and plasma testosterone of the tortoise, *Testudo hermanni hermanni* Gmeln, *Gen. Comp. Endocrinol.*, 44, 194, 1981.

Lance, V.A., and Rostal, D., The annual reproductive cycle of the male and female desert tortoise: Physiology and endocrinology, *Chel. Cons. Biol.*, 4, 302, 2002.

LeBuff, C., *The Loggerhead Turtle in the Eastern Gulf of Mexico*, Sanibel, FL: Caretta Research, 1990.

Lee, P.L.M., and Hays, G.C., Polyandry in a marine turtle: Females make the best of a bad job, *Proc. Natl. Acad. Sci. USA*, 101, 6530, 2004.

Legler, J.M., Observations on the sexual behavior of captive turtles, *Lloydia*, 18, 95, 1955.

Legler, J.M., Extra-uterine migration of ova in turtles, *Herpetologica*, 14, 49, 1958.

Legler, J.M., Natural history of the ornate box turtle, *Terrapene ornata ornata* Agassiz, *Univ. Ks. Pub. Mus. Nat. Hist.*, 11, 527, 1960a.

Legler, J.M., Remarks on the natural history of the Big Bend slider, *Pseudemys scripta gaigeae* Hartweg, *Herpetologica*, 16, 139, 1960b.

Legler, J.M., Cloacal gills in Australian chelid turtles (Video and Abstract), Salt Lake City, UT: University of Utah Educational Media Service, 1979.

Legler, J.M., Australian chelid turtles: Reproductive patterns in wide-ranging taxa, in *Biology of Australasian Frogs and Reptiles*, G. Grigg, R. Shine, and H. Ehmann (eds.), New South Wales, Australia: Royal Zoological Society, 1985, 117.

Legler, J.M., and Cann, J., A new genus and species of chelid turtle from Queensland, Australia, *Contrib. Sci. Natur. Hist. Mus. LA Co.*, 324, 1, 1980.

Licht, P., Endocrinology of Reptilia—the pituitary system, in *Chemical Zoology, Vol. IX: Amphibia and Reptilia*, M. Florkin and B. Scheer (eds.), New York: Academic Press, 1974, 399.

Licht, P., Evolutionary and functional aspects of pituitary gonadotropins in the green turtle *Chelonia mydas*, *Am. Zool.*, 20, 565, 1980.

Licht, P., Reptiles, in *Marshall's Physiology of Reproduction, Vol. 1: Reproductive Cycles of Vertebrates*, G.E. Lamming (ed.), Edinburgh, UK: Churchill Livingstone, 1984, 206.

Licht, P., and Crews, D., Gonadotropin stimulation of in vitro progesterone production in reptilian and amphibian ovaries, *Gen. Comp. Endocrinol.*, 29, 141, 1976.

Licht, P., Rainey, W., and Cliffton, K., Serum gonadotropin and steroids associated with breeding activities in the green sea turtle *Chelonia mydas*. II. Mating and nesting in natural populations, *Gen. Comp. Endocrinol.*, 40, 116, 1980.

Licht, P., Wood, J., Owens, D.W., and Wood, F., Serum gonadotropin and steroids associated with breeding activities in the green sea turtle *Chelonia mydas*. I. Captive animals, *Gen. Comp. Endocrinol.*, 39, 274, 1979.

Licht, P., Wood, J., and Wood, F., Annual and diurnal cycles in plasma testosterone and thyroxine in the male green sea turtle *Chelonia mydas*, *Gen. Comp. Endocrinol.*, 57, 335, 1985.

Limpus, C.J., A study of the loggerhead seaturtle, *Caretta caretta,* in eastern Australia, Ph.D. thesis, University of Queensland, Brisbane, Australia, 1985.

Limpus, C.J., Puberty and first breeding in *Caretta caretta*, in *Proceedings of the Tenth Annual Workshop on Sea Turtle Biology and Conservation*, T.H.C. Richardson, J.I.C Richardson, and M.C. Donnelly (eds.), NOAA Technical Memorandum NMFS-SEFC-278, 1990, 81.

Limpus, C.J., Estimation of tag loss in marine turtle research, *Aust. Wild. Res.*, 19, 457, 1992.

Limpus, C.J., The green turtle, *Chelonia mydas*, in *Queensland: Breeding Males in the Southern Great Barrier Reef*, Wildlife Research, 20, 513, 1993.

Limpus, C.J., and Limpus, D.J., Biology of the loggerhead turtle in western south Pacific Ocean foraging areas, in *Loggerhead Sea Turtles*, A. Bolten and B. Witherington (eds.), Washington, DC: Smithsonian Institution Press, 2003, 93.

Limpus, C.J., and Nicholls, N., The southern oscillation regulates the annual numbers of green turtles (*Chelonia mydas*) breeding around northern Australia, *Aust. Wildl. Res.*, 15, 157, 1988.

Limpus, C.J., and Nicholls, N., ENSO regulation of Indo-Pacific green turtle populations, in *Applications of Seasonal Climate Forecasting in Agricultural and Natural Ecosystems—The Australian Experience*, G.L. Hammer, N. Nicholls, and C. Mitchell (eds.), The Netherlands: Kluwer Academic, 2000, 399.

Limpus, C.J., and Reed, P.C., Green turtles, *Chelonia mydas*, stranded by Cyclone Kathy on the south-western coast of the Gulf of Carpentaria, *Aust. Wildl. Res.* 12, 523, 1985a.

Limpus, C.J., and Reed, P.C., The green turtle, *Chelonia mydas*, in Queensland: Population structure in a coral reef feeding ground, in *Biology of Australasian Frogs and Reptiles*, G. Grigg, R. Shine, and H. Ehmann (eds.), Sydney: Surrey Beatty & Sons, 1985b, 47.

Lofts, B., Patterns of testicular activity, in *Perspectives in Endocrinology, Hormones in the Lives of Lower Vertebrates*, E.J. Barrington and C.B. Jorgensen (eds.), New York: Academic Press, 1968, 239.

Lofts, B., Seasonal cycles in reptilian testes, *Gen. Comp. Endocrinol.*, Suppl. 2, 146, 1969.

Lofts, B., and Boswell, C., Seasonal changes in the distribution of the testis lipids of the Caspian terrapin *Clemmys caspica*, *Proc. Zool. Soc. Lond.*, 136, 581, 1961.

Lofts, B., and Tsui, H.W., Histological and histochemical changes in the gonads and epididymides of the male soft shelled turtle *Trionyx sinensis*, *J. Zool.*, 1181, 57, 1977.

Long, D.R., and Rose, F.L., Pelvic girdle size relationships in three turtle species, *J. Herpetol.*, 23, 315, 198■.

Lovich, J.E., *Macroclemys, M. temminckii*, in *Catalogue of American Amphibians and Reptiles*, SSAR, 562. , 1993.

Loyez, M., Recherches sur le développement ovarian des oeufs meroblastiques, *Archs. Anat. Microsc.*, 8, 69, 190■.

Lutz, P., and Musick, J.A., *The Biology of Sea Turtles, Vol. I*, Boca Raton, FL: CRC Press, 1997.

Lutz, P., Musick, J.A., and Wyneken, J., *The Biology of Sea Turtles, Vol. II*, Boca Raton, FL: CRC Press, 200■.

Mahmoud, I., Courtship behavior and sexual maturity in four species of kinosternid turtles, *Copeia*, 1967, 314, 1967.

Mahmoud, I., and Klicka, J., Seasonal gonadal changes in kinosternid turtles, *J. Herpetol.*, 6, 183, 1972.

Mahmoud, I.Y., Hess, G.L., and Klicka, J., Normal embryonic stages of the western painted turtle, *Chrysemys picta belli*, *J. Morphol.*, 141, 269, 1973.

Màrques, M.R., Villanueva, O.A., and Penaflores, S.C., Sonopsis de Datos Biologicos Sobre la Tortuga Golfina *Lipidochelys olivacea* (Eschscholtz 1829), Sinopsis Sobre la Pesca (INP/S2, SAT) INP 2, 1, 1976.

Mason, R.T., Reptilian pheromones, in *Biology of the Reptilia, Vol. 18: Physiology*, C. Gans and D. Crews (eds.), New York: Academic Press, 1992, 114.

McDowell, S.B., On the major arterial canals in the ear-region of testudinoid turtles and the classification of the Testudinoidea, *Bull. Mus. Comp. Zool. Harvard*, 125, 23, 1961.

McDowell, S.B., The genus *Emydura* (Testudines: Chelidae) in New Guinea with notes on the penial morphology of Pleurodira, in *Advances in Herpetology and Evolutionary Biology*, A.G.J. Rhodin and K. Miyata (eds.), Cambridge, MA: Museum of Comparative Zoology, 1983, 169.

McPherson, R.J., and Marion, K.R., Seasonal testicular cycle of the stinkpot turtle (*Sternotherus odoratus*) in central Alabama, *Herpetologica*, 37, 33, 1981.

McPherson, R.J., and Marion, K.R., Reproductive variation between two populations of *Sternotherus odoratus* in the same geographic area, *J. Herpetol.*, 17, 181, 1983.

McPherson, R.J., Boots, L.R., MacGregor, R., and Marion, K.R., Plasma steroids associated with seasonal reproductive changes in a multiclutched freshwater turtle, *Sternotherus odoratus*, *Gen. Comp. Endocrinol.*, 48, 440, 1982.

Miller, J.D., Criteria for staging reptilian embryos, in *Biology of Australasian Frogs and Reptiles*, G. Grigg, R. Shine, and H. Ehmann (eds.), Aust. Zool. Soc. NSW, Chipping, NSW, Australia: Surrey Beatty & Sons, 1985a, 305.

Miller, J.D., Embryology of marine turtles, in *Biology of the Reptilia, Vol. 14: Development*, C. Gans, F. Billett, and P.F.A. Maderson (eds.), New York: John Wiley & Sons, 1985b, 269.

Miller, J.D., Reproduction in sea turtles, in *The Biology of Sea Turtles*, P.L. Lutz and J.A. Musick (eds.), Boca Raton, FL: CRC Press, 1997, 51.

Miller, J.D., and Limpus, C.J., Incubation period and sexual differentiation in the green turtle, *Chelonia mydas* L., in *Proc. Melbourne Herpetol. Symp.*, C. Banks and A. Martin (eds.), Royal Melbourne Zoological Gardens, Melbourne, 1981, 66.

Miller J.D., and Limpus, C.J., Ontogeny of marine turtle gonads, in *The Biology of Sea Turtles, Vol. II*, P. Lutz, J. Musick, and J. Wyneken (eds.), Boca Raton, FL: CRC Press, 2003, 199.

Miller, J.D., Limpus, C.J., and Godfrey, M.H., Nest site selection, oviposition, eggs, development, hatching and emergence of loggerhead turtles, in *The Biology of the Loggerhead Turtle*, A. Bolten and B. Witherington (eds.), Washington, DC: Smithsonian Institution Press, 2003, 125.

Miller, K., Packard, G.C., and Packard, M.J., Hydric conditions during incubation influence locomotor performance of hatchling snapping turtles, *J. Exp. Biol.*, 127, 401, 1987.

Milstead, W.W., and Tinkle, D.W., *Terrapene* of western Mexico, with comments on the species groups within the genus, *Copeia*, 1967, 180, 1967.

Minnich, J., The use of water, in *Biology of the Reptilia, Vol. 12: Physiological Ecology*, C. Gans and F.H. Pough (eds.), New York: Academic Press, 1982, 325.

Moll, D., The ecology of sea beach nesting in slider turtles (*Trachemys scripta venusta*) from Caribbean Costa Rica, *Chel. Cons. Biol.*, 1, 107, 1994.

Moll, D., and Moll, E.O., *The Ecology, Exploitation, and Conservation of River Turtles*, New York: Oxford University Press, 2004.

Moll, E.O., Latitudinal and intersubspecific variation in reproduction of the painted turtle, *Chrysemys picta*, *Herpetologica*, 29, 307, 1973.

Moll, E.O., Reproductive cycles and adaptations, in *Turtles: Perspectives and Research*, M. Harless and H. Morlock (eds.), Malabar, FL: John Wiley & Sons, 1979, 305.

Moll, E.O., and Legler, J.M., The life history of a neotropical slider turtle, *Pseudemys scripta* (Schoepff), in Panama, *Bull. LA Co. Mus. Nat. Hist.*, 11, 1, 1971.

Moore, M.C., and Lindzey, J., The physiological basis of sexual behavior in male reptiles, in *Biology of the Reptilia, Vol. 18: Physiology*, G. Gans and D. Crews (eds.), Chicago: University of Chicago Press, 1992, 70.

Morreale, S.J., Ruiz, C.J., Spotila, J.R., and Standora, E.A., Temperature-dependent sex determination: Current practices threaten conservation of sea turtles, *Science*, 216, 1245, 1982.

Mortimer, J.A., The influence of beach sand characteristics on the nesting behavior and clutch survival of green turtles (*Chelonia mydas*), *Copeia*, 1990, 802, 1990.

Motz, V.A., and Callard, I.P., Seasonal variation in the oviductal morphology of the painted turtle, *Chrysemys picta*, *J. Morphol.*, 207, 59, 1991.

Mrosovsky, N., Mechanism of nocturnal emergence of hatchling sea turtles: Control by thermal inhibition of activity, *Nature*, 220, 1338, 1968.

Mrosovsky, N., Pivotal temperatures for loggerhead turtles (*Caretta caretta*) from northern and southern nesting beaches, *Can. J. Zool.*, 66, 661, 1988.

Mrosovsky, N., Dutton, P.H., and Whitmore, C.P., Sex ratios of two species of sea turtle nesting in Suriname, *Can. J. Zool.*, 62, 2227, 1984a.

Mrosovsky, N., Hopkins-Murphy, S.R., and Richardson, J., Sex ratio of sea turtles: Seasonal changes, *Science*, 225, 739, 1984b.

Mrosovsky, N., and Yntema, C.L., Temperature dependence of sexual differentiation in sea turtles: Implications for conservation practices, *Biol. Cons.*, 18, 271, 1980.

Munson, J.P., Researches on the oogenesis of the tortoise, *Clemmys marmorata*, *Am. J. Anat.*, 3, 311, 1904.

Nicholson, F.A., and Risley, P.L., A study of the urogenital systems of *Emys blandingii*, with observations on the occurrence of the Mullerian ducts in males, *Proc. Iowa Acad.* Sci., 47, 343, 1941.

Noble, G.A., and Noble, E.R., *A Brief Anatomy of the Turtle*, Stanford, CA: Stanford University Press, 1940, 48.

O'Steen, S., Embryonic temperature influences juvenile temperature choice and growth rate in snapping turtles *Chelydra serpentina*, *J. Exp. Biol.*, 201, 439, 1998.

O'Steen, S., and Janzen, F.J., Embryonic temperature affects metabolic compensation and thyroid hormones in hatchling snapping turtles, *Physiol. Biochem. Zool.*, 72, 520, 1999.

Obst, F.J., *Turtles, Tortoises and Terrapins*, Melbourne: Macmillan, 1986.

Oliver, J., *North American Amphibians and Reptiles*, Princeton, NJ: Van Nostrand, 1955.

Owens, D.W., The comparative reproductive physiology of sea turtles, *Am. Zool.*, 20, 549, 1980.

Owens, D.W., Hormones in the life history of sea turtles, in *The Biology of Sea Turtles, Vol. I*, P. Lutz, and J. Musick (eds.), Boca Raton, FL: CRC Press, 1997, 315.

Owens, D.W., and Morris, Y.A., The comparative endocrinology of sea turtles, *Copeia*, 1985, 723, 1985.

Packard, G.C., The physiological and ecological importance of water to embryos of oviparous reptiles, in *Egg Incubation: Its Effects on Embryonic Development in Birds and Reptiles*, D.C. Deeming and M.W.J. Ferguson (eds.), Cambridge, UK: Cambridge University Press, 1991, 213.

Packard, G.C., Temperatures during winter in nests with hatchling painted turtles (*Chrysemys picta*), *Herpetologica*, 53, 89, 1997.

Packard, G.C., Fasano, S.L., Attaway, M.B., Lohmiller, L.D., and Lynch, T.L., Thermal environment for overwintering hatchlings of the painted turtle (*Chrysemys picta*), *Can. J. Zool.*, 75, 401, 1997a.

Packard, G.C., Lang, J.W., Lohmiller, L.D., and Packard. M.J., Cold tolerance in hatchling painted turtles (*Chrysemys picta*): Supercooling or freeze tolerance?, *Physiol. Zool.*, 70, 670, 1997b.

Packard, G.C., and Packard, M.J., Patterns of nitrogen excretion by embryonic softshelled turtles (*Trionyx spiniferus*) developing in cleidoic eggs, *Science*, 221, 1049, 1983.

Packard, G.C., and Packard, M.J., Coupling of physiology of embryonic turtles to the hydric environment, in *Respiration and Metabolism of Embryonic Vertebrates*, R.S. Seymour (ed.), Dordrecht, The Netherlands: W. Junk Publishers, 99, 1984.

Packard, G.C., and Packard, M.J., The physiological ecology of reptilian eggs and embryos, in *Biology of the Reptilia, Vol. 16*, C. Gans and R.B. Huey (eds.), New York: Alan R. Liss, 1988a, 523.

Packard, G.C., and Packard, M.J., Water relations of embryonic snapping turtles (*Chelydra serpentina*) exposed to wet or dry environments at different times in incubation, *Physiol. Zool.*, 61, 95, 1988b.

Packard, G.C., and Packard, M.J., Growth of embryonic softshell turtles is unaffected by uremia, *Can. J. Zool.*, 68, 841, 1990.

Packard, G.C., and Packard, M.J., Delayed inoculative freezing is fatal to hatchling painted turtles (*Chrysemys picta*), *Cryo-Lett.*, 14, 273, 1993a.

Packard, G.C., and Packard, M.J., Hatchling painted turtles (*Chrysemys picta*) survive exposure to subzero temperatures during hibernation by avoiding freezing, *J. Comp. Physiol. B*, 163, 147, 1993b.

Packard, G.C., and Packard, M.J., Type of soil affects survival by overwintering hatchlings of the painted turtle, *J. Therm. Biol.*, 22, 53, 1997.

Packard, G.C., and Packard, M.J., The overwintering strategy of hatchling painted turtles, or how to survive in the cold without freezing, *BioScience*, 51, 199, 2001.

Packard, G.C., and Packard, M.J., Cold acclimation enhances cutaneous resistance to inoculative freezing in hatchling painted turtles, *Chrysemys picta*, *Funct. Ecol.*, 17, 94, 2003a.

Packard, G.C., and Packard, M.J., Natural freeze-tolerance in hatchling painted turtles?, *Comp. Biochem. Physiol.*, 134, 233, 2003b.

Packard, G.C., Packard, M.J., Miller, K., and Boardman, T. J., Influence of moisture, temperature, and substrate on snapping turtle eggs and embryos, *Ecology*, 68, 983, 1987.

Packard, G.C., Packard, M.J., and Benigan, L., Sexual differentiation, growth, and hatching success by embryonic painted turtles incubated in wet and dry environments at fluctuating temperatures, *Herpetologica*, 47, 125, 1991.

Packard, G.C., Packard, M.J., and Boardman, T.J., Influence of hydration of the environment on the pattern of nitrogen excretion by embryonic snapping turtles (*Chelydra serpentina*), *J. Exp. Biol.*, 108, 195, 1984.

Packard, G.C., Packard, M.J., Lang, J.W., and Tucker, J.K., Tolerance for freezing in hatchling turtles, *J. Herpetol.*, 33, 536, 1999.

Packard, G.C., Packard, M.J., and McDaniel, L.L., Seasonal change in the capacity for supercooling by neonatal painted turtles, *J. Exp. Biol.*, 204, 1667, 2001.

Packard, G.C., Packard, M.J., McDaniel, P.L., and McDaniel. L.L., Tolerance of hatchling painted turtles to subzero temperatures, *Can. J. Zool.*, 67, 828, 1989.

Packard, G.C., Packard, M.J., Miller, K., and Boardman, T.J., Effects of temperature and moisture during incubation on carcass composition of hatchling snapping turtles (*Chelydra serpentina*), *J. Comp. Physiol. B*, 158, 117, 1988.

Packard, G.C., and Phillips, J.A., The importance of the physical environment for the incubation of reptilian eggs, in *Captive Management and Conservation of Amphibians and Reptiles*, J.B. Murphy, K. Adler, and J.T. Collins (eds.), SSAR, 1994, 195.

Packard, G.C., Taigen, T.L., Packard, M.J., and Boardman, T.J., Water relations of pliable-shelled eggs of common snapping turtles (*Chelydra serpentina*), *Can. J. Zool.*, 58, 1404, 1980.

Packard, G.C., Taigen, T.L., Packard, M.J., and Boardman, T.J., Changes in mass of eggs of softshell turtles (*Trionyx spiniferus*) incubated under hydric conditions simulating those of natural nests, *J. Zool.*, 193, 81, 1981.

Packard, M.J., and Clark, N.B., Aspects of calcium regulation in embryonic lepidosaurs and chelonians and a review of calcium regulation in embryonic archosaurs, *Physiol. Zool.*, 69, 435, 1996.

Packard, M.J., and DeMarco, V.G., Eggshell structure and formation in eggs of oviparous reptiles, in *Egg Incubation: Its Effects on Embryonic Development in Birds and Reptiles*, D.C. Deeming and M.W.J. Ferguson (eds.), Cambridge, UK: Cambridge University Press, 1991, 53.

Packard, M.J., and Packard, G.C., Effect of water balance on growth and calcium mobilization of embryonic painted turtles (*Chrysemys picta*), *Physiol. Zool.*, 59, 398, 1986.

Packard, M.J., and Packard, G.C., Environmental modulation of calcium and phosphorus metabolism in embryonic snapping turtles (*Chelydra serpentina*), *J. Comp. Physiol.*, 159B, 501, 1989.

Packard, M.J., and Packard, G.C., Sources of calcium, magnesium, and phosphorus for embryonic softshell turtles (*Trionyx spiniferus*), *J. Exp. Zool.*, 258, 151, 1991.

Packard, M.J., and Packard, G.C., Accumulation of lactate by frozen painted turtles (*Chrysemys picta*) and its relationship to freeze tolerance, *Physiol. Biochem. Zool.*, 77, 433, 2004.

Palmer, B.D., and Guillette, L.J. Jr., Histology and functional morphology of the female reproductive tract of the tortoise *Gopherus polyphemus*, *Am. J. Anat.*, 183, 200, 1988.

Palmer, B.D., and Guillette, L.J. Jr., Morphological changes in the oviductal endometrium during the reproductive cycle of the tortoise *Gopherus polyphemus*, *J. Morphol.*, 204, 323, 1990.

Palmer, B.D., and Guillette, L.J. Jr., Oviductal proteins and their influence on embryonic development in birds an reptiles, in Egg Incubation: Its Effects on Embryonic Development in Birds and Reptiles, D.C. Deeming and M.W.J. Ferguson (eds.), Cambridge, UK: Cambridge University Press, 1991, 29.

Parker, G.H., The direction of the ciliary currents in the oviducts of vertebrates, *Am. J. Physiol.*, 87, 93, 1928.

Parker, G.H., The passage of sperms and of eggs through the oviducts in terrestrial vertebrates, *Phil. Trans.*, 219, 381, 1931.

Parmenter, C.J., Reproduction and survivorship of *Chelodina longicollis* (Testudinata: Chelidae), in *Biology of Australasian Frogs and Reptiles*, G. Grigg, R. Shine, and H. Ehmann (eds.), *Roy. Soc. NSW*, 1985, 53.

Pasteels, J., Une table analytique du développement des reptiles. I. Stades de gastrulation chez chéloniens et lacertiliens, *Ann. Soc. Roy. Zool.*, Belgique 87, 217, 1957.

Pasteels, J., Développement embryonnaire, in *Traite De Zoologie, Vol. 14*, P.P. Grasst (ed.), Paris: Masson and Co., 1970, 893.

Patterson, R., The role of urination in egg predator defense in the desert tortoise (*Gopherus agassizi*), *Herpetologica*, 27, 197, 1971.

Paukstis, G.L., and Janzen, F.J., Sex determination in reptiles: Summary of effects of constant temperatures of incubation on sex ratios of offspring, *Smith. Herpetol. Inform. Serv.*, 83, 1, 1990.

Paukstis, G.L., Gutzke, W.H.N., and Packard, G.C., Effects of substrate water potential and fluctuating temperatures on sex ratios of hatchling painted turtles (*Chrysemys picta*), *Can. J. Zool.*, 62, 1491, 1984.

Pearse, D.E., and Avise, J.C., Turtle mating systems: Behavior, sperm storage, and genetic paternity, *J. Hered.*, 92, 206, 2001.

Pearse, D.E., Janzen, F.J., and Avise, J.C., Genetic markers substantiate long-term storage and utilization of sperm by female painted turtles, *Heredity*, 86, 378, 2001.

Pearse, D.E., Janzen, F.J., and Avise, J.C., Multiple paternity, sperm storage, and reproductive success of female and male painted turtles (*Chrysemys picta*) in nature, *Behav. Ecol. Sociobiol.*, 51, 164, 2002.

Pellegrini, G., Sur la cellule interstitielle du testicule, *C.R. Assoc. Anat.*, 20, 314, 1925a.

Pellegrini, G., Sulle modificazione degli elementi intersterstiziale del testicolo negli animali ad attività sessuale periodica, *Arch. Ital. anat. Embriol.*, 22, 550, 1925b.

Peterson, C.C., Anhomeostasis: Seasonal water and solute relations in two populations of the desert tortoise (*Gopherus agassizii*) during chronic drought, *Physiol. Zool.*, 69, 1324, 1996.

Pieau, C., Modalities of the action of temperature on sexual differentiation in field-developing embryos of the European pond turtle *Emys obicularia* (Emydidae), *J. Exp. Zool.*, 220, 353, 1982.

Plotkin, P.T., Byles, R.A., Rostal, D.C., and Owens, D.W., Independent verses socially facilitated oceanic migrations of the olive ridley, *Lepidochelys olivacea*, *Mar. Biol.*, 122, 137, 1995.

Plummer, M.V., Some aspects of nesting success in the turtle, *Trionyx muticus*, *Herpetologica*, 32, 353, 1976.

Plummer, M.V., Notes on the courtship and mating behavior of the soft-shelled turtle, *Trionyx muticus* (Reptilia, Testudines, Trionychidae), *J. Herpetol.*, 11, 90, 1977.

Polisar, J., Reproductive biology of a flood-season nesting freshwater turtle of the northern neotropics: *Dermatemys mawii* in Belize, *Chel. Cons. Biol.*, 2 13, 1996.

Powell, C.B., Female sexual cycles of *Chrysemys picta* and *Clemmys insculpta* in Nova Scotia, *Can. Fid. Nat.*, 81, 134, 1967.

Pritchard, P.C.H., Notes on Persian turtles, *Br. J. Herpetology*, 3, 271, 1966.

Pritchard, P.C.H., *Living Turtles of the World*, Neptune City, NJ: T.H.F. Publications, 1967.

Pritchard, P.C.H., The leatherback or leathery turtle, *Dermochelys coriacea*, Monogr. 1. Int. Union Conserv. Natur. Res., Morges, Switzerland, 1971.

Pritchard, P.C.H., *Encyclopedia of Turtles*, Neptune, NJ: T.F.H. Publications, 1979.

Pritchard, P.C.H., *The Alligator Snapping Turtle, Biology and Conservation*, Malabar, FL: Kreiger Publishing, 2006.

Pritchard, P.C.H., and Trebbau, P., *The Turtles of Venezuela*, SSAR, 1984.

Ragotzkie, R.A., Mortality of loggerhead turtle eggs from excessive rainfall, *Ecology*, 40, 303, 1959.

Ratterman, R.J., and Ackerman, R.A., The water exchange and hydric microclimate of painted turtle (*Chrysemys picta*) eggs incubating in field nests, *Physiol. Zool.*, 62, 1059, 1989.

Raynaud, A., and Pieau, C., Embryonic development of the genital system, in *Biology of the Reptilia, Vol. 15*, C. Gans and F. Billett (eds.), New York: John Wiley & Sons, 1985, 149.

Reese S.A., Crocker, C.A., Jackson, D.A., and Ultsch, G.R., The physiology of hibernation among painted turtles: The midland painted turtle (*Chrysemys picta marginata*), *Respir. Physiol.*, 124, 43, 2000.

Reese S.A., Jackson, D.C., and Ultsch, G.R., Lactate accumulation and glycogen depletion in hatchling turtles (*Chrysemys picta bellii*, *Chelydra serpentina*, and *Graptemys geographica*) submerged at 3°C, *Integr. Comp. Biol.*, 42, 1299, 2002a.

Reese S.A., Jackson, D.C., and Ultsch, G.R., The physiology of overwintering in a turtle that occupies multiple habitats, the common snapping turtle (*Chelydra serpentina*), *Physiol. Biochem. Zool.*, 75, 432, 2002b.

Reese S.A., Jackson, D.C., and Ultsch, G.R., Hibernation in freshwater turtles: Softshell turtles (*Apalone spinifera*) are the most intolerant of anoxia among North American species, *J. Comp. Physiol. B*, 173, 263, 2003.

Renous, S., Rimblot-Baly, F., Fretey, J., and Pieau, C., Caractéristiques du développement embryonnaire de la tortue Luth, *Dermochelys coriacea* (Vandelli, 1761), *Ann. Sci. Nat. Zool. Paris*, 10, 197, 1989.

Restrepo, A., Piñeros, V., and Páez, V., Nest site selection by Colombian slider turtles, *Trachemys callirostris callirostris* (Testudines: Emydidae), in the Mompos Depression, Colombia, *Chel. Cons. Biol.*, 5 , 249, 2006.

Rhodin, A.G.J., Mittermeier, R.A., and Hall, P.M., Distribution, osteology, and natural history of the Asian giant softshell turtle, *Pelochelys bibroni*, in Papua New Guinea, *Chel. Cons. Biol.*, 1, 19, 1993.

Rimkus, T.A., Hruska, N., and Ackerman, R.A., Separating the effects of vapor pressure and heat exchange on water exchange by snapping turtle (*Chelydra serpentina*) eggs, *Copeia*, 2002, 706, 2002.

Risley, P.L., Contributions on the development of the reproductive system in the musk turtle, *Sternotherus odoratus* (Latreille). II. Gonadogenesis and sex differentiation, *Z. Zellforsch.*, 18, 493, 1933.

Risley, P.L., The spermatogenic and seasonal cycle of the testis of the musk turtle, *Scernotherus odoratus* (Latreille), *Anat. Rec.*, 57, suppl. 78, 1933b.

Roosenburg, W.M., Maternal condition and nest site choice: An alternative for the maintenance of environmental sex determination, *Am. Zool.*, 36, 157, 1996.

Roques, S., Diaz-Panaiagua, C., and Andreu, A.C., Microsatellite markers reveal multiple paternity and sperm storage in the Mediterranean spur-thighed tortoise, *Testudo gracea*, *Can. J. Zool.*, 82, 153, 2004.

Roques, S., Diaz-Panaiagua, C., Portheault, A., Perez-Santigosa, N., and Hidalgo-Vila, J., Sperm storage and low incidence of multiple paternity in the European pond turtle, *Emys orbicularis*: A secure but costly strategy? *Biol. Cons.*, 129, 236, 2006.

Rose, F., Drotman, R., and Weaver, W., Electrophoresis of chin gland extracts of *Gopherus* (Tortoises). *Comp. Biochem. Physiol.*, 29, 847, 1969.

Rostal, D., Seasonal reproductive biology of the Kemp's Ridley (*Lepidochelys kempii*): Comparison of captive and wild populations, *Chel. Cons. Biol.*, 4, 788, 2005.

Rostal, D., Wibbels, T., Grumbles, J.S., Lance, V.A., and Spotila, J.R., Chronology of sex determination in the desert tortoise (*Gopherus agazzizii*), *Chel. Cons. Biol.*, 4, 313, 2002.

Rowe, J.W., Egg size and shape variation within and among Nebraskan painted turtle (*Chrysemys picta bellii*) populations: Relationships to clutch and maternal body size, *Copeia*, 1994, 1034, 1994.

Roze, J.A., Pilgrim of the river. Life cycle of the Orinoco river turtle has many unusual features, *Nat. Hist.*, 73, 34, 1964.

Salhanick, A.R., Voto, C.C., Fox, T.O., and Callard, I.P., Estrogen binding proteins in the oviduct of the turtle, *Chrysemys picta*: Evidence for a receptor species, *Endocrinology,* 105, 1388, 1979.

Salih, M., Remarks on the morphology of the digestive, respiratory and female urogenital systems of the soft-shelled turtle *Trionyx euphraticus* (Daudin), *Bull. Iraq. Nat. Hist. Mus.*, 3, 1, 1965.

Sarkar, S,. Sarkar, N.K., and Maiti, B.R., Histological and functional changes of oviductal endometrium during seasonal reproductive cycle of the soft-shelled turtle, *Lissemys punctata punctata*, *J. Morphol.*, 224, 1, 1995.

Sarkar, S., Sarkar, N.K., and Maiti, B.R., Seasonal pattern of ovarian growth and interrelated changes in plasma steroid levels vitellogenesis and oviductal function in the adult female soft-shelled turtle, *Lissemys punctata punctata*, *Can. J. Zool.*, 74, 303, 1996.

Sarkar, S., Sarkar, N.K., and Maiti, B.R., Oviductal sperm storage structure and their changes during the seasonal (dissociated) reproductive cycle in the soft-shelled turtle *Lissemys punctata punctata*, *J. Exp. Zool.*, 295A, 1, 83, 2003.

Schleich, H.H., and Kästle, W., *Reptile Eggshells*, SEM Atlas, Stuttgart: Gustav Fischer Verlag, 1988.

Schramm, B.G., Lance, V.A., and Casares, M., Reproductive cycles of male and female giant tortoises (*Geochelone nigra*) on the Galápagos Islands by plasma steroid analysis and ultrasound scanning, Linnaeus Fund Research Report, *Chel. Cons. Biol.*, 3, 523, 1999.

Schumacher, G.H., The head muscles and hyolaryngeal skeleton of turtles and crocodilians, in *Biology of the Reptilia, Vol. 4*, C. Gans and T.S. Parsons (eds.), New York: Academic Press, 1973, 101.

Schwarzkopf, L., and Brooks, R.J., Sex determination in northern painted turtles: Effect of incubation at constant and fluctuating temperatures, *Can. J. Zool.*, 63, 2543, 1985.

Schwarzkopf, L., and Brooks, R.J., Nest-site selection and offspring sex ratio in painted turtles, *Chrysemys picta*, *Copeia*, 1987, 53, 1987.

Seidel, M.E., and Dreslik, M.J., *Pseudemys concinna* (LeConte) River Cooter, in *Catalogue of American Amphibians and Reptiles*, SSAR, 626.1, 1996.

Seidel, M.E., and Ernst, C.H., *Trachemys scripta* (Schoepff) Pond Slider, in *Catalogue of American Amphibians and Reptiles,* SSAR, 831.1, 2006.

Sen, M., and Maiti, B.R., Histomorphological changes of the ovary and oviduct during sexual maturity in the soft-shelled turtle, *Lissemys punctata punctata*, *Zool. Anz.*, 225, 391, 1990.

Sever, D., and Hamlett, W., Female sperm storage in reptiles, *J. Exp. Zool.*, 292, 187, 2002.

Seymour, R.S., Physiological adaptations to aquatic life, in *Biology of the Reptilia, Vol. 13*, Physiology. C. Gans and F.H. Pough (eds.), New York: Academic Press, 1982, 1.

Shah, R.V., The musculature of a cryptodire (*Deirochelys*) and a pleurodire (*Chelodina*) compared, *Bull. Mus. Comp. Zool.* (Harvard), 129, 343, 1963.

Shine, R., The reproductive biology of Australian reptiles: A search for general patterns, in *Biology of Australian Frogs and Reptiles*, G. Grigg, R. Shine, and H. Ehmann (eds.), Rcy. Soc. NSW, 297, 1985.

Simkiss, K., The sources of calcium for the ossification of the embryos of the giant leathery turtle, *Comp. Biochem. Physiol.*, 7, 71, 1962.

Singh, D.P., Annual sexual rhythm in relation to environmental factors in a tropical pond turtle, *Lissemys punctata granosa*, *Herpetologica*, 33, 190, 1977.
Spencer, R., Growth patterns of two widely distributed freshwater turtles and a comparison of common methods used to estimate age, *Aust. J. Zool.*, 50, 477, 2002.
Spencer, R., and Thompson. M.B., The significance of predation in nest site selection of turtles: An experimental consideration of macro- and microhabitat preferences, *Oikos*, 102, 592, 2003.
Spencer, R.J., Thompson, M.B., and Banks, P.B., Hatch or wait? A dilemma in reptilian incubation, *Oikos*, 93, 401, 2001.
Spotila, J.R., Zimmerman, L.C., Binkley, C.A., Grumbles, J.S., Rostal, D.C., List, A. Jr., Beyer, E.C., Phillips, K.M., and Kemp, S.J., Effects of incubation conditions on the sex determination, hatching success, and growth of hatchling desert tortoises, *Gopherus agassizii*, *Herpetol. Monog.*, 8, 103, 1994.
Sprando, R.L., and Russell, L.D., Spermatogenesis in the red-eared slider (*Pseudemys scripta*) and the domestic fowl (*Gallus domesticus*): A study of cytoplasmic events including cell volume changes and cytoplasmatic elimination, *J. Morphol.*, 198, 95, 1988.
Steen, J.B., *Comparative Physiology of Respiratory Mechanisms*, New York: Academic Press, 1971.
Stefan, Y., Réaction des canaux de Müller des jeunes males de *Emys leprosa* S. Non encore différences a l'action du propionate d'androsténediol, *C.R. Soc. Biol. Paris*, 153, 1234, 1959.
Stefan, Y., Contributions a l'étude experimentale de l'intersexualité chez un chélonian *Emys leprosa*, *Bull. Biol. Fr. Belg.*, 97, 363, 1963.
Stephens, P.R., and Wiens, J.J., Ecological diversity and phylogeny of emydid turtles, *Biol. J. Linn. Soc.*, 79, 577, 2003.
Stephens, P.R., and Wiens, J.J., Ecological structure of emydid turtle communities: The effects of phylogeny and dispersal, *Am. Nat.*, 164, 244, 2004.
Storey K.B., and Storey, J.M., Freeze tolerance in animals, *Physiol. Rev.*, 68, 27, 1988.
Storey K.B., Storey, J.M., Brooks, S.P.J., Churchill, T.A., and Brooks, R.J., Hatchling turtles survive freezing during winter hibernation, *Proc. Natl. Acad. Sci. USA*, 85, 8350, 1988.
Swingland, I.R., Reproductive effort and life history strategy of the Aldabra giant tortoise, *Nature*, 269, 402, 1977.
Swingland, I.R., and Coe, M.J., The natural regulation of giant tortoise populations on Aldabra Atoll: Reproduction, *J. Zool. Lond.*, 186, 285, 1978.
Swingland, I.R., and Coe, M.J., The natural regulation of giant tortoise populations on Aldabra Atoll: Recruitment, *Phil. Trans. Roy. Soc. B*, 286, 177, 1979.
Swingland, I.R., and Klemens, M., The Conservation Biology of Tortoises, Occasional Papers of the IUCN Species Survival Commission No. 5., IUCN, Gland, Switzerland, 1989.
Thing, A., The formation and structure of the zona pellucida in the ovarian eggs of turtles, *Am. J. Anat.*, 23, 237, 1918.
Thompson, M.B., The physiology and ecology of the eggs of the pleurodiran tortoise *Emydura macquarrii* (Gray, 1831), *Aust. Wild. Res.*, 10, 363, 1983.
Thompson, M.B., Water exchange in reptilian eggs, *Physiol. Zool.*, 60, 1, 1987.
Thompson, M.B., Nest temperatures in the pleurodiran turtle, *Emydura macquarii*, *Copeia*, 1988, 996, 1988.
Thompson M.B., and Speake, B.K., Egg morphology and composition, in *Reptilian Incubation: Environment, Evolution and Behaviour*, D.C. Deeming (ed.) Nottingham, UK: Nottingham University Press, 2004, 45.
Thomson, J.S., The anatomy of the tortoise, *Sci. Proc. Roy. Dublin Soc.*, 20, 359, 1932.
Thornhill, C.M., Comparative reproduction of the turtle, *Chrysemys scripta elegans*, in heated and natural lakes, *J. Herpetol.*, 16, 347, 1982.
Tinkle, D.W., Additional remarks on the extra-uterine migration of ova in turtles, *Herpetologica*, 15, 161, 1959.
Tinkle, D.W., Geographic variation in reproduction, size, sex ratio, and maturity of *Sternotherus odoratus* (Testudinata: Chelydridae), *Ecology*, 42, 68, 1961.
Tinkle, D.W., Congdon, J.D., and Rosen, P.C., Nesting frequency and success: Implications for the demography of painted turtles, *Ecology*, 62, 1426, 1981.
Tiwari, M., and Bjorndal, K.A., Variation in morphology and reproduction in loggerheads, *Caretta caretta*, nesting in the United States, Brazil, and Greece, *Herpetologica*, 56, 343, 2000.
Tokita, M., and Kuratani, S., Normal embryonic stages of the Chinese softshelled turtle *Pelodiscus sinensis* (Trionychidae), *Zool. Sci. (Tokyo)*, 18, 705, 2001.
Tucker, J.K., Body size and migration of hatchling turtles: Inter- and intraspecific comparisons, *J. Herpetol.*, 34, 541, 2000.
Tucker, J.K., Age, body size, and reproductive patterns in the red-eared slider (*Trachemys scripta elegans*) in west-central Illinois, *Herpetol. Nat. Hist.*, 8, 181, 2001.

Tucker, J.K., and Janzen, F.J., Order of oviposition and egg size in the red-eared slider turtle (*Trachemys scripta elegans*), *Can. J. Zool.*, 76, 377, 1998.

Tucker, J.K., Janzen, F.J., and Paukstis, G.L., Response of embryos of the red-eared turtle (*Trachemys scripta elegans*) to experimental exposure to water saturated substrates, *Chel. Cons. Biol.*, 2, 345, 1997.

Ultsch, G.R., Ecology and physiology of hibernation and overwintering among freshwater fishes, turtles, and snakes, *Biol. Rev.*, 64, 435, 1989.

Ultsch, G.R., and Jackson, D.C., Long-term submergence at 3°C of the turtle *Chrysemys picta bellii* in normoxic and severely hypoxic water III. Effects of changes in ambient pO_2 and subsequent air breathing, *J. Exp. Biol.*, 97, 87, 1982a.

Ultsch, G.R., and Jackson, D.C., Long-term submergence at 3°C of the turtle, *Chrysemys picta bellii*, in normoxic and severely hypoxic water I. Survival, gas exchange and acid-base status, *J. Exp. Biol.*, 96, 11, 1982b.

Ultsch, G.R., and Jackson, D.C., Acid-base status and ion balance during simulated hibernation in freshwater turtles from the northern portions of their ranges, *J. Exp. Biol.*, 273, 482, 1995.

Ultsch, G.R., and Reese, S.A., Ecology and physiology of overwintering, in *The Biology of the Snapping Turtle*, R.J. Brooks, A.C. Steyermark, and M.S. Finkler (eds.), Washington, DC: Smithsonian Institute Press, in press.

Ultsch, G.R., and Wasser, J.S., Plasma ion balance of North American freshwater turtles during prolonged submergence in normoxic water, *Comp. Biochem. Physiol. A*, 97, 505, 1990.

Ultsch, G.R., Carwile, M.E., Crocker, C.E., and Jackson, D.C., The physiology of hibernation among painted turtles: The Eastern painted turtle *Chrysemys picta picta*, *Physiol. Biochem. Zool.*, 72, 493, 1999.

Valenzuela, N., Multiple paternity in side-neck turtles *Podocnemius expansa*: Evidence from microsatellite DNA data, *Mol. Ecol.*, 99, 2000.

Valenzuela, N., Temperature-dependant sex determination, in *Reptilian Incubation: Environment, Evolution and Behaviour*, D.C. Deeming (ed.), Nottingham, UK: Nottingham University Press, 2004, 211.

Valenzuela, N., and Janzen, F.J., Nest-site philopatry and the evolution of temperature-dependent sex determination, *Evol. Ecol. Res.*, 3, 779, 2001.

Van Abbema, J. (ed.), Proceedings: Conservation, Restoration, and Management of Tortoises and Turtles, an International Conference, New York Turtle and Tortoise Society, New York, 1997.

Viosca, P., The *Pseudemys troosti-elegans* complex, a case of sexual dimorphism, *Copeia,* 1933, 208, 1933.

Vogt, R.C., Freshwater turtle nesting and reproduction in the neotropics, in ASIH/ HL/ SSAR/ AFS-ELHS/ AES/ GIS Joint Meetings, Seattle, WA, 301, 1997.

Vogt, R.C., and Bull, J.J., Ecology of hatchling sex ratio in map turtles, *Ecology*, 65, 582, 1984.

Walker, W.F., The locomotor apparatus of testudines, in *Biology of the Reptilia, Vol. 4*, C. Gans and T.S. Parsons (eds.), New York: Academic Press, 1973, 1.

Walker, W.F., Locomotion, in *Turtles, Perspectives and Research*, M. Harless and H. Morlock (eds.), New York: John Wiley, 1979, 435.

Wallace, B.P., Sotherland, P.R., Spotila, J.R., Reina, R.D., Franks, B.E., and Paladino, F.V., Biotic and abiotic factors affect the nest environment of embryonic leatherback turtles, *Dermochelys coriacea*, *Physiol. Biochem. Zool.*, 77, 423, 2004.

Weaver, W., Courtship and combat behavior in *Gopherus berlandieri*, *Bull. Fl. St. Mus.*, 15, 1, 1970.

Webb, G.J.W., Choquenot, D., and Whitehead, P.J., Nests, eggs, and embryonic development of *Carettochelys inscuipta* (Chelonia: Carettochelidae) from northern Australia, *J. Zool B*, 1, 521, 1986.

Webb, R.G., Size at sexual maturity in the male softshell turtle, *Trionyx ferox emoryi*, *Copeia*, 1956, 121, 1956.

Webb, R.G., Observations on the life histories of turtles (genus *Pseudemys* and *Graptemys*) in Lake Texoma, Oklahoma, *Am. Midl. Nat.*, 65, 193, 1961.

Webb, R.G., North American recent soft-shelled turtles (Family Trionychidae), *Univ. Ks. Pubis. Mus. Nat. Hist.*, 13, 429, 1962.

Webb, R.G., *Trionyx* Geoffroy Saint-Hilaire, Softshell Turtles, in *Catalogue of American Amphibians and Reptiles*, SSAR, 487.1, 1990.

White, J.B., and Murphy, C.G., The reproductive cycle and sexual dimorphism of the common snapping turtle, *Chelydra serpentina serpentina*, *Herpetologica*, 29, 240, 1973.

Wibbels, T., Owens, D., Limpus, C.J., and Amoss, M.S., Seasonal changes in serum gonadal steroids associated with migration, mating, and nesting in loggerhead sea turtles, *Gen. Comp. Endocrin.*, 79, 154, 1990.

Wibbels, T., Owens, D.W., and Rostal, D.C., Soft plastra of adult male sea turtles: An apparent secondary sexual characteristic, *Herpetol. Rev.*, 22, 47, 1991.

Wibbels, T., Wilson, C., and Crews, D., Mullerian duct development and regression in a turtle with temperature-dependent sex determination, *J. Herpetol.*, 33, 149, 1999.

Wilbur, H.M., The evolutionary and mathematical demography of the turtle *Chrysemys picta*, *Ecology*, 56, 64, 1975a.

Wilbur, H.M., A growth model for the turtle *Chrysemys picta*, *Copeia*, 1975, 337, 1975b.

Wilbur, H.M., and Morin, P.J., Life history evolution in turtles, in *Biology of the Reptilia, Vol. 16, Ecology. B. Defense and Life History*, C. Gans and R.B. Huey (eds.), New York: Alan R. Liss, 1988, 387.

Wilhoft, D.C., Hotaling, E., and Franks, P., Effects of temperature on sex determination in embryos of the snapping turtle, *Chelydra serpentina*, *J. Herpetol.*, 17, 38, 1983.

Willard R., Packard, G.C., Packard, M.J., and Tucker, J.K., The role of the integument as a barrier to penetration of ice into overwintering hatchlings of the painted turtle (*Chrysemys picta*), *J. Morphol.*, 246, 150, 2000.

Williams, E.E., Variation and selection in the cervical central articulations in living turtles, *Bull. Amer. Mus. Nat. Hist.*, 94, 505, 1950.

Wilson, D., Nest-site selection: Microhabitat variation and its effects on the survival of turtle embryos, *Ecology*, 79, 1884, 1998.

Witzell, W.N., Synopsis of the biological data on the hawksbill turtle, *Eretmochelys imbricata* (Linneaus 1766), *FAO Fish. Synop.*, 137, 1, 1983.

Wolverton, E., Winter survival of hatchling painted turtles in northern Minnesota, *Copeia*, 1961, 109, 1961

Wood, J.R., and Wood, F.E., Reproductive biology of captive green sea turtles *Chelonia mydas*, *Am. Zool.*, 20, 499, 1980.

Wyneken, J., The anatomy of sea turtles, US Dept. of Commerce NOAA Technical Memorandum NMFS-SEFSC-470, 1, 2001.

Yamada, F., Studies on the ciliary movements in the oviduct, *Jap. J. Physiol.*, 1-2, 194, 1952.

Yntema, C.L., A series of stages in the embryonic development of *Chelydra serpentina*, *J. Morphol.*, 125, 219, 1968.

Yntema, C.L., Effects of incubation temperatures on sexual differentiation in the turtle, *Chelydra serpentina*, *J. Morphol.*, 150, 453, 1976.

Yntema, C.L., Incubation times for eggs of the turtle *Chelydra serpentina* (Testudinidae: Chelydridae) at various temperatures, *Herpetologica*, 34, 274, 1978.

Yntema, C.L., Temperature levels and periods of sex determination during incubation of eggs of *Chelydra serpentina*, *J. Morphol.*, 159, 17, 1979.

Zangerl R., The turtle shell, in *Biology of the Reptilia*, C. Gans, A. Bellairs, and T. Parsons (eds.), New York: Academic Press, 1969, 311.

Zug, G.R., The penial morphology and the relationships of cryptodiran turtles, *Occ. Pap. Mus. Zool. Mich.*, 647, 1, 1966.

Zweifel, R.G., Long-term ecological studies on a population of painted turtles, *Chrysemys picta*, on Long Island, New York, *Am. Mus. Nov.*, 2952, 1, 1989.

11 Mixed and Uniform Brood Sex Ratio Strategy in Turtles: *The Facts, the Theory, and Their Consequences*

Vincent Hulin, Marc Girondot, Matthew H. Godfrey, and Jean-Michel Guillon

CONTENTS

11.1 INTRODUCTION

Temperature-dependent sex determination (TSD) occurs when males or females are differentially produced according to the incubation temperature (Bull, 1983). Since the discovery of TSD in a squamate by Charnier (1966), this pattern of sex determination has been described in various reptiles: all crocodilians (Deeming, 2004), tuataras (Nelson et al., 2004), some squamates (Harlow, 2004), and 64 out of the 79 studied turtle species (Ewert et al., 2004). Other reptile species exhibit genotypic sex determination (GSD), where sexual phenotype is independent of embryonic incubation temperature. GSD in reptiles is sometimes linked with heteromorphic sex chromosomes, with males or females being the heterogametic sex. However, many species with GSD do not exhibit strong differentiation of sex chromosomes. Overall, the presence of dimorphic sex chromosomes is not necessarily mutually exclusive of TSD, as has been demonstrated in various amphibians (Chardard et al., 2004) and one lizard (Shine et al., 2002).

Three distinct patterns of TSD are observed in reptiles but only two are present in turtles. Pattern TSD Ia or MF is observed in many turtles and is characterized by the production of males at lower incubation temperatures and females at higher temperatures (Ewert et al., 1994). For pattern TSD Ib or FM, females are produced at lower incubation temperatures and males at higher temperatures. This pattern is observed in some lizards (Viets et al., 1994) but was originally described in crocodilians. The recent availability of more complete data for some crocodile species has revealed that many species actually exhibit TSD II or FMF, where females are produced at low and high temperatures and males at intermediate ones (Lang & Andrews, 1994). Note that some turtle species also exhibit pattern II (Ewert et al., 1994). The relationships among these patterns are subject to debate. It has been proposed that FMF is the general pattern for reptiles and that FM or MF are simply observed because extreme incubation temperatures have not been adequately studied or because sufficiently lower or higher incubation temperatures are not conducive to successful incubation (Pieau et al., 1995).

The precise timing when sex determination is sensitive to temperature during development has been studied in various reptiles. The timing is always linked with the first stages of gonadal development until the end of the second third of development (Pieau & Dorizzi, 2004). This homogeneity among various reptilian orders as well as recent phylogenetic analyses (Janzen & Krenz, 2004) suggest a common origin for TSD in this class.

The selective forces explaining the prevalence of TSD in turtles remain elusive. The most-studied hypothesis was formulated by Charnov and Bull (1977). According to this theoretical model, environmental sex determination (ESD) should be favored over GSD when offspring develop in a spatially heterogeneous (patchy) environment for one parameter, this parameter influencing the fitness of sexes differently. Parents and offspring should also have no control over which patch type offspring develop in, and mating should take place among individuals coming from different patches. Its application to reptiles posits differential fitness for sexual phenotypes depending on a parameter correlated with their incubation temperature. Whereas these conditions indeed select for environmental sex determination in a theoretical model (Bull, 1981), they have never been conclusively demonstrated in reptiles. Though not yet validated, alternative models bring new perspective on this subject (Hulin & Guillon, 2007; Julliard, 2000; Reinhold, 1998; Roosenburg, 1996).

During extreme climatic events, greater numbers of unisex nests can be produced. If these conditions persist in the long term, the population sex ratio would become highly biased and could present an evolutionary drawback of TSD in turtles. This evolutionary question is today of great importance because of predicted rapid climate change and associated global warming (IPCC, 2001). To assess the evolutionary significance of TSD in turtles, we tried to answer three questions. First, we theoretically compared strategies producing unisex or mixed-sex ratio within a nest and their contributions to the population. Second, we looked for the risk of extinction of turtle populations according to the brood sex ratio strategy (mixed or unisex nests) used by individuals. Third, to compare our theoretical predictions with real-world scenarios we reviewed the literature and calculated the proportion of unisex nests in different turtle populations. Our results are relevant to the discussion on the evolution of TSD and its consequences on turtle populations.

11.2 UNISEX OR MIXED SEX RATIO STRATEGY IN A NEST: WHICH IS BEST?

Let us take first a simple model of population dynamics with constant population size N and constant average brood sex ratio in the population sr measured in male frequency. Two strategies will be examined: a unisex sex ratio strategy (USRS), where clutches produce all males with a frequency sr', or all females with a frequency $1 - sr'$, and a mixed sex ratio strategy (MSRS) where the male frequency within a clutch is sr' and the female frequency is $1 - sr'$. During its lifetime, an individual will produce K eggs in L clutches. The brood size is then K/L. The distribution of sex ratios produced by this individual is obtained from a binomial distribution with the total number of events being the total number of clutches for USRS (because a clutch will be all male or all female) or the total number of eggs for MSRS (because an egg is either male or female). Note that these two strategies are the two extremes of a continuum. The distribution of sex ratios is then $B(L, sr')$ for the individuals that use USRS and $B(K, sr')$ for the individuals that use MSRS (Figure 11.1). Note that the total number of eggs in both cases is K, and therefore the strategy does not influence the global output of juveniles.

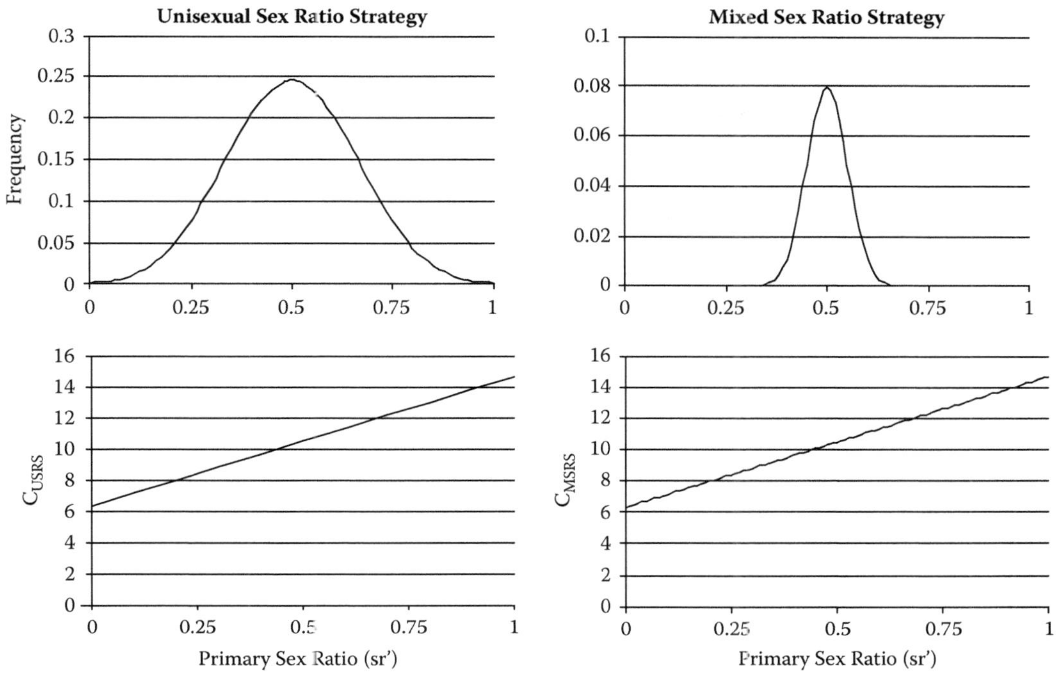

FIGURE 11.1 Distribution of primary sex ratios produced during an individual's lifetime that uses unisex or mixed sex ratio strategy ($sr' = 0.5$, $N = 1000$, $K = 100$, $L = 10$). The contribution to the next generation measured in number of juveniles produced for each combination is shown in the lower row for $sr = 0.3$, $N = 1000$, $K = 100$, and $L = 10$.

The M male progeny of an individual will compete with other males of the population to reproduce, and the F females will compete with other females of the population. Thus, the contribution of an individual to the next generation will be $M/_{Nsr}$ by the way of its male progeny and $F/_{N(1\text{-}sr)}$ by the way of its female progeny (Shaw & Mohler, 1953).

The contribution of one individual using USRS who produces i unisex male clutches among the L clutches she produces during her lifetime is

$$\frac{iK}{NsrL} + \frac{(L - i)K}{N(1 - sr)L}$$

An individual using MSRS produces i male eggs among the K eggs she produces during her lifetime. The contribution of this individual to the next generation is therefore

$$\frac{i}{Nsr} + \frac{K - i}{N(1 - sr)}$$

Then the average contribution C of strategy MSRS and USRS are estimated as the sum of the contribution of each brood composition weighted by its frequency among the possible broods,

$$C_{MSRS} = \sum_{i=0}^{K} \binom{i}{K} (sr')^{i} (1 - sr')^{K-i} \left(\left(\frac{i}{N\,sr} \right) + \left(\frac{K-i}{N(1-sr)} \right) \right)$$

As

$$\sum_{i=0}^{x} \binom{i}{x} (p)^{i} (1-p)^{x-i}\, i = E\left(\mathrm{B}(x,p)\right) = x.p$$

and

$$\sum_{i=0}^{x} \binom{i}{x} (p)^{i} (1-p)^{x-i} (x-i) = E\left(\mathrm{B}(x,1-p)\right) = x.(1-p)$$

then

$$C_{MSRS} = C_{USRS} = \frac{K}{N} \left(\frac{sr'}{sr} + \frac{1-sr'}{1-sr} \right)$$

Overall, the choice of using a mixed or unisex sex ratio strategy has no influence on the contribution to the next generation and is therefore essentially neutral (Figure 11.1).

Whereas the use of mixed or unisex sex ratio strategy is not under selection, there is still one potential difference in the probability of extinction of the population when all individuals use USRS versus MSRS. If mortality occurs mainly at the level of the nest, such as egg destruction during incubation (i.e., a nest is destroyed or not, Eckrich & Owens, 1995; Girondot et al., 2002), each year very few nests may effectively contribute to the population. Under the USRS scenario, juveniles that survive for a particular year have a higher probability of being of the same sex. Consider the case when only one nest escapes destruction each year and adults reproduce *Y* years in this population. Then the probability that simply by chance the population becomes unisex is $sr'^{Y} + (1 - sr')^{Y}$. The first occurrence of a unisex outcome for the population will follow a geometric distribution with parameter $p = sr'^{Y} + (1 - sr')^{Y}$ and with mean $\frac{1}{p}$.

Hence, for $Y = 20$ and $sr' = 0.5$ (i.e., half of the nests are all male producing and the other half are all female producing), a unisex outcome for the population will occur once every ~500,000 years, on average (Figure 11.2). Although this may appear to be a relatively rare event, one should recall that TSD is an ancient character in reptiles, having appeared between 100 and 300 million years ago (Janzen & Krenz, 2004). In the case of a MSRS, the probability that simply by chance the population becomes unisex is $sr'^{Y.K/L} + (1 - sr')^{Y.K/L}$. For $Y = 20$, $sr' = 0.5$, and $K/L = 10$, a unisex outcome for the MSRS population is expected to occur once every ~10^{60} years, on average (Figure 11.2).

When mortality occurs at the scale of the whole nest, a lineage using the USRS should face a higher probability of extinction over the long term, so only those using the MSRS should have been able to survive until now. This type of group-selection argument is similar to explanations for the long-term advantage of sex (Gouyon et al., 1989; Nunney, 1989).

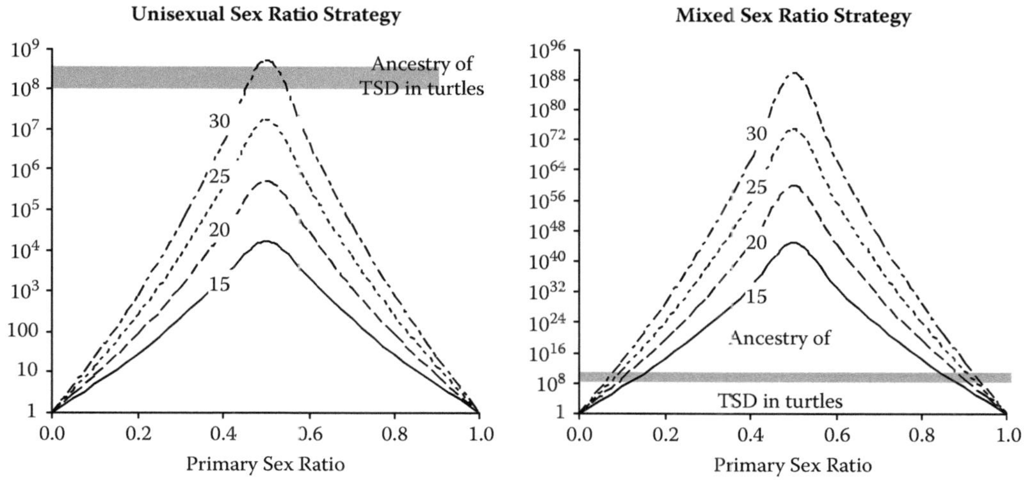

FIGURE 11.2 Mean time in years to arrive at unisex populations in species with TSD when one nest per year escapes destruction. The number of cohorts contributing to reproduction is shown on the top of each curve. Note that the placement of the bar "Ancestry of TSD in turtles" (estimated time since TSD has appeared in turtles) is the same in both graphs although the scale is different. Mixed sex ratio strategy: $K / L = 10$.

11.3 REVIEW OF THE PROPORTION OF UNISEX NESTS FROM FIELD STUDIES

11.3.1 Sampling Requirements

Ideally, to assess the exact proportion of unisex nests in turtle species it would be necessary to classify the sex of all hatchlings in all nests deposited by each turtle of the considered population during its total lifespan—a logistical impossibility. Therefore, for our purposes we reviewed the published literature for estimates of offspring sex ratios in turtles. We consider the following components of study design as minimal requirements for adequately assessing the level of unisex nests in a turtle population.

11.3.1.1 Fair Spatial Sampling

For turtle species with TSD, the sex ratio in a nest is dependent on the thermal conditions where the nest develops. As most nesting areas cannot be considered thermally homogeneous (Hays et al., 1995; Mrosovsky et al., 1984a), the spatial location of a nest has an impact on its hatchling sex ratio. An estimation of the proportion of unisex nests is then representative only of the part of the nesting area where nests have been sampled. For example, in sea turtle species it is known that females may nest on several beaches (Eckert et al., 1989); these beaches can be thermally heterogeneous (shade due to the vegetation, composition of the sand, cooling effect of the tide, and so on). Therefore, studies aiming to estimate the proportion of unisex nests at the scale of the nesting beach must sample nests in the different areas of the beach. In addition, studies aiming to estimate this proportion at the scale of a geographic area must sample nests in each nesting beach.

11.3.1.2 Fair Temporal Sampling

Nest sex ratio is influenced by the seasonality of nesting. Thermal conditions vary at intra- and inter-annual scales, causing the nest sex ratios to vary during and between nesting seasons (Godfrey & Mrosovsky, 1999). Therefore, an estimation of the proportion of unisex nests can be biased if the field study is constrained either to a fraction of the entire nesting season or to a single nesting season.

11.3.1.3 Accurate Classification of Sex

The sex of turtle hatchlings can be determined by several methods. Due to a lack of external morphological differences between male and female hatchlings, direct observation of gonadal structure has been considered to be the most accurate method of classifying sex (Mrosovsky & Godfrey, 1995). However, direct observation usually requires the sacrifice of hatchlings being studied. As many turtle species are protected, some authors have used indirect estimates of nest sex ratio. These methods are based on models using temperature, or a proxy of temperature (e.g., duration of incubation), to estimate the sex ratio of the nest. However, because of differences between individuals in the consequences of thermal conditions on sex, these indirect methods are imprecise and must be interpreted with caution. The radioimmunoassay (RIA) of testosterone is another indirect method to classify the sex of hatchlings without killing them, but it must be parameterized for each species to which it is applied. To date, this method (Lance & Valenzuela, 1992) has been successfully used only for *Podocnemis expansa* (Valenzuela, 2001a; Valenzuela et al., 1997). An early report of the effectiveness of RIA in sexing loggerhead sea turtle hatchlings by Crain et al. (1995) has not been successfully replicated (Merchant-Larios, 1999).

The primary method then remains the direct observation of hatchlings' gonads. Most studies focused only on a small sample of hatchlings or eggs to limit the consequences for the population. When small samples are used to estimate the nest sex ratio, another potential bias could arise if the sample is not representative of all hatchlings within a clutch. Indeed, thermal conditions are known to vary within the nest causing male and female hatchlings to be more or less frequent depending on the position in the nest (Georges, 1992; Godfrey et al., 1997).

11.3.2 Data from the Literature

For the purposes of our study, we estimated the proportion of unisex nests in different turtle populations based on data from studies where sexual phenotype was determined by the structure of the gonad or by RIA of testosterone, and sex ratio values are independently given for each natural nest. (Note that we excluded studies that focused on nests that had been manipulated, such as by relocation to a protected hatchery.) Thirty-three studies were retained using these criteria (Table 11.1). Twenty-two of these also give the exact number of hatchlings sexed for each nest. As a turtle population is difficult to delimit, especially in the case of sea turtles, in the present section the term "population" is used to designate geographically distinct nesting areas that may not always fit the theoretical concept in population biology.

From these 33 studies, we compiled data for 25 populations of 13 species (Table 11.1). The number of populations for each species varies from one (for seven species) to six (for *Caretta caretta*). For each population, we have data covering 1 (for 13 cases) to 6 years (for *Chrysemys picta* in Illinois and *Dermochelys coriacea* in French Guiana and Suriname), with data for at least two nesting seasons for other 11 populations. Some studies focused on small numbers of nests (Table 11.1). Three were conducted on only one nest and no other studies concerned the same species or the same population—Bull & Vogt (1979) on *Trionyx spiniferus*, Demuth (2001) on *Gopherus polyphemus*, and Dalrymple et al. (1985) on *Eretmochelys imbricata* in Florida. One population was represented by only three nests—Alho (1985) on *Podocnemis expansa* in Brazil—and one by only two nests—Kaska et al. (1998) on *Caretta caretta* in Cyprus). Six studies focused on one to five nests but concerned populations that were also studied in other years. We sought to include as much data as possible to facilitate our analyses without sacrificing the essential criteria set out previously.

As for all meta-analyses, we observed a large heterogeneity in the quality of data. For instance, inter-seasonal and intra-seasonal temporal variability is differentially described in all populations. For example, half of the studies (24 studies of 33) sampled nests at different days encompassing an important proportion of the nesting season (Table 11.1). Also, the spatial variability was different, depending on the study (Table 11.1): only six studies sampled nests in different nesting beaches/

TABLE 11.1
Description of the Studies Used for the Review*

Species	Geographic Area	Year(s)	Temporal Sampling	Spatial Sampling	Sexual Phenotype Identification	Hatchlings per Nest	Nests
Caretta caretta	Cyprus	1995[5]	10-Jun and 27-Jul	2 beaches (Akdeniz and Karpaz)	Eggs sampled at different depths in the nest a few days before emergence and artificially incubated thereafter	9–21	2
	Espirito Santo and Bahia (Brazil)	1996[6] and 1997[6] (data of both years mixed together)		2 beaches (Comboios and Praia do Forte)	Hatchlings sampled during emergence	10	51
	Cape Canaveral (Florida)	1986[3]	Daily every 2 weeks from 17-May to 16-Aug		Hatchlings sampled during emergence	10	45
		1987[4] and 1988[4]	Daily every 2 weeks from 17-May to 16-Aug		Hatchlings sampled during emergence	10	76
	Natal (South Africa)	1984[2] and 1985[2]	Daily from 27-Oct 1984 to 01-Jan-1985		Eggs sampled in the nest at the end of the thermosensitive period and artificially incubated thereafter	20–25	17
	South Carolina	1979[1], 1980[1], and 1982[1]		3 islands (Sand island, South island, and Kiawah island)	Hatchlings sampled during emergence	10	18
	Turkey	1995[5]	Daily from 8-Jun to 2-Aug	4 beaches (Dalyan, Fethiye, Patara, and Kizilot)	Hatchlings found in the nest a few days after emergence or eggs sampled at different depths in the nest a few days before emergence and artificially incubated thereafter	9–21	6
		2000[7], 2001[7], and 2002[7]	Daily between Jun and Jul		Hatchlings found in the nest a few days after emergence or eggs sampled at different depths in the nest a few days before emergence and artificially incubated	2–6	21
Carettochelys insculpta	Northern Territory (Australia)	1986[8]	Daily between 6-Aug and 15-Dec		Hatchlings sampled in nests a few days before emergence	4–11	6
		1996[9], 1997[9], and 1998[9]			Eggs sampled in nests a few days before emergence and artificially incubated	—	140

(continued)

TABLE 11.1 (continued)
Description of the Studies Used for the Review*

Species	Geographic Area	Year(s)	Temporal Sampling	Spatial Sampling	Sexual Phenotype Identification	Hatchlings per Nest	Nests
Chelonia mydas	Ascension Island	1999[13]	11 days between 03-Jan and 08-Apr	3 beaches (South West beach, Long beach, and North East beach)	Hatchlings sampled in the nest after emergence	3–9	20
	Caribbean coast (Costa Rica)	1980[11]		Tortuguero beach divided into high-, mid-, and low-beach	Eggs sampled in the nest a few days before emergence and artificially incubated or hatchlings sampled during the emergence	20	15
	Cyprus	1995[5]	5 days between 8-Jun and 25-Jul	2 beaches (Akdeniz and Karpaz)	Eggs sampled at different depths in the nest a few days before emergence and artificially incubated	9–21	5
		1998[12]	Daily or every 2 days between May and Oct	2 beaches (Alagadi and west coast of the island)	Dead hatchlings found in the nest a few days after emergence	1–10	48
	Suriname	1980[10]	Daily from 27-May to 9-Jun		Hatchlings sampled during emergence	10	12
		1981[15] and 1982[15]	Every 2–4 days from 7-Mar to 12-Aug		Hatchlings sampled during emergence	10	113
		1993[14]	Nearly every day between 3-Mar and 1-Sep	Beach divided into vegetation area, vegetation line, and sand area	Hatchlings sampled during emergence	3–10	79
Chelydra serpentina	Mississippi River (Illinois)	1999[16]			Hatchlings sampled during emergence	10	14
Chrysemys picta	Ontario (Canada)	1984[17]			Hatchlings sampled in nests a few days before emergence	—	21

	Mississippi River (Illinois)	1980[21]	Nearly every day from 31-May to 28-Jun		Hatchlings sampled in nests a few days before emergence	—	2
		1989[18], 1990[18], 1991[18], and 1992[18]	Almost daily in Jun		Hatchlings sampled in nests a few days before emergence	10	116
		1995[19]	Almost daily in Jun	Nesting area divided into low, intermediate, and high vegetation cover	Living or dead hatchlings remaining in nests after emergence	6–14	10
		1995[20]	Almost daily in Jun		Hatchlings remaining in nests after emergence	> 6	157
Dermochelys coriacea	French Guiana/ Suriname	1981[22], 1983[22], 1984[22], and 1985[22]	Daily between 14-Jun and 31-Jul		Hatchlings sampled during emergence	8–40	34
		1982[15]	Every 2–4 days from 7-Mar to 12-Aug		Hatchlings sampled during emergence	10	29
		1993[14]	Nearly every day between 3-Mar and 1-Sep	Beach divided into vegetation area, vegetation line, and sand area	Hatchlings sampled during emergence	3–10	27
	Playa Grande (Pacific coast of Costa Rica)	1994[23]			Hatchlings sampled in nests laid below the high tide line		18

(continued)

TABLE 11.1 (continued)
Description of the Studies Used for the Review*

Species	Geographic Area	Year(s)	Temporal Sampling	Spatial Sampling	Sexual Phenotype Identification	Hatchlings per Nest	Nests
Eretmochelys imbricata	Soldier Key (Dade County, Florida)	1982[24]	One nest		70 hatchlings among the 141 hatchlings of the nest (all hatchlings were dead)	70	1
	Buck Island	1994[25]			Dead hatchlings found in the nest a few days after emergence	574 from 51 nests	51
Graptemys spp.	Mississippi River (Illinois)	1978[27]			Hatchlings sampled in nests a few days before emergence	—	202
		1980[21]	Nearly every day from 31-May to 28-Jun		Hatchlings sampled during emergence	—	75
		1982[26]	Nearly every day from Jun to Sept		Hatchlings sampled in nests a few days before emergence	At least 8	14
Gopherus polyphemus	Cape Canaveral (Florida)	—[28]	One nest		Hatchlings sampled in the nest before emergence	5	1
Macrochelys temminckii	Apalachicola River (Florida)	1990[29] and 1991[29]			Hatchlings remaining in nests after emergence	6	47
Podocnemis expansa	Pará State (Brazil)	1984[30]			Hatchling sampled during emergence	—	3
	Middle Caquetá River (Colombia)	1993[31] and 1994[31]		3 beaches (Yarumal, Centro, and Guadual)	RIA of testosterone for all living hatchlings sampled between hatching and emergence	19–42	6
		1997[32] and 1998[32]			RIA of testosterone for living hatchlings sampled between hatching and emergence	30	3

Podocnemis unifilis	Guapore River (Rondonia, Brazil)	1990[33]	Daily from Aug to Dec		Hatchlings remaining in nests after emergence	10	15
Trionyx spiniferus	Mississippi River (Illinois)	1978[21]	Nearly every day from 31-May to 28-Jun		Hatchlings sampled in nests before emergence	At least 8	1

* *In the absence of notation, live hatchlings were sampled. The classification of hatchling was done by observation of the histological structure of the gonads unless otherwise noted. A cell with no data means that the information was not found in the corresponding article. A blank cell in the "Spatial sampling" column means that the study focused on only one beach/nesting area. Nesting beach/area is named in the "Spatial Sampling" column when listed in the source.*

[1]*Mrosovsky et al. (1984);* [2]*Maxwell et al. (1988);* [3]*Mrosovsky & Provancha (1989);* [4]*Mrosovsky & Provancha (1992);* [5]*Kaska et al. (1998);* [6]*Mrosovsky et al. (1999);* [7]*Kaska et al. (2006);* [8]*Georges (1992);* [9]*Doody et al. (2004);* [10]*Mrosovsky (1982);* [11]*Spotila et al. (1987);* [12]*Broderick et al. (2000);* [13]*Godley et al. (2002);* [14]*Godfrey et al. (1996);* [15]*Mrosovsky et al. (1984);* [16]*Kolbe & Janzen (2002);* [17]*Schwarzkopf & Brooks (1985);* [18]*Janzen (1994b);* [19]*Weisrock & Janzen (1999);* [20]*Valenzuela & Janzen (2001);* [21]*Vogt & Bull (1984);* [22]*Rimblot-Baly et al. (1987);* [23]*Binckley et al.(1998);* [24]*Dalrymple et al. (1985);* [25]*Wibbels et al. (1999);* [26]*Bull (1985);* [27]*Bull & Vogt (1979);* [28]*Demuth (2001);* [29]*Ewert (1994);* [30]*Alho et al. (1985);* [31]*Valenzuela et al. (1997);* [32]*Valenzuela (2001a);* [33]*De Souza & Vogt (1994).*

areas and four sampled nests in different sub-areas of the beach (according to the surrounding vegetation or the position in the beach).

The collection methods used in selected studies to obtain the hatchlings also differ greatly. In the majority of the cases, authors sampled from 10 to 20 hatchlings per clutch during emergence. Some other studies sampled embryos in the nest a few days before the estimated end of incubation and let embryos complete incubation in the laboratory (Doody et al., 2004; Kaska et al., 1998; Maxwell et al., 1988; Spotila et al., 1987). Because sexual differentiation of hatchling sex is thermally sensitive during the second third of incubation, we assumed that altering thermal conditions during the end of incubation did not alter the natural sex ratio. These two sampling methods result in an unbiased estimate of the offspring sex ratio if sampling is representative of eggs or hatchlings from different positions in the nest. Except for Kaska et al. (1998), the precision about the sampling within the nest was not given in the article, although some authors stated that they randomly sampled from all emergent hatchlings. Finally, some studies used dead hatchlings found in the nest after emergence had been observed (Broderick et al., 2000; Kaska et al., 2006; Wibbels et al., 1999). Although avoiding the problem of sacrificing living hatchling, this method may not represent an unbiased sample of hatchlings. Indeed, it is possible that dead hatchlings remaining in the nest were not randomly distributed in the nest and thus had experienced a different thermal regime of incubation when compared to the hatched eggs. For example, the thermal conditions at the bottom of the nest tend to be cooler than at the top of the nest (Kaska et al., 1998).

11.3.3 Analyses of the Sex Ratio Data

Overall, the proportion of unisex turtle nests was 0.65 and of these unisex nests, a proportion of 0.72 was all female. The proportion of unisex nests was also calculated independently for each population and year. A hierarchical model was built to test for the origin of heterogeneity (species, population, and year) in the proportion of unisex nests. For this test, *Trionyx spiniferus* was removed, as this species possesses GSD. Data were divided into groups of unisex male, unisex female, and mixed sex nests. The likelihood of the observations was calculated using a multinomial model. In short, the –ln likelihood of an observation of m_i unisex male, f_i unisex female, and u_i mixed sex ratio nests was calculated as

$$L_i = -Ln\left(\frac{(m_i + f_i + u_i)!}{m_i!f_i!u_i!} pm_i^{m_i} pf_i^{f_i} pu_i^{u_i}\right)$$

where pm_i, pf_i, and pu_i respectively are the expected proportion of males, females, and mixed sex ratio nests. These expected proportions were calculated taking into account a species, *species + population* or *species + population + year* effect. The various likelihoods were compared using likelihood ratio test where $-2(L_A - L_B)$ was distributed as a χ^2 with a degree of freedom equal to the difference in the number of parameters between model A and B. We found a highly significant species effect (LRT = 280.46, 22 df, $p = 10^{-46}$), a population within species effect (LRT = 224.18, 44 df, $p = 10^{-35}$), and a year within population and species effect (LRT = 275.62, 58 df, $p = 10^{-30}$). Similar results are found if nests were grouped in unisex or mixed sex ratio nests; there was a highly significant species effect (LRT = 161.47, 11 df, $p = 10^{-29}$), a population within species effect (LRT = 144.53, 11 df, $p = 10^{-25}$), and a year within population and species effect (LRT = 95.63, 29 df, $p = 10^{-9}$). These effects were significant whether including the population where only one nest was sampled or not.

Indeed, the proportion of unisex nests differed greatly between species, between populations, and between years (Table 11.2; Figure 11.3). For example, for *Caretta caretta* the proportion of unisex nests varied from 0.07 in Turkey to 0.73 in Brazil, and just within Turkey the proportion varied

TABLE 11.2
Proportion of Unisex Nests for Different Populations*

	Population	Prop. of Unisex Nests	No. of Nests
Caretta caretta	South Carolina	0.33	18
	Cape Canaveral (Florida)	0.32	121
	Natal (South Africa)	0.53	17
	Espirito Santo and Bahia (Brazil)	0.73	51
	Turkey	0.04	27
	Cyprus	0.50	2
Carettochelys insculpta	Northern Territory (Australia)	0.78	146
Chelonia mydas	Cyprus	0.85	52
	Caribbean coast (Costa Rica)	0.33	15
	Ascension Island	0.52	21
	Suriname	0.31	204
Chelydra serpentina	Mississippi River (Illinois)	0.86	14
Chrysemys picta	Wolf Howl Pond (Canada)	0.71	21
	Mississippi River (Illinois)	0.67	129
Dermochelys coriacea	Suriname	0.59	90
	Pacific coast (Costa Rica)	1.00	18
Eretmochelys imbricata	Soldier Key (Florida)	1.00	1
	Buck Island	0.96	51
Gopherus polyphemus	Cape Canaveral (Florida)	0.00	1
Graptemys spp.	Mississippi River (Illinois)	0.78	291
Macrochelys temminckii	Apalachicola River (Florida)	0.51	47
Podocnemis expansa	Pará State (Brazil)	0.00	3
	Tamanco (Colombia)	0.11	9
Podocnemis unifilis	Guapore River (Brazil)	0.20	15
Trionyx spiniferus	Mississippi River (Illinois)	0.00	1

* *Note that Trionyx spiniferus is a species with GSD.*

from 0.00 in 2000, 2001, and 2002 to 0.17 in 1995. The variability of the proportion of unisex nests reported in the literature is in accordance with the theoretical prediction of our model. Indeed, we predicted that unisex and mixed sex strategies are evolutionarily neutral.

Most studies based their estimates of nest sex ratios on a sample of hatchlings. If the sampled nest was mixed, the probability of observing a unisex sample of hatchlings is inversely proportional to the number of hatchlings sampled. Except in the rare cases when all hatchlings from a nest were sexed, most of the studies had relatively small sample sizes and thus it was generally not possible to reject at the 5% significance level the hypothesis that a nest was actually mixed when the observed sample showed only one sex. Therefore, the proportion of mixed nests reported in the literature may be underestimated.

The overall sex ratio is biased toward females (66%) but mean sex ratios in nests were also calculated independently for each population (Table 11.3) and year (not shown). A hierarchical analysis (see previous paragraph) showed that there are significant differences between species (LRT = 330.51, 10 df, $p = 10^{-65}$), populations of the same species (LRT = 728.44, 17 df, $p = 10^{-153}$) and years within a population (LRT = 875.50, 37 df, $p = 10^{-172}$). Only a few populations showed a higher proportion of males, and the single population where only males were reported in the nest is

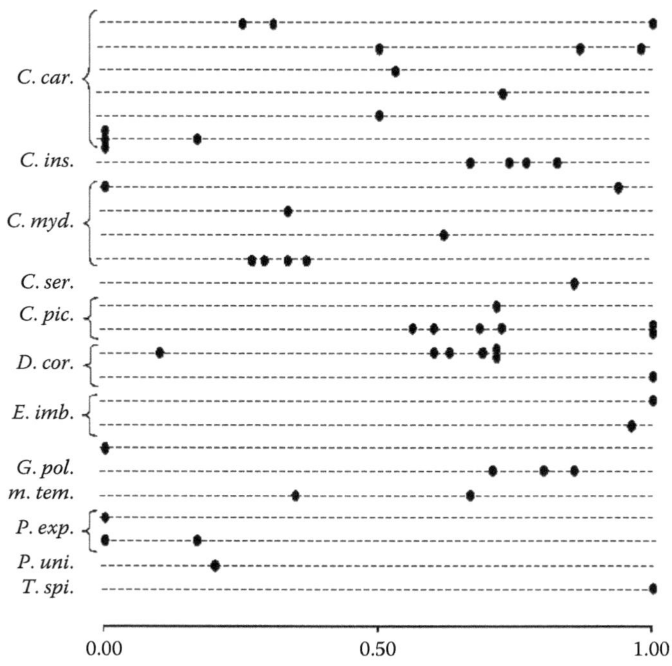

FIGURE 11.3 Proportion of unisex nests for all sampled populations of all species. Each line represents a population. Each dot on the line represents the proportion of unisex nests for a given year. Note that *Trionyx spinifera* is a species with GSD.

atypical (Dalrymple et al., 1985). More generally, studies where nest sex ratios are skewed relative to 0.5 are also those studies with the lowest number of sampled nests. Nest sex ratios can be considered as biased toward females because 18 populations (including studies where the number of studied nests was high) out of 22 showed a female biased mean sex ratio. This strong bias of sex ratios toward females is at the limit of evolutionary possibility (Charnov & Dawson, 1989) and may suggest that primary sex ratio of some populations is not at the evolutionary equilibrium. One possible cause of this general bias toward females in populations of all species could be global warming.

11.4 GLOBAL WARMING AND THE FATE OF TSD TURTLES

Here we focus on threats specific to reptile species with TSD, leaving aside more general threats concerning cold-blooded vertebrates, temperate biodiversity, or aquatic animals. Present scenarios of greenhouse gas emissions predict a doubling of atmospheric CO_2 to be reached between years 2050 and 2100 (IPCC, 2001). According to simulations, this should be followed by a 2 to 11°C increase in global temperatures (Stainforth et al., 2005). This warming is expected to have a profound impact on the offspring sex ratio of turtle populations, as higher temperatures could impede the production of males (Janzen, 1994a). Reptiles with TSD have already been confronted to major climatic deteriorations (Rage, 1998) and yet have survived until now, although present warming is believed to be much more sudden than past episodes. A rapid evolutionary response would be necessary for species to "catch up" to the speed of climate change. For an appropriate response to occur, there must be either heritable behavioral/physiological variation among individuals in a population or the adaptive plasticity of a phenotypic trait. The frequency-dependent selection for the production of males could then result in various outcomes, as follows.

TABLE 11.3
Mean Sex Ratios (percentage of males) and Variance among Nests and Number of Studied Nests of Populations*

	Population	Sex Ratio	Variance	No. of Nests
Caretta caretta	South Carolina	0.47	0.007	18
	Cape Canaveral (Florida)	0.09	0.0004	121
	Natal (South Africa)	0.64	0.011	17
	Espirito Santo and Bahia (Brazil)	0.34	0.088	51
	Turkey	0.33	0.002	27
	Cyprus	0.09	0.007	2
Carettochelys insculpta	Northern Territory (Australia)	0.30	0.028	6
Chelonia mydas	Cyprus	0.04	0.002	52
	Caribbean coast (Costa Rica)	0.57	0.009	15
	Ascension Island	0.19	0.004	21
	Suriname	0.34	0.001	204
Chelydra serpentina	Mississippi River (Illinois)	0.09	0.005	14
Chrysemys picta	Wolf Howl Pond (Canada)	0.13	0.002	21
	Mississippi River (Illinois)	0.46	0.001	129
Dermochelys coriacea	Suriname	0.38	0.003	90
	Pacific coast (Costa Rica)	0.00	0	18
Eretmochelys imbricata	Soldier Key (Florida)	1.00		1
Gopherus polyphemus	Cape Canaveral (Florida)	0.20		1
Graptemys spp.	Mississippi River (Illinois)	0.40	0.001	291
Macrochelys temminckii	Apalachicola River (Florida)	0.34	0.003	47
Podocnemis expansa	Pará State (Brazil)	0.03	0.001	3
	Tamanco (Colombia)	0.25	0.011	9
Podocnemis unifilis	Guapore River (Brazil)	0.58	0.007	15
Trionyx spiniferus	Mississippi River (Illinois)	0.54		1

* *Note that Trionyx spiniferus is a species with GSD.*

First, TSD could be lost in favor of GSD, thereby ensuring an equilibrated primary sex ratio. Although this transition probably occurred in the past of six turtle lineages (Janzen & Krenz, 2004), it is unlikely that many turtle species could respond in this way in a short time. The transition from TSD to GSD would imply that genetic variation for sex determination already includes individuals with GSD instead of TSD, or that TSD may mutate to GSD in the next generations. To date, only one lizard species is reported to exhibit both GSD and TSD (Shine et al., 2002); this has not been discovered in any turtle species so far.

Second, TSD could evolve by a displacement of the pivotal temperature (toward a higher value) or an extension of the transitional range of temperature (a flattening of the response to temperature, especially in higher temperatures). This mechanism would allow the production of males in the future at temperatures that are now completely feminizing. There is ample evidence for the existence of clutch effects on the sex ratio of laboratory-incubated eggs, which may be the expression of genetic or maternal effects (Ewert et al., 1994; Mrosovsky, 1988). Assuming the absence of maternal effects, two studies have found high heritable genetic variation for sex ratio in a population of map turtles (*Graptemys ouachitensis*: $h^2 = 0.82$; Bull et al., 1982a) and in a population of snapping turtles (*Chelydra serpentina*: $h^2 = 0.34$ to 0.76; Janzen, 1992). This indicates the potential for evolutionary change in response to sex ratio selection in TSD turtles. A simulation model of sex ratio evolution in *Chrysemys picta* in response to climate change parameters also favored this scenario (Morjan, 2003b).

However, environmental variability could override the expression of individual variation and thus hamper the selection on sex ratio. In the case of TSD, individual variation could be overridden if eggs in natural nests were incubated under conditions that would always give males or always give females, regardless of genetic factors. Then the reaction norm to temperature during sex determination would have little influence on the sex that is finally expressed. The effective heritability of the reaction norm could thus be very low (Bull et al., 1982a). Comparing the response to temperature between populations has revealed no consistent trend of higher pivotal temperatures in warmer climate for turtles (Bull et al., 1982a; Ewert et al., 2004; Ewert et al., 1994, 2005; Mrosovsky, 1988; Vogt & Flores-Villela, 1992) and for a TSD lizard (Doody et al., 2006). Geographic variation in nest-site choice compensating for temperature variation seems to explain this lack of a trend (Doody et al., 2006; Ewert et al., 1994, 2005; Morjan, 2003b). This has led to the proposal that adaptation to increasing temperature could evolve more easily by the evolution of nest site choice (Bulmer & Bull, 1982; Janzen & Morjan, 2001). The presence of other maternal effects, such as the possible influence of yolk hormones (Bowden et al., 2000), could also prevent the action of selection on genetic variation for the sex ratio (St. Juliana et al., 2004).

Third, as lower temperatures are usually masculinizing in turtles, females could select cooler places for nesting. Micro-environmental heterogeneity of temperatures is well documented on nesting grounds—nests under vegetation or closer to water are usually cooler, giving the females the opportunity to select a specific temperature regime for their brood (Janzen, 1994b; Kamel & Mrosovsky, 2005; Morjan, 2003b; Mrosovsky et al., 1986[a]). As noted previously, it has been observed that female turtles tend to nest in shadowed places in warmer climates (Ewert et al., 1994, 2005; Morjan, 2003b). Bulmer and Bull (1982) suggested that if nesting behavior is heritable, nest temperature could change faster than thermal response of offspring sex. Indeed, repeatability of nest site choice by individual females has been found in marine turtles (Kamel & Mrosovsky, 2004, 2005) and freshwater turtles (Janzen & Morjan, 2001). Repeatability is a prerequisite for heritability but is not enough to conclude that the nesting behavior is heritable. Furthermore, individual preferences could be overridden if micro-environmental variation of temperatures between nests was of minor importance compared to climatic variation of temperatures (within the nesting season or across years). This would result in low repeatability in nest sex ratios (Valenzuela & Janzen, 2001) and low effective heritability for nest temperatures (Morjan, 2003a). Without heritability, nest site selection could still evolve as a plastic trait. According to this hypothesis, female turtles would modify the placement of their nest depending on the thermal environment encountered in any given season. For this to occur, it would imply that females use environmental cues at the time of nesting that are correlated with temperatures during the thermosensitive period of incubation. This eventuality has not been properly evaluated, but hatchling sex ratios are often reported to be variable among years at the level of a population (Godfrey et al., 1996; Janzen, 1994b). This suggests that potential plasticity of the nesting behavior—including nest-site choice, nesting phenology, and nest depth—cannot fully compensate for climatic inter-annual variations. Alternatively, it has been proposed that the nest site may be inherited by imprinting (Julliard, 2000; Reinhold, 1998), i.e., nesting females would tend to return to the place or environment where they incubated and hatched themselves. As a result, the nest site would behave as a female-transmitted trait, which could constrain sex ratio evolution and possibly hinder the production of males (Freedberg & Wade, 2001).

Species could also evolve to nest at higher latitudes to compensate for the climatic change. This possibility is restricted to species that are able to undertake long-distance migrations, principally sea turtles. Over evolutionary time, it is likely that this mechanism has provided new nesting beaches for refuge—during ice-age episodes, for example. Sea turtles are known to exhibit fidelity to nesting beaches or groups of nesting beaches (Bjorndal et al., 1985; Carr & Carr, 1972; Gyuris & Limpus, 1988; Mortimer & Portier, 1989; Schulz, 1975) but imperfect homing may allow shifts between nesting areas and occasional colonization events (Bowen et al., 1989; Meylan et al., 1990). However, it is difficult to assess whether this change could occur fast enough in response to global warming, and the increasing human development of shorelines for recreational, urban, or industrial activities could be an obstacle to the colonization of new nesting areas by sea turtles.

Fourth, breeding and nesting phenology could evolve to compensate for global warming. Seasonal shifts in the sex ratio from male-biased to female-biased have been reported in field studies (Doody et al., 2004; Maxwell et al., 1988; Rimblot-Baly et al., 1987). Nesting earlier (or later) in the season could ensure then that incubation takes place during a time when relatively cool temperatures still allow the production of males. It is known that turtle populations from different latitudes usually differ in the date of onset of their reproductive activity (Ewert et al., 2005). There is currently no evidence for heritable variability in nesting phenology. However, plastic behavioral response could occur via heat accumulation, as suggested for temperate freshwater turtles (Bull et al., 1982b; Congdon et al., 1983; Doody et al., 2004; Iverson et al., 1997; Obbard & Brooks, 1987) and loggerhead sea turtles (Weishampel et al., 2003).

Fifth, female turtles could modify their behavior by digging deeper (cooler) nests (Glen & Mrosovsky, 2004; Valenzuela, 2001b; Wilhoft et al., 1983). Individual variation has here been reported but is supposed to result mostly from differences in body size, with bigger (older) females digging deeper nests (Iverson et al., 1997; Morjan, 2003b; Vogt & Bull, 1982). However, in *Chrysemys picta* it seems that nest depth varies between populations, independent of female size (Morjan, 2003b). No study has yet addressed if there is size-independent variation of nest depth among individuals within a population or if females can modify this behavior in response to climatic fluctuations. In any case, the size of the turtle puts a limit on the depth of the nest, so that nests could go beyond that limit only by selection for a bigger size.

11.5 UNISEX VERSUS MIXED BROODS AND THE EVOLUTION OF TSD

The question of whether brood sex ratios are mixed or unisex in natural nests is relevant to the response of TSD to climate change. Natural selection can operate on a trait depending on its level of variation and heritability in a population. Among the main traits of TSD that show or may show heritability in TSD reptiles are the sex determination of embryos by temperature and the choice of nesting site. What does the proportion of mixed sex broods in natural nests tell us about individual variation? There are various possible reasons for mixed offspring sex ratios. The thermal environment may be variable within the nest, ranging from masculinizing to feminizing temperatures. The sex of the embryo would then depend on its position within the nest. The larger the clutch, the bigger the nest chamber, so thermal conditions may differ from the bottom (usually cooler) to the top (usually warmer) and with higher diel variation of temperatures (Georges, 1992; Hanson et al., 1998; Wilhoft et al., 1983). In large clutches of sea turtles, metabolically generated heat of developing eggs is also responsible for higher temperatures in the core of the nest than at the side (Broderick et al., 2001; Godfrey et al., 1997; Maxwell et al., 1988). The heterogeneity of temperatures in the nest may thus favor the production of mixed sexes.

Alternatively, if temperatures do not vary much with the position in the nest, a mixed sex ratio may be the expression of genetic variation between embryos (Girondot et al., 1994). Indeed, each offspring inherits only one allele from its mother and one allele from its father, with multiple paternity further increasing the genetic diversity of the clutch (Pearse & Avise, 2001). At intermediate temperatures, genetic variation would result in different sexes being expressed by embryos developing at the same temperature. A mixed brood may also be expected if sex determination is not a deterministic process but rather a probabilistic process: any embryo would have a probability $p(T,$

x) of becoming male or female, depending on the temperature (T) and genetic factors (x). Under the latter hypothesis, mixed sex ratios could arise without temperature or even genetic variation but genetic factors would modify the probability of being a male or a female at a given temperature.

In any case, mixed sex ratios favor the heritability of sex determination. For individual variation in TSD to be effectively inherited, the variation must be expressed by embryos during sex determination. In the first case, when nest conditions vary from the masculinizing to feminizing temperatures, some proportion of the embryos must develop under intermediate temperatures that allow the expression of genetic variation. In the second case, when temperature does not vary within the nest, the observation of mixed sex ratios is direct or indirect evidence for the expression of genetic factors: if sex determination is a deterministic process, then mixed sex ratios can only result from the expression of genetic variation; if sex determination is a probabilistic process, then mixed sex ratios indicate either a temperature where any small genetic variation would have a high influence on sex ratios—e.g., T such that $0 < p(T, x) < p(T, x+\delta) < 1$—or the expression of high genetic variation within the clutch—e.g., (x_i, x_j) such that $p(T, x_i) \approx 0$ and $p(T, x_j) \approx 0.6$. Similarly, the fact that all clutches do not produce only females or only males at a given time favors the hypothesis that nest site choice has an influence on sex ratio, and could be effectively heritable.

Overall, prospects for TSD turtles may depend on the proportion of mixed and unisex broods detected in populations. Where unisex broods predominate, the thermal environment may severely constrain the evolution of sex ratios, with potentially grim consequences on the future of the population. This is especially true in populations already showing a large excess of all-female nests, an excess that is expected to increase as global climate continues to warm (e.g., *Caretta caretta* in Florida). Where mixed broods predominate (e.g., *Caretta caretta* in Turkey or *Chelonia mydas* in Suriname), traits like sex determination or nest site selection are more likely to evolve to equilibrate the sex ratio. Yet maternal effects may complicate the picture and make the future less predictable. In addition, the rapidity of temperature rise together with other human-related threats (poaching, incidental capture in fisheries, pollution, habitat loss) challenge the quiet rhythm of natural selection in long-lived turtles.

REFERENCES

Alho, C.J.R., Danni, T.M.S., and Padua, L.F.M., Temperature-dependent sex determination in *Podocnemis expansa* (Testudinata: Pelomedusidae), *Biotropica*, 17, 75–78, 1985.

Binckley, C.A., Spotila, J.R., Wilson, K.S., and Paladino, F.V., Sex determination and sex ratios of pacific leatherback turtles, *Dermochelys coriacea*, *Copeia*, 1998, 291–300, 1998.

Bjorndal, K.A., Carr, A., Meylan, A., and Mortimer, J.A., Reproductive-biology of the hawksbill *Eretmochelys imbricata* at Tortuguero, Costa Rica, with notes on the ecology of the species in the Caribbean, *Biol. Cons.*, 34, 353–368, 1985.

Bowden, R.M., Ewert, M.A., and Nelson, C.E., Environmental sex determination in a reptile varies seasonally and with yolk hormones, *Proc. Roy. Soc. Lond.*, 267, 1745–1749, 2000.

Bowen, B.W., Meylan, A., and Avise, J.C., An odyssey of the green sea turtle: Ascension Island revisited, *Proc. Nat. Acad. Sci. USA*, 86, 573–576, 1989.

Broderick, A.C., Godley, B.J., and Hays, G.C., Metabolic heating and the prediction of sex ratios for green turtles (*Chelonia mydas*), *Physiol. Biochem. Zool.*, 74, 161–170, 2001.

Broderick, A.C., Godley, B.J., Reece, S.E., and Downie, J.R., Incubation periods and sex ratios of green turtles: Highly female biased hatchling production in the eastern Mediterranean, *Mar. Ecol. Prog. Ser.*, 202, 273–281, 2000.

Bull, J.J., Evolution of environmental sex determination from genotypic sex determination, *Heredity*, 47, 173–184, 1981.

Bull, J.J., *Evolution of Sex Determining Mechanisms*, Menlo Park, CA: Benjamin/Cummings Publishing, 1983.

Bull, J.J., Sex ratio and nest temperature in turtles: Comparing field and laboratory data, *Ecology*, 66, 1115–1122, 1985.

Bull, J.J., and Vogt, R.C., Temperature-dependent sex determination in turtles, *Science*, 206, 1186–1188, 1979.

Bull, J.J., Vogt, R.C., and Bulmer, M.G., Heritability of sex ratio in turtles with environmental sex determination, *Evolution*, 35, 333–341, 1982a.
Bull, J.J., Vogt, R.C., and McCoy, C.J., Sex determining temperatures in turtles: A geographic comparison, *Evolution*, 36, 13–26, 1982b.
Bulmer, M.G., and Bull, J.J., Models of polygenic sex determination and sex ratio control, *Evolution*, 36, 13–26, 1982.
Carr, A., and Carr, M.H., Site fixity in the Caribbean green turtle, *Ecology*, 53, 425–429, 1972.
Chardard, D., Penrad-Mobayed, M., Chesnel, A., Pieau, C., and Dournan, C., Thermal sex reversals in amphibians, in *Temperature-Dependent Sex Determination in Vertebrates*, N. Valenzuela and V.A. Lance (eds.), Washington, DC: Smithsonian Books, 2004, 59–70.
Charnier, M., Action de la température sur la sex-ratio chez l'embryon d'Agama agama (Agamidae, Lacertilien). *C.R. Séances Soc. l'Ouest Afr.*, 160, 620–622, 1966.
Charnov, E.L., and Bull, J.J., When is sex environmentally determined?, *Nature*, 266, 828–830, 1977.
Charnov, E.L., and Dawson, T.E., Environmental sex determination with overlapping generations, *Am. Nat.*, 134, 806–816, 1989.
Congdon, J.D., Tinkle, D.W., Breitenbach, G.L., and van Loben Sels, R.C., Nesting ecology and hatching success in the turtle *Emydoidea blandingii*, *Herpetologica*, 39, 417–429, 1983.
Crain, A.D., Bolten, A.B., Bjorndal, K.A., Guillette, L.J., and Gross, T.S., Size-dependent, sex-dependent and seasonal changes in insulin-like growth factor I in the loggerhead sea turtle (*Caretta caretta*), *Gen. Compar. Endocr.*, 98, 219–226, 1995.
Dalrymple, G.H., Hampp, J.C., and Wellins, D.J., Male-biased sex ratio in a cold nest of a hawksbill sea turtle (*Eretmochelys imbricata*), *J. Herpetol.*, 19, 158–159, 1985.
De Souza, R.R., and Vogt, R.C., Incubation temperature influences sex and hatchling size in the neotropical turtle *Podocnemis unifilis*. *J. Herpetol.*, 28, 453–464, 1994.
Deeming, D.C., Prevalence of TSD in crocodilians, in *Temperature-Dependent Sex Determination in Vertebrates*, N. Valenzuela and V.A. Lance (eds.), Washington, DC: Smithsonian Books, 2004, 33–41.
Demuth, J.P., The effects of constant and fluctuating incubation temperatures on sex determination, growth, and performance in the tortoise *Gopherus polyphemus*, *Can. J. Zool.*, 79, 1609–1620, 2001.
Doody, S., Georges, A., and Young, J. E., Determinants of reproductive success and offspring sex in a turtle with environmental sex determination, *Biol. J. Linn.* Soc., 81, 1–16, 2004.
Doody, S., Guarino, E., Georges, A., Corey, B., Murray, G., and Ewert, M. A., Nest site choice compensates for climate effects on sex ratio in a lizard with environmental sex determination, *Evol. Ecol.*, 20, 307–330, 2006.
Eckert, K.L., Eckert, S.A., Adams, T.W., and Tucker, A.D., Inter-nesting migrations by leatherback sea turtles (*Dermochelys coriacea*) in the west-Indies, *Herpetologica*, 45, 190–194, 1989.
Eckrich, C.E., and Owens, D.W., Solitary versus arribada nesting in the olive ridley sea turtles (*Lepidochelys olivacea*): A test of the predator-satiation hypothesis, *Herpetologica*, 51, 349–354, 1995.
Ewert, M.A., Etchberger, C.R., and Nelson, C.E., Turtle sex-determination modes and TSD patterns, in *Temperature-Dependent Sex Determination in Vertebrates*, N. Valenzuela and V.A. Lance (eds.), Washington, DC: Smithsonian Books, 21–32, 2004.
Ewert, M.A., and Jackson, D.R., Program Final Report, 1994.
Ewert, M.A., Jackson, D.R., and Nelson, C.E., Patterns of temperature-dependent sex determination in turtles, *J. Exp. Zool.*, 270, 3–15, 1994.
Ewert, M.A., Lang, J.W., and Nelson, C.E., Geographic variation in the pattern of temperature-dependent sex determination in the American snapping turtle (*Chelydra serpentina*), *J. Zool. Lond.*, 265, 81–95, 2005.
Freedberg, S., and Wade, M.J., Cultural inheritance as a mechanism for population sex-ratio bias in reptiles, *Evolution*, 55, 1049–1055, 2001.
Georges, A., Thermal characteristics and sex determination in field nests of the pig-nosed turtle, *Carettochelys insculpta* (Chelonia: Carettochelydidae), from northern Australia, *Austral. J. Zool.*, 40, 511–521, 1992.
Girondot, M., Tucker, A.D., Rivalan, P., Godfrey, M.H., and Chevalier, J., Density-dependent nest destruction and population fluctuations of Guianan leatherback turtles, *Anim. Conserv.*, 5, 75–84, 2002.
Girondot, M., Zaborski, P., Servan, J., and Pieau, C., Genetic contribution to sex determination in turtles with environmental sex determination, *Genet. Res.*, 63, 117–127, 1994.
Glen, F., and Mrosovsky, N., Antigua revisited: the impact of climate change on sand and nest temperatures at a hawksbill turtle (*Eretmochelys imbricata*) nesting beach, *Glob. Change Biol.*, 10, 2036–2045, 2004.
Godfrey, M.H., and Mrosovsky, N., Estimating hatchling sex ratios, in *Research and Management Techniques for the Conservation of Sea Turtles*, K.L. Eckert, K.A. Bjorndal, F.A. Abreu-Grobois, and M. Donnelly (eds.), IUCN/SSC Marine Turtle Specialist Group Publication, 1999, 136–138.

Godfrey, M.H., Barreto, R., and Mrosovsky, N., Estimating past and present sex ratios of sea turtles in Suriname, *Can. J. Zool.*, 74, 267–277, 1996.

Godfrey, M.H., Barreto, R., and Mrosovsky, N., Metabolically-generated heat of developing eggs and its potential effect on sex ratio of sea turtle hatchlings, *J. Herpetol.*, 31, 616–619, 1997.

Godley, B.J., Broderick, A.C., Glen, F., and Hays, G.C., Temperature-dependent sex determination of Ascension Island green turtles, *Mar. Ecol. Prog. Ser.*, 226, 115–124, 2002.

Gouyon, P.-H., Gliddon, C.J., and Couvet, D., The evolution of reproductive systems: A hierarchy of causes, in *Plant Population Ecology*, A.J. Davy, M.J. Hutchings, and A. Watkinson (eds.), *Brit. Ecol. Soc. Symp.* 28, Oxford, UK, 1989, 23–33.

Gyuris, E., and Limpus, C.J., The loggerhead turtle *Caretta caretta* in Queensland: Population breeding structure, *Austral. Wild. Res.*, 15, 197–209, 1988.

Hanson, P.E., Wibbels, T., and Martin, R.E., Predicted female bias in sex ratios of hatchling loggerhead sea turtles from a Florida nesting beach, *Can. J. Zool.*, 76, 1850–1861, 1998.

Harlow, P.S., TSD in lizards, in *Temperature-Dependent Sex Determination in Vertebrates*, N. Valenzuela and V.A. Lance (eds.), Washington, DC: Smithsonian Books, 42–52, 2004.

Hays, G.C., Adams, C.R., Mortimer, J.A., and Speakman, J.R., Inter- and intra-beach thermal variation for green turtle nests on Ascension Island, South Atlantic, *J. Mar. Biol. Assoc. UK*, 75, 405–411, 1995.

Hulin, V., and Guillon, J.-M., Female philopatry in a heterogeneous environment: Ordinary conditions leading to extraordinary ESS sex ratios, *B.M.C. Evol. Biol.*, 7, 2007.

IPCC, Past and future CO2 atmospheric concentrations, in *Climate Change 2001: Synthesis Report*, 2001.

Iverson, J.B., Higgins, H., Sirulnik, A., and Griffiths, C., Local and geographic variation in the reproductive biology of the snapping turtle (*Chelydra serpentina*), *Herpetologica*, 53, 96–117, 1997.

Janzen, F.J., Heritable variation for sex ratio under environmental sex determination in the common snapping turtle (*Chelydra serpentina*), *Genetics*, 131, 155–161, 1992.

Janzen, F.J., Climate change and temperature-dependent sex determination in reptiles, *Proc. Nat. Acad. Sci. USA*, 91, 7487–7490, 1994a.

Janzen, F.J., Vegetational cover predicts the sex ratio of hatchling turtles in natural nests, *Ecology*, 75, 1593–1599, 1994b.

Janzen, F.J., and Krenz, J.G., Phylogenetics: Which was first, TSD or GSD?, in *Temperature-Dependent Sex Determination in Vertebrates*, N. Valenzuela and V.A., Lance (eds.), Washington, DC: Smithsonian Books, 2004, 121–130.

Janzen, F.J., and Morjan, C.L., Repeatability of microenvironment-specific nesting behaviour in a turtle with environmental sex determination, *Anim. Behav.*, 62, 73–82, 2001.

Julliard, R., Sex-specific dispersal in spatially varying environments leads to habitat-dependent evolutionarily stable offspring sex ratios, *Behav. Ecol.*, 11, 421–428, 2000.

Kamel, S.J., and Mrosovsky, N., Nest site selection in leatherbacks, *Dermochelys coriacea*, individual patterns and their consequences, *Anim. Behav.*, 68, 357–366, 2004.

Kamel, S.J., and Mrosovsky, N., Repeatability of nesting preferences in the hawksbill sea turtle, *Eretmochelys imbricata*, and their fitness consequences, *Anim. Behav.*, 70, 819–828, 2005.

Kaska, Y., Downie, J.R., Tippett, R., and Furness, R.W., Natural temperature regimes for loggerhead and green turtle nests in the eastern Mediterranean, *Can. J. Zool.*, 76, 723–729, 1998.

Kaska, Y., Ilgaz, C., Ozdemir, A., Baskale, E., Turkozan, O., Baran, I., and Stachowitsch, M., Sex ratio estimations of loggerhead sea turtle hatchlings by histological examination and nest temperatures at Fethiye beach, Turkey, *Naturwissenschaft*, 93, 338–343, 2006.

Kolbe, J.J., and Janzen, F.J., Impact of nest-site selection on nest success and nest temperature in natural and disturbed habitats, *Ecology*, 83, 269–281, 2002.

Lance, V.A., and Valenzuela, N., A hormonal method to determine the sex of hatchling giant river turtles, *Podocnemis expansa*. Application to endangered species research, *Am. Zool.*, 32, 16, 1992.

Lang, J.W., and Andrews, H.V., Temperature-dependent sex determination in crocodilians, *J. Exp. Zool.* 270, 28–44, 1994.

Maxwell, J.A., Motara, M.A., and Frank, G.H., A micro-environmental study of the effect of temperature on the sex ratios of the loggerhead turtle, *Caretta caretta*, from Tongaland, Natal, *S. Afr. J. Zool.*, 23, 342–350, 1988.

Merchant Larios, H., Determining hatchling sex, in *Research and Management Techniques for the Conservation of Sea Turtles*, K.L. Eckert, K.A. Bjorndal, F.A. Abreu-Grobois, and M. Donnelly (eds.), ICUN/SSC Marine Turtle Specialist Group Publication, 1999, 130–135.

Meylan, A., Bowen, B.W., and Avise, J.C., A genetic test of the natal homing versus social facilitation models for green turtle migration, *Science*, 248, 724–727, 1990.
Morjan, C.L., How rapidly can maternal behavior affecting primary sex ratio evolve in a reptile with environmental sex determination?, *Am. Nat.*, 162, 205–219, 2003a.
Morjan, C.L., Variation in nesting patterns affecting nest temperatures in two populations of painted turtles (*Chrysemys picta*) with temperature-dependent sex determination, *Behav. Ecol. Sociobiol.*, 53, 254–261, 2003b.
Mortimer, J.A., and Portier, K.M., Reproductive homing and internesting behaviour of the green turtle (*Chelonia mydas*) at Ascension Island, South Atlantic Ocean, *Copeia*, 1989, 962–977, 1989.
Mrosovsky, N., Sex ratio bias in hatchling sea turtles from artificially incubated eggs, *Biol. Cons.*, 23, 309–314, 1982.
Mrosovsky, N., Pivotal temperatures for loggerhead turtles (*Caretta caretta*) from northern and southern nesting beaches, *Can. J. Zool.*, 66, 661–669, 1988.
Mrosovsky, N., and Godfrey, M.H., Manipulating sex ratios: Turtle speed ahead!, *Chel. Cons. Biol.*, 1, 238–240, 1995.
Mrosovsky, N., and Provancha, J., Sex ratio of loggerhead sea turtles hatching on a Florida beach, *Can. J. Zool.*, 67, 2533–2539, 1989.
Mrosovsky, N., and Provancha, J., Sex ratio of hatchling loggerhead sea turtles: Data and estimates from a 5-year study, *Can. J. Zool.*, 70, 530–538, 1992.
Mrosovsky, N., Baptistotte, C., and Godfrey, M.H., Validation of incubation duration as an index of the sex ratio of hatchling sea turtles, *Can. J. Zool.*, 77, 831–835, 1999.
Mrosovsky, N., Dutton, P.H., and Whitmore, C.P., Sex-ratios of 2 species of sea turtle nesting in Suriname, *Can. J. of Zool.*, 62, 2227–2239, 1984a.
Mrosovsky, N., Hopkins-Murphy, S.R., and Richardson, J.E., Sex ratios of sea turtles: Seasonal changes, *Science*, 225, 739–741, 1984b.
Nelson, N.J., Cree, A., Thompson, M.B., Keall, S.N., and Daugherty, C.H., TSD in tuataras, in *Temperature-Dependent Sex Determination in Vertebrates*, N. Valenzuela and V.A. Lance (eds.), Washington, DC: Smithsonian Books, 2004, 53–58.
Nunney, L., The maintenance of sex by group selection, *Evolution*, 43, 245–257, 1989.
Obbard, M.E., and Brooks, R.J., Prediction of the onset of the annual nesting season of the common snapping turtle (*Chelydra serpentina*), *Herpetologica*, 43, 324–328, 1987.
Pearse, D.E., and Avise, J.C., Turtle mating systems: Behavior, sperm storage, and genetic paternity, *J. Hered.*, 92, 206–211, 2001.
Pieau, C., and Dorizzi, M., Oestrogens and temperature-dependent sex determination in reptiles: All is in the gonads, *J. Endocrinol.*, 181, 367–377, 2004.
Pieau, C., Girondot, M., Desvages, G., Dorizzi, M., Richard-Mercier, N., and Zaborski, P., Temperature variation and sex determination in reptilia, *Exper. Med.*, 13, 516–523, 1995.
Rage, J.C., Latest cretaceous extinctions and environmental sex determination in reptiles, *Bull. Soc. Géolog. Fr.*, 169, 479–483, 1998.
Reinhold, K., Nest-site philopatry and selection for environmental sex determination, *Evol. Ecol.*, 12, 245–250, 1998.
Rimblot-Baly, F., Lescure, J., Fretey, J., and Pieau, C., Sensibilité à la température de la différenciation sexuelle chez la tortue luth, *Dermochelys coriacea* (Vandelli, 1761); application des données de l'incubation artificielle à l'étude de la sex-ratio dans la nature, *Ann. Sci. Nat. Zool.*, 8, 277–290, 1987.
Roosenburg, W.M., Maternal condition and nest site choice: An alternative for the maintenance of environmental sex determination?, *Am. Zool.*, 36, 157–168, 1996.
Schulz, J.P., Sea turtles nesting in Surinam, *Zool. Verhand.*, 143, 1–143, 1975.
Schwartzkopf, L., and Brooks, R.J., Sex determination in northern painted turtles: Effect of incubation at constant and fluctuating temperatures, *Can. J. Zool.*, 63, 2543–2547, 1985.
Shaw, R.F., and Mohler, J. D., The selective significance of the sex ratio, *Am. Nat.*, 87, 337–342, 1953.
Shine, R., Elphick, M.J., and Donnellan, S., Co-occurrence of multiple, supposedly incompatible modes of sex determination in a lizard population, *Ecol. Lett.*, 5, 486–489, 2002.
Spotila, J.R., Standora, E.A., Morreale, S.J., and Ruiz, G.J., Temperature dependent sex determination in the green turtle (*Chelonia mydas*): Effects on the sex ratio on a natural nesting beach, *Herpetologica*, 43, 74–81, 1987.

St. Juliana, J.R., Bowden, R.M., and Janzen, F.J., The impact of behavioral and physiological maternal effects on offspring sex ratio in the common snapping turtle, *Chelydra serpentina*, *Behav. Ecol. Sociobiol.*, 56, 270–278, 2004.

Stainforth, D., Aina, T., Christensen, C., Collins, M., Faull, N., Frame, D.J., Kettleborough, J.A., Knight, S., Martin, A., Murphy, J.M., Piani, C., Sexton, D., Smith, L.A., Spicer, R.A., Thorpe, A.J., and Allen, M.R., Uncertainty in predictions of the climate response to rising levels of greenhouse gases, *Nature*, 433, 403–406, 2005.

Valenzuela, N., Constant, shift, and natural temperature effects on sex determination in *Podocnemis expansa* turtles, *Ecology*, 82, 3010–3024, 2001a.

Valenzuela, N., Genetic differentiation among nesting beaches in the highly migratory giant river turtle (*Podocnemis expansa*) from Colombia, *Herpetologica*, 57, 48–57, 2001b.

Valenzuela, N., Botero, R., and Martinez, E., Field study of sex determination in *Podocnemis expansa* from colombian Amazonia, *Herpetologica*, 53, 390–398, 1997.

Valenzuela, N., and Janzen, F.J., Nest-site philopatry and the evolution of temperature-dependent sex determination, *Evol. Ecol. Res.*, 3, 779–794, 2001.

Viets, B.E., Ewert, M.A., Talent, G., and Nelson, C.E., Sex-determining mechanisms in squamate reptiles, *J. Exp. Zool.*, 270, 45–56, 1994.

Vogt, R.C., and Bull, J.J., Temperature controlled sex-determination in turtles: Ecological and behavioral aspects, *Herpetologica*, 38, 156–164, 1982.

Vogt, R.C., and Bull, J.J., Ecology of hatchling sex ratio in map turtles, *Ecology*, 65, 582–587, 1984.

Vogt, R.C., and Flores-Villela, O., Effects of incubation temperature on sex determination in a community of Neotropical freshwater turtles in southern Mexico, *Herpetologica*, 48, 265–270, 1992.

Weishampel, J., Bagley, D.A., Ehrhart, L.M., and Rodenbeck, B., Spatiotemporal patterns of annual sea turtle nesting behaviors along an East Central Florida beach, *Biol. Cons.*, 110, 295–303, 2003.

Weisrock, D.W., and Janzen, F.J., Thermal and fitness-related consequences of nest location in painted turtles (*Chrysemys picta*), *Funct. Ecol.*, 13, 94–101, 1999.

Wibbels, T., Hillis-Starr, Z.-M., and Phillips, B., Female-biased sex ratios of hatchling hawksbill sea turtles from a Caribbean nesting beach, *J. Herpetol.*, 33, 142–144, 1999.

Wilhoft, D.C., Hotaling, E., and Franks, P., Effects of temperature on sex determination in embryos of the snapping turtle, *Chelydra serpentina*, *J. Herpetol.*, 17, 38–42, 1983.

12 The Physiology and Anatomy of Anoxia Tolerance in the Freshwater Turtle Brain

Sarah L. Milton

CONTENTS

12.1 INTRODUCTION

The vertebrate nervous system is composed of the central nervous system (CNS), the brain and spinal cord, and the peripheral nervous system (PNS), which is responsible for relaying signals back and forth between the CNS and the rest of body. Both the CNS and the PNS contain two main cell types, the neurons (nerve cells) themselves and a number of supporting cells called glia. Glia in mammalian brains are about ten times as numerous as neurons and are responsible for such diverse functions as the physical and biochemical support of the neurons, axonal guidance during development, and the recycling of neurotransmitters (Nicholls, 1994). The brain itself is comprised of three major divisions: the caudal brainstem includes the medulla, midbrain, and cerebellum; the diencephalon consists of the thalamus, hypothalamus, and pituitary; and the rostral cerebral hemispheres contain the striatum, olfactory bulb, neocortex, hippocampus, and dentate gyrus (Nicholls, 1994). Communication between the neurons and between neurons and other cells depends primarily on the release of chemical messengers called neurotransmitters from the axonal bulbs of one cell (the presyanptic cell) that bind to receptors on the post-synaptic cell. Whereas more than 100 neuroactive compounds have been discovered, they are generally divided into three categories, including: Type I neurotransmitters, such as glutamate and gamma-aminobutyric acid (GABA) that may account for up to 90% of synaptic transmission; the "classical" Type II neurotransmitters, such as acetylcholine and the catecholamines like dopamine and serotonin; and the neuropeptides (Type III neurotransmitters) that serve primarily as neuromodulators rather than acting primarily at synapses (Nicholls, 1994). The transmission of electrical signals within the nervous system (which depends on ion fluctuations across cell membranes) and the re-uptake and recycling of neurotransmitters, are both energetically expensive, and this high-requisite ATP consumption makes the brains of most animals critically dependent on a constant supply of oxygen.

However, some animals have evolved to withstand extended periods without oxygen; those that can survive extended periods without oxygen (anoxia) without apparent damage are considered to be "facultative anaerobes," of which the best described are several species of freshwater turtle and a northern European fish, the Crucian carp (*Carassius carassius*). A number of animals can actually survive very low oxygen conditions (hypoxia) for extended periods of time, such as hibernating frogs (Boutilier, 2001; Ultsch et al., 2004), crayfish (McMahon, 2001), and a variety of fish such as Tilapia (Mohamed, 1981; van Ginneken et al., 1995), and many of these can go several hours without oxygen (Knickerbocker & Lutz, 2001; Cowan & Storey, 2001; Fujimori & Abe, 2002; Milton et al., 2003; Wu & Storey, 2005; Warren & Jackson, 2005). Hypoxia-tolerant animals are generally those that regularly experience low oxygen conditions, including burrowers, hibernators, and divers (Ramirez et al., 2007). For example, sea turtles spend only 3 to 6% of their time at the surface (Lutcavage & Lutz, 1997), and have been recorded during the winter remaining submerged for as long as 7 hours (Hochscheid et al., 2005) or possibly longer (Felger et al., 1976; Ogren & McVea, 1995), although most dives are shorter and the animals remain aerobic (Southwood et al., 1999). However, true facultative anaerobes exhibit truly astounding abilities to tolerate complete anoxia. The musk turtle (*Sternotherus odoratus*) can remain submerged in anoxic waters for an average of 21.6 days at 3°C, the map turtle (*Graptemys geographica*) for a mean of 45 days, and the painted turtle (*Chrysemys picta picta*) for 150 days (Ultsch, 2006). The ability of some turtles to withstand such extended periods without oxygen is not apparently due to significant variations in brain complexity or structures, but rather due to specific adaptations at the physiological and molecular level, the topic of discussion of this chapter.

Whereas there are of course differences in organization between the more phylogenetically primitive brain of reptiles and the more complex mammalian brains, the general functional similarities outweigh specific differences in anatomical structure and sophistication (Figure 12.1a). Physiological and anatomical evidence suggests that the turtle brain in particular has retained features that were probably present in the brains of the reptilian ancestors of mammals (Hall & Ebner, 1970), and histochemical applications have established a number of homologies of structures within the vertebrate brain (Parent, 1979) and have been used to examine evolutionary development (Zhu et al., 2005). Of course, there are distinct differences in complexity (and size), in particular in the development of the cortex and the connections between the cortex and other parts of the brain. In reptiles, the cortex (pallium) consists of three cellular layers that form the medial and dorsomedial complex (homologous to the mammalian hippocampus), dorsal cortex (equivalent to the mammalian isocortex), and the lateral cortex (piriform cortex) (Zhu et al., 2005; Figure 12.1b). Layers of the mammalian isocortex contain numerous groups of transmitter-specific neurons, including those for cholecystokinin-8 (CCK8), acetylcholine, substance P, GABA, and glutamate (GLUT). In turtles, neurons of the dorsal cortex are positive only for those compounds that are found in large numbers of neurons in layers V-VI in mammalian isocortex (i.e., substance P, GABA, and glutamate) (Reiner, 1991). In addition, neurons labeled for markers of CCK8, acetylcholine, and other compounds found mainly in layers II-IV of the mammalian isocortex, are absent or extremely rare in the turtle dorsal cortex (Reiner, 1991). Thus, the dorsal cortex in turtles appears to lack several of the major cell types characteristic of layers II-IV of the mammalian isocortex but possesses cell types characteristic of layers V-VI, and one major step in the evolution of the reptilian cortex into mammalian cortex must have been the addition of the types of neurons found in the granular and supragranular layers of mammalian isocortex (Reiner, 1993). The major groups of neurons containing these neurotransmitters in turtle dorsal cortex are reported to be very similar in morphology to their counterparts

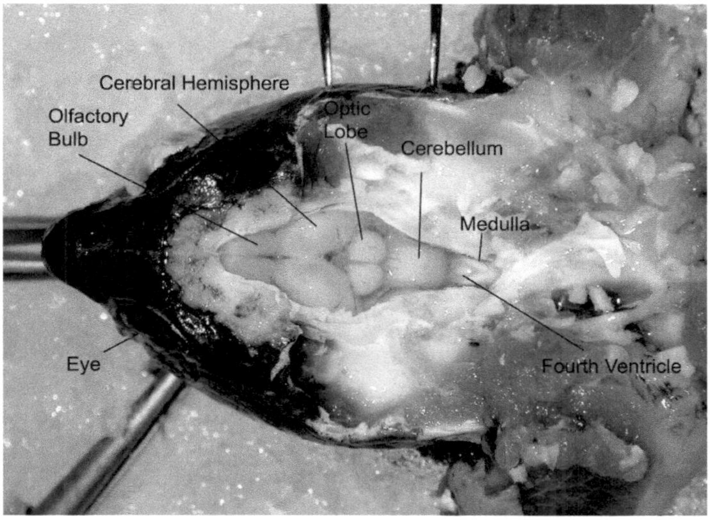

(a)

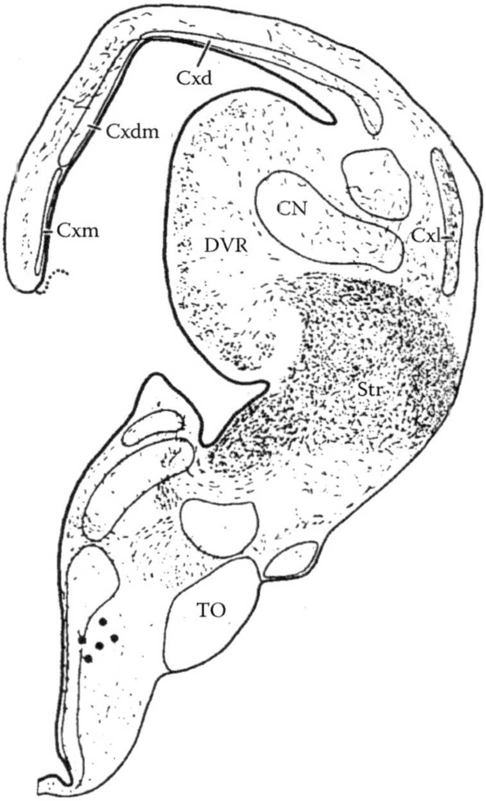

(b)

FIGURE 12.1 Anatomy of the turtle brain: (a) external anatomy, dorsal view; (b) coronal section to identify regions of interest described in this chapter. CN, core nucleus of DVR; dCx, dorsal cortex; dmCx, dorsomedial cortex; mCx, medial cortex; DVR, dorsoventricular ridge; LCx, lateral cortex; Str, striatum; TO, optic tract. Photo (a) courtesy of J. Wyneken.

in mammalian isocortex (Reiner, 1991), although they do appear to have reduced surface area and spines compared to typical mammalian cells (Schmidt-Kastner, personal communnication).

However, the links between anatomy and function in reptiles have been far less studied when compared to what is known about mammalian brains, with the exception of a few regions, most notably the olfactory system (Halpern, 1991), turtle retina and retinal connections to the basal optic nucleus (BON), and the pineal/parietal system (Underwood, 1992). The BON is part of the turtle accessory optic system, functionally equivalent to the medial terminal nucleus of the accessory optic system in mammals, and is involved in the reflex arc that contributes to retinal image stability (Rosenberg & Ariel, 1996). The visual system of the turtle has been heavily investigated because it retains functionality even after several days *in vitro*, thus making it an easily accessible and easily manipulated part of the CNS.

Other targets of particular interest are those areas of the brain thought to play significant roles in sexual differentiation. As turtles, like many other reptiles, exhibit temperature-dependent sex-determination, interactions between the gonads and developing brain have also been examined (Jeysuria & Place, 1998; Salame-Mendez et al., 1998). However, both the visual system and current paradigms of sexual differentiation have been widely described elsewhere (e.g., Kogo & Ariel, 1997; Kogo et al., 2002; Crews, 2003) and for the most part are beyond the scope of this chapter. The focus here will be on one highly significant difference between the brains of mammals and those of certain freshwater turtles (e.g., *Chrysemys picta*, *Trachemys scripta*): whereas the mammalian brain begins to die within minutes if deprived of oxygen, some freshwater turtles can withstand and recover from complete anoxia of hours to days at room temperature to as long as weeks or months at 3°C (Reese et al., 2001; Reese et al., 2002; Reese et al., 2004; Warren et al., in press). Freshwater turtles that have been investigated include both North and South American species, though the length of time that the animals can withstand and recover from anoxia varies from species to species and ranges from hours to days at warmer temperatures to weeks or months during winter hibernation. The two best studied species renowned for anoxia tolerance are the red-eared pond slider (*Trachemys scripta*) and the painted turtle (*Chrysemys picta*); other species are known to tolerate anoxia but have been less well investigated, including the map turtle (*Graptemys geographica*), snapping turtle (*Chelydra serpentina*), a South American slider (*Trachemys dorbigni*), and the musk turtle (*Sternotherus odoratus*) (Ultsch, 2006).

As turtles are ectothermic vertebrates, there is already an increased tolerance to a lack of oxygen associated with overall lower metabolic rates. In an early comparative study (Belkin, 1963), reptiles were categorized by their tolerance to anoxia: of those species examined, all the snakes, lizards, sea turtles, and crocodilians survived on average 45 minutes of anoxia, whereas the freshwater turtles as a group were generally able to withstand 12 hours of anoxia. When corrected for temperature, the brains of other ectotherms like the rainbow trout are as equally susceptible to anoxia as the mammalian brain (Nilsson et al., 1993). The remarkable tolerance of certain turtle species to anoxia is not then merely a side effect of ectothermy (Figure 12.2) but is the result of specific adaptations of the turtle brain at the physiological and molecular level that do not occur in other reptiles (Nilsson et al., 1991). These adaptations allow the turtle to enter a state of deep, reversible hypometabolism, with an overall reduction in metabolic rate to only 10 to 15% of basal (Jackson, 1968; Doll et al., 1994). With this significant reduction in metabolic rate, energy demand is reduced to match the reduced energy supplied by anaerobic glycolysis such that there is no long-term mismatch between energy supply and demand that results in the catastrophic failure of homeostasis typical of mammalian cells. However, the reduction in metabolism is not necessarily a generalized, global occurrence across the brain (nor indeed, across the whole body) but is instead a regionally specialized, tightly regulated suite of adaptations that may permit the functional downregulation of certain parts of the brain while activity is maintained in others (Table 12.1).

Of all mammalian tissues, the brain is the most energetically demanding, accounting for 20% of total body oxygen consumption, and thus neurons are considered to be the most sensitive of cells to low oxygen (hypoxia). In mammalian research, the levels of oxygen used to make an animal

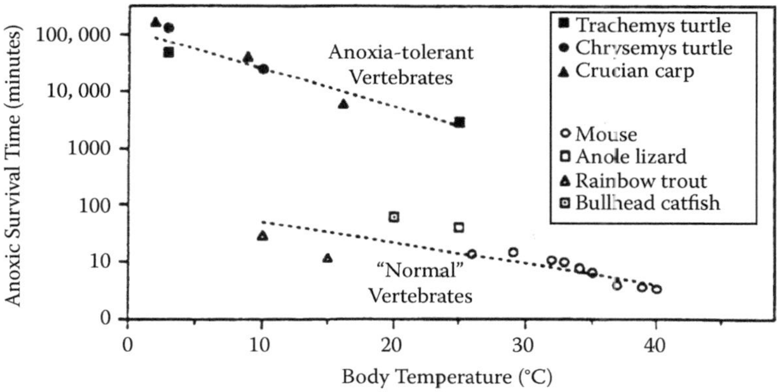

FIGURE 12.2 Anoxic survival time of vertebrates at different temperatures. Anoxia-tolerant vertebrates show a more than 1000-fold higher capacity to survive anoxia than most vertebrates, even at similar body temperatures. Reprinted from Lutz et al. (2003), with kind permission from Springer Science and Business Media.

TABLE 12.1
Summary of the Events Resulting in Anoxic/Ischemic Death in the Mammalian Brain, and Adaptations of the Anoxia-Tolerant Turtle that Allow It to Survive without Oxygen

Death Cascade in the Anoxia-Sensitive Mammalian Brain	Adaptations for Survival in the Anoxia-Tolerant Turtle Brain
⇩ Oxygen supply	Lower oxidative metabolism (⇩ cytochrome oxidase, ⇩ citrate synthase)
→ Glycolysis	High glycolytic capacity (⇧ hexokinase, ⇧ LDH)
⇩ ATP	ATP levels maintained
⇩ ATP dependent processes	Controlled hypometabolism, glycolysis
→ Ion imbalances -> membrane depolarization	⇩ Membrane ion flux (VGCC, Na^+, K-ATP, NMDA, Kv1) ⇩ Na^+/K^+ pumping; membrane potential maintained
⇧ Release of excitoxins (glutamate, dopamine) (⇧ release, ⇩ reuptake)	⇩ Glutamate release, DA/glu reuptake continue ⇧ Neuronal inhibition (AD, GABA, delta-opioids)
Glutamate stimulation of NMDA receptors → ⇧ Intracellular calcium	
→ Activation of cell death pathways	⇧ Activation of cell protection pathways (HSP)
Restoration of oxygen	
⇧ Reactive oxygen species (ROS)	⇧ Antioxidant capacity (ascorbate, GSH) Suppression of ROS production
→ Activation of cell death pathways	→ Activation of cell protection pathways (HSP, Ngb)
Limited neurogenesis	Potential for neurogenesis

hypoxic is in the range of 10 to 14% O_2 (versus the normal 21%), resulting in an arterial oxygen partial pressure of 40 to 50 mmHg versus normoxic arterial P_{O2} of 100 mmHg (Huang et al., 1994; Fokkelman et al., 2007). By contrast, normal arterial oxygen levels in the freshwater turtle *Trachemys scripta* are only 60 to 70 mmHg, and anaerobic metabolism—defined by a drop in oxygen consumption rates and increases in plasma lactate—begins at about 22 mmHg (Milton, 1994). Recent studies have shown a wide variation in the capacity of neurons to tolerate hypoxia, reflective of the function and the degree of hypoxia normally encountered (Milton & Prentice, in press). In the mammalian brain, the hippocampus, corpus striatum, and Purkinje cells are the most vulnerable to hypoxic or ischemic damage; within the hippocampus, CA1 neurons are more sensitive than cells of the CA3 region (Pulsinelli, 1985). For comparison, let us look first at the sequence of events in the mammalian brain that occur when oxygen is lacking—for example, as might happen in the event of a cardiac arrest, or more locally in the event of a stroke.

12.2 THE BRAIN IN CRISIS

As the end electron acceptor that makes mitochondrial oxidative phosphorylation possible, oxygen is critical to nearly all forms of life on earth. Severe hypoxia or cerebral ischemia forces tissues to rely on the ATP supplied by anaerobic glycolysis, which supplies only 2 moles of ATP per mole of glucose rather than the far greater amounts of ATP generated from oxidative phosphorylation. In theory, each mole of glucose supplies 36 moles of ATP by oxidative phosphorylation, though empirical measurements indicate that true ATP production is closer to 25 moles/mole glucose (Hochachka & Somero, 2002). When forced to rely on less than 1/10th of its normal energy supply, cellular ATP demand rapidly outstrips ATP supply and ATP-driven processes begin to fail. In the neurons, failure of ATP-dependent ion transporters leads to the breakdown of membrane potential; K^+ leakage out of the cells triggers membrane depolarization and the release of excitatory amino acids (EAA) such as aspartate and glutamate (Lutz et al., 2003). Elevated levels of glutamate, due to both increased release and decreased reuptake, are generally regarded as a significant factor leading to hypoxic or ischemic cell death. High extracellular glutamate stimulates NMDA receptors and permits an influx of calcium from the extracellular space (Harukuni & Bhardwaj, 2006). Increased intracellular Ca^{2+} is thought to lead to neurodegenerative events including the generation of excess levels of reactive oxygen species (ROS), lipid peroxidation, and the activation of executioner caspases that trigger apoptosis (programmed cell death). The activation of glutamate receptors (Dugan et al., 1995; Sharp et al., 2005; Kahlert et al., 2005) and elevated intracellular calcium (Sharikabad et al., 2004; Nagy et al., 2004) are associated with free-radical formation, and the direct activation of NMDA receptors results in a massive release of hydroxyl radicals (OH) (Lancelot et al., 1998; Laplanche et al., 2003). High intracellular calcium also activates proteases, lipases, and endonucleases, which in turn destroy cellular integrity (Lutz et al., 2003).

In addition to glutamate related damage, hypoxic or ischemic events cause the release of the monoamine dopamine into the extracellular space; however, in contrast to other excitotoxins, increases are seen before cellular depolarization, even under conditions of mild hypoxia (Globus et al., 1988; Huang et al., 1994) and may be a major cause of hypoxic/ischemic damage (Mitsuyo et al., 2003). Dopamine is thought to contribute to neuronal damage by increasing the release of excitatory neurotransmitters, through the production of oxygen free radicals, by inhibiting Na^+/K^+ ATPase and by uncoupling glucose metabolism from cerebral blood flow (Lutz et al., 2003). Following hypoxia/ischemia, additional neuronal damage occurs during reperfusion, thought to be caused by the post-ischemic release of oxygen radicals, the synthesis of nitric oxide, inflammation, and an imbalance between the excitatory and inhibitory neurotransmitter systems (Berger et al., 2002); this damage can continue for days after the initial hypoxic/ischemic insult.

12.3 THE ANATOMY AND PHYSIOLOGY OF ANOXIC SURVIVAL

By contrast, certain species of freshwater turtle can survive extended periods of anoxia and fully recover (Ultsch, 2006), as can the Crucian carp (*Carassius carassius*), a northern European fish that likewise winters in iced-over, anoxic, ponds (Lutz et al., 2003). Extended anoxic survival involves the downregulation of energy-demanding processes and those that are potentially damaging to cellular survival, and a concurrent upregulation of a variety of protective mechanisms that decrease energy demand and promote cell survival. For example, factors that are downregulated include the enzymes of both glycolysis and oxidative phosphorylation (Brooks & Storey, 1988; Brooks & Storey, 1989), an overall decrease in protein synthesis (Fraser et al., 2001), the release of excitotoxic neurotransmitters (Nilsson & Lutz, 1991; Milton & Lutz, 1998; Milton et al., 2002; Milton & Lutz, 2005), and the activity of ion channels, all of which decrease electrical activity (Fernandes et al., 1997) and thus lower both energy demand and overall metabolic rate.

Conversely, a number of pro-survival mechanisms are either already constitutively elevated or are clearly upregulated, including increasing glycolytic energy supply through high constitutive tissue levels of glycogen (McDougal et al., 1968; Clark & Miller, 1973), increases in plasma glucose (Milton, 1994), increased brain blood flow (Hylland et al., 1994), the release of inhibitory neuromodulators such as adenosine (Nilsson & Lutz, 1992) and GABA (Nilsson & Lutz, 1991), as well as the maintenance of (Lutz & Manuel, 1999) or increase in (Lutz & Leone-Kabler, 1995) neurotransmitter receptor function. Changes at the molecular level also occur, including increased expression or activation of heat shock proteins (Prentice et al., 2004), members of the mitogen activated kinase families (Greenway & Storey, 2000; Milton, unpublished data), and the recently discovered neuroglobin (Milton et al., 2006).

12.3.1 Energy Supply and the Enzymes of Anaerobic Glycolysis

The upregulation of energy conserving processes, coupled to the downregulation of energy expenditures, leaves the turtle in a state of profound, reversible, metabolic depression. The key to survival is to lower energy demand to meet the reduced energy supplied by anaerobic glycolysis; thus, the ability to carry out widespread glycolysis is critical. A study by Suarez et al. (1989) comparing glucose use in rats and turtles showed that after accounting for temperature differences, there was at least a sixfold difference in metabolic rates between rats and turtles, suggesting that lower rates of ATP use in turtles are inherently protective in anoxia as they are more easily supported by anaerobic glycolysis. However, lower metabolic rates and temperature differences alone are clearly not sufficient for anoxia survival, as the brains of other poikilotherms respond to anoxia in a manner similar to mammals. Remarkably, at their respective biological temperatures, the glycolytic capacity of rats and turtles is similar, as determined by the activities of hexokinase and lactate dehydrogenase (Suarez et al., 1989). This indicates not only a greater reliance on glycolysis by turtles even under normoxic conditions but a ready capacity for anaerobic glycolysis in anoxia. Key enzymes that have been studied in relation to brain metabolism include both those critical to oxidative phosphorylation, primarily cytochrome c oxidase, and those of the glycolytic pathway including hexokinase, lactate dehydrogenase, and glycogen phosphorylase. As is also seen in mammals, there is a detectable general pattern of differences in metabolism between gray and white matter in the turtle brain, such as gray/white matter differences in glucose metabolism indicated by 2-deoxyglucose use and citrate synthase (Sokoloff et al., 1977; Suarez et al., 1989). These differences result in a repeatedly observed rostral-to-caudal gradation of metabolic indicators, including the enzymes of energy production and antioxidants, that in general declines in parallel with the increase in white matter content of the more caudal brain regions (Oke et al., 1987), perhaps because the metabolic rate of the soma (and greater numbers of energy consuming dendrites and synapses) is greater than that of axons and a greater proportion of gray matter is somata.

As the terminal oxidase of the electron transport chain, the activity of cytochrome oxidase (CO) is directly linked to oxygen consumption. As neural activity is constrained, at least in part, by the availability of cellular ATP, and the activity of CO in aerobic metabolism determines the level of ATP available to the cell, then CO activity in turn can serve as a marker of metabolic capacity (Sakata et al., 2005). Hence, CO histochemistry has been used extensively to follow changes in neural metabolism after experimental manipulations; for example, significant decreases in CO activity have been shown to follow experimentally induced decreases in excitatory afferent input in mammalian cells (Hevner et al., 1992; Hevner & Wong-Riley, 1993; Wong-Riley et al., 1998a). On a finer scale, increased CO activity indicative of elevated metabolic capacity in the hypothalamus, nucleus sphericus, and septum is associated with heightened aggressiveness and territoriality in both male and female leopard geckos (Coomber et al., 1997; Sakata et al., 2002). In general, CO activity closely follows that of Na^+/K^+ ATPase, rather than being limited to a particular neurotransmitter or cell signaling system (Wong-Riley et al., 1998b; Sakata et al., 2005). When Xia et al. (1992) compared CO activity in the adult and newborn rat and the turtle brain, activity levels were found to be highest in the rostral brain areas in both adult rat and turtle compared to the brain stem and spinal cord. Although overall activity in the turtle brain was only 20 to 30% that of the rat brain, in keeping with their overall lower oxygen demand, the trend was similar, with CO activity in the turtle brain highest in the cortex, dorsal ventricular ridge, and the paleostriatum augmentum (Figure 12.3).

A similar pattern was reported by Suarez et al. (1989) for citrate synthase (CS), an enzyme of the citric acid cycle. In the rat, CS activity was highest in the cortex and striatum and lowest in the pons-medulla and hypothalamus. In the turtle, activity was highest in the cerebellum and dorsal ventricular ridge and lowest in the medulla, although activity was also low in the olfactory lobe. Overall oxidative capacity was two- to three-fold lower in turtles than rats (Suarez et al., 1989). As is the general pattern for CO, CS activity could be linked to Na^+/K^+ ATPase activity; both were highest in the cortex of the rat and lowest in the hypothalamus, whereas both were highest in the dorsal ventricular ridge (DVR) of the turtle—the reptilian structural analogue to the mammalian hippocampus—and lowest in the olfactory bulb (Suarez et al., 1989).

Alternatively, activity of the glycolytic enzyme hexokinase (HK) is actually higher in some parts of the turtle brain compared to the rat, with a very different distribution from that of CO (Xia et al., 1992). In the adult rat, HK levels are still higher in the cortex than the brainstem, as with CO, but there is comparatively less heterogeneity. Cortex levels of HK activity in the turtle brain are indistinguishable from the rat despite innate differences in metabolic demand, whereas levels in the

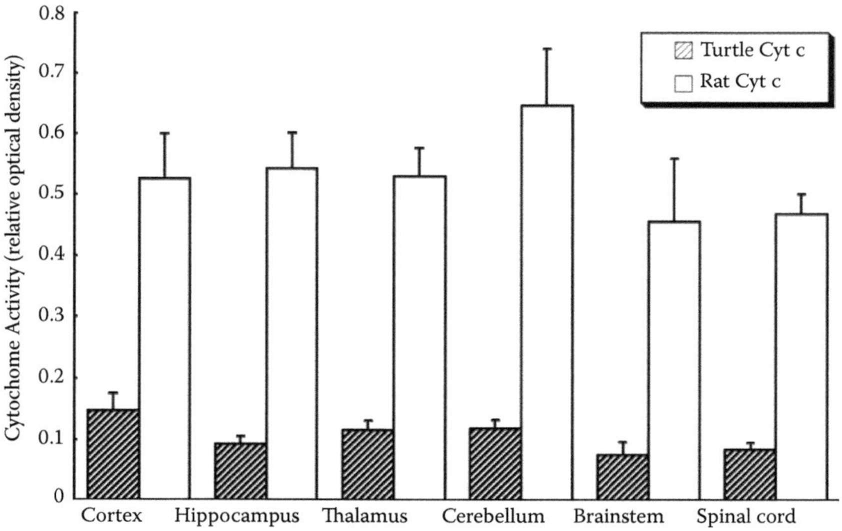

FIGURE 12.3 Cytochrome c oxidase activities in the CNS of the adult rat and the freshwater turtle *Trachemys scripta*. Values are means ± SD of relative optical densities for $n = 9$ corresponding brain sections in three animals per group. CO activity in all turtle sections is significantly lower than that seen in rat CNS. Data adapted from Xia et al. (1992).

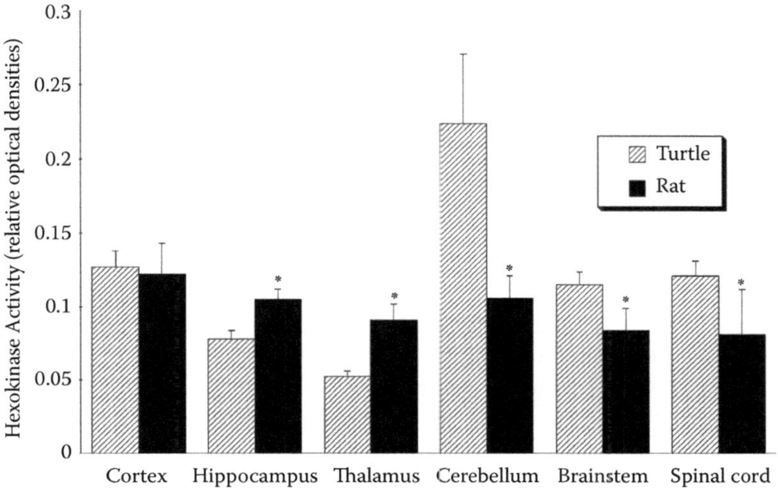

FIGURE 12.4 Hexokinase activities in the CNS of the adult rat and the freshwater turtle *Trachemys scripta*. Values are means ± SD of relative optical densities for $n = 12$ corresponding brain sections in four animals per group. The asterisk indicates a significant difference between rat and turtle. Data adapted from Xia et al. (1992).

hippocampus, thalamus, and hypothalamus are still approximately half that of the rat (Figure 12.4). Importantly, HK activity in the turtle CNS ranges from 37% to more than 100% higher in the spinal cord, brain stem, and cerebellum compared to activity levels in the adult rat (Xia et al., 1992), presumably conferring a greater glycolytic capacity to these areas and to the turtle brain overall.

Lactate dehydrogenase (LDH) activities follow the same rostral-to-caudal pattern in the rat brain as other metabolic enzymes, being highest in the cortex and striatum and lowest in the pons-medulla and hypothalamus, whereas turtle LDH activity was highest in the thalamus and optic lobe and lowest in the olfactory lobe and medulla (Suarez et al., 1989). However, when normalized for temperature turtles appear to have more HK and LDH activity overall, such that even lower rates in the medulla are still high compared to the rat (Suarez et al., 1989). High rates of glycolysis in the hindbrain could then support, for example, the continued functioning in anoxia of the Na^+/K^+ ATPase (Hylland et al., 1997), and thus continued functioning of the lower brain regions. Compared to mammals, anoxia-tolerant turtles have a high brain glycogen content (McDougal et al., 1968; Partata & Marques, 1994), with glycogen located primarily in the neurons rather than in the glia as is seen in mammals (Partata & Achaval-Elena, 1995). Anoxia also produces a marked hyperglycemia in turtles, with plasma glucose increasing from basal values of 3 mM to as high as 25 mM at room temperature (Penney, 1974; Keiver et al., 1992).

The idea that high glycolytic capacity is indicative of continued function during anoxia is supported by the work of Partata et al. (1999). They examined the distribution of both glycogen phosphorylase (GP) and CO in the South American turtle *Trachemys dorbigni* to create a metabolic map of the turtle brain, finding that GP and CO activities were neither distributed diffusely nor associated exclusively with a particular functional system. As in studies of the rat, the patterns of enzyme distribution partially overlapped with activity of both enzymes high in discrete portions of the cortex (cortex medialis, cortex dorsomedialis), in the striatum and substantia nigra, and in the primordium hippocampi (Partata et al., 1999). In the rat, GP is found primarily in glial cells (Swanson & Choi, 1993), whereas in turtles it is localized to the neurons (Partata & Achaval-Elena, 1995). Of course, as GP is part of the glycogen utilization pathway, the authors correlate GP activity with anaerobic metabolism, and thus suggest that in some parts of the brain both aerobic and anaerobic metabolism contribute to functional energy demand, whereas in other regions one pathway is more dominant. However, an alternative explanation is that it is more critical to maintain function in some regions of the brain during anoxia than in others. For example, in the striatum the release and reuptake of neurotransmitters continues during long-term anoxia (Milton & Lutz, 1998; Milton et al., 2002) despite energetic costs of reuptake transport mechanisms of as high as 1.5 ATP/molecule (Swanson & Duan, 1999). The continued release and re-uptake of these compounds implies a likely function for the neurotransmitters during anoxia, perhaps to maintain neuronal networks or prepare for the eventual upregulation of metabolism during recovery (Milton et al., 2002), but the energetic demands of transport also means a continuous energy supply is required. Thus, the high activity levels of both GP and CO would allow aerobic metabolism under basal conditions, with a change to anaerobiosis during anoxia at levels sufficient to maintain neurotransmitter homeostasis, albeit at a reduced level. Along these lines, Partata et al. (1999) note the remarkable overlap in distribution of GP and CO in the medulla, where HK activity was also found to be high (Xia et al., 1992). The authors also suggested that the CO and HK overlap is critical as CO may regulate the onset of glycolysis; lactate production is known to begin before CO is fully reduced in the anoxic turtle brain (Lutz et al., 1984; Lutz et al., 1985). By contrast, the dorsal ventricular ridge and most layers of the optic tectum expressed significant GP levels but not CO activity, indicating some continued function for the eye during anoxia (Rosenberg & Ariel, 1990; Fan et al., 1995, 1997). However, the BON showed detectable CO but not GP (Partata et al., 1999), though continued normal electrical activity has been shown to occur in the accessory optic system for several days *in vitro* (Ariel & Fan, 1993; Johnson et al., 1998). The sum of these metabolic enzymes studies then implies that the continued function of the brainstem in anoxia is critical to anoxic survival or recovery.

Alternatively, these studies did not measure enzyme activity levels during anoxia, and it has been known for many years that enzymes of the glycolytic pathway as well as those of aerobic metabolism are strongly downregulated in anoxia both in the brain and in other organs (Brooks & Storey, 1988, 1989). In turtles, as in other good facultative anaerobes, glycolytic activation is attenuated after the first hour or two of anoxia; as the rate of energy consumption gets more and more depressed, so does the rate of energy production (Kelly & Storey, 1988). Despite the widespread presence of glycogen phosphorylase reported by Partata et al. (1999), GP activity in the brain is reduced by 70% over the initial hours of anoxia (Brooks & Storey, 1988). Phosphofructokinase (PFK) shows increased inhibition as well, reflecting its role as the primary locus of glycolytic rate depression (Brooks & Storey, 1988). In most animals, hypoxia induces a sharp increase in anaerobic energy production (the Pasteur Effect) as metabolic supply attempts to keep pace with metabolic demand. The Pasteur Effect results primarily from the release of metabolic control at the PFK locus (Storey, 1988); in the turtle, covalent modification of enzymes results rather in a depression of metabolic rate (Brooks & Storey, 1988, 1989).

Energy production then continues in the anoxic turtle brain, albeit at greatly reduced rates that suffice to meet the reduced energy demands of the hypometabolic state, with some sections of the brain metabolically equipped for extended glycolysis. The distribution of glycolytic enzymes within the brain is likely to indicate critical areas that maintain function even in extended anoxia; the hypometabolism of anoxia is not simply a "shutting down" of the brain but is a selective process presumed to maximize energy savings while allowing critical processes to continue. Studies have revealed that whereas many processes are minimized, others continue in anoxia, including the continued release and reuptake of neurotransmitters, the upregulation of protective molecular pathways, and selective electrical activity (Lutz & Milton, 2004).

12.3.2 Ion Channels

As with all cells of the body, neurons maintain their ion balance across the cell membrane with a higher concentration of K^+ ions inside the cell compared to the outside, whereas Na^+ ion gradients are the reverse. These ion differences are used by the neuron to generate action potentials, and the restoration of ion gradients after an action potential—plus counteracting continuous leak channels within the membrane—consumes enormous amounts of energy. Within the brain, as much as 50 to 60% of the total normoxic ATP demand is spent by the Na^+-K^+-ATPase alone to maintain ion homeostasis (Erecsinska & Silver, 1989), thus any reduction in ion flux across the membrane is likely to have a profound effect on survival by decreasing energy demand. For example, numerous studies indicate that an influx of Ca^{2+} from the extracellular space into the cytoplasm is the key event in a variety of neurodegenerative processes; cell death results from the calcium-related generation of ROS (Sharikabad et al. 2004; Nagy et al. 2004), lipid peroxidation, and the activation of executioner caspases (Lee et al., 2005). Disruption of cellular calcium homeostasis has been proposed as a critical event in both apoptosis and necrosis, and increases in cytoplasmic free calcium levels precede apoptotic cell death under a variety of circumstances (Kruman & Mattson, 1999), including glutamate stimulation (Stout et al., 1998) and superoxide exposure.

Suarez et al. (1989) examined the abundance of voltage gated calcium channels (VGCC) in the turtle brain. They report no consistent trends in distribution between the rat and turtle brain (Figure 12.5); while VGCC density was only half that in the turtle thalamus compared to the rat, it was a third greater in the turtle cerebellum (Suarez et al., 1989). In the rat, VGCCs were most abundant in the hippocampus, striatum, and cortex and lowest in the hypothalamus and pons, thus following to a major extent the established trend in CO, CS, and glycolytic enzyme activity (Suarez et al., 1989; Partata et al., 1999). By contrast, the abundance of the VGCC in the turtle was highest in the olfactory lobe (which has low metabolic enzyme activity) and the DVR (which has high enzyme activity); VGCCs were lowest in the turtle thalamus (high LDH activity) and the medulla (low LDH and HK activity) (Suarez et al., 1989). Thus, simple differences in VGCC abundance is not the basis for differences between the mammalian and turtle brain in either anoxia tolerance or hypoxia-sensitivity, nor does it appear that VGCC distribution can be linked in the turtle to regional metabolic capacity.

The distributions of relatively few other ion channels have been described in the turtle brain (Xia & Haddad, 1991, 1993), with some additional studies restricted to very limited regions of the brain, e.g., the brainstem and spinal cord (Keifer & Carr, 2000). Xia and Haddad (1993) compared the kinetics and distribution of voltage-sensitive Na^+ channels in the adult rat and turtle brain using

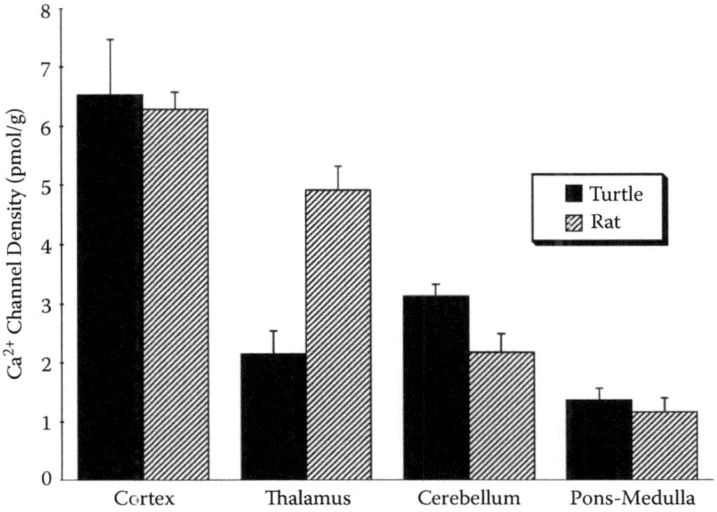

FIGURE 12.5 Voltage-dependent calcium channel densities (pmol/g) in homologous regions of the turtle and rat brains at 25°C. Data are means ± SE, $n = 5$ and 4 in the rat and turtle brains, respectively. Data adapted from Suarez et al. (1989).

autoradiography of bound saxitoxin. As with the voltage-gated calcium channels, Na^+ channels in the rat generally followed established metabolic gradients, with a very heterogeneous distribution. Saxitoxin binding in the rat brain was highest in the cortex, hippocampus, amygdala, and cerebellum and low in the brainstem and spinal cord (Xia & Haddad, 1993). Whereas saxitoxin in the turtle had similar binding properties as in the rat, overall binding density was much lower, especially in the cerebellum and rostral areas such as the cortex. Very few areas, including the medullary raphe, had similar or higher channel density in the turtle compared to the rat (Xia & Haddad, 1993). However, the differences in receptor density are unlikely to explain the approximately 100-fold difference in anoxia sensitivity between the turtle and mammalian brains, a conclusion also reached by Edwards et al. (1989). That study found that turtle synaptosomes had a voltage-gated Na^+ channel density approximately a third that of rat synaptosomes—but again, this alone is of little significance when compared to the enormous differences in anoxia tolerance (Edwards et al., 1989).

Yet another difference in ion channels between turtles and rats lies in the density of ATP-sensitive potassium channels (K-ATP). K-ATP channels were first discovered in cardiac myocytes (Noma, 1983) and have subsequently been found in many other excitable cell types (Ashcroft, 1988), including neurons (Mourre et al., 1989; Krnjevic, 1993). These channels are normally inhibited by physiological levels of ATP but open when ATP levels decrease (Ben Ari et al., 1990; Cameron & Baghdady, 1994). As the open probability of K-ATP channels is thus directly linked to intracellular energy (ATP/ADP) levels, the channels link the metabolic status of a cell to its electrical activity (Liss & Roeper, 2001). Sulfonylurea binding studies indicate the K-ATP are widely expressed throughout different regions of the brain, including pyramidal and striatal neurons, various nuclei of the hypothalamus, and GABAergic and dopaminergic neurons of the substantia nigra (Liss & Roeper, 2001).

K-ATP channels are thought to play a critical protective role in the early stages of brain ischemia in mammals (Ben Ari et al., 1990) through hyperpolarization of the membrane, by increasing brain blood flow (and thus glucose delivery) (Tomiyama et al., 1999), and decreasing the release of excitatory neurotransmitters (Tanaka et al., 1996; Wind et al., 1997). Blockade of K-ATP in the anoxic rat brain increases neuronal depolarization, thus suggesting that open K-ATP can dampen the anoxia-induced depolarization of CNS neurons (Jiang & Haddad, 1991). By contrast, K-ATP activation does not have any significant effect on K^+ flux in newborn rats or adult turtles, suggesting that these channels are not a major route of K^+ loss in more hypoxia-tolerant animals (Jiang & Haddad, 1991). Using the binding of glibenclamide, a specific inhibitor that binds to the sulfonylurea subunit of K-ATP channels, Xia and Haddad (1991) compared the binding properties and distribution of K-ATP in the rat and turtle brains with autoradiography. High- and low-affinity binding channels were revealed in both, but the rat had far higher binding densities and a more heterogeneous distribution. In the rat, there was a more than a ten-fold difference between the highest and lowest densities, with highest K-ATP levels in the substantia nigra, hippocampus, cerebellum, and a few thalamic nuclei. Intermediate levels were found in the basal ganglia, septum, thalamus, and hypoglossal nuclei, whereas densities were lowest in the cerebellum, brainstem, and spinal cord (Xia & Haddad, 1991). The cortex itself is highly variable; K-ATP are present at high densities in the hippocampal CA3 and CA4 regions and the dentate gyrus but are only intermediate in the CA1 and CA2 regions. Interestingly, the CA3 region has proven to be less sensitive to hypoxic-ischemic brain damage than the CA1 area (Pulsinelli, 1985); perhaps this is related to the protection afforded by higher densities of these ATP-sensitive channels.

In the rat brain, we again see a general rostral-to-caudal gradient, though there is less specific correlation between K-ATP channels and the enzymes of metabolism, indicating perhaps a greater association of K-ATP with specific pathways and nuclei rather than with generalized brain metabolic activity. For example, high densities in the substantia nigra could be related to the theoretical role of K-ATP channels in depressing the excessive release of excitotoxins such as dopamine in the face of an energy deficit; the substantia nigra has a high density of dopaminergic neurons that terminate in the striatum (Parent, 1979). As an aside, stimulation of these channels to increase hypoxic/ischemic neuronal survival in the mammalian brain has been an interest for nearly two decades with little result.

However, in contrast to high K-ATP channel density in the adult rat, binding densities were very low in the newborn rat and turtle brains, both more hypoxia-tolerant than the adult rat. Average densities of the K-ATP channels were 25 and 36 fmol/mg protein in the turtle and newborn rat, respectively, versus a mean 219 fmol/mg in the adult rat (Xia & Haddad, 1991). Also in sharp contrast to the adult rat was the overall homogeneity of binding density in turtles: densities in most brain regions ranged from 7 to 25 fmol/mg protein in turtles, whereas in the adult rat values ranged from 46 fmol/mg in the spinal cord to around 309 fmol/mg in the substantia nigra. Whereas densities were lower and more homogeneous in the newborn rat than the adult, this trend was even greater in the turtle brain, especially in the more caudal areas. Xia and Haddad (1991) reported densities too low to measure in the spinal cord, intermediate binding in the rostral and cerebellar regions, and values as high as 40 fmol/ng in the striatum. K-ATP densities in the turtle brain are thus about 1/10th the values of the rat brain, in keeping with their lower overall metabolic intensity; the authors conclude that K-ATP channel density then could not be the critical difference that allows for extended anoxia in turtles (Xia & Haddad, 1991).

However, more recent work has shown that even if K-ATP density is not high in the turtle brain, K-ATP channel function is of critical importance in anoxia tolerance, performing many of the same protective functions as have been reported in mammalian neurons (although apparently with more success). In mammalian cells, the opening of K-ATP channels in response to low intracellular ATP is in part protective because the K^+ outflow temporarily hyperpolarizes the cell. It is critical to realize that if ATP levels do not recover, the increased conductance will result in a more rapid efflux of K^+ and shorten the time to depolarization (Lutz et al., 2003); indeed, the K-ATP channels are thought to be a major route of K^+ efflux during anoxic depolarization in mammals (Xia & Haddad, 1991), though they appear of less importance in the depolarizing turtle brain (Pek-Scott & Lutz, 1998). In turtles, there is much evidence that membrane potential is maintained during anoxia (Lutz et al., 2003), and anoxia is associated with a decrease in K^+ efflux that can be initially blocked by K-ATP blockers (Pek-Scott & Lutz, 1998). Whereas the addition of glibenclamide blocks this anoxia-induced reduction in K^+ efflux during the first hour, the effect is diminished by 2 hours anoxia and completely disappears by 4 hours anoxia; these differences are thought to be associated with a temporary fall and subsequent recovery in cellular energy stores (Pek-Scott & Lutz, 1998). The temporary opening of K-ATP channels is protective in the turtle brain during a period of temporary energy crisis: open K-ATP are associated with the decreased release of both dopamine (Milton & Lutz, 2005) and glutamate (Milton et al., 2002) during a period when the turtle brain has perhaps less energy to spare on reuptake transport mechanisms.

However, whereas the distribution and densities of ion channels may not reveal differences between anoxia-tolerant and hypoxia-sensitive animals, or determine which parts of the brain are more or less active in anoxia, changes in ion channel activity are more revealing. Modulating the conductance of ion channels can result in significant alterations to metabolism. For example, increasing Na^+ channel activity with veratridine increases energy consumption by 20% in rat and turtle synaptosomes (Edwards et al., 1989). Conversely, blocking voltage-dependent Na^+ channels with tetrodotoxin causes a 22% decrease in the energy consumption of turtle synaptosomes (Edwards et al., 1989) and a 20% decrease in the *in situ* perfused rat brain (Xie et al., 1994). Thus, the ability to alter ion conductance could be a powerful mechanism to decrease metabolic costs by reducing the costs of ion pumping. Extensive evidence shows that ion flux through membrane channels indeed decreases in the anoxic turtle brain ("channel arrest"), including decreased K^+ flux (Pek-Scott & Lutz, 1998) and a reduction in the density of voltage-gated Na^+ channels (Perez-Pinzon et al., 1992). In the isolated turtle cerebellum, 4 hours anoxia produced a 42% decrease in the density of voltage-gated Na^+ channels (Perez-Pinzon et al., 1992).

Anoxia-induced decreases in Ca^{2+} influx through the glutamate-responsive N-methyl-D-aspartate (NMDA) receptors of the turtle cerebrocortex have also been well described (Bickler &

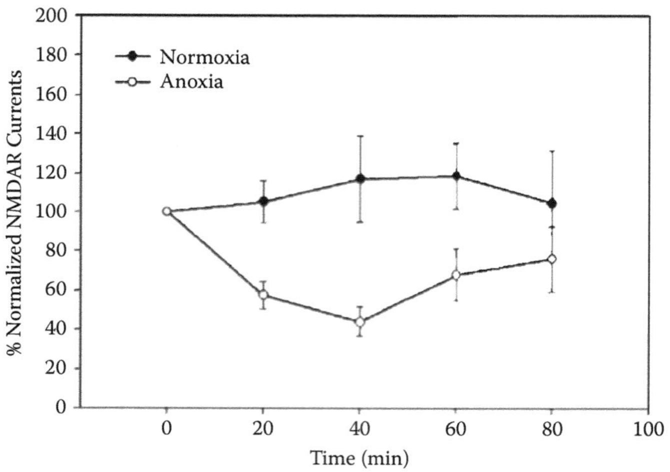

FIGURE 12.6 Normalized whole-cell NMDA currents in turtle cortical neurons. Data are mean ± SEM of 5 to 11 independent experiments. Reprinted from Shin et al. (2005), with permission from Elsevier.

Gallego, 1993; Shin et al., 2005) (Figure 12.6). Ionized calcium in the cerebrospinal fluid increases five- to six-fold during prolonged anoxia, greatly increasing the potential for Ca^{2+} influx and neurotoxicity. In cortical sheets dissected out from turtles submerged in anoxic water (2 to 4°C) for 2 hours to 6 weeks, the Ca^{2+} influx via NMDA channels decreased 30 to 40% within the first 2 hours of anoxia and remained stable over 5 further weeks of anoxia (Bickler, 1998). As NMDA receptors are inhibited by both acidity (Traynelis & Cull-Candy, 1990) and high magnesium (Ascher & Nowak, 1987), and anoxia induces both a five-fold increase in Mg^{2+} and a significant drop in pH, these changes will partly offset the long-term increase in calcium (Bickler, 1998) but do not explain the decrease in NMDA receptor activity within 2 hours of anoxic submergence.

A recent study by Prentice et al. (2003) indicated one mechanism by which the downregulation of ion channels may occur, in demonstrating that transcription of the voltage-dependent K^+ channel Kv1 is reversibly regulated by oxygen supply. Four hours of anoxia reduced Kv1 transcription to less than 20% of basal rates; mRNA levels were restored following 4 hours reoxygenation *in vivo* (Prentice et al., 2003). However, neither actin nor hypoxia-inducible factor (HIF-1) transcription were altered by 4 hours anoxia, indicating that the down-regulation of ion channel transcription is a targeted adaptation rather than a general effect of anoxia.

Of course, other mechanisms may also play a role in channel arrest, such as phosphorylation and dephosphorylation; covalent modification of proteins is one of the simplest methods to control activity, as occurs with metabolic enzymes (Brooks & Storey, 1988). Reversible protein phosphorylation has been shown to play a role in hypometabolism in a variety of enzyme systems and in different animal phyla, indicating its widespread phylogenetic conservation of the mechanism (MacDonald & Storey, 1999; Ramnanan & Storey, 2006). For example, NMDA receptor activity (the primary glutamate receptor of the CNS) is also decreased in the turtle brain during anoxia, associated with a decrease in Ca^{2+} permeability (Bickler, 1998; Bickler et al., 2000). Bickler and coworkers have recently described several mechanisms by which NMDA receptors are silenced that work at different times during anoxia in the turtle brain, with dephosphorylation occurring in the first few minutes of anoxia; over days to weeks, receptors are actually removed from the cell membrane (Bickler et al., 2000). However, none of these studies on ion channel activity examined regional differences in the brain, so it is unknown what role relative changes in channel arrest play in altering the activity of specific areas of the CNS.

The decreased ion permeability reported in the turtle brain is paralleled by both a dramatic down-regulation of the EEG (Fernandes et al., 1997) and reductions in Na^+/K^+ ATPase activity (Hylland et al., 1997). During the initial hour of transition to the anoxic hypometabolic state, there is a dramatic decrease in brain electrical activity in turtles, though the pattern of changes suggests a systematic shutdown rather than simply a global depression of nervous activity (Fernandes et al., 1997). In essence, the turtle enters a reversible coma, though the significantly depressed EEG is regularly interrupted (every 0.5 to 2 min) by short bursts (3 to 15 sec) of mixed frequency activity, lending support to the hypothesis that some activities in the brain continue even in prolonged anoxia (Lutz & Milton, 2004). Again, the EEG work by Fernandes et al. (1997) was at the whole brain level, such that the relative activity of different regions, as suggested by metabolic indicators and ion channel distribution, could not be determined. It has been noted that turtle brainstem neurons *in vitro* have less spontaneous spike activity than is seen in homologous structures in mammals *in vivo* (Rosenberg & Ariel, 1990; Fan et al., 1995, 1997), though whether this is an effect of lower body temperatures or an adaptation to conserve energy is not known; clearly, lower spike activity would conserve energy and reduce excitotoxic damage. Buck and Bickler (1998) found that anoxia induces increases in the threshold potential of isolated cortical sheets in *Chrysemys picta*, leading to energy savings by increasing the synaptic input needed to trigger an action potential ("spike arrest") and thus reducing electrical activity in the neuron.

However, depressed ion channel conductance and reduced electrical activity are reflected in alterations in Na^+/K^+ ATPase activity that are indeed different in individual regions of the turtle brain. Na^+/K^+ ATPase activity is reduced by 31% in the turtle cortex (telencephalon) after 24 hours anoxia and by 34% in the cerebellum (Hylland et al., 1997). Na^+/K^+ ATPase activity in the turtle brainstem is slightly (though not significantly) higher than in the telencephalon or cerebellum under basal conditions; activity also drops by 24 hours anoxia but the difference is not significant (Figure 12.7). Despite the lower densities of ion channels, measures of Na^+/K^+ ATPase activity that are initially higher and decrease less in the brain stem support the conclusions suggested by the distribution of the enzymes of anaerobic metabolism—that is, that the brainstem in the anoxic turtle remains active. Continued activity may reflect a role in maintaining basal physiological functions such as circulatory control, or perhaps this area is responsible for taking the turtle out of its comatose state upon reoxygenation (Hylland et al., 1997). As an aside, Na^+/K^+ ATPase activity in the brain of the anoxia-tolerant Crucian carp (*Carassius carassius*) does not decrease in anoxia (Hylland et al., 1997); there is also evidence that rates of ion flux are maintained (Johansson & Nilsson, 1995).

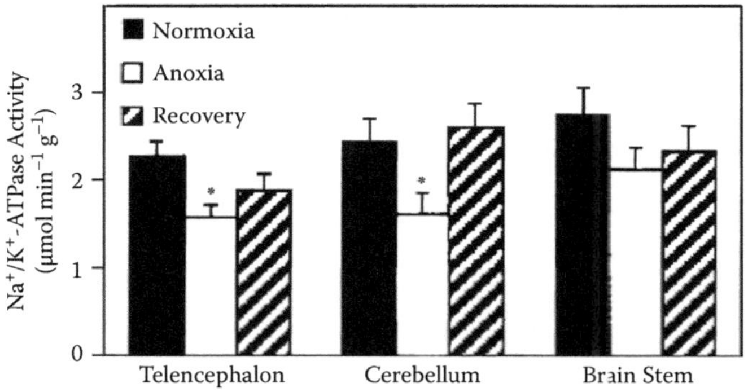

FIGURE 12.7 Na^+/K^+ ATPase activity in three parts of the normoxic, 24 anoxic, and recovered *T. scripta* brain. Data are mean ± SEM of $n = 6$ animals. The asterisk indicates $P < 0.05$. Reprinted from Hylland et al. (1997), with permission from Elsevier.

The downregulation of ion channels should not be taken as a general response to anoxia in the turtle brain. Whereas channels that would promote excitation are downregulated, there is an upregulation of receptors for inhibitory compounds that promote the hypometabolic state. One well-known inhibitory compound in the mammalian brain is gamma-aminobutyric acid (GABA). The binding of GABA to the mammalian $GABA_A$ receptor, a transmembrane chloride channel, opens the channel and triggers a transient increase in chloride flux that keeps the membrane at the resting potential or even leads to hyperpolarization (Kardos, 1993), thus countering the effects of excitatory events. $GABA_B$ receptors are metabotropic, linked through G-proteins to K^+ channels such that GABA binding opens K^+ channels and decreases neuronal excitability. This same basic response also occurs in turtles: the focal application of GABA to the somata of pyramidal cells in the turtle cortex leads to an increase in membrane conductance and hyperpolarization (Kriegstein & Connors, 1986). Autoradiographic studies have shown that the distribution of $GABA_A$ and $GABA_B$ receptors in the turtle cortex and cerebellum is similar to that seen in mammals (Schlegel & Kriegstein, 1987; Albin & Gilman, 1992). In the turtle forebrain, benzodiazepine (a GABA agonist) binding is greatest in the anterior olfactory nucleus, the lateral and dorsal cortices, and the DVR (Schlegel & Kriegstein, 1987). Not only is the distribution pattern similar but unlike many receptors, the constitutive levels of $GABA_A$ receptors are very high in the turtle brain and indeed are comparable to the rat brain. The B_{max} for the specific binding of the benzodiazepine ligand [^{3}H]flunitrazepam to the cerebral hemispheres of normoxic turtles (2404 fmol/mg protein) is in the same range as the values of homologous regions of the rat brain: 2583 fmol/mg protein in the cortex and 1886 fmol/mg in the striatum (Ninomiya et al., 1988).

In contrast to glutamate, studies indicate that binding to $GABA_A$ receptors is enhanced in some areas of the brain in anoxia. Sakurai et al. (1993) reported enhanced binding of [^{3}H]flunitrazepam at 2 and 6 hours anoxia in the *Trachemys* DVR and at 2 hours (but not 6 hours) in the striatum, suggesting an enhanced inhibitory response in the turtle forebrain during anoxia. Lutz and Leone-Kabler (1995) extended this study, showing that the affinity of specific binding (K_d) did not change in the turtle brain over 24 hours anoxia, but that B_{max} increased to 29% over control (21% over control by 12 hours anoxia). Thus, the increases seen in the Sakurai et al. (1993) study were due to an upregulation of the receptor itself rather than to changes in affinity. In concert with these alterations in $GABA_A$ receptor binding is an increase in both whole tissue GABA (Nilsson et al., 1991) and an increased release of GABA into the extracellular space (Nilsson & Lutz, 1991).

Muscarinic (acetylcholine) receptors are also widely distributed in the CNS as well as in the autonomic nervous system ganglia and seem to play a role in learning and memory in mammals and also in turtles (Petrillo et al., 1994). In the turtle forebrain, muscarinic receptors have a similar distribution to GABA receptors, with receptors most densely located in the striatum, nucleus accumbens, DVR, lateral geniculate nucleus, and anterior olfactory nucleus (Schlegel & Kriegstein, 1987). Along with any direct role of acetylcholine as a neurotransmitter, muscarinic receptors could also play a potential inhibitory role in these highly active regions of the turtle brain, as stimulation of the receptors in the mammal brain causes a striking depression of N-, P-/Q-, and L-type calcium currents (although R-type calcium currents and spikes are enhanced by muscarinic receptor agonists) (Tai et al., 2006), and muscarinic autoreceptors in rats are able to suppress neurotransmitter release through the opening of K^+ channels (Drukarch et al., 1990). However, the role of acetylcholine in the turtle brain has not been studied except as a stimulator of nitric oxide (NO) production. NO is an endogenous vasodilator in turtles; acetylcholine increases cerebral blood flow in the normoxic turtle brain but not in the anoxic brain (Hylland et al., 1996). Acetylcholine does play an important role in the turtle heart, capable of triggering right to left shunting (Hicks & Malvin, 1992) and inducing bradycardia in warm but not cold-acclimated turtles (Hicks & Farrell, 2000), and has been studied as a neurotransmitter of the turtle visual system (Vigh & Witkovsky, 2004).

12.3.3 Neurotransmitters and Neuromodulators

12.3.3.1 Inhibitory Compounds

Changes in neurotransmitter release are of course key factors in the regulation of brain activity, and this is even more true in matters of anoxia tolerance as the increased release of some neuroactive compounds (GABA, glycine, adenosine) provide protection during periods of oxygen deprivation, whereas other compounds (glutamate, dopamine) are toxic when released in excess (Lutz et al., 2003).

As noted previously, GABA increases in both whole brain tissues and in the extracellular space in the anoxic turtle. As the most abundant inhibitory neurotransmitter in the vertebrate nervous system, GABA increases in the brain even in mammals faced with hypoxia (Nilsson & Lutz, 1993), most likely because the conversion of glutamate to GABA is anaerobic, whereas its breakdown is oxygen dependent (Lutz et al., 2003). In the turtle, GABA increases 45 to 60% over 2 to 4 hours anoxia to as high as 127% over basal by 12 hours anoxia at room temperature (Hitzig et al., 1985; Lutz et al., 1985; Nilsson et al., 1990). Of course, intracellular neurotransmitter increases will have little effect on brain metabolism unless released into the extracellular space where they can impact receptors. And indeed, extracellular GABA increases from a basal level of 0.3 μM to a mean of 27 μM in the striatum of the anoxic turtle over a period of hours; the rise in extracellular GABA is thus greater than the intracellular increase (Nilsson & Lutz, 1991). However, the increase in extracellular GABA does not begin until about 100 minutes of anoxia, just about the time that adenosine, another neuroprotective compound, has begun to decline.

Adenosine (AD) results from the breakdown of the high energy phosphate compounds ATP, ADP, and AMP, and has been widely accepted as a neuroprotective compound in the hypoxic/ischemic brain. Declining ATP levels and the resulting increase in adenosine would then indicate a loss of energy stores, and in the brain this results in increased brain blood flow (Collis, 1989), glycogenolysis (Magistretti et al., 1986), and the suppression of excitatory neurotransmitter release (Stone, 1991; Prince & Stevens, 1992), all of which serve to decrease neuronal energy demand while concurrently increasing energy supply. However, AD receptors are negatively impacted by hypoxia/ischemia; ischemia results in a rapid depletion of AD A_1 receptors in the gerbil (Onodera & Kogure, 1985) and rat brain (Lee et al., 1986; Nagasawa et al., 1994), whereas as little as 2 minutes of anoxia leads to the persistent down-regulation of hippocampal receptors (Aden et al., 1994). Whereas AD may then aid hypoxic survival, its effects are clearly limited in the mammalian brain. However, in anoxia-tolerant animals AD release plays a number of critical protective roles that result in extended survival (Lutz et al., 2003). Anoxia results in a marked, though temporary, rise in extracellular AD to approximately ten-fold over basal in the turtle striatum (Nilsson & Lutz, 1992). Nilsson and Lutz (1992) proposed that the temporary fall in ATP known to occur during the initial hour of anoxia (when ATP use initially outstrips ATP supply) results in AD increases that signal insufficient energy and mediate some of the changes that then reduce metabolic rate. Extracellular AD in the anoxic turtle brain has since been shown to increase cerebral blood flow (Hylland et al., 1994), to play a critical role in channel arrest as a mediator of NMDA receptor and K^+ channel down-regulation (Buck & Bickler, 1998; Pek & Lutz, 1997), and to inhibit the release of excitatory neurotransmitters including glutamate (Milton et al., 2002) and dopamine (Milton & Lutz, 2005).

However, as with GABA, increases in extracellular levels of AD would mean little if receptor populations were down-regulated as in the rat and gerbil. As is seen with ion channels, AD A_1 receptor density in *Trachemys* is only 1/10th that of the rat brain, in keeping with their overall lower metabolic rates, and has a lower binding affinity (Lutz & Manuel, 1999). Receptor densities in the turtle forebrain (cerebral hemispheres) is similar to the hindbrain (rest of brain minus the brainstem), with approximately 100 fmol/mg protein versus 1400 fmol/mg protein in the rat fore- and hindbrain. The turtle forebrain has a higher affinity but is still less than half that of the rat forebrain (Lutz & Manuel, 1999). However, in sharp contrast to the rat or gerbil brain receptor function is maintained in the turtle forebrain over 24 hours anoxia, whereas in the hindbrain the

K_d (dissociation constant) significantly decreases by 6 hours anoxia; that higher affinity is maintained at 12 and 24 hours anoxia (Lutz & Manuel, 1999). As there is some evidence that AD continues to be released periodically throughout long-term anoxia (Lutz & Kabler, 1997), maintenance of receptor function would be critical for AD to continue its protective role. For example, the stimulation of AD A_1 and GABA receptors inhibits glutamate release during long-term anoxia (Thompson et al., 2007). The increased A_1 receptor sensitivity in the hindbrain could thus be a means to increase its sensitivity (to lower AD levels) and hence its effectiveness.

Another family of receptors that are likewise inhibitory in the brain, like $GABA_A$ receptors, have been reported to exist at far higher densities than would be expected from the general pattern of low receptor/ion channels densities in turtle versus mammalian brains. Xia and Haddad (2001) compared the affinity, density, and distribution of δ and μ-opioid receptors in the turtle and rat brain. δ-opioid receptor activation protects neurons from glutamate and hypoxia-induced injury in mammals (Zhang et al., 2000, 2001), and they hypothesized that an upregulation could be one mechanism that protects the turtle brain in anoxia. Using receptor binding assays and autoradiography, the authors report both a much higher density and lower dissociation constant of δ-opioid receptors in the turtle (Xia & Haddad, 2001). As noted in the case of metabolic enzymes, the distribution was not uniform in either the turtle or the rat brain (Figure 12.8). In the rat, the highest density of receptors was in the forebrain in the cortex and caudate putamen, with intermediate levels in the amygdala and low levels everywhere else including the hippocampus, thalamus, and hypothalamus. The now familiar rostral-caudal gradient also occurs in the turtle, with binding density much higher in the forebrain than the brainstem, and highest in the DVR. The cortex had a lower B_{max} than the DVR but was still more than 30% higher than other brain regions, and more than four times as dense as the rat cortex (Xia & Haddad, 2001). Densities were also higher than the rat brain in the hippocampus, medulla, and spinal cord.

By contrast, μ-opioid receptors were higher in the rat than in turtles, but μ- and κ-opioid receptors have little to no role in neuroprotection from glutamate or hypoxic stress (Zhang et al., 2000, 2001). Those areas, which in the mammal have the highest metabolic activity and are most sensitive to hypoxia or glutamate damage (the cortex, pyramidal cells, and striatum), in turtles also have the most protection; GABA and AD increase, binding is maintained or increases, K-ATP channels open, and there is a high density of δ-opioid receptors. Clearly, not all processes are simply shut down in the anoxic turtle brain, nor is the process homogenous across the brain; inhibitory processes are seen to be upregulated, whereas excitatory processes are inhibited, and these events occur to differing degrees in the fore- and hindbrain.

12.3.3.2 Excitatory Neurotransmitters

Because of glutamate's role in excitotoxic cell death through stimulation of the NMDA receptor, distribution of glutamatergic pathways and receptors are of particular interest in the study of anoxia tolerance. Glutamate is the dominant excitatory neurotransmitter in the vertebrate brain and spinal cord, having effects through both ionotropic and metabotropic glutamate receptors (Gardoni & Di Luca, 2006; Camacho & Massieu, 2006). Keifer and Carr (2000) examined the regional distribution of both ionotropic and metabotropic glutamate receptors in the cerebellum and brain stem of *T. scripta*. Where their interest was focused on the cerebellorubral circuit (a major descending motor system found in most vertebrates that contains the red nucleus, lateral reticular nucleus, and cerebellum), immunocytochemistry and light microscopy revealed intense staining for the $NMDAR_1$ receptor subunit in the thalamus (dorsomedial nucleus anterior), the periventricular nucleus, lateral geniculate nucleus, optic tectum, substantia nigra, in the Purkinje and granular cell layers of the cerebellum, and in all the nuclei associated with spinal nerves (Keifer & Carr, 2000). Apparently the glutamate receptor distribution in the brain stem and cerebellum of turtles is similar to that of rats (Keifer & Carr, 2000); glutamate receptors are also detectable in cortical cells and pyramidal cells (Blanton & Kriegstein, 1992), and in the optic system (Kogo et al., 2002), whereas Sakurai

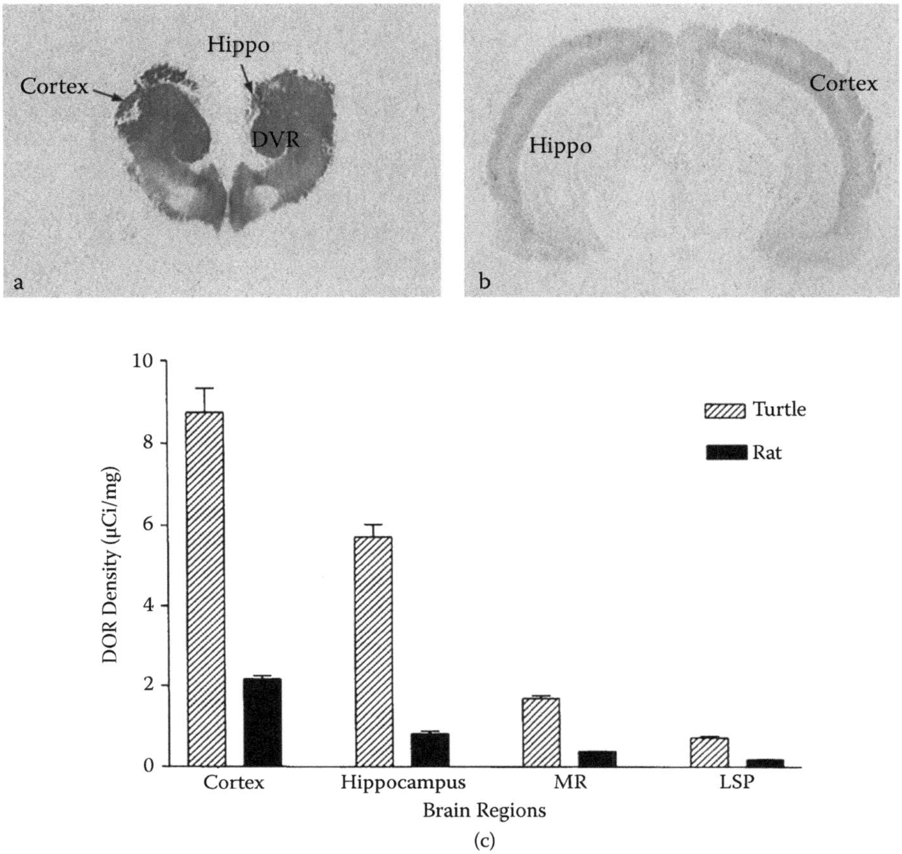

FIGURE 12.8 Delta-opioid receptor density and distribution in representative coronal sections of the (a) turtle and (b) rat forebrain, and (c) quantitative differences between the turtle and rat. Hippo, hippocampus; MR, reticular formation of medulla; LSP, lateral spinal nucleus. From Xia and Haddad (2001), with permission.

et al. (1993) reported abundant binding of the glutamate agonist [^{3}H]MK 801 in the dorsal cortex, hippocampus, and DVR of *Trachemys*, with lower levels in the striatum. As in the mammal, glutamate receptors are generally excitatory: glutamate caused excitation of the "molecular layers" of the pyramidal cells of the cortex (Kriegstein & Connors, 1986), whereas both glutamate and NMDA produce powerful activation of nonpyramidal cells in the turtle (Kriegstein & Connors, 1986; Russo & Velluti, 1992). Clearly, differences in anoxia sensitivity between turtles and other vertebrates are not due to significant differences in neuronal pathways or to differences in neurotransmitter or receptor subpopulations.

Interest among many researchers has focused on the cerebellorubral pathway because it has similar circuitry to that of the rat without the overlying influence of the motor cortex—which is not present in turtles (Keifer & Lustig, 2000), although recent work on corticohypothalamic connections has suggested that a homologous region may be embedded within the medial cortex (Zhu et al., 2005). However, in terms of anoxia tolerance the striatum is of greater interest due to its heavy innervation by dopaminergic and glutamatergic neurons. As discussed previously, excess glutamate and dopamine are thought to be responsible for the majority of hypoxic/ischemic and reoxygenation damage in the CNS. Therefore, it could be expected that the heavy glutamatergic and dopaminergic innervation of the substantia nigra and striatum of the turtle, as that in mammalian brains, would result in significant cellular damage in anoxia/reoxygenation. In contrast to mammalian cells, neurons of freshwater turtles like *T. scripta* do not lose ion balance in anoxia: extracellular K^+ levels remain low (Sick et al., 1982), intracellular calcium does not increase (Bickler, 1992), and there are no extracellular increases in excitotoxic compounds like glutamate (Nilsson & Lutz, 1991) or dopamine (Milton & Lutz, 1998). And whereas glutamate levels remain fairly constant in the brains of mammals (for the short time that they survive) (Siesjo, 1978), glutamate levels actually decline in the whole brain of the anoxic turtle (Nilsson et al., 1990); the decline may be related to the fact that neuronal glutamate synthesis is oxygen dependent. However, of more importance is the lack of increase in extracellular glutamate levels in the anoxic turtle brain (Nilsson & Lutz, 1991) when compared to the excitotoxic surge that follows hypoxia or ischemia in the mammalian brain (Lutz et al., 2003).

Long thought to be the result of simply preventing release into the extracellular space, glutamate homeostasis has recently been shown to result from a combined strategy of decreased release and continued reuptake by both glial and neuronal cells (Milton et al., 2002). By 4 hours anoxia, glutamate release is decreased by 47% relative to the normoxic animal modulated by both adenosine receptors and K-ATP channels (Milton et al., 2002); there is a further decrease as anoxia continues that is linked to GABA receptors (Thompson et al., 2007). However, active reuptake processes continue albeit at significantly reduced rates, suggesting that the animal may be attempting to retain neuronal function while simultaneously reducing energy expenditures (Milton et al., 2002). A need to maintain neuronal networks and synaptic function, or perhaps to keep the brain "on" in preparation for recovery, may underlie the continued release and reuptake of neurotransmitters, as well as the burst electrical activity apparent in the EEG (Fernandes et al., 1997), thus implying a critical role in maintaining these processes despite the energy cost. Hints of this need to maintain function also appear in the mammalian literature, where a great deal of research on potential therapeutic interventions for stroke has focused on blocking the release or ultimate effects of glutamate. Whereas laboratory studies have been promising, therapeutic interventions have failed completely in clinical trials (Hoyte et al., 2004), suggesting a critical role for at least a minimum of glutamate cycling to maintain normal brain function. One recent study of traumatic head injury in mice in fact reported significant attenuation of neural deficits and cognitive performance following glutamate receptor stimulation, rather than NMDA blockade, for 24 and 48 hours after the initial blunt trauma (Biegon et al., 2004). Dopamine has also been shown to have protective as well as pathological effects on brain cells (Rosin et al., 2005).

Like glutamate, dopamine continues to be released and taken back up by specific transport mechanisms in the anoxic turtle brain (Milton & Lutz, 1998; Milton & Lutz, 2005). As with glutamate pathways and receptors, dopaminergic structures in the reptilian and mammalian brains are generally homologous; the morphological organization of central monoamine neurons as a whole in *Chrysemys picta* resembles in many respects the well-documented arrangement of mammalian monoaminergic neurons (Parent, 1979). The numerous catecholamine-containing somata also relate to well-characterized catecholamine groups in mammals (Parent, 1979). And also like mammals, the basal ganglia structures (striatum and palladial structures) in both geckos (*Gekko gecko*) and *Trachemys scripta* display the highest immunoreactivity for proteins associated with D1-type dopamine receptors as well as tyrosine hydroxylase (TH, the key enzymes for dopamine synthesis) (Smeets et al., 2001). Biochemical studies have indicated that the ventral striatum of *Chrysemys picta* contains the highest concentrations of dopamine and serotonin in the brain, with more than four times the average whole brain dopamine concentration and approximately twice the serotonin content of whole brain extracts (Maickel et al., 1968; Welch & Welch, 1969). The areas most densely innervated by TH- immunoreactive fibers in *Trachemys scripta* include the striatum and amygdaloid complex, the substantia nigra, and the ventromedial part of the rhombencephalon (Smeets et al., 2003). Many dopaminergic neurons of the substantia nigra terminate in the striatum, which has been extensively studied in terms of neurotransmitter homeostasis during anoxic survival in turtles. The work by Smeets et al. (2003) and others that demonstrates that the distribution of key dopaminergic pathways in the basal ganglia of reptiles then largely resembles that of other amniotes, again emphasizing that differences in anoxia tolerance are not due to significant underlying differences between mammalian and turtle brains in terms of structure or function but instead are due to specific adaptations to common pathways that permit anoxic survival.

Dopamine (DA), like glutamate, is released in excess in the mammalian brain when oxygen deprivation results in decreased energy supplies. However, even mild hypoxia in the mammalian brain significantly increases extracellular levels of dopamine (Huang et al. 1994), which is readily oxidized to form ROS (Obata, 2002); dopamine also contributes to neuronal damage (Mitsuyo et al., 2003) by modulating the release of excitatory amino acids (especially glutamate) by inhibiting Na^+/K^+ ATPase and by uncoupling glucose metabolism from cerebral blood flow (Lutz et al., 2003). As with glutamate, dopamine homeostasis in anoxia is maintained by both decreasing release and through active reuptake, though these two strategies are used at different timepoints (Milton & Lutz, 2005). Cellular energy stores temporarily decrease during the initial hour of anoxia (Lutz et al., 1984; Buck et al., 1998), and this drop in ATP opens K-ATP channels for 1 to 2 hours, after which ATP levels are restored and the channels close (Pek-Scott & Lutz, 1998). As occurs with glutamate, this opening of K-ATP channels blocks the release of DA in the anoxic turtle brain such that extracellular concentrations of the monoamine remain at basal levels (Milton & Lutz, 2005). By 4 hours anoxia, K-ATP channels are closed and no longer have a physiological effect (Pek-Scott & Lutz, 1998); DA during long-term anoxia is then released into the extracellular space, with homeostasis maintained by continued active reuptake (Milton & Lutz, 1998). Whereas neurotransmitter homeostasis has not been specifically examined in other regions of the brain, it is clear from studies of both glutamate and dopamine (and EEG patterns) that the brain is not completely "shut off" even as channel arrest and decreased synaptic transmission greatly lower electrical activity and overall energy demand. Some functions and areas of the brain, such as the brainstem and striatum, maintain more function than others.

12.3.3.3 Nitric Oxide

Since the initial discovery that nitric oxide (NO) has biological activity as a vasodilator (Ignarro et al., 1987; Palmer et al., 1987), evidence has accumulated that this small, gaseous free radical plays a role in a number of physiological systems (Moncada & Higgs, 2006), including a number of aspects of neuronal communication in the brain (Bredt et al., 1990; Bohme et al., 1991), in macrophage function (Lancaster & Hibbs, 1990; Thomsen et al., 1990), platelet aggregation (Bassenge et al., 1989), and penile erection (Ignarro et al., 1990). Control of NO is at the level of biosynthesis, as it easily penetrates biological membranes and thus cannot be stored in vesicles. Nitric oxide synthase (NOS) occurs in both a constitutive and inducible forms, with the constitutive form able to generate only small amount of NO for short periods, whereas the more slowly activated inducible form can generate large amounts for longer periods. However, all isoforms catalyze the same reaction:

$$\text{L-arginine} + O_2 \xrightarrow{NOS} \text{NO} + \text{L-citrulline}$$

As the reaction consumes molecular oxygen, NO cannot be produced in the fully anoxic brain but has been shown to have both protective and pathophysiological roles in hypoxia and ischemia. For instance, the increase in intracellular calcium due to glutamate stimulation of NMDA receptors stimulates neuronal NOS; the resulting increase in NO has both positive effects (increased brain blood flow by vasodilation) and detrimental ones (glutamate release, lipid peroxidation, and protein damage) (Dawson, 1994). The greater levels of NO produced by the later-acting inducible NOS have been reported to impair mitochondrial function (Dawson, 1994) and DNA synthesis (Kwon et al., 1991; Henry et al., 1993), as well as potentially impairing glycolysis by stimulating the inactivation of glyceraldehydes-3-phosphate dehydrogenase (Zhang & Snyder, 1992). Alternatively, NO has also been shown to play a role in preconditioning in the brain (Schroter et al., 2005; Yamada et al., 2006; Yuan et al., 2006) and would thus be protective.

NO is detectable in many areas of the reptilian brain and spinal cord. Bruning et al. (1994) investigated the distribution of NOS in the *Trachemys* brain using NADPH-diaphorase histochemistry. In the forebrain, neurons in the olfactory tubercle, the basal ganglia complex, the basal amygdaloid nucleus, suprapeduncular nucleus, and the posterior hypothalamic area stain intensely for NOS. Tracts were observed connecting the basal amygdaloid nucleus with the hypothalamus (corresponding to the stria terminalis), whereas tectothalamic and thalamotectal fibers run along the ventromedial edge of the optic tract to cross in the supraoptic decussation. Strongly NOS-positive neurons are present in the substantia nigra and in several nuclei of the midbrain, including the optic tectum. In the cerebellum, strong staining was confined to bundles of afferent fibers running in the lower molecular and in the Purkinje cell layer that appear to include ascending projections from the dorsal funicular nucleus or the spinal cord. NOS-positive cells were also found in various nuclei of the caudal brainstem, including the cerebellar nuclei, in the superior vestibular nucleus, and in the reticular nuclei (Bruning et al., 1994).

With such a widespread yet distinct distribution of NOS, it is likely that NO also acts as a messenger molecule in different parts of the reptilian brain, with the pattern of expression appearing to have evolved in certain areas before the radiation of present mammalian, avian, and reptilian species (Bruning et al., 1994). However, the only role of NO in the turtle brain determined so far is as a vasodilator. NO appears to be an endogenous vasodilator in both the turtle (Hylland et al., 1996) and the Crucian carp (Hylland & Nilsson, 1995); NOS inhibitors are able to block the increase in cerebral blood flow induced by acetylcholine during normoxia. However, NOS inhibitors have no effect on the anoxic increases in blood flow in either animal, which are induced instead by increases in adenosine (Hylland et al., 1994). As the synthesis of NO requires oxygen, the lack of NO effects in anoxia is not surprising; however, the NO donor sodium nitroprusside (SNP) also loses its ability to stimulate cerebral blood flow in the anoxic animal (Hylland et al., 1996). In light of NO's assorted pathophysiological effects during hypoxia/ischemia, Lutz et al. (2003) suggested that perhaps the neural responsiveness to NO is somehow down-regulated in good facultative anaerobes. Considering the widespread distribution of NOS in the turtle brain, the ability to avoid detrimental impacts from NO in the initial hypoxic stages prior to full anoxia or upon reperfusion would clearly increase neuronal survival, especially in those areas with known continued functioning, e.g., the visual system and striatum.

12.4 THE VISUAL SYSTEM

Strangely enough, one such place that appears to have continued activity is the eye. The optic system of the turtle is one brain region that has been specifically investigated and visual responses are described by numerous researchers, both in the basal optic nucleus (BON) of the visual accessory system (for reviews, see *Biology of the Reptilia*, v. 17, Gans & Ulinski, eds., 1992; also papers coauthored by M. Ariel) and the parietal eye/pineal system (Owens et al., 1980; Vivien-Roels et al., 1988). The BON has been extensively studied in terms of basic visual neuronal connections and the interactions between inhibitory and excitatory signaling (e.g., Fan et al., 1995; Kogo & Ariel, 1997). While that research is beyond the scope of this chapter (and worthy of entire chapters in its own right), the basic evidence at least demonstrates again that there is nothing unusual about the optic system of the turtle in terms of composition and neurotransmitter function. The retina of the turtle is set on the vertebrate plan with a complement of photoreceptors; bipolar, horizontal, and amacrine cells; and ganglion cells. Of the 40 or so chemicals thought to be neurotransmitters, turtle retinal cells have been found to respond to almost two dozen (Granda & Sisson, 1992). Developmental studies confirm that the eye is part of the CNS; it is therefore a readily accessible part of the CNS that has been particularly utilized in turtles because their "robust cells" maintain functionality *in vitro* for extended periods of time (Granda & Sisson, 1992; Ariel & Fan, 1993; Johnson et al., 1998).

For example, extracellular recordings of neurons *in vitro* show that the accessory optic system remains tuned for specific directions of visual pattern motion presented to the contalateral eye even after several days *in vitro* (Rosenberg & Ariel, 1990). Studies confirm that GABAergic neurons and both $GABA_A$ and $GABA_B$ receptors are present, with GABA also appearing to act on inhibitory presynaptic receptors (Martin & Ariel, 2005). Almost all bipolar cells of *T. scripta* are strongly immunoreactive for glutamate (Ehinger et al., 1988); there is little by way of NMDA-receptor mediated visual response, but glutamate does appear to be responsible for activity in the BON via AMPA receptors (Kogo et al., 2002). Of the catecholamines, dopamine is the most common in the retina, located in *T. scripta* almost exclusively in amacrine cells and acting via D_1 and D_2 receptors (Witkovsky et al., 1984, 1987), where it appears to drive light adaptation.

However, despite the large body of work utilizing the "robust" turtle retina, there has been virtually no work looking at the responses of the visual system to anoxia. In a study of the anoxia tolerant Crucian carp, evoked potentials in the retina and optic tectum in response to light flashes were examined. Anoxia rapidly causes an approximate 80% suppression of the light evoked potentials in both the retina and the optic tectum (Johansson et al., 1997); sound-evoked electrical activity is similarly depressed in the anoxic goldfish (Suzue et al., 1987; Fay & Ream, 1992), leaving the anoxic fish essentially deaf and blind. Curiously, glycolysis is enhanced in the anoxic carp, allowing it to maintain some degree of activity during anoxia (Lutz & Nilsson, 1997); thus, the sensory shutdown may be a way to save energy in temporarily unnecessary parts of the brain (Lutz et al., 2003). By contrast, a similar study in the turtle indicated that electrical activity in the retina is maintained during anoxia (Stenslokken et al., unpublished data) despite the fact that the turtles are mostly "comatose" in anoxia (Lutz et al., 2003). Perhaps the turtles maintain some function in the visual system in part to maintain circannual rhythms or photoperiodic responses and restore function in the spring—anecdotal evidence tells of turtles seen swimming under the ice prior to the spring thaw, and some studies indicate the occurrence of both circannual and circadian rhythms in turtles (Mahapatra et al., 1988; Mahmoud & Licht, 1997). It would be interesting to examine neurotransmitter balance in the visual system of anoxic turtles as a way to compare an "active" system to parts of the brain that are presumably inactive.

Ariel (2006) did investigate the effects of adenosine on spontaneous and evoked potentials in the basal optic nucleus of *T. scripta* to determine if AD mediated hypoxia responses in the visual system. Using an *in vitro* isolated brain preparation with the eyes attached, he carried out both extracellular and intracellular recordings of spike activity in normoxia, hypoxia, and with and without the general AD antagonist, theophylline. Hypoxia alone resulted in very small decreases in spike activity, but hypoxia in the presence of theophylline increased spike rates by 3 to 300%; increases in spike rates occurred for both spontaneous activity and that evoked by visual stimulation. A small—though not statistically significant—increase in the visual response also occurred in normoxia with theophylline (although spike activity actually decreased), indicating that AD may play some role in maintaining the low spontaneous spike rates recorded even during normoxia (Rosenberg & Ariel, 1990; Fan et al., 1995, 1997) as well as protecting brainstem viability during hypoxia without reducing visually mediated brainstem reflex control (Ariel, 2006). In all cases, BON neurons exhibited direction sensitivity without a change in the preferred direction; that is, even during brainstem hypoxia with or without AD blockade, BON cells can still function. So despite the apparent continued functioning of the visual system in the anoxic turtle, its utility in an otherwise quiescent animal is unknown; vision may play a role in detecting "spring" before water temperatures have warmed significantly, but surprisingly little information exists on the control of seasonal rhythms and hibernation/emergence in reptiles, though there have been investigations into emergence behavior and reproductive state (Underwood, 1992) as well as freeze/thaw tolerance in frogs and turtles (Storey & Storey, 1994; Storey, 2006). A few studies have examined photoperiodism and hormonal control in freshwater turtles (Ganzhorn & Licht, 1983; Mahapatra et al., 1988; Mendoco & Licht, 2005) and sea turtles (Wibbels et al., 1990), whereas soil temperature appears to be the primary determinant of hatchling and adult emergence in the spring after overwintering belowground (Tucker 1999; Spencer et al., 2001; Nagle et al., 2004). Emergence in adult freshwater turtles that hibernate for the winter underwater has yet to be investigated.

Interestingly, melatonin is also found in the retina, although it is secreted in the dark (Cardinali & Rosner, 1971) and inhibits DA release from amacrine cells (Dubocovich, 1986). In this way, melatonin is implicated in the regulation of dark and light adaptation, and is thus complementary to the role of melatonin in circadian rhythms. Daily rhythms are evident in light/dark-exposed Testudines, including *Testudo hermanni* (Hermann's tortoise) (Vivien-Roels, 1983, 1985), the green sea turtle, *Chelonia mydas* (Owens et al., 1980), and the eastern box turtle, *Terrapene carolina* (Vivien-Roels et al., 1988). The pineal organ is present in turtles, and like all reptile pineals contain photosensory cells, supporting cells, and neurons, though the number and ultrastructure of these cells varies among species (Underwood, 1992). Neurophysiological investigations confirm the morphological evidence of photoreception by the pineal, with afferent nerves that project to the pretectal and tegmental areas of the brain (Hamaski & Eder, 1977; Quay, 1979). In lizards and turtles, the pineal organs are clearly organized for heavy secretory activity, with generally well-developed Golgi complexes, large numbers of free ribosomes, and both rough and smooth endoplasmic reticulum (Quay, 1979; Collin & Oksche, 1981). Examinations of the ability of light to penetrate to the base of the brain in a variety of vertebrates have shown that longer wavelengths (700 to 750 nm) penetrate more effectively than light of shorter wavelengths (Hartwig & van Veen, 1979)—the same wavelengths that also penetrate more deeply through the water and thus could potentially influence the visual system of hibernating turtles.

Pineal and blood melatonin levels in the box turtle (*Terrapene carolina*) are affected by both light and temperature (Vivien-Roels et al., 1988), but the studies of activity patterns that have been done in reptiles have focused primarily on the circadian rhythms of daily activity rather than on circannual rhythms (Ganzhorn & Licht, 1983; Mahapatra et al., 1988; Mendoco & Licht, 2005). A study by Vivien-Roels et al. (1979) did find that the maximum concentration of melatonin and the amplitude of circadian fluctuations in the tortoise *Testudo hermanni* increased during the breeding season, and that circadian rhythms disappeared completely during winter and hibernation. Whereas in mammals melatonin receptors are largely restricted to the pineal gland and hypothalamus (Morgan et al., 1994), there is a more widespread distribution in non-mammalian vertebrates, with concentrated densities in the visual areas of the brain as well as in the pineal and hypothalamic areas (Siuciak et al., 1991; Wiechmann & Wirsig-Weichmann, 1992). Larson-Prior et al. (1996) used autoradiography to determine areas of high melatonin receptor density in *C. picta*, finding that the greatest binding was found within specific regions associated with the visual system. The pineal gland was the most heavily labeled structure of the CNS; label was also found in the supraoptic nucleus and optic tectum in addition to targets of the optic tectum, such as the tectothalamic tract and the dorsal part of the anterior DVR (Larson-Prior et al., 1996). Interestingly, there was no label found in the reticular or raphe nuclei, nor in the rhombencephalon, despite known projections from the optic tectum to the medulla; as the point of the study was to identify those areas of the brain that could serve as substrates for the integration of photic and hormonal information, it is curious that an area likely to be critical for arousal from hibernation is not apparently linked to the pineal system. Alternatively, the only non-visual structures found to bind 2-[^{125}I]iodomelatonin were the striatum, the habencular nucleus, and one of its target nuclei, the interpeduncular nucleus (Larson-Prior et al., 1996). The habenacular system is involved in the control of both autonomic and endocrine functions, and projections from it access the hypothalamus, striatum, and portions of the reticular formation involved in arousal (Herkenham & Nauta, 1979). It is of note that melatonin is secreted by parts of the brain that potentially remain active in anoxia because melatonin is also a potent antioxidant (Nogues et al., 2006) and is linked to increases in antioxidant enzymes (Esparza et al., 2005; Eskiocak et al., in press). While not examined in turtles, melatonin administration in hypoxia-tolerant goldfish did decrease lipid damage in muscles in fish subjected to hypoxia-reoxygenation stress, though it did not alter mortality related to oxidative stress (Lopez-Olmeda et al., 2006).

12.5 ANTIOXIDANTS

If the cells survive the initial anoxia/ischemia event, further damage occurs upon reoxygenation/reperfusion, as this results in a rapid transient increase in reactive oxygen species (ROS) that can damage cellular proteins, lipids, and nucleotides (Christophe & Nicolas, 2006). ROS species include superoxide radicals (O_{2-}), H_2O_2, and hydroxyl radicals (OH^-). As ROS are continuously produced as a by-product of oxidative metabolism even under basal conditions, animals are generally supplied with innate antioxidant defenses including a variety of enzymatic and non-enzymatic compounds such as catalase, glutathione-S-transferases, and the superoxide dismutases (SOD). It is only when ROS production exceeds defenses, as may occur after an ischemic/reperfusion episode, that free radical damage occurs. Increased ROS production has been linked to cell death associated with acute cerebral damage such as ischemia-reperfusion injury (Cao et al., 1988; Delbarre et al., 1992; Lei et al., 1997), as well as with long-term degenerative disorders such as Parkinson's Disease (Ebadi et al., 1996; Cohen et al., 1997; Tatton & Chalmers-Redman, 1998) and amyotrophic lateral sclerosis (ALS) (Rosen et al., 1993). ROS-induced damage occurs when inherent defense mechanisms are overwhelmed; the brain is particularly vulnerable to oxidative damage because it is rich in easily oxidized unsaturated fatty acids and iron but relatively poor in antioxidant defenses (Floyd & Carney, 1992). Treatment with various antioxidant compounds has been shown to decrease free radical accumulation and cell death both *in vivo* and *in vitro* following oxygen deprivation/reoxygenation in mammalian brains (Sheng et al., 2002; Wang et al., 2003).

Because the turtle is routinely exposed to hypoxia/reoxygenation events, it is clearly able to either prevent excess ROS formation or to counteract these highly reactive compounds with sufficient antioxidants. And indeed, measurements of lipid peroxidation damage products such as thiobarbituric acid reactive substances (TBARS) show minimal changes in turtle tissues following anoxic submergence and recovery (Willmore & Storey, 1997a). Low levels of oxidative damage are due no doubt in part to the presence in turtles of constitutively high levels of antioxidant enzymes, including catalase, superoxide dismutase (SOD), and alkyl hydroperoxide reductase (Willmore & Storey, 1997a, 1997b). Glutathione peroxidase (GPOX) and SOD activities in turtle livers are actually comparable to mammalian enzyme activity (Willmore & Storey, 1997b) despite the much lower overall metabolic rate of turtles.

Whereas the majority of studies did not examine antioxidant levels in specific regions of the brain, certain protective compounds show clear patterns of distribution that are likely related to activity levels and the potential for oxidative damage. Two of these compounds are the antioxidants ascorbate and glutathione (GSH), the most abundant low molecular mass antioxidants in the CNS; both compounds are effective scavengers of peroxyl and hydroxyl radicals, superoxide anions, and singlet oxygen (Lyrer et al., 1991). Ascorbate and GSH are generally localized in the intracellular compartment (Schenk et al., 1982; Rice & Nicholson, 1987) but are lost into the extracellular space when mammalian cells undergo anoxic depolarization (Hillered et al., 1988; Landolt et al., 1992). In mammalian cells, ascorbate is found at ten-fold higher concentrations in neurons versus glia (Rice & Russo-Menna, 1998); as ascorbate is taken up from the CSF after crossing the blood-brain barrier, expression of ascorbate transporter SVCT2 (sodium-dependent vitamin C transporter type 2) is likewise greater in neurons than glial cells (Tsukaguchi et al., 1999; Berger & Hediger, 2000). By contrast, glial cells have higher levels of GSH than are found in neurons (Slivka et al., 1987; Raps et al., 1989; Makar et al., 1994; Rice & Russo-Menna, 1998). Prolonged ischemia leads to a dramatic decrease in mammalian tissue levels of ascorbate and GSH as well as other antioxidants (Lyrer et al., 1991), and this presumably increases susceptibility to reperfusion damage. By contrast, not only do turtle cells resist anoxic depolarization, but ascorbate levels are maintained *in vitro* without an external supply even during prolonged hypoxia (Rice & Cammack, 1991), indicative of some other means of homeostasis in addition to uptake from the circulatory system via the CSF (ascorbate is synthesized by the reptilian kidney and the mammalian liver). Interestingly, it has also been suggested that ascorbate and GSH may help limit excitotoxicity from ischemic glutamate release

(Rothman & Olney, 1986), perhaps because calcium effects are through second messenger systems that may in turn act through free radicals.

Turtle brain ascorbate levels overall are several times higher than in anoxia-intolerant species, being 100% higher than levels in the rat brain and 200% above levels in guinea pig brain (Rice et al., 1995). These high ascorbate levels, like anoxia tolerance in general, are not a side-benefit of ectothermy; whereas the hypoxia-resistant box turtle and garter snake have higher ascorbate levels than the rat, if not as high as the pond turtle, the hypoxia-intolerant frog (*Xenopus laevis*) does not (Rice et al., 1995). However, both ascorbate and GSH levels do show an anterior-posterior gradient across the turtle brain (Figure 12.9), as has also been noted in the mammalian CNS (Milby et al., 1982; Oke et al., 1987; Rice et al., 1995). Highest ascorbate values are found in the olfactory bulb, cortex, and dorsal ventricular ridge. The optic tectum and cerebellum have intermediate levels, with the lowest ascorbate in the brainstem, spinal cord, and optic nerve (Rice et al., 1995; Perez-Pinzon

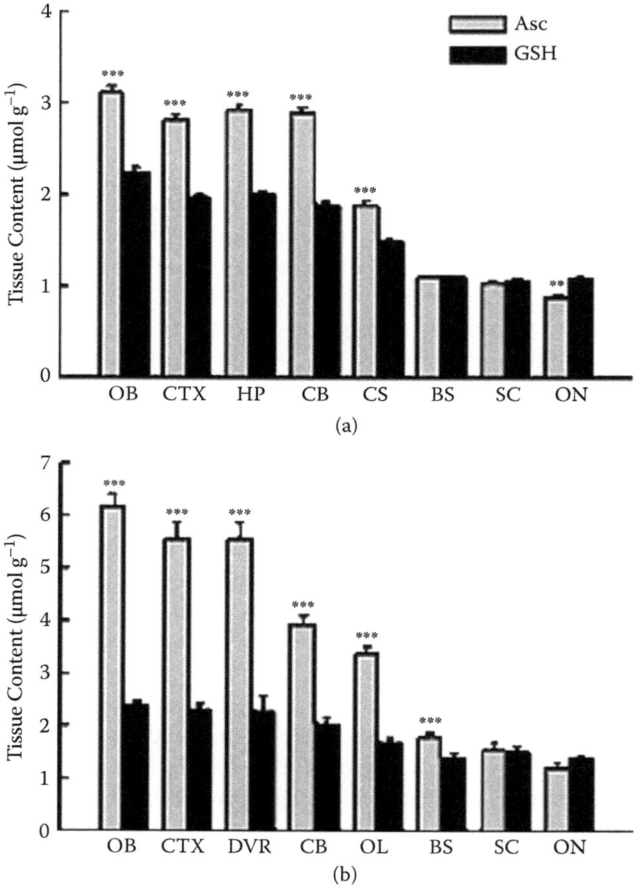

FIGURE 12.9 Regional distribution of ascorbate and GSH in the (a) rat and (b) turtle (pooled data from *T. scripta* and *C. picta*). Ascorbate levels are generally higher than GSH in both animals (significant differences: two asterisks indicates $P < 0.01$, three asterisks indicates $P < 0.001$), with ascorbate higher in the turtle than the rat (note different scales of y-axes) in most regions. OB, olfactory bulb; CTX, cortex; HP, hippocampus; CB, cerebellum; CS, superior colliculus; BS, brainstem; SC, spinal cord; ON, optic nerve; DVR, dorsal ventricular ridge; OL, optic lobe. Data are mean ± SEM from n = 5 to 31 animals. Reprinted from Rice et al. (2002), with permission from Elsevier.

& Rice, 1995). While only 10 to 20% higher in the turtle than the rat, GSH follows a similar anterior-posterior gradient, with levels highest in the olfactory bulb, cortex, and DVR (Rice et al., 1995; Perez-Pinzon & Rice, 1995). These regional patterns are maintained in both winter and summer, though overall ascorbate and GSH levels are 20 to 40% lower in winter than summer (Perez-Pinzon & Rice, 1995).

As with the enzymes of metabolism, it has been suggested that the anterior to posterior decrease in ascorbate reflects the increase in white matter content of the more caudal brain regions (Oke et al., 1987), perhaps because the metabolic rate of the soma is greater than that of axons. For example, the general pattern of high ascorbate in gray matter compared to white matter also follows gray/white matter differences in glucose metabolism, as indicated by 2-deoxyglucose utilization and citrate synthase (Sokoloff et al., 1977; Suarez et al., 1989), although this trend breaks down for more specific regional differences (Perez-Pinzon & Rice, 1995). This general pattern of ascorbate and GSH distribution also mirrors the pattern of cytochrome c oxidase (CO) in the turtle CNS.

Another protective compound and potential antioxidant with differential distribution in the brain is pituitary adenylate cyclase activating polypeptide (PACAP) (Reglodi et al., 2001). While two isoforms occur, PACAP-27 and PACAP-38, PACAP-38 is the primary isoform, found in nearly invariant form in every vertebrate so far examined, as well as in the earthworm (Reglodi et al., 2000c; Somogyvari-Vigh et al., 2000) and tunicate (Arimura, 1998). A peptide so highly conserved over species separated by 700 million years of evolution presumably is of critical importance, and PACAP is thought to have numerous actions as a hormone, neurohormone, neurotransmitters, and trophic factor (Arimura, 1998; Dedja et al., 2005). *In vitro*, PACAP has been shown to be protective against developmental apoptosis in cultured cerebellar granule cells (Vaudry et al., 2000) and protects as well against apoptotic cell death from ethanol (Vaudry et al., 2005). PACAP inhibits H_2O_2-induced apoptosis and inhibits its deleterious effects on mitochondrial membrane potential and DNA fragmentation (Vaudry et al., 2002) as well as reduces glutamate-induced neuronal cell death in pure rat cortical cultures (Mario et al., 1996). *In vivo*, PACAP is also protective under hypoxic/ischemic conditions, attenuating cell death in the CA1 region of the rat hippocampus after global ischemia (Uchida et al., 1996) and decreasing infarct size after focal cerebral ischemia (Reglodi et al., 2000a). PACAP appears to act through PAC1 receptors that lead to the activation of MAPK pathways and phosphorylation of extracellular regulated kinase (ERK) pathways. ERK activation is generally considered to be protective following ischemia (Ferrer et al., 2003; Park et al., 2004); PACAP may also suppress presumptive pro-death pathways such as jun-activated kinase (JNK) (Glazner et al., 1999; Racz et al., 2006).

In the turtle, a PACAP-related protective response in anoxia was demonstrated in the retina by Rabl et al. (2002), where the responses to light flashes of horizontal cells from the eyecup were examined over time. The response amplitudes to light of cells containing PACAP-38 in the medium were higher than control amplitudes at all anoxic time points: the amplitude of responses in control cells decreased by about one half to two thirds over 46 hours anoxia, versus an approximate one third decrease in amplitude below basal (in normoxic cells) in PACAP-enhanced cells (Rabl et al., 2002).

Compared to mammalian brain, PACAP-38 levels are extraordinarily high in the turtle (PACAP-27 is nearly undetectable), with 18- to 60-fold higher concentrations than in the rat (Reglodi et al., 2001) and 10 to 100 times higher levels than are reported for the human brain (Palkovits et al., 1995). Even relatively low concentration areas of the turtle brain (cortex and striatum) were 18- and 19-fold higher than analogous regions in the rat (Figure 12.10). Again, this is not a general phenomenon common to ectotherms; PACAP-38 levels in the frog and fish (*Anguilla anguilla*) are in the mammalian range (Yon et al., 1993; Montero et al., 1998). Reglodi et al. (2001) report very high PACAP-38 levels in the turtle *Trachemys scripta* in almost all areas of the brain examined, with the highest levels in the brainstem, hypothalamus, tectum, median pallidum (hippocampus), and spinal cord. Concentrations were lowest in the optic nerve, pituitary gland, choroids plexus, and in the retina. By contrast, other peptides without neuroprotective effects have the same or reduced levels in the turtle compared to mammalian brains (Reiner, 1991). It is interesting to note

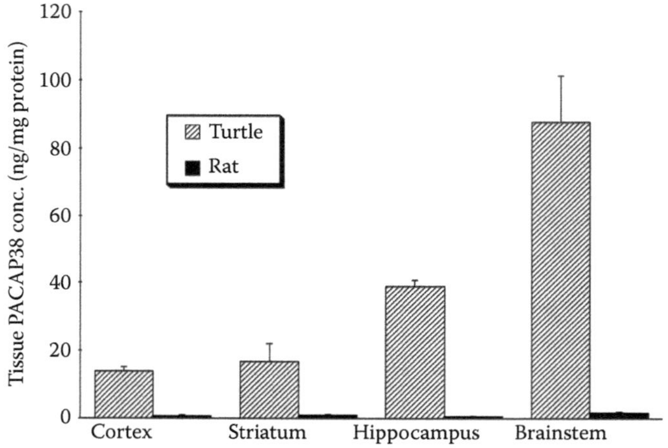

FIGURE 12.10 Immunoreactive PACAP-38 in the CNS of the turtle (*T. scripta*) and rat. Note that unlike other antioxidants and metabolic enzymes, PACAP increases on the rostral-caudal gradient. Data adapted from Reglodi et al. (2001).

that in contrast to ascorbate, PACAP levels do not show a rostral to caudal concentration gradient, and are highest in white matter, perhaps to some extent equalizing protection across the brain. In the rat, PACAP levels are high both in regions corresponding to parts of the turtle brain where the peptide is elevated, and is also found in regions highly sensitive to hypoxia (cortex, hippocampus, and striatum) (Reglodi et al., 2001). Of course, whereas PACAP has been shown to have protective effects against oxidative stress *in vitro* (Vaudry et al., 2002) and helps maintain neuronal function in isolated turtle cells (Rabl et al., 2002), the functional role of such high levels in the turtle brain has yet to be determined. High levels in white matter may serve as yet unknown purposes in addition to neuroprotection.

The apparent "stockpiling" of PACAP-38 in the turtle brain has been reported with other neuroprotective compounds. In addition to the previously mentioned high constitutive levels of multiple antioxidants, Prentice et al. (2004) reported high basal levels of heat shock proteins in the brains of *Trachemys scripta*; in the mammalian brain, Hsp72 is virtually undetectable in unstressed animals. As molecular chaperones that guide the folding of nascent proteins, and the refolding or chaperoning to sites of degradation of damaged proteins, heat shock proteins (HSPs) in the 70 kDa family in particular are thought to provide protection against a variety of stress related damage (Snoeckx et al., 2001), including cerebral ischemia (Blanco et al., 2006). Along with potential protective roles during anoxia itself, there is also evidence from mammalian studies that increased HSPs protect against ROS damage (Plumier et al., 1995). In mammalian cells, Hsp72 is nearly undetectable under basal conditions but highly inducible by stressors including ischemia and hypoxia (Murphy et al., 1999; Snoeckx et al., 2001), whereas Hsp73 is considered a constitutive protein little induced by stress (Snoeckx et al., 2001). By contrast, both Hsp73 and Hsp72 are present at relatively high levels in the normoxic turtle brain (Figure 12.11), and both are induced over 12 hours anoxia, although a differential pattern of induction suggests that the HSPs may play differing roles in short- and long-term anoxia (Prentice et al., 2004).

Neuroglobin is another compound suggested to have antioxidant properties. Only recently discovered in the mammalian brain (Burmester et al., 2000), neuroglobin has since been reported in fish, amphibians, and birds (Burmester & Hankeln, 2004), and also in the brains of *T. scripta* (Milton et al., 2006). Globins are generally oxygen transport or storage proteins, and a role as a "neuronal myoglobin" has been suggested but not proven for neuroglobin. Other potential roles for neuroglobin include acting as a terminal oxidase to regenerate NAD+ under anaerobic conditions

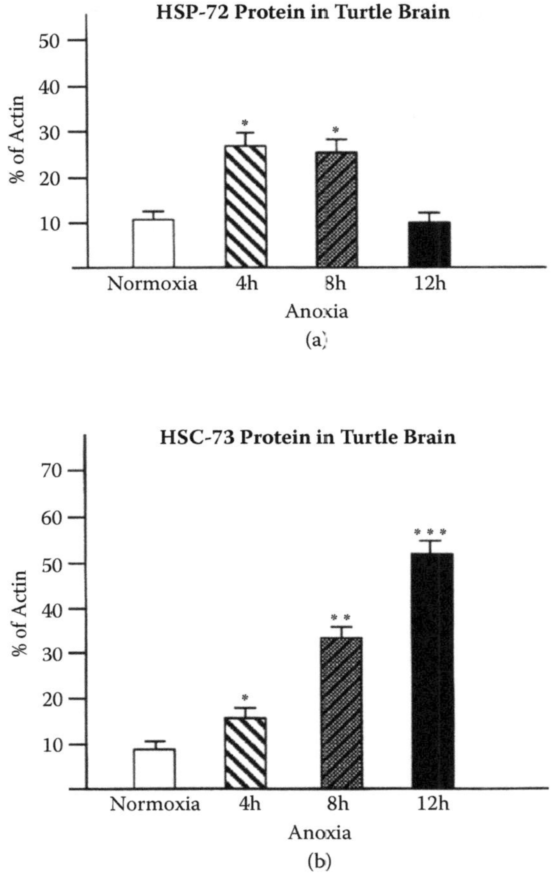

FIGURE 12.11 Heat shock proteins (a) Hsp72 and (b) Hsc73 are upregulated in anoxia in the brain of *T. scripta*, although with different patterns of expression, implying perhaps different roles in early and late anoxia. Both Hsp72 and Hsc73 are present at detectable levels even in the normoxic brain. Levels of Hsp proteins are expressed as a percentage of the Western signal intensity for actin. Turtles were maintained in normoxia or exposed to 4, 8, or 12 hours 99.99% N_2. One or two indicates $P < 0.05$, three indicates $P < 0.01$ compared with normoxic control levels, with $n = 4$ animals per timepoint. Data adapted from Prentice et al. (2004).

(Sowa et al., 1998; Kriegl et al., 2002), as a sensor to detect oxygen levels, or aiding in the detoxification of reactive oxygen species (Fordel et al., 2006). In the turtle, neuroglobin is significantly upregulated (3.5 times) over 4 hours hypoxia, with a progressive but less significant increase (to 200% of basal) of 4 hours anoxia but a greater increase upon reoxygenation following anoxia, to 4.7 times over control (Milton et al., 2006). The greater effects of hypoxia and reoxygenation on neuroglobin transcript levels compared to anoxia indicate a more likely role as an ROS scavenger than as an oxygen carrier. However, as with the HSPs, regional differences in neuroglobin expression were not investigated.

12.6 NEUROGENESIS

Even with the hypometabolism of anoxia and hypothermia of winter hibernation, many turtles still fail to survive the season, implying that months of anoxia can result in severe damage despite their tolerance (Ultsch, 2006). Research has shown that one factor limiting winter anoxic survival in freshwater turtles may be their capacity to buffer lactic acid (Jackson, 2004; Jackson et al., 2006), which can increase from basal levels of about 1 mmol/l to as high as 150 to 200 mmol/l. The large, mineralized shells of turtles such as *Chrysemys* and *Trachemys* buffer lactic acid through the release of calcium and magnesium carbonate into the extracellular fluid to neutralize acid as well as by the uptake and storage of calcium in the shell (Jackson, 1997) and long bones (Jackson et al., 2000; Jackson et al., 2006). It has been suggested that differences in the ability to buffer lactate may account for differences in survival time between those turtles that can survive "only" 25 to 50 days of anoxia at 3°C (e.g., *T. scripta* or the map turtle, *Graptemys geographica*) versus those that tolerate as many as 100 to 125 days of anoxia (e.g., *Chrysemys picta* and the snapping turtle, *Chelydra serpentina*). Even with the ability to buffer enormous acid loads, pH over weeks to months of anoxia falls to 7.1, near the limit for survival at 3°C (Ultsch & Jackson, 1982; Reese et al., 2001; Reese et al., 2004; Warren et al., in press). Of course, low pH can irreversibly damage proteins and other cellular structures and result eventually in death even in animals able to tolerate the anoxia itself.

The replacement of irreversibly damaged neurons following recovery would then allow turtles to survive repeated winters even with considerable damage, depending on the extent and location of impacted cells. Much recent research in representatives of all major vertebrate taxa supports the concept that some neurons and glial cells in certain parts of the CNS continue to be produced throughout life (Gross, 2000; Alvarez-Buylla et al., 2001). Studies concerned with the occurrence of immature proliferating cells in the CNS of vertebrates have focused primarily on birds and mammals, and show that in adult mammals at least neurogenesis appears to be limited to a few areas of the brain (Gould et al., 1999a; Peretto et al., 1999; Hastings et al., 2001). However, all reptiles examined thus far continue to add neurons at a high rate and in many regions of the adult brain (Font et al., 2001). The continued addition of neurons and glial cells in adults may explain the age-related increase in brain size (Lopez-Garcia et al., 1984); in addition, some lizards have been shown to regenerate large portions of damaged cerebral cortex (Font et al., 1991, 1997). This "adult neurogenesis" has been described in numerous areas of the brain in lizards as well as the turtle, but there are regional differences as well as interspecific variations in neurogenic capacity (Font et al., 2001). Brain areas of recruitment or proliferation in the turtle *T. scripta* include the main and accessory olfactory bulbs, rostral forebrain, all cortical areas, anterior dorsal ventricular ridge, septum, striatum, and cerebellum (Perez-Canellas & Garcia-Verdugo, 1992; Perez-Canellas et al., 1997). A temperature-related response has been reported in adult lizards (Ramirez et al., 1997; Penafiel et al., 2001), whereas recent work has expanded these findings to turtles, demonstrating the presence of recently divided cells positive for neuronal markers in the spinal cord of juvenile *Chrysemys d'orbigny* (Fernandez et al., 2002; Russo et al., 2004). New neurons are born in the ependyma of the ventricular walls and migrate through the brain parenchyma to their final destinations (Lopez-Garcia et al. 1988a, 1990a; Perez-Canellas & Garcia-Verdugo, 1996) and the data suggest that most regions of the ventricular zone are neurogenic in the turtle (Font et al., 1991). Unlike in the lizards, adult neurogenesis in *T. scripta* also gives rise to both neurons and free glial cells, although glial cell production appears to be limited to the striatum (Perez-Canellas et al., 1997). Whereas the functional consequences of the continuous generation and migration of new neurons in reptiles is still a matter of conjecture, with suggested functions in learning and memory, the stronger proliferation of cells under warm conditions than at colder temperatures in turtles (Radmilovich et al. 2003) could also provide a potential means during the warmer months to replace anoxia-damaged cells (Milton & Prentice, in press). Ischemic insult has been shown to increase neurogenesis in the adult mammalian brain (Liu et al., 1998); the proliferating cells arise from neuronal stem or progenitor cells (Sakakibara et al., 1996) that differentiate into neurons in later weeks (Yagita et al., 1999). While Font et al. (2001) comment that "it is difficult to imagine that reptiles or other vertebrates might endure sublethal injuries to specific portions of their brain with such high frequency that selection for neuronal regeneration plays a significant role," that is clearly not the case for turtles.

12.7 CONCLUSIONS

It is clear from a large body of research that differences in anoxic survival between mammals (and nearly all other vertebrates, for that matter) and the anoxia-tolerant turtles are not due to significant differences in overall brain structure, neuronal pathways, or the presence or absence of particular neurotransmitters. Indeed, the protective pathways thus far examined in the turtle are not unique in and of themselves but occur in mammalian systems as well— however, they are expressed more robustly, and more successfully, in turtles. A great deal of research has revealed the physiological—and increasingly, the molecular mechanisms—behind the remarkable anoxia tolerance of these animals: channel arrest, the enhancement of protective pathways such as GABA and adenosine, and the stockpiling of some protective compounds, coupled to the suppression of pathological events like increases in extracellular glutamate and dopamine, or ROS production. However, many of these studies were restricted to a particular region of the brain or, like measures of EEG activity, are global responses. More effective correlations between anatomy and the physiology of anoxia tolerance could be constructed with additional studies comparing specific brain structures of known function under anoxic and normal conditions.

Some notable trends do emerge from the literature, beginning with the general metabolic rostral-caudal gradient detectable in turtles as well as in mammalian brains. In contrast to mammals, the gradient in turtles is less well defined in general, although is quite distinctive for some compounds—an anatomical difference with significant physiological implications. For instance, the rostral-caudal gradient is most pronounced for the enzymes of oxidative phosphorylation as well as some antioxidants, reflecting the higher metabolic rates of gray matter. Interestingly, it is echoed by gradients in receptor density for inhibitory compounds like GABA and the δ-opioid receptors. In turtles, it then seems that whereas higher brain functions consume the most energy under normoxic conditions, these are also the areas most strongly suppressed during anoxia. By contrast, some glycolytic enzymes increase as one moves towards the hindbrain and brainstem of the turtle, which likely reflects—or contributes to—differences in activity during anoxia. Certain antioxidants, such as PACAP-38, are likewise present at higher levels in the hindbrain than the turtle forebrain, perhaps "evening out" overall antioxidant capacity and ensuring sufficient global protection against oxidative stress upon recovery. Despite the large overall metabolic suppression, the turtle brain is clearly not "shut off" but instead is down-regulated in a controlled manner that allows basal functions to continue, and perhaps maintains a state of minimal readiness for the eventual return of oxygen. The turtle brain's ability to withstand and recover from long-term anoxia is not a matter of passive tolerance but the result of interlocking adaptations concerning energy production, neurotransmitter balance, and the regulation of ion flux that produce a state of deep hypometabolism overall while apparently maintaining basal functions in critical areas. But whereas certain turtle species are far more resistant to anoxia not only with respect to mammals but also when compared to other reptiles, there has been no systematic study of specific areas of the brain to determine if all parts survive equally well or if some regions, as in the mammal, are more vulnerable than others. Clearly, more investigation is needed to identify and correlate regional differences in potentially protective mechanisms with both normal function and anoxic activity. Functioning at a mere 10 to 15% of normal metabolic rates, one can assume that the anoxic turtle brain is "stripped down" to the minimum of absolutely necessary functions to support life and prepare for recovery. Not only might this provide greater insight into the anatomy and physiology of anoxia tolerance itself, but knowledge of which pathways are strongly downregulated and in what regions maintaining function is critical could in turn point to avenues of therapeutic intervention in the hypoxic/ischemic mammalian brain.

ACKNOWLEDGMENTS

Research referenced in this chapter by S.L. Milton, P.L. Lutz, and H.M. Prentice was funded by the National Science Foundation, American Heart Association, National Institute of Health, and the Florida Atlantic University Foundation. This chapter was inspired by the work of and is dedicated to the memory of Peter Lutz. Special thanks to R. Schmidt-Kastner and an anonymous reviewer for their constructive comments on the manuscript.

REFERENCES

Aden, U., Linstrom, K., Bona, E., Hagberg, H., and Fredholm, B.B., Changes in adenosine receptors in the neonatal rat brain following hypoxic ischemia, *Mol. Brain Res.*, 23, 254, 1994.

Albin, R.L., and Gilman, S., $GABA_A$, $GABA_B$, benzodiazepine binding sites in the cerebellar cortex of the red-eared turtle (*Pseudemys scripta*), *Brain Res.*, 595, 164, 1992.

Alvarez-Buylla, A., Garcia-Verdugo, J.M., and Tramontin, A.D., A unified hypothesis on the lineage of neural stem cells, *Nat. Rev. Neurosci.*, 2, 287, 2001.

Ariel, M.M., Modulation of visual inputs to accessory optic system by theophylline during hypoxia, *Exp. Brain Res.*, 172, 351, 2006.

Ariel, M., and Fan, T.X., Electrophysiological evidence for a bisynaptic retinocerebellar pathway, *J. Neurophysiol.*, 69,1323, 1993.

Arimura, A., Perspectives on pituitary adenylate cyclase activating polypeptide (PACAP) in the neuroendocrine, endocrine, and nervous systems, *Jap. J. Physiol.*, 48, 301, 1998.

Ashcroft, F.M., Adenosine 5′-triphosphate-sensitive potassium channels, *Ann. Rev. Neurosci.*, 11, 97, 1988.

Ascher, P., and Nowak, L., Electrophysiological studies of NMDA receptors, *Trends Neurol. Sci.*, 10, 284, 1987.

Bassenge, E., Busse, R., and Pohl, U., Inhibition of thrombocyte aggregation and adhesion by endothelium-derived relaxant factor (EDRF) and their pathophysiologic significance, *Z. Kardiol.*, 78 Suppl. 6, 54, 1989.

Belkin, D.A., Anoxia: Tolerance in reptiles, *Science*, 139, 492, 1963.

Ben Ari, Y., Krnjevic, K., and Crepel, V., Activators of ATP-sensitive K^+ channels reduce anoxic depolarization of CA3 hippocampal neurons, *Neurosci.*, 37, 55, 1990.

Berger, U.V., and Hediger, M.A., The vitamin C transporter SVCT2 is expressed by astrocytes in culture but not in situ, *Neuoreport*, 11, 1395, 2000.

Berger, R., Garnier, Y., and Jensen, A., Perinatal brain damage: underlying mechanisms and neuroprotective strategies, *J. Soc. Gynecol. Investig.*, 9, 319, 2002.

Bickler, P.E., Cerebral anoxia tolerance in turtles: Regulation of intracellular calcium and pH, *Am. J. Physiol.*, 263, R1298, 1992.

Bickler, P.E., Reduction of NMDA receptor activity in cerebrocortex of turtles (*Chrysemys picta*) during 6 wk of anoxia, *Am. J. Physiol. Reg. Integr. Comp. Physiol.*, 275, 86, 1998.

Bickler, P.E., and Gallego, S.M., Inhibition of brain calcium channels by plasma proteins from anoxic turtles, *Am. J. Physiol.*, 265, R277, 1993.

Bickler, P.E., Donohoe, P.H., and Buck, L.T., Hypoxia-induced silencing of NMDA receptors in turtle neurons, *J. Neurosci.*, 20, 3522, 2000.

Biegon, A., et al., Dynamic changes in N-methyl-D-aspartate receptors after closed head injury in mice: Implications for treatment of neurological and cognitive deficits, *Proc. Natl. Acad. Sci. USA*, 101, 5117, 2004.

Blanco, M., et al., Ischemic preconditioning: A novel target for neuroprotective therapy, *Cerebrovasc. Dis.*, 21 Suppl. 2, 38, 2006.

Blanton, M.G., and Kriegstein, A.R., Properties of amino acid neurotransmitter receptors of embryonic cortical neurons when activated by exogenous and endogenous agonists, *J. Neurophysiol.*, 67, 1185, 1992.

Bohme, G.A., et al., Possible involvement of nitric oxide in long-term potentiation, *Eur. J. Pharmacol.*, 199(3), 379, 1991.

Boutilier, R.G., Mechanisms of metabolic defense against hypoxia in hibernating frogs, *Respir. Physiol.*, 128(3), 365, 2001.

Bredt, D.S., Hwang, P.M., and Snyder, S.H., Localization of nitric oxide synthase indicating a neural role for nitric oxide, *Nature.*, 347(6295), 768, 1990.

Brooks, S.P.J., and Storey, K.B., Anoxia brain function: Molecular mechanisms of metabolic depression, *FEBS*, 232, 214, 1988.

Brooks, S.P.J., and Storey, K.B., Regulation of glycolytic enzymes during anoxia in the turtle *Pseudemys scripta*, *Am. J. Physiol.*, 257, R278, 1989.

Bruning, G., Wiese, S., and Mayer, B., Nitric oxide synthase in the brain of the turtle *Pseudemys scripta elegans*, *J. Comp. Neurol.*, 348, 183, 1994.

Buck, L.T., and Bickler, P.E., Adenosine and anoxia reduce N-methyl-D-aspartate receptor open probability in turtle cerebrocortex, *J. Exp. Biol.*, 201, 289, 1998.

Buck, L.T., et al., Reversible decreases in ATP and PCr concentrations in anoxic turtle brain, *Comp. Biochem. Physiol. A*, 120, 633, 1998.

Burmester, T., and Hankeln, T., Neuroglobin: A respiratory protein of the nervous system, *News Physiol Sci.*, 19, 110, 2004.

Burmester, T., et al., A vertebrate globin expressed in the brain, *Nature*, 407, 520, 2000.

Cameron, J.S., and Baghdady, R., Role of ATP sensitive potassium channels in long-term adaptation to metabolic stress, *Cardiovasc. Res.*, 28, 788, 1994.

Cao, W., et al., Oxygen free radical involvement in ischemia and reperfusion injury to brain, *Neurosci. Lett.*, 88, 233, 1988.

Camacho, A., and Massieu, L., Role of glutamate transporters in the clearance and release of glutamate during ischemia and its relation to neuronal death, *Arch. Med. Res.*, 37, 11, 2006.

Cardinali, D.P., and Rosner, J.M., Metabolism of serotonin by the rat retina *in vitro*, *J. Neurochem.*, 18, 1769, 1971.

Christophe, M., and Nicolas, S., Mitochondria: A target for neuroprotective interventions in cerebral ischemia-reperfusion, *Curr. Pharm. Des.*, 12, 739, 2006.

Clark, V.M., and Miller, A.T., Studies on anaerobic metabolism in the freshwater turtles (*Pseudemys scripta elegans*), *Comp. Biochem. Physiol.*, 44A, 55, 1973.

Cohen, G., Farooqui, R., and Kesler, N., Parkinson disease: A new link between monoamine oxidase and mitochondrial electron flow, *Proc. Nat. Acad. Sci. USA*, 94, 4890, 1997.

Collin, J.P., and Oksche, A., Structural and functional relationships in the nonmammalian pineal gland, in *The Pineal Gland: Anatomy and Biochemistry*, R.J. Reiter (ed.), Boca Raton, FL: CRC Press, 1981, 27–67.

Collis, M.G., The vasodilator role of adenosine, *Pharmac. Ther.*, 41, 143, 1989.

Coomber, P., Crews, D., and Gonzalez-Lima, F., Independent effects of incubation temperature and gonadal sex on the volume and metabolic capacity of brain nuclei in the leopard gecko (*Eublepharis macularis*), a lizard with temperature-dependent sex determination, *J. Comp. Neurol.*, 380, 409, 1997.

Cowan, K.J., and Storey, K.B., Protein kinase and phosphatase responses to anoxia in crayfish, *Orconectes virilis*: Purification and characterization of cAMP-dependent protein kinase, *Comp. Biochem. Physiol. B Biochem. Mol. Biol.*, 130(4), 565, 2001.

Crews, D., Sex determination: Where environment and genetics meet, *Evol. Dev*, 5(1), 50, 2003.

Dawson, D.A., Nitric oxide and focal ischemia: Multiplicity of actions and diverse outcome, *Cerebrovasc. Brain Metab. Rev.*, 6, 299, 1994.

Dedja, A., Sokolowska, P., and Nowak, J.Z., Neuroprotective potential of three neuropeptides PACAP, VIP, and PHI, *Pharmacol. Rep.*, 57, 307, 2005.

Delbarre, B., Delbarre, G., and Calinon, F., Free radicals, glutamate, dopamine, and serotonin after ischemia-reperfusion insult in aging gerbil brain, *Drug Dev. Res.*, 26, 419, 1992.

Doll, C.J., Hochachka, P.W., and Hand, S.C., A microcalorimetric study of turtle cortical slices: Insight into brain metabolic depression, *J. Exp. Biol.*, 191, 141, 1994.

Drukarch, B., Schepens, E., and Stoof, J.C., Muscarinic receptor activation attenuates D2 dopamine receptor mediated inhibition of acetylcholine release in rat striatum: Indications for a common signal transduction pathway, *Neurosci.*, 37, 1, 1990.

Dubocovich, M.L., Modulation of dopaminergic activity by melatonin in retina, in *Pineal and Retinal Relationships*, P.J. O'Brian and D.C. Klein (eds.), New York: Academic Press, 1986, 239–252.

Dugan, L.L., et al., Mitochondrial production of reactive oxygen species in cortical neurons following exposure to N-methyl-D-aspartate, *J. Neurosci.*, 15, 6377, 1995.

Ebadi, M., Srinivasan, S.K., and Baxi, M.D., Oxidative stress and antioxidant therapy in Parkinson's disease, *Prog. Neurobiol.*, 48, 1, 1996.

Edwards, R., Lutz, P.L., and Baden, D., Relationship between energy expenditure and ion channel function in the rat and turtle brain, *Am. J. Physiol.*, 255, R1345, 1989.

Ehinger, B., Bipolar cells in the turtle retina are strongly immunoreactive for glutamate, *Proc. Natl. Acad. Sci. USA*, 85, 8321, 1988.

Erecsinska, M., and Silver, I.A., ATP and brain function, *J. Cereb. Blood Flow Metab.*, 9, 2, 1989.

Eskiocak, S., The effect of melatonin on protein oxidation and nitric oxide in the brain tissue of hypoxic neonatal rats, *Brain Dev.*, 29(1), 19, 2007.
Esparza, J.L., et al., Melatonin reduces oxidative stress and increases gene expression in the cerebral cortex and cerebellum of aluminum-exposed rats, *J. Pineal Res.*, 39, 129, 2005.
Fan, T.X., et al., Visual responses and connectivity in the turtle pretectum, *J. Neurophysiol.*, 73, 2507, 1995.
Fan, T.X., Scudder, C., and Ariel, M., Neuronal responses to turtle head rotation in vitro, *J. Neurobiol.*, 33, 99, 1997.
Fay, R.R., and Ream, T.J., The effects of temperature change and transient hypoxia on auditory nerve fibre response in the goldfish (*Carassius auratus*), *Hearing Res.*, 58, 9, 1992.
Felger, R.S., Cliffton, K., and Regal, P.J., Winter dormancy in sea turtles: Independent discovery and exploitation in the Gulf of California by two local cultures, *Science*, 191, 283, 1976.
Fernandes, J.A., et al., Electroencephalogram activity in the anoxic turtle brain, *Am. J. Physiol.*, 273, R911, 1997.
Fernandez, A., Radmilovich, M., and Trujillo-Cenoz, O., Neurogenesis and gliogenesis in the spinal cord of turtles, *J. Comp. Neuro.*, 453, 131, 2002.
Ferrer, I., et al., Early modifications in the expression of mitogen-activated protein kinase (MAPK/ERK), stress-activated kinases SAPK/JNK and p38, and their phosphorylated substrates following focal cerebral ischemia, *Acta Neuropathol.*, 105, 425, 2003.
Floyd, R.A., and Carney, J.M., Free radical damage to protein and DNA: Mechanisms involved and relevant observations on brain undergoing oxidative stress, *Lab. Invest.*, 47, 412, 1992.
Fokkelman, K., et al., Tissue-specific changes in glutathione content of hypoxic newborn pigs reoxygenated with 21% or 100% oxygen, *Eur. J. Pharmacol.*, 562(1-2), 132, 2007.
Font, E., et al., Neuron regeneration reverses 3-acetylpyridine-induced cell loss in the cerebral cortex of adult lizards, *Brain Res.*, 551, 230, 1991.
Font, E., et al., 3-Acetylpyridine-induced degeneration and regeneration in the adult lizard brain: A qualitative and quantitative analysis, *Brain Res.*, 754, 245, 1997.
Font, E., et al., Neurogenesis and neuronal regeneration in the adult reptilian brain, *Brain Behav. Evol.*, 58, 276, 2001.
Fordel, E., et al., Neuroglobin and cytoglobin overexpression protects human SH-SY5Y neuroblastoma cells against oxidative stress-induced cell death, *Neurosci. Lett.*, 410(2), 146, 2006.
Fraser, K.P., et al., Complete suppression of protein synthesis during anoxia with no post-anoxia protein synthesis debt in the red-eared slider turtle *Trachmeys scripta*, *J. Exp. Biol.*, 204, 4353, 2001.
Fujimori, T., and Abe, H., Physiological roles of free D- and L-alanine in the crayfish *Procambarus clarkii* with special reference to osmotic and anoxic stress responses, *Comp. Biochem. Physiol. A Mol. Integr. Physiol*, 131(4), 893, 2002.
Ganzhorn, D., and Licht, P., Regulation of seasonal gonadal cycles by temperature in the painted turtle *Chrysemys picta*, *Copeia*, 1983, 347, 1983.
Gardoni, F., and Di Luca, M., New targets for pharmacological intervention in the glutamatergic synapse, *Eur. J. Pharmacol.*, 545, 2, 2006.
Glazner, G.W., et al., Activity-dependent neurotrophic factor: A potent regulator of embryonic growth and development, *Anat. Embryol. (Berl)*, 200, 65, 1999.
Globus, M.Y.-T., et al., Effects of ischemia on the in vivo release of stratal dopamine, glutamate, and alpha-amino-butyric acid studied by intracerebral microdialysis, *J. Neurochem.*, 51, 1455, 1988.
Gould, E., et al., Learning enhances adult neurogenesis in the hippocampal formation, *Nat. Neurosci.*, 2, 260, 1999a.
Granda, A.M., and Sisson, D.F., Retinal function in turtles, in *Biology of the Reptilia*, v. 17, C. Gans and P.S. Ulinski (eds.), Chicago: University of Chicago Press, 1992, 136–174.
Greenway, S.C., and Storey, K.B., Mitogen-activated protein kinases and anoxia tolerance in turtles, *J. Exp. Zool.*, 287, 477, 2000.
Gross, C.G., Neurogenesis in the adult brain: Death of a dogma, *Nat. Rev. Neurosci.* 1, 67, 2000.
Hall, W.C., and Ebner, F.F., Thalamotelencephalic projections in the turtle (*Pseudemys scripta*), *J. Comp. Neurol.*, 140, 101, 1970.
Halpern, M., Nasal and chemical senses in reptiles: Structure and function, in *Hormones, Brain, and Behavior: Biology of the Reptilia*, v. 18, C. Gans and D. Crews (eds.), Chicago: University of Chicago Press, 1991, 423–523.
Hamaski, G.I., and Eder, D.J., Adaptive radiation of the pineal system, in *Handbook of Sensory Physiology*, F. Crescitelli (ed.), New York: Springer-Verlag, 1977, 497–548.

Hartwig, H.G., and van Veen, T., Spectral characteristics of visible radiation penetrating into the brain and stimulating extraretinal photoreceptors, *J. Comp. Physiol.*, 130, 277, 1979.
Harukuni, I., and Bhardwaj, A., Mechanisms of brain injury after global cerebral ischemia, *Neurol Clin.*, 24(1), 1, 2006.
Hastings, N.B., Tanapat, P., and Gould, E., Neurogenesis in the adult mammalian brain, *Clin. Neurosci. Res.*, 1, 175, 2001.
Henry, Y., et al., EPR characterization of molecular targets for NO in mammalian cells and organelles, *FASEB J.*, 7, 1124, 1994.
Herkenham, M., and Nauta, W.J.H., Efferent connections of the habernacular nuclei in the rat, *J. Comp. Neurol.*, 187, 19, 1979.
Hevner, R.F., and Wong-Riley, M.T., Mitochondrial and nuclear gene expression for cytochrome oxidase subunits are disproportionately regulated by functional activity in neurons, *J. Neurosci.*, 13, 1805, 1993.
Hevner, R.F., Duff, R.S., and Wong-Riley, M.T., Coordination of ATP production and consumption in brain: Parallel regulation of cytochrome oxidase and Na^+, K^+-ATPase, *Neurosci. Lett.*, 138, 188, 1992.
Hicks, J.W., and Malvin, G.M., Mechanism of intracardiac shunting in the turtle *Pseudemys scripta*, *Am. J. Physiol.*, 262, R986, 1992
Hicks, J.W., and Farrell, A.P., The cardiovascular responses of the red-eared slider (*Trachemys scripta*) acclimated to either 22 or 5°C. II. Effects of anoxia on adrenergic and cholinergic control, *J. Exp. Biol.*, 203, 3775, 2000.
Hillered, L., et al., Increased extracellular levels of ascorbate in the striatum after middle cerebral artery occlusion in the rat monitored by intracerebral microdialysis, *Neurosci. Lett.*, 95, 286, 1988.
Hitzig, B.M., et al., Brain amino acid concentrations during diving and acid-base stress in turtles, *J. Appl. Physiol.*, 58, 1751, 1985.
Hochachka, P.W., and Somero, G.N., *Biochemical Adaptation*, New York: Oxford University Press, 2002.
Hochscheid, S., Bentivegna, F., and Hays, G.C., First records of dive durations for a hibernating sea turtle, *Biol. Lett.*, 1, 82, 2005.
Hoyte, L., et al., The rise and fall of NMDA antagonists for ischemic stroke, *Curr. Mol. Med.*, 4, 131, 2004.
Huang, C.C., et al., Relationship of extracellular dopamine in striatum of newborn piglets to cortical oxygen pressure, *Neurochem. Res.*, 9(6), 649, 1994.
Hylland, P., and Nilsson, G.E., Evidence that acetylcholine mediates increased cerebral blood flow velocity in crucian carp through a nitric oxide-dependent mechanism, *J. Cereb. Blood Flow Metab.*, 15, 519, 1995.
Hylland, P., Nilsson, G.E., and Lutz, P.L., Time course of anoxia induced increase of cerebral blood flow rate in turtles: Evidence for a role of adenosine, *J. Cereb. Blood Flow Metab.*, 16, 290, 1994.
Hylland, P., Nilsson, G.E., and Lutz, P.L., Role of nitric oxides in the elevation of cerebral blood flow induced by acetylcholine in the turtle, *J. Cereb. Blood Flow Metab.*, 16, 290, 1996.
Hylland, P., et al., Brain Na^+/K^+-ATPase activity in two anoxia tolerant vertebrates: Crucian carp and freshwater turtle, *Neurosci. Lett.*, 235, 89, 1997.
Ignarro, L.J., et al., Endothelium-derived relaxing factor produced and released from artery and vein is nitric oxide, *Proc. Natl. Acad. Sci. USA*, 84(24), 9265, 1987.
Ignarro, L.J., et al., Nitric oxide and cyclic GMP formation upon electrical field stimulation cause relaxation of corpus cavernosum smooth muscle, *Biochem. Biophys. Res. Commun.*, 170(2), 843, 1990.
Jackson, D.C., Metabolic depression and oxygen depletion in the diving turtle, *J. Appl. Physiol.*, 24, 503, 1968.
Jackson, D.C., Lactate accumulation in the shell of the turtle *Chrysemys picta bellii*, during anoxia at 3 and 10°C, *J. Exp. Biol.*, 200, 2295, 1997.
Jackson, D.C., Surviving extreme lactic acidosis: The role of calcium lactate formation in the anoxic turtle, *Respir. Physiol. Neurobiol.*, 144, 173, 2004.
Jackson, D.C., Crocker, C.E., and Ultsch, G.R., Bone and shell contribution to lactic acid buffering of submerged turtles *Chrysemys picta bellii* at 3°C, *Am. J. Physiol. Regul. Integr. Comp. Physiol.*, 278, R1564, 2000.
Jackson, D.C., et al., Comparative shell buffering properties correlate with anoxia tolerance in freshwater turtles, *Am. J. Physiol. Regul. Integr. Comp. Physiol.*, 292(2), R1008, 2006.
Jeyasuria, P., and Place, A., The brain-gonadal embryonic axis in sex determination of reptiles: A role for cytochrome P450 aromatase, *J. Exp. Zool.*, 281, 428, 1998.
Jiang, C.S., and Haddad, G.G., Effect of anoxia on intracellular and extracellular potassium activity in hypoglossal neurons in-vitro, *J. Neurophysiol.*, 66, 103, 1991.
Johansson, D., and Nilsson, G.E., Roles of energy status KATP channels and channel arrest in fish brain K^+ gradient dissipation during anoxia, *J. Exp. Biol.*, 198, 2575, 1995.

Johansson, D., Nilsson, G.E., and Doving, K.B., Anoxic depression of light-evoked potentials in retina and optic tectum of crucian carp, *Neurosci. Lett.*, 237, 73, 1997.

Johnson, S.M., Johnson, R.A., and Mitchell, G.S., Hypoxia, temperature, and pH/CO_2 effects on respiratory discharge from a turtle brainstem preparation, *J. Appl. Physiol.*, 84, 649, 1998.

Kahlert, S., Zundorf, G., and Reiser, G., Glutamate-mediated influx of extracellular Ca^{2+} is coupled with reactive oxygen species generation in cultured hippocampal neurons but not in astrocytes, *J. Neurosci. Res.*, 79, 262, 2005.

Kardos, J., The $GABA_A$ receptor channel mediated chloride ion translocation through the plasma membrane: New insights from 36 Cl^- flux measurements, *Synapse*, 13, 74, 1993.

Keifer, J., and Carr, M.T., Immunocytochemical localization of glutamate receptor subunits in the brain stem and cerebellum of the turtle *Chrysemys picta*, *J. Comp. Neurol.*, 427, 455, 2000.

Keifer, J., and Lustig, D.G., Comparison of cortically and subcortically controlled motor systems. II. Distribution of anterogradely labeled terminal boutons on intracellularly filled rubrospinal neurons in rat and turtle, *J. Comp. Neurol.*, 416, 101, 2000.

Keiver, K.M., Weinberg, J., and Hochachka, P.W., The effect of anoxic submergence and recovery on circulating levels of catecholamines and corticosterone in the turtle, *Chrysemys picta*, *Gen. Comp. Endocrinol.*, 85, 308, 1992.

Kelly, D.A., and Storey, K.B., Organ-specific control of glycolysis in anoxic turtles, *Am. J. Physiol.*, 255, R774, 1988.

Knickerbocker, D.L., and Lutz, P.L., Slow ATP loss and the defense of ion homeostasis in the anoxic frog brain, *J. Exp. Biol.*, 204 (Pt 20), 3547, 2001.

Kogo, N., and Ariel, M., Membrane properties and monosynaptic retinal excitation of neurons in the turtle accessory optic system, *J. Neurophysiol.*, 78, 614, 1997.

Kogo, N., Fan, T.X., and Ariel, M., Synaptic pharmacology in the turtle accessory optic system, *Exp. Brain Res.*, 147, 464, 2002.

Kriegl, J.M., et al., Ligand binding and protein dynamics in neuroglobin, *Proc. Natl. Acad. Sci. USA*, 99, 7992, 2002.

Kriegstein, A.R., and Connors, B.W., Cellular physiology of the turtle visual cortex: Synaptic properties and intrinsic circuitry, *J. Neurosci.*, 6, 178, 1986.

Krnjevic, K., Membrane current activation during hypoxia in hippocampal neurons, in *Surviving Hypoxia: Mechanisms of Control and Adaptation*, P.W. Hochachka, P.L. Lutz, T. Sick, M. Rosenthal, and G. van den Thilart (eds.), Boca Raton, FL: CRC Press, 1993, 365.

Kruman, I.I., and Mattson, M.P., Pivotal role of mitochondrial calcium uptake in neural cell apoptosis and necrosis, *J. Neurochem.*, 72, 529, 1999.

Kwon, N.S., Stuehr, D.J., and Nathan, C.F., Inhibition of tumor cell ribonucleotide reductase by macrophage-derived nitric oxide, *J. Exp. Med.*, 174, 761, 1991.

Lancaster, J.R. Jr., and Hibbs, J.B. Jr., EPR demonstration of iron-nitrosyl complex formation by cytotoxic activated macrophages, *Proc. Natl. Acad. Sci. USA*, 87(3), 1223, 1990.

Lancelot, E., et al., A microdialysis study investigating the mechanisms of hydroxyl radical formation in rat striatum exposed to glutamate, *Brain Res*, 809, 294, 1998.

Landolt, H., et al., Extracellular antioxidants and amino acids in the cortex of the rat: Monitoring by microdialysis of early ischemic changes, *J. Cereb.Blood Flow Metab.*, 12, 96, 1992.

Laplanche, L., et al., Hydroxyl radicals release in rat striatum involves metabotropic glutamate receptors, *Brain Res. Bull.*, 61, 453, 2003.

Larson-Prior, L.J., Siuciak, J.A., and Dubocovich, M.L., Localization of 2-[^{125}I]iodomelatonin binding sites in visual areas of the turtle brain, *Eur. J. Pharmacol.*, 297, 181, 1996.

Lee, K.S., Tetzlaffand, W., and Kreutzberg, G.W., Rapid downregulation of hippocampal adenosine receptors following brief anoxia, *Brain Res.* 380, 155, 1986.

Lee, H.J., Ban, J.Y., and Seong, Y.H., Blockade of 5-HT(3) receptor with MDL7222 and Y25130 reduces hydrogen peroxide-induced neurotoxicity in cultured rat cortical cells, *Life Sci.*, 78(3), 294, 2005.

Lei, B., Adachi, N., and Arai, T., The effect of hypothermia on H_2O_2 production during ischemia and reperfusion: A microdialysis study in the gerbil hippocampus, *Neurosci. Lett.*, 222, 91, 1997.

Liss, B., and Roeper, J., Molecular physiology of neuronal K-ATP channels (review), *Molec. Mem. Biol.*, 18, 117, 2001.

Liu, J., et al., Increased neurogenesis in the dentate gyrus after transient global ischemia in gerbils, *J. Neurosci.*, 18, 7768, 1998.

Lopez-Garcia, C., Tineo, P.L., and del Corral, J., Increase of the neuron number in some cerebral cortical areas of a lizard, *Podarcis hispanica*, (Steind, 1870), during postnatal periods of life, *J. Hirnforsch.*, 25, 255, 1984.

Lopez-Garcia, C., et al., Delayed postnatal neurogenesis in the cerebral cortex of lizards, *Dev. Brain Res.*, 43, 167, 1988a.

Lopez-Garcia, C., et al., Late generated neurons in the medial cortex of adult lizards send axons that reach the Timm-reactive zones, *Dev. Brain Res.*, 57, 249, 1990a.

Lopez-Olmeda, J.F., et al., Effects of melatonin administration on oxidative stress and daily locomotor activity patterns in goldfish, *J. Physiol. Biochem.*, 62, 17, 2006.

Lutcavage, M.E., and Lutz, P.L., Diving physiology, in *The Biology of Sea Turtles*, P.L. Lutz and J.A. Musick (eds.), Boca Raton, FL: CRC Press, 1997, 277–296.

Lutz, P.L., and Kabler, S.A., Release of adenosine and ATP in the brain of the freshwater turtle (*Trachemys scripta*) during long-term anoxia, *Brain Res.*, 769, 281, 1997.

Lutz, P.L., and Leone-Kabler, S.A., Upregulation of $GABA_A$ receptor during anoxia in the turtle brain, *Am. J. Physiol.*, 37, R1332, 1995.

Lutz, P.L., and Nilsson, G.E., Contrasting strategies for anoxic brain survival—glycolysis on or off, *J. Exp. Biol.*, 200, 411, 1997.

Lutz, P.L., and Manuel, L., Maintenance of adenosine A_1 receptor function during long term anoxia in the turtle brain, *Am. J. Physiol.*, 276, R633, 1999.

Lutz, P.L., and Milton, S.L., Negotiating brain anoxia survival in the turtle, *J. Exp. Biol.*, 207, 3141, 2004.

Lutz, P.L., et al., Relationships between aerobic and anaerobic energy production in turtle brain in situ, *Am. J. Physiol.*, 247, R740, 1984.

Lutz, P.L., Edwards, R., and McMahon, P., GABA concentrations are maintained in the anoxic turtle brain, *Am. J. Physiol.*, 249, R372, 1985.

Lutz, P.L., Nilsson, G.E., and Prentice, H., *The Brain without Oxygen: Causes of Failure Molecular and Physiological Mechanisms for Survival*, 3rd ed., Dordrecht: Kluwers Press, 2003.

Lyrer, P., et al., Levels of low molecular weight scavengers in the brain during focal ischemia, *Brain Res.*, 567, 317, 1991.

MacDonald, J.A., and Storey, K.B., Regulation of ground squirrel Na^+K^+-ATPase by reversible phosphorylation during hibernation, *Biochem. Biophys. Res. Comm.*, 254, 424, 1999.

Magistretti, P.J., Hof, P.R., and Martin, J.L., Adenosine stimulates glycogenolysis in mouse cerebral cortex: A possible coupling mechanisms between neuronal activity and energy metabolism, *J. Neurosci.*, 6, 2553, 1986.

Mahapatra, M.S., Mahata, S.K., and Maiti, B.R., Circadian rhythms and influence of light on serotonin, norepinephrine, and epinephrine contents in the pineal-paraphyseal complex of soft-shelled turtles (*Lissemys punctata punctata*), *Gen. Comp. Endocrinol.*, 71(1), 183, 1988.

Mahmoud, I.Y., and Licht, P., Seasonal changes in gonadal activity and the effects of stress on reproductive hormones in the common snapping turtle, *Chelydra serpentina*, *Gen. Comp. Endocrinol.*, 107(3), 359, 1997.

Maickel, P.R., et al., A method for the determination of serotonin and norepinephrine in discrete areas of the rat brain, *Int. J. Neuropharmac.*, 7, 275, 1968.

Makar, T.K., et al., Vitamin E, ascorbate, glutathione, glutathione disulfide, and enzymes of oxidative metabolism in cultures of chick astrocytes and neurons: Evidence that astrocytes play an important role in oxidative processes in the brain, *J. Neurochem.*, 62, 45, 1994.

Mario, H., et al., Pituitary adenylate-cyclase activating polypeptide protects neurons from glutamate-induced cytotoxic injury, *Brain Res.*, 741, 82, 1996.

Martin, J., and Ariel, M., Localization of GABA markers in the turtle's basal optic nucleus, *Brain Res.*, 1066, 109, 2005.

McDougal, D.B. Jr., et al., The effects of anoxia upon energy sources and selected metabolic intermediates in the brains of fish, frog, and turtle, *J. Neurochem.*, 15, 577, 1968.

McMahon, B.R., Respiratory and circulatory compensation to hypoxia in crustaceans, *Respir. Physiol.*, 128(3), 349, 2001.

Mendoco, M.T., and Licht, P., Photothermal effects on the testicular cycle in the musk turtle, *Sternotherus odoratus*, *J. Exp. Zool.*, 239, 117, 2005.

Milby, K., Oke, A., and Adams, R.N., Detailed mapping of ascorbate distribution in rat brain, *Neurosci. Lett.*, 28, 15, 1982.

Milton, S.L., The physiology of hypoxia and anoxia tolerance in three species of turtles: The loggerhead sea turtle (*Caretta caretta*), green sea turtle (*Chelonia mydas*) and freshwater *Trachemys scripta*, Ph.D. thesis, University of Miami, Miami, FL, 1994.

Milton, S.L., and Lutz, P.L., Low extracellular dopamine levels are maintained in the anoxic turtle brain, *J. Cereb. Blood Flow Metab.*, 18, 803, 1998.

Milton, S.L., and Lutz, P.L., Adenosine and ATP sensitive potassium channels modulate dopamine release in the anoxic turtle (*Trachemys scripta*) striatum, *Am. J. Physiol.*, 289, R77, 2005.

Milton, S.L., and Prentice, H.M., Beyond anoxia: The physiology of metabolic downregulation and recovery in the anoxia-tolerant turtle, *Comp. Biochem. Physiol. A*, 147(2), 277, 2007.

Milton, S.L., Thompson, J.W., and Lutz, P.L., Mechanisms for maintaining extracellular glutamate in the anoxic turtle striatum, *Am. J. Physiol.*, 282, R1317, 2002.

Milton, S.L., Manuel, L., and Lutz, P.L., Slow death in the leopard frog *Rana pipiens*: Neurotransmitters and anoxia tolerance, *J. Exp. Biol.*, 206 (Pt 22), 4021, 2003.

Milton, S.L., et al., The regulation of neuroglobin gene transcription in hypoxia and anoxia in the brain of the anoxia-tolerant turtle *Trachemys scripta*, *J. Biomed. Sci.*, 13, 509, 2006.

Mitsuyo, T., et al., Facilitation of ischemia-induced release of dopamine and neuronal damage by dexamethasone in the rat striatum, *Eur. J. Pharmacol.*, 465(3), 267, 2003.

Mohamed, M.P., Metabolism of *Tilapia mossambica* (Peters) with emphasis on hypoxia, *Ind. J. Exp. Biol.*, 19(11), 1098, 1981.

Moncada, S., and Higgs, E.A., The discovery of nitric oxide and its role in vascular biology, *Br. J. Pharmacol.*, 147 Suppl. 1, S193, 2006.

Montero, M., et al., Distribution, characterization, and growth-hormone-releasing activity of pituitary adenylate cyclase activating polypeptide in the European eel, *Anguilla anguilla*, *Endocrinol.*, 139, 4300, 1998.

Morgan, P.J., et al., Melatonin receptors: Localization, molecular pharmacology and physiological significance, *Neurochem. Int.*, 24, 101, 1994.

Mourre, C., et al., Antidiabetic sulfonylureas: Localization of binding sites in the brain and effects on the hyperpolarization induced by anoxia in hippocampal slices, *Brain Res.*, 486, 159, 1989.

Murphy, S.J., et al., Regional expression of the heat shock protein 72 mRNA following mild and severe hypoxia in the neonatal piglet brain, *Adv. Exp. Med. Biol.*, 471, 155, 1999.

Nagasawa, H., Araki, T., and Kogure, K., Alteration of adenosine A_1 receptor in the post-ischemic rat brain, *Euroreport*, 5, 1453, 1994.

Nagle, R.D., Lutz, C.L., and Pyle, A.L., Overwintering in the nest by hatchling map turtles (*Graptemys geographica*), *Can. J. Zool.*, 82(8), 1211, 2004.

Nagy, K., et al., Diazoxide preconditioning protects against neuronal cell death by attenuation of oxidative stress upon glutamate stimulation, *J. Neurosci. Res.*, 76, 697, 2004.

Nicholls, D.G., *Proteins, Transmitters, and Synapses*, Boston: Blackwell Scientific Publications, 1994.

Nilsson, G.E., and Lutz, P.L., Release of inhibitory neurotransmitters in response to anoxia in turtle brain, *Am. J. Physiol.*, 261, R32, 1991.

Nilsson, G.E., and Lutz, P.L., Adenosine release in the anoxic turtle brain as a mechanism for anoxic survival, *J. Exp. Biol.*, 162, 345, 1992.

Nilsson, G.E., and Lutz, P.L., Role of GABA in hypoxia tolerance, metabolic depression, and hibernation—possible links to neurotransmitter evolution, *Comp. Biochem. Physiol.*, 105C, 329, 1993.

Nilsson, G.E., Alfaro, A.A., and Lutz, P.L., Changes in turtle brain neurotransmitters and related substances during anoxia, *Am. J. Physiol.*, 259, R376, 1990.

Nilsson, G.E., Lutz, P.L., and Jackson, T.L., Neurotransmitters and anoxic survival in the brain: A comparison between anoxia tolerant and anoxia intolerant vertebrates, *Physiol. Zool.*, 64, 638, 1991.

Nilsson, G.E., et al., Brain sensitivity to anoxia in fish as reflected by changes in extracellular K^+ activity, *Am. J. Physiol.*, 264, R250, 1993.

Ninomiya, H., et al., Increased binding of [^{3}H]muscimol and [^{3}H]flunitrazepam in the rat brain under hypoxia, *J. Neurochem.*, 51, 1111, 1988.

Nogues, M.R., et al., Melatonin reduces oxidative-stress in erythrocytes and plasma of senescence-accelerated mice, *J. Pineal Res.*, 41, 142, 2006.

Noma, A., ATP-regulated K^+ channels in cardiac muscle, *Nature*, 305, 147, 1983.

Obata, T., Dopamine efflux by MPTP and hydroxyl radical generation, *J. Neural Transm.*, 109, 1159, 2002.

Ogren, L., and McVea, C.J., Apparent hibernation by sea turtles in North American waters, in *Biology and Conservation of Sea Turtles*, K.A. Bjorndal (ed.), Washington, DC: Smithsonian Institution Press, 1995, 127–132.

Oke, A.F., May, L., and Adams, R.N., Ascorbic acid distribution patterns in human brain: A comparison with non-mammalian species, *Ann. NY Acad. Sci.*, 498, 1, 1987.

Onodera, H., and Kogure, K., Autoradiographic visualization of adenosine A_1 receptors in the gerbil hippocampus: Changes in the receptor density after transient ischemia, *Brain Res.*, 345, 406, 1985.

Owens, D.W., Gern, W.A., and Ralph, C.L., Melatonin in the blood and cerebrospinal fluid of the green sea turtle (*Chelonia mydas*), *Gen. Comp. Endocrinol.*, 40, 180, 1980.

Palkovits, M., Somogyvari-Vigh, A., and Arimura, A., Concentrations of pituitary adenylate cyclase activating polypeptide (PACAP) in human brain nuclei, *Brain Res.*, 699, 116, 1995.

Palmer, R.M.J., Ferrige, A.G., and Moncada, S., Nitric oxide accounts for the biological activity of endothelium derived relaxing factor, *Nature*, 327, 524, 1987.

Parent, A., Monaminergic systems of the brain, in *Biology of the Reptilia, v. 10, Neurobiology*, C. Gans (ed.), New York: Academic Press, 1979, 247–286.

Park, E.M., et al., A neuroprotective role of extracellular signal-regulated kinase in N-acetyl-O-methyldopamine-treated hippocampal neurons after exposure to in vitro and in vivo ischemia, *Neurosci.*, 123, 147, 2004.

Partata, W.A., and Marques, M., Effects of fasting and seasonal variation in brain glycogen disposition in the turtle *Chrysemys dorbigni*, *Comp. Biochem. Physiol.*, 107A, 727, 1994.

Partata, W.A., and Achaval-Elena, M., Glycogen distribution in the central nervous system of the turtle *Chrysemys dorbigni*, *Rev. Bras. Cien Morfol.*, 12, 3, 1995.

Partata, W.A., et al., Distribution of glycogen phosphorylase and cytochrome oxidase in the central nervous system of the turtle *Trachemys dorbigni*, *Comp. Biochem. Physiol.*, 124A, 113, 1999.

Pek, M., and Lutz, P.L., Role for adenosine in "channel arrest" in the anoxic turtle brain, *J. Exp. Biol.*, 200, 1913, 1997.

Pek-Scott, M., and Lutz, P.L., K^+ATP channel activation provides transient protection in anoxic turtle brain, *Am. J. Physiol.*, 275, R2023, 1998.

Penafiel, A., Temperature affects adult neurogenesis in the lizard brain, *Int. J. Dev. Biol.*, 45, 83, 2001.

Penney, D.G., Effects of prolonged diving anoxia on the turtle, *Pseudemys scripta elegans*, *Comp. Biochem. Physiol.*, 47A, 933, 1974.

Peretto, P., et al., The subependymal layer in rodents: A site of structural plasticity and cell migration in the adult mammalian brain, *Brain Res. Bull.*, 49, 221, 1999.

Perez-Canellas, M.M., and Garcia-Verdugo, J.M., Adult neurogenesis in reptiles: A comparative study using [^{3}H]thymidine autoradiography, *Eur. J. Neurosci.*, S5, 294, 1992.

Perez-Canellas, M.M., and Garcia-Verdugo, J.M., Adult neurogenesis in the telencephalon of a lizard: A [^{3}H]thymidine autoradiographic and bromodeoxyuridine immunocytochemical study, *Dev. Brain Res.*, 93, 49, 1996.

Perez-Canellas, M.M., Font, E., and Garcia-Verdugo, J.M., Postnatal neurogenesis in the telencephalon of turtles: Evidence for non-radial migration of new neurons from distant proliferative ventricular zones to the olfactory bulbs, *Dev. Brain Res.*, 101, 125, 1997.

Perez-Pinzon, M.A., and Rice, M.E., Seasonal- and temperature-dependent variation in CNS ascorbate and glutathione levels in the anoxia-tolerant turtles, *Brain Res.*, 705, 45, 1995.

Perez-Pinzon, M.A., et al., Membrane and synaptic activity during anoxia in the isolated turtle cerebellum, *J. Comp. Physiol.*, 16, 345, 1992.

Petrillo, M., Ritter, C.A., and Powers, A.S., A role for acetylcholine in spatial memory in turtles, *Physiol. Behav.*, 56, 135, 1994.

Plumier, J.C., et al., Transgenic mice expressing the human heat shock protein 70 have improved post-ischemic myocardial recovery, *J. Clin. Invest.*, 95, 1854, 1995.

Prentice, H.M., et al., Gene transcription of brain voltage-gated potassium channels is reversibly regulated by oxygen supply, *Am. J. Physiol. Regul. Integr. Comp. Physiol.*, 285, R1317, 2003.

Prentice, H.M., et al., The upregulation of cognate and inducible heat shock proteins in the anoxic turtle brain, *J. Cereb. Blood Flow Metab.*, 24, 826, 2004.

Prince, D.A., and Stevens, C.F., Adenosine decreases neurotransmitter release at central synapses, *Proc. Natl. Acad. Sci. USA*, 89, 8585, 1992.

Pulsinelli, W.A., Selective neuronal vulnerability: Morphological and molecular characteristics, in *Molecular Mechanisms of Ischemic Brain Damage: Progress in Brain Research*, K. Kogure, K.A. Hossman, and B.K. Siesjo (eds.), Amsterdam: Elsevier Science, 1985, 29–37.

Quay, W.B., The parietal eye-pineal complex, in *Biology of the Reptilia*, v.9, C. Gans, R.G. Northcutt, and P. Ulinski (eds.), New York: Academic Press, 1979, 245–406.

Rabl, K., et al., PACAP inhibits anoxia-induced changes in physiological responses in horizontal cells in the turtle retina, *Regul. Pept.*, 109(1-3), 71, 2002.

Racz, B., et al., The neuroprotective effects of PACAP in monosodium glutamate-induced retinal lesion involve inhibition of proapoptotic signaling pathways, *Regul. Pept.*, 137(1-2), 20, 2006.

Radmilovich, M., Fernandez, A., and Trujillo-Cenoz, O., Environment temperature affects cell proliferation in the spinal cord and brain of juvenile turtles, *J. Exp. Biol.*, 206, 3085, 2003.

Ramirez, J.M., Folkow, L.P., and Blix, A.S., Hypoxia tolerance in mammals and birds: From the wilderness to the clinic, *Ann. Rev. Physiol.*, 69, 113, 2007.

Ramirez, C., et al., Photo-period temperature and neuroblast proliferation-migration in lizard cortex, *Neuroreport*, 7, 1257, 1997.

Ramnanan, C.J., and Storey, K.B., Suppression of Na^+K^+-ATPase activity during estivation in the land snail *Otala lacteal*, *J. Exp. Biol.*, 209, 677, 2006.

Raps, S.P., et al., Glutathione is present in high concentrations in cultured astrocytes but not in cultured neurons, *Brain Res.*, 493, 398, 1989.

Reese, S.A., et al., The physiology of hibernation in common map turtles (*Graptemys geographica*), *Comp. Biochem. Physiol. A*, 130, 331, 2001.

Reese, S.A., Jackson, D.C., and Ultsch, G.R., The physiology of overwintering in a turtle that occupies multiple habitats, the common snapping turtle (*Chelydra serpentina*), *Physiol. Biochem. Zool.*, 75, 432, 2002.

Reese, S.A., et al., Geographic variation of the physiological response to overwintering in the painted turtle (*Chrysemys picta*), *Physiol. Biochem. Zool.*, 77, 619, 2004.

Reglodi, D., et al., Delayed systemic administration of PACAP38 is neuroprotective in transient middle cerebral artery occlusion in the rat, *Stroke*, 31, 1411, 2000a.

Reglodi, D., et al., Neuroprotective effects of PACAP38 in a rat model of transient focal ischemia under different experimental conditions, *Ann. NY Acad. Sci.*, 921, 119, 2000b.

Reglodi, D., et al., PACAP38-like immunoreactivity in the nervous system of oligochaeta, *Peptides*, 21, 183, 2000c.

Reglodi, D., et al., Pituitary adenylate cyclase activating polypeptide is highly abundant in the nervous system of anoxia-tolerant turtle *Pseudemys scripta elegans*, *Peptides*, 22, 873, 2001.

Reiner, A., A comparison of neurotransmitter-specific and neuropeptide-specific neuronal cell types present in the dorsal cortex in turtles with those present in the isocortex in mammals: Implications for the evolution of isocortex, *Brain Behav. Evol.*, 38, 53, 1991.

Reiner, A., Neurotransmitter organization and connections of turtle cortex: Implications for the evolution of mammalian isocortex, *Comp. Biochem. Physiol. Comp. Physiol.*, 104, 735, 1993.

Rice, M.E., and Nicholson, C., Interstitial ascorbate in turtle brain is modulated by release and extracellular volume change, *J. Neurochem.*, 49, 1096, 1987.

Rice, M.E., and Cammack, J., Anoxia-resistant turtle brain maintains ascorbic acid content in vitro, *Neurosci. Lett.*, 132, 141, 1991.

Rice, M.E., and Russo-Menna, I., Differential compartmentalization of brain ascorbate and glutathione between neurons and glia, *Neurosci.*, 82, 1213, 1998.

Rice, M.E., Lee, E.J.K., and Choy, Y., High levels of ascorbic acid, not glutathione, in the CNS of anoxia-tolerant reptiles contrasted with levels in anoxia-intolerant species, *J. Neurochem.*, 64, 1790, 1995.

Rice, M.E., et al., Brain antioxidant regulation in mammals and anoxia-tolerant reptiles: Balanced for neuroprotection and neuromodulation, *Comp. Biochem. Physiol. C*, 133, 515, 2002.

Rosen, D.R., et al., Mutations in Cu/Zn SOD gene are associated with familial amyotrophic lateral sclerosis, *Nature*, 362, 59, 1993.

Rosenberg, A.F., and Ariel, M., Visual-response properties of neurons in turtle basal optic nucleus in vivo, *J. Neurophysiol.*, 63, 1033, 1990.

Rosenberg, A.F., and Ariel, M., A model for optokinetic eye movements in turtles that incorporates properties of retinal slip neurons, *Vis. Neurosci.*, 13, 375, 1996.

Rosin, C., et al., Dopamine D2 and D3 receptor agonists limit oligodendrocyte injury caused by glutamate oxidative stress and oxygen/glucose deprivation, *Glia*, 52(4), 336, 2005.

Rothman, S.M., and Olney, J.W., Glutamate and the pathophysiology of hypoxic-ischemic brain damage, *Ann. Neurol.*, 19, 105, 1986.

Russo, R.E., and Velluti, J.C., Inhibitory effects of excitatory amino acids on pyramidal cells of the in vitro turtle medial cortex, *Exp. Brain Res.*, 92, 85, 1992.

Russo, R.E., et al., Functional and molecular clues reveal precursor-like cells and immature neurons in the turtle spinal cord, *J. Physiol.*, 560, 831, 2004.

Sakakibara, S., et al., Mouse-Musashi-1, a neural RNA-binding protein highly enriched in the mammalian CNS stem cell, *Dev. Biol.*, 176, 230, 1996.

Sakata, J.T., et al., Social experience affects territorial and reproductive behaviors in male leopard geckos, *Eublepharis macularis*, *Anim. Behav.*, 63, 487, 2002.

Sakata, J.T., Crews, D., and Gonzalez-Lima, F., Behavioral correlates of differences in neural metabolic capacity, *Brain Res. Rev.*, 48, 1, 2005.

Sakurai, S.Y., et al., Unchanged [^{3}H]MK-801 binding and increased [^{3}H]flunitrazepan binding in the turtle forebrain during anoxia, *Brain Res.*, 625, 181, 1993.

Salame-Mendez, A., et al., Response of diencephalons but not the gonad to female-promoting temperature with elevated estradiol levels in the sea turtle *Lepidochelys olivacea*, *J. Exp. Zool.*, 280, 304, 1998.

Schenk, J.O., et al., Homeostatic control of ascorbate concentration in CNS cf extracellular fluid, *Brain Res.*, 253, 353, 1982.

Schlegel, J.R., and Kriegstein, A.R., Quantitative autoradiography of muscarinic and benzodiazepine receptors in the forebrain of the turtle, *Pseudemys scripta*, *J. Comp. Neurol.*, 264, 521, 1987.

Schroter, A., et al., Nitric oxide applications prior and simultaneous to potentially excitotoxic NMDA-evoked calcium transients: cell death or survival. *Brain Res.*, 1060(1–2), 1, 2005.

Sharikabad, M.N., Ostbye, K.M., and Brors, O., Effect of hydrogen peroxide on reoxygenation-induced Ca^{2+} accumulation in rat cardiomyocytes, *Free Radic. Biol. Med.*, 37, 531, 2004.

Sharp, C.D., et al., N-methyl-D-aspartate receptor activation in human cerebral endothelium promotes intracellular oxidant stress, *Am. J. Physiol. Heart Circ. Physiol.*, 288, H1893, 2005.

Sheng, H., et al., Effects of metalloporphyrin catalytic antioxidants in experimental brain ischemia, *Free Rad. Biol. Med.*, 33(7), 947, 2002.

Shin, D.S., et al., Calcium and protein phosphatase 1/2A attenuate N-methyl-D-aspartate receptor activity in the anoxic turtle cortex, *Comp. Biochem. Physiol. A Mol. Integr. Physiol.*, 142, 50, 2005.

Sick, T.J., et al., Brain potassium ion homeostasis, anoxia, and metabolic inhibition in turtles and rats, *Am. J. Physiol.* 243, R281, 1982.

Siesjo, B.K., *Brain Energy Metabolism*, New York: John Wiley & Sons, 1978.

Siuciak, J.A., Krause, D.N., and Dubocovich, M.L., Quantitative pharmacological analysis of 2-[^{125}I]-iodomelatonin binding sites in discrete areas of the chicken brain, *J. Neurosci.*, 11, 2855, 1991.

Slivka, A., Mytilineou, C., and Cohen, G., Histochemical evaluation of glutathione in brain, *Brain Res.*, 409, 275, 1987.

Smeets, W.J.A.J., Lopez, J.M., and Gonzalez, A., Immunohistochemical localization of DARPP-32 in the brain of the lizard *Gekko gecko*: Co-occurrence with tyrosine hydroxylase, *J. Comp. Neurol.*, 435, 194, 2001.

Smeets, W.J.A.J., Lopez, J.M., and Gonzalez, A., Immunohistochemical localization of DARPP-32 in the brain of the turtle, *Pseudemys scripta elegans*: Further assessment of its relationship with dopaminergic systems in reptiles, *J. Chem. Neuroanat.*, 25, 83, 2003.

Snoeckx, L.H., et al., Heat shock proteins and cardiovascular pathophysiology, *Physiol. Rev.*, 81, 1461, 2001.

Sokoloff, L., et al., The [^{14}C]deoxyglucose method for the measurement of local cerebral glucose utilization: Theory, procedure, and normal values in the conscious and anesthetized albino rat, *J. Neurochem.*, 28, 897, 1977.

Somogyvari-Vigh, A., et al., Tissue distribution of PACAP27 and 38 in the Oligochaeta: PACAP27 is the dominant form in the nervous system of *Lumbricus polyphemus*, *Peptides*, 21, 1185, 2000.

Southwood, A.L., et al., Heart rates and diving behavior of leatherback sea turtles in the eastern Pacific Ocean, *J. Exp. Biol.*, 202 (Pt 9), 1115, 1999.

Sowa, A.W., et al., Altering hemoglobin levels changes energy status in maize cells under hypoxia, *Proc. Natl. Acad. Sci. USA*, 95, 10317, 1998.

Spencer, R.J., Thomas, M.B., and Banks, P.B., Hatch or wait? A dilemma in reptilian incubation, *Oikos*, 93, 401, 2001.

Stone, T.W., Receptors for adenosine and adenine nucleotides, *Gen. Pharmacol.*, 22, 25, 1991.

Storey, K.B., Suspended animation: The molecular basis of metabolic depression, *Can. J. Zool.*, 66, 124, 1988.

Storey, K.B., Reptile freeze tolerance: Metabolism and gene expression, *Cryobiol.*, 52, 1, 2006.

Storey, K.B., and Storey, J.M., Metabolic rate depression in animals: Transcriptional and translational controls, *Biol. Rev. Camb. Phil. Soc.*, 79, 207, 2004.

Stout, A.K., et al., Glutamate-induced neuron death requires mitochondrial calcium uptake, *Nat. Neurosci.*, 1, 366, 1998.

Suarez, R.K., et al., Turtles and rats: A biochemical comparison of anoxia-tolerant and anoxia-sensitive brains, *Am. J. Physiol.*, 257, R1083, 1989.

Suzue, T., Wu, G.-B., and Furukawa, T., High susceptibility to hypoxia of afferent synaptic transmission in the goldfish sacculus, *J. Neurophysiol.*, 58, 1066, 1987.
Swanson, R.A., and Choi, D.W., Glial glycogen stores affect neuronal survival during glucose deprivation in vitro, *J. Cereb. Blood Flow Metab.*, 13, 162, 1993.
Swanson, R.A., and Duan, S., Regulation of glutamate homeostasis in the brain, *J. Exp. Biol.*, 200, 401, 1999.
Tai, C., Kuzmiski, J.B., and MacVictor, B.A., Muscarinic enhancement of R-type calcium currents in hippocampal CA1 pyramidal neurons, *J. Neurosci.*, 26, 6249, 2006.
Tanaka, T., et al., ATP-sensitive K^+ channel openers block sulpiride-induced dopamine release in rat striatum, *Eur. J. Pharmacol.*, 297, 35, 1996.
Tatton, W.G., and Chalmers-Redman, R.M., Mitochondria in neurodegenerative apoptosis: An opportunity for therapy?, *Ann. Neurol.*, 44 (3 Suppl 1), S134, 1998.
Thompson, J.W., Prentice, H.M., and Lutz, P.L., Regulation of extracellular glutamate levels in the long-term anoxic turtle striatum: coordinated activity of glutamate transporters, adenosine, K(ATP) (+) channels and GABA. *J. Biomed. Sci.*, in press, 2007.
Thomsen, L.L, Ching, L.M., and Baguley, B.C., Evidence for the production of nitric oxide by activated macrophages treated with the antitumor agents flavone-8-acetic acid and xanthenone-4-acetic acid, *Cancer Res.*, 50(21), 6966, 1990.
Tomiyama, Y., Brian, J.E. Jr., and Todd, M.M., Cerebral blood flow during hemodilution and hypoxia in rats: Role of ATP-sensitive potassium channels, *Stroke*, 30, 1942, 1999.
Traynelis, S.F., and Cull-Candy, S.G., Proton inhibition of N-methyl-D-aspartate receptors in cerebellar neurons, *Nature*, 345, 347, 1990.
Tsukaguchi, H., et al., A family of mammalian Na^+-dependent L-ascorbic acid transporters, *Nature*, 399, 70, 1999.
Tucker, J.K., Environmental correlates of hatchling emergence in the red-eared turtle, *Trachemys scripta elegans*, in Illinois, *Chel. Cons. Biol.*, 3(3), 401, 1999.
Uchida, D., Prevention of ischemia-induced death of hippocampal neurons by pituitary adenylate cyclase activating polypeptide, *Brain Res.*, 736, 280, 1996.
Underwood, H., Endogenous rhythms, in *Hormones, Brain and Behavior: Biology of the Reptilia*, v. 18, C. Gans and D. Crews (eds.), Chicago: University of Chicago Press, 1992, 229–297.
Ultsch, G.R., The ecology of overwintering among turtles: Where turtles overwinter and its consequences, *Biol. Rev. Camb. Philos. Soc.*, 81, 339, 2006.
Ultsch, G.R., and Jackson, D.C., Long-term submergence at 3°C of the turtle *Chrysemys picta bellii*, in normoxic and severely hypoxic water. I. Survival, gas exchange and acid-base status, *J. Exp. Biol.*, 96, 11, 1982.
Ultsch, G.R, Reese, S.A., and Stewart, E.R., Physiology of hibernation in *Rana pipiens*: Metabolic rate, critical oxygen tension, and the effects of hypoxia on several plasma variables, *J. Exp. Zoolog. A Comp. Exp. Biol.*, 301(2), 169-76, 2004.
Underwood, H., Endogenous rhythms, in *Hormones, Brain, and Behavior: Biology of the Reptilia*, v. 18, C. Gans and D. Crews (eds.), Chicago: University of Chicago Press, 1992, 229–297.
Van Ginneken, V., et al., Fish muscle energy metabolism measured during hypoxia and recovery: An in vivo ^{31}P-NMR study, *Am. J. Physiol.*, 268(5 Pt 2), R1178, 1995.
Vaudry, D., et al., The neuroprotective effect of pituitary adenylate cyclase activating polypeptide on cerebellar granule cells is mediated through inhibition of the CED3-related cysteine protease caspase-3/CPP32, *Proc. Natl. Acad. Sci. USA*, 97, 13390, 2000.
Vaudry, D., et al., PACAP protects cerebellar granule neurons against oxidative stress-induced apoptosis, *Eur. J. Neurosci.*, 15, 1451, 2002.
Vaudry, D., et al., Endogenous PACAP acts as a stress response peptide to protect cerebellar neurons from ethanol or oxidative insult, *Peptides*, 26, 2518, 2005.
Vigh, J., and Witkovsky, P., Neurotransmitter actions on transient amacrine and ganglion cells of the turtle retina, *Vis. Neurosci.*, 21, 1, 2004.
Vivien-Roels, B., The pineal gland and the integration of environmental information: Possible role of hydroxy- and methoxyindoles, *Molec. Physiol.*, 4, 331, 1983.
Vivien-Roels, B., Interaction between photoperiod, temperature, pineal, and seasonal reproduction in nonmammalian vertebrates, in *The Pineal Gland: Current State of Pinela Research*, B. Mess, C. Ruzsas, L. Tima, and P. Pevet (eds.), Amsterdam: Elsevier Press, 1985, 187-209.
Vivien-Roels, B., Arendt, J., and Bradtke, J., Circadian and circannual fluctuations of pineal indoleamines (serotonin and melatonin) in *Testudo hermanni gmelin* (reptilia, chelonia), *Gen. Comp. Endocrinol.*, 37, 197, 1979.

Vivien-Roels, B., Pevet, P., and Claustrat, B., Pineal and circulating melatonin rhythms in the box turtles *Terrapene carolina triungus*: Effect of photoperiod, light pulse, and environmental temperature, *Gen. Comp. Endocrinol.*, 69, 163, 1988.
Wang, H., et al., Over-expression of antioxidant enzymes protects cultured hippocampal and cortical neurons from necrotic insults, *J. Neurochem.*, 87(6), 1527, 2003.
Warren, D.E., and Jackson, D.C., The role of mineralized tissue in the buffering of lactic acid during anoxia and exercise in the leopard frog *Rana pipiens*, *J. Exp. Biol.*, 208(Pt 6), 1117, 2005.
Warren, D.E., Reese, S.A., and Jackson, D.C., The factors that limit survival of red-eared slider turtles, *Trachemys scripta*, during long-term anoxic submergence at 3°C, *Physiol. Biochem. Zool.*, in press.
Warren, J.S., et al., ^{31}P-NMR measurements of intracellular pH and high energy phosphate concentrations in isolated, perfused, working turtle hearts during anoxia and acidosis, *Am. J. Physiol. Regul. Integr. Comp. Physiol.*, 259, R521, 1990.
Wibbels, T., et al., Seasonal changes in serum gonadal steroids associated with migration, mating, and nesting in the loggerhead sea turtle (*Caretta caretta*), *Gen. Comp. Endocrinol.*, 79(1), 154, 1990.
Wiechmann, A.F., and Wirsig-Wiechmann, C.R., Asymmetric distribution of melatonin receptors in the brain of a lizard *Anolis caroliensis*, *Brain Behav. Evol.*, 43, 26, 1992.
Willmore, W.G., and Storey, K.B., Glutathione systems and anoxia tolerance in turtles, *Am. J. Physiol.*, 273, R219, 1997a.
Willmore, W.G., and Storey, K.B., Antioxidant systems and anoxia tolerance in a freshwater turtle *Trachemys scripta elegans*, *Mol. Cell. Biochem.*, 170, 177, 1997b.
Wind, T., et al., Activation of ATP-sensitive potassium channels decreases neuronal injury caused by chemical hypoxia, *Brain. Res.*, 751, 295, 1997.
Witkovsky, P., Eldred, W., and Karten, H.J., Catecholamine- and indoleamine-containing neurons in the turtle retina, *J. Comp. Neurol.*, 228, 217, 1984.
Witkovsky, P., Alones, V., and Piccolino, M., Morphological changes induced in turtle retinal neurons by exposure to 6-hydroxydopamine and 5,6-dihydroxytryptamine, *J. Neurocytol.*, 16, 55, 1987.
Wong-Riley, M.T., et al., Neurochemical organization of the macaque striate cortex: correlation of cytochrome oxidase with Na^+K^+ ATPase, NADPH-diaphorase, nitric oxide synthase, and N-methyl-D-aspartate receptor subunit I, *Neurosci.*, 83, 1025, 1998a.
Wong-Riley, M.T., et al., Neurochemical organization of the macaque retina: Effect of TTX on levels and gene expression of cytochrome oxidase and nitric oxide synthase and on the immunoreactivity of Na^+K^+ ATPase and NMDA receptor subunit I., *Vos. Res.*, 38, 1455, 1998b.
Wu, S., and Storey, K.B., Up-regulation of acidic ribosomal phosphoprotein P0 in response to freezing or anoxia in the freeze tolerant wood frog, *Rana sylvatica*, *Cryobiology*, 50(1), 71, 2005.
Xia, Y., and Haddad, G.G., Major differences in CNS sulfonylurea receptor distribution between the rat (newborn, adult) and turtle, *J. Comp. Neurol.*, 113, 37, 1991.
Xia, Y., and Haddad, G.G., Neuroanatomical distributions and binding properties of saxitoxin sites in the rat and turtle CNS, *J. Comp. Neurol.*, 330, 363, 1993.
Xia, Y., and Haddad, G.G., Major difference in the expression of delta- and mu-opioid receptors between turtle and rat brain, *J. Comp. Neurol.*, 436, 202, 2001.
Xia, Y., Jiang, C., and Haddad, G.G., Oxidative and glycolytic pathways in rat (newborn and adult) and turtle brain: Role during anoxia, *Am. J. Physiol.*, 262, R595, 1992.
Xie, Y., et al., Effects of the sodium channel blocker tetrodotoxin (TTX) on cellular ion homeostasis in rat brain subject to complete ischemia, *Brain Res.*, 652, 216, 1994.
Yagita, Y., et al., Difference of cell proliferation profiles in the rat various regions after transient global ischemia, *J. Cereb. Blood Flow Metab.*, 19 (Suppl. 1), S667, 1999.
Yamada, T., et al., Nitric oxide produced during sublethal ischemia is crucial for the preconditioning-induced down-regulation of glutamate transporter GLT-1 in neuron/astrocyte co-cultures, *Neurochem. Res.*, 31(1), 49, 2006.
Yon, L., et al., Neuroanatomical and physiological evidence for the involvement of pituitary adenylate cyclase activating polypeptide in the regulation of the distal lobe of the frog pituitary, *J. Neuroendocrinol.*, 5, 289, 1993.
Yuan, H.B., et al., Hypothermic preconditioning reduces Purkinje cell death possibly by preventing the over-expression of inducible nitric oxide synthase in rat cerebellar slices after an in vitro simulated ischemia, *Neurosci.*, 142(2), 381, 2006.
Zhang, J., and Snyder, S.H., Nitric oxide stimulates auto-ADP-ribosylation of glyceradehyde-3-phosphate dehydrogenase, *Proc. Natl. Acad. Sci. USA*, 89, 9382, 1992.

Zhang, G.H., Haddad, G.G., and Xia, Y., δ-, but not μ- and κ-opioid receptor activation protects neocortical neurons from glutamate induced excitotoxic injury, *Brain Res.*, 852, 143, 2000.
Zhang, G.H., Gibney, G.T., and Xia, Y., The role of opioid receptors in neuroprotection against hypoxic insult, *Soc. Neurosci. Abs.*, 2001.
Zhu, D., et al., Thalamocortical connections in the pond turtle *Pseudemys scripta elegans*, *Brain Behav. Evol.*, 65, 278, 2005.

13 The Relationships of Turtles within Amniotes

Olivier Rieppel

CONTENTS

13.1 INTRODUCTION

Turtles (Testudines) are characterized by a highly derived body plan rendering the analysis of their relationships with other reptiles within amniotes difficult. The most salient feature of the turtle body plan is the dermal armor composed of a dorsal carapace and a ventral plastron. The development of the turtle shell is correlated with intricate changes most prominently affecting the dorsal axial skeleton and the pectoral girdle (Romer, 1956). The turtle carapace combines endoskeletal (ribs and vertebral neural arches) and exoskeletal (dermal) components in an entirely novel, indeed unique, structure (Goette, 1899; Völker, 1913; Versluys, 1914; Zangerl, 1939, 1969; Vallén, 1942; Kälin, 1945; Burke, 1989; Rieppel & Reisz, 1999; Rieppel, 2001). The development of the carapace involves a redirection of sclerotome cell migration to form the ribs in a more superficial position, i.e., within the dermal carapacial disk (Burke, 1989; Gilbert et al., 2001; Gilbert et al., 2004; Kuraku et al., 2005), and an anterior shift of the neural arches relative to the centra in the dorsal vertebral column by half a segment (Goette, 1899; Rieppel, 2001). The plastron of turtles includes the modified dermal components of the pectoral girdle (interclavicle and clavicles forming the epi- and entoplastral plates respectively (Zangerl, 1939, 1969). More posterior parts of the plastron appear to be derived from dermal gastral ribs (Zangerl, 1939). The endoskeletal pectoral girdle is located between carapace and plastron and forms a tri-radiate structure that comprises the scapulo-coracoid and a large, medioventrally directed acromion process. Because the ribs are embedded in the costal plates of the carapace and the scapulo-coracoid located between carapace and plastron, turtles are unique among tetrapods in having a scapula that is located at a morphological level deep to the ribs.

The profound structural changes that are correlated with the development of the turtle shell render the skeletal morphology of turtles difficult to compare with that of other reptiles. There are also no fossil intermediates known that link turtles with a more generalized group of reptiles. This not only renders the analysis of the phylogenetic relationships of turtles difficult but also engenders debates about primary homology assessments (Rieppel, 1996; Hill, 2005; for discussion, see Harris et al., 2007). In the present contribution, a brief historical survey of the changing ideas about turtle origins will be followed by a summary of the current status of the debate. A concluding section will discuss the ecological setting of turtle origins.

13.2 EARLY IDEAS ON TURTLE ORIGINS

Following initial doubts and debate (Baur, 1889, 1895; Cope, 1893), the classic studies by Osborn (1903) and Williston (1917) firmly established the configuration of the temporal region in the skull as the key character for higher-level reptile classification (Romer, 1956; Kuhn-Schnyder, 1980). On that basis, turtles were recognized as anapsid reptiles, characterized by the absence of any temporal fenestration, and thus as the most primitive lineage of all extant reptiles. The search for turtle ancestry consequently focused on Paleozoic stem reptiles. The most comprehensive early analysis of turtle relationships by Gregory (1946) established the herbivorous pareiasaurs from the Paleozoic as the group most closely related to turtles, whereas Olsen (1947) found turtles to be more closely related to the Paleozoic diadectomorphs, now no longer included in amniotes (Laurin & Reisz, 1997).

Continuing the tradition of seeking turtle ancestry amongst Paleozoic stem-reptiles, Romer (1968) proposed the procolophonids as their closest relatives. Gaffney (1980) was perhaps the first author to address the issue of turtle relationships using the logic of cladistic phylogeny reconstruction as employed in the analysis of three-taxon statements. His conclusion was that turtles are the sister-group to a taxon that includes synapsids and diapsids. This hypothesis, based on the presence of Jacobson's organ in mammals and diapsids but its absence in turtles, stands in marked contrast with Goodrich's (1916) subdivision of amniotes into Theropsida (mammals) and Sauropsida (reptiles and birds), based as it was on the aortic arch system. In an earlier paper, Gaffney and McKenna (1979) had recognized the captorhinomorphs as the closest relatives of turtles (see also Gaffeny, 1990; Gaffney & Meylan, 1988; Gauthier et al., 1988), with the implication that captorhinomorphs themselves, together with turtles, become the sister-group of a synapsid–diapsid clade.

With the advent of computer-supported broad-scale cladistic analysis, reptiles were found to subdivide into two major clades, the Paleozoic Parareptilia, sister-group of the Eureptilia (Gauthier et al., 1988; Laurin & Reisz, 1995). Pareiasaurs as well as procolophonids were recognized as parareptilian clades, whereas the captorhinomorphs were recognized as the stem of Eureptilia. The competing hypotheses of turtle relationships outlined previously thus placed turtles alternatively in the Parareptilia or the Eureptilia, respectively. However, computer-aided broad-scale analyses of turtle relationships continued to focus on Paleozoic parareptiles, finding again procolophonids (Figure 13.1; Reisz & Laurin, 1991; Laurin & Reisz, 1995) or pareiasaurs (Figure 13.1; Lee 1993, 1995, 1996, 1997a) as the sister-group of turtles. The emerging consensus was that turtles are nested

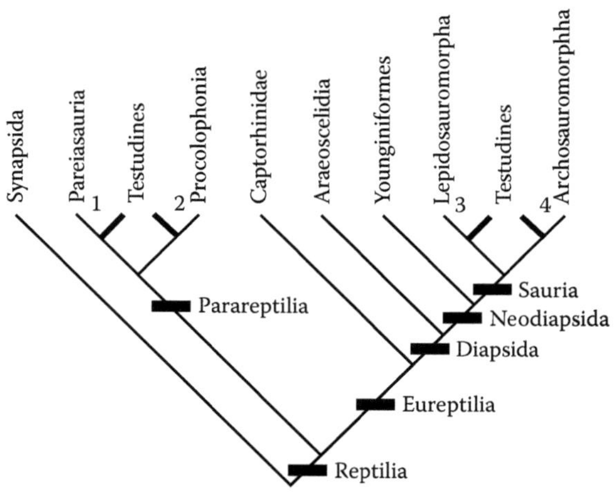

FIGURE 13.1 Alternative hypotheses of phylogenetic relationships of turtles (Testudines). 1: Reisz & Laurin (1991); Laurin & Reisz, (1997); 2: Lee (1995, 1997); 3: Hill (2005); 4: Merck (1997), and molecular analyses.

within parareptiles, but it is important to recognize that all of the aforementioned cladistic analyses were based on the presupposition that turtles are stem reptiles. The consequence was that if any diapsid was included in the analysis, it was the Paleozoic stem-diapsid *Petrolacosaurus* (Reisz, 1981). This implicitly placed greater weight on "anapsid" characters at the cost of characters that turtles might share with crown-group diapsids (Sauria).

However, there had been a number of earlier authors who had questioned the anapsid status of turtles and who considered them to be diapsids instead. Goodrich (1916, 1930) was one of the first to point out that reptiles would not form a natural group unless they were grouped with birds within Sauropsida. He thought turtles ought to be included in this group as well, and on this basis concluded that the temporal region in the turtle skull must have been secondarily closed. Boulenger (1918) added to the debate his observations that revealed differences of detail in the structure of the temporal region of the skull in the captorhinomorphs and in the oldest known fossil turtle, *Proganochelys*, from the early Late Triassic of Germany. Broom (1924) emphasized the striking similarities in the pelvic girdle and hind limbs that are shared by turtles and diapsid lepidosaurs such as *Sphenodon* (Rieppel, 1995). Lakjer (1926) believed the lower temporal arch to have been reduced to the quadrato-maxillary ligament in the modified diapsid skull of squamates and considered turtles to be modified diapsids as well, based on the presence of a quadrato-maxillary ligament in that group. In contrast, Rieppel (1990) found no support for diapsid affinities of turtles in the structure of the trigeminal jaw adductor musculature. DeBeer (1937) used developmental data concerning the occipito-atlantal joint to justify his hypothesis that turtles are related to (diapsid) archosaurian reptiles. Hofsten (1941) seconded that view and postulated a relationship of turtles with crocodiles and birds on the basis of the presence of a secondary subclavian artery. The hypothesis of archosaurian affinities of turtles earned significant subsequent support, again mostly on the basis of cladistic argumentation (Ax, 1984; Gardiner, 1993; Hennig, 1983; Lovtrup, 1977, 1985; Remane, 1959; see the discussion in Rieppel, 2000a). It is interesting to note that a diapsid status of turtles was accepted almost exclusively by zoologists, whereas their status as anapsid reptiles remained the consensus view among paleontologists (Romer, 1956, 1968; Kuhn-Schnyder, 1980).

13.3 THE RECENT REVIVAL OF THE DEBATE

The issue of turtle relationships within amniotes was brought back to center stage by Lee's studies (1993, 1995, 1996, 1997a), who found turtles to be nested within the pareiasaurs and, on that basis, developed an evolutionary scenario for the gradual derivation of the turtle body plan from that of a derived pareiasaur such as *Anthodon*. Key elements of that evolutionary scenario are the derivation of the turtle shell from osteoderms and a posterior shift of the pectoral girdle that would result in a morphological position of the scapula inside the rib cage. Whereas plausible from a gradualistic-phylogenetic perspective, this scenario does not explain the morphological complexity of the turtle carapace (Goette, 1899; Burke, 1989; Rieppel & Reisz, 1999; Rieppel, 2001; Gilbert et al., 2001; Gilbert et al., 2004; Kuraku et al., 2005), nor is it supported by turtle ontogeny, which shows no indication of a backward shift of the pectoral girdle (Ruckes, 1929; Burke, 1991; Rieppel, 1996). The conclusion had to be that "developmental specialization does not provide any hint as to the way in which this pattern evolved phylogenetically" (Carroll, 1988).

Lee's (1995, 1997a) phylogenetic hypothesis of preiasaurian relationships of turtles (Figure 13.1) and the evolutionary scenario based on it (Lee, 1993, 1996) were challenged by the finding that turtles could be the sister-group of a clade of Triassic marine reptiles, the Sauropterygia (Rieppel, 2000b), both groups nesting at the base of the lepidosauromorph lineage of the diapsid tree (Rieppel, 1994; Rieppel & DeBraga, 1996; DeBraga & Rieppel, 1997; Rieppel & Reisz, 1999). These results conflict with those of Merck (1997; see also the discussion in Brochu, 2001), whose analysis showed turtles again to group with Triassic marine reptiles (in this case ichthyosaurs, sauropterygians, and thalattosaurs) but all of them nesting at the base of the archosauromorph clade of the diapsid radiation (Figure 13.1).

The increased interest in turtle relationships as reflected by these morphological studies triggered a series of molecular studies that all confirmed the archosaurian affinities of turtles (Platz & Conlon, 1997; Gorr et al., 1998; Kirsch & Mayer, 1998; Zardoya & Meyer, 1998; Hedges & Poling, 1999; Kumazawa & Nishida, 1999; Mannen & Li, 1999; Mindell et al., 1999; Cao et al., 2000; Page, 2000; Janke et al., 2001; Li & Tsoi, 2002; Rest et al., 2003; Iwabe et al., 2005). However, there was variation in the relative relationships of turtles, crocodiles, and birds as indicated by DNA data: some studies found crocodiles to be the sister taxon of turtles (supported primarily by nuclear data), whereas in other analyses turtles were found to be the sister-group of a clade comprising crocodiles and birds, supported primarily by mitochondrial genes (Iwabe et al., 2005).

Problems of the molecular analyses of turtle relationships, possibly responsible for the partial incongruence of the signals obtained, were identified as long-branch attraction, taxon sampling, and rate heterogeneity among the genes employed (Brochu, 2001). Incongruent signals for turtle relationships based on morphological data were attributed to different philosophies "regarding how different types of data should be handled in phylogeny reconstruction" (Hill, 2005; for further discussion of that issue, see Rieppel & Kearney, 2006; Harris et al., 2007). Lee (2001) presented a simultaneous analysis (Nixon & Carpenter, 1996) that combined osteological data with characters of the soft anatomy taken from Gauthier et al. (1988) and with the mitochondrial DNA data from Zardoya and Meyer (1998; data modified as described by Lee, 2001). Whereas the analysis of the molecular data alone showed turtles to group with archosaurs (Lee, 2001, fig. 5; congruent with Zardoya & Meyer, 1998), as did the combined analysis when fossils were excluded, the combined analysis that included fossils recovered a monophyletic clade of Diapsida that excluded turtles. Lee's (2001) study documents a dominance of morphological data coded for extant and fossil amniotes over the molecular data that were employed.

The soft anatomical characters employed by Lee (2001) in his combined analysis and their potential phylogenetic significance were scrutinized by Rieppel (2000a). The osteological characters and their coding used in the analyses of turtle relationships within amniotes (Lee, 1995, 1997a, 1997b; DeBraga & Rieppel, 1997; Rieppel & Reisz, 1999), in turn, were re-evaluated by Hill (2005), whose study is particularly important as it includes a broader taxon sampling than did previous studies and adds a suite (71) of new characters relating to the histology and osteological morphology of the integument and osteoderms. The significance of these added characters lies is the fact that they remove much of the ambiguity that was obtained when "the presence of dorsal dermal ossifications of any kind (trunk osteoderms or carapacial elements) [is interpreted] as a potential synapomorphy linking turtles with parareptiles" (Hill, 2005; compare also Lee, 1993, with Rieppel, 2001). On the basis of his data set, Hill (2005) obtained diapsid affinities of turtles again, yet in a novel tree topology that shows turtles to be the sister-group of lepidosaurs (Figure 13.1). This result reflects back on Broom's (1924) comments on the striking similarities in the morphology of the pelvis and hind limbs that turtles share with *Sphenodon* (Rieppel, 1995).

On the whole, a general consensus seems to emerge that places turtles among the diapsid reptiles, but it remains to be seen whether they will be found to be more closely related to the lepidosauromorph or the archosauromorph clade among the Sauria. One possible reason for that remaining uncertainty may lie in the fact that the basal split of Sauria into a lepidosauromorph and an archosauromorph lineage is less well constrained than is currently acknowledged. However, most recently the occurrence of a cleithrum has been reported in the dermal pectoral girdle of the turtle *Kayentachelys aprix* from the Lower Jurassic of Arizona (Joyce et al., 2006). The cleithra identified in *Kayentachelys* and other basal fossil turtles such as *Proganochelys* and *Proterochersys* have alternatively been interpreted as dorsal epiplastral processes (Gaffney, 1990). Cleithra are reduced and lost early within reptiles, such that their presence in basal fossil turtles would conflict with a position of turtles among crown-group diapsids (Sauria) (Joyce et al., 2006).

The claim that a cleithrum is present in fossil basal turtles goes back to Jaekel's (1915) description of the Upper Triassic turtle *Stegochelys dux*, now considered a junior synonym of *Proganochelys quenstedti*. Jaekel (1915) based his identification of a cleithrum in *Proganochelys* on its close proximity to the upper end of the scapular process. But as noted by Gaffney (1990), the Berlin specimen of *Proganochelys* available to Jaekel (1915) did not preserve the anterior part of the plastron. Jaekel (1915) consequently did not recognize the cleithrum as a dorsal epiplastral process.

The homology of the epiplastron of turtles with the clavicle of other amniotes is undisputed (Jaekel, 1915; Zangerl, 1939, 1969; Gaffney, 1990; Joyce et al., 2006). Comparison of the pectoral girdle of *Proganochelys* with that of *Captorhinus* led Gaffney (1990) to homologize the dorsal epiplastral process of *Proganochelys* with the dorsal process of the clavicle in *Captorhinus* that is applied to the anterior edge of the scapular blade. To consider the dorsal epiplastral process a cleithrum would mean that the clavicle of fossil basal turtles lost all dorsal expansion as it was transformed into the epiplastron. The key feature used by Joyce et al. (2006) to claim the presence of a cleithrum in *Kayentachelys* is that some—but not all—of the specimens available appear to document a developmental origin of the dorsal epiplastral process that is independent from the epiplastron proper. The absence of sutures or sutural facets in *Proganochelys*, *Proterochersis*, and other specimens of *Kayentachelys* is claimed to document the ontogenetic fusion of the cleithrum with the epiplastron (clavicula). Given the nature of fossil material in general, and the frequent difficulty to distinguish sutures from breaks, the claim that a cleithrum is present in *Kayentachelys*, and by implication in other fossil basal turtles that have a dorsal epiplastral process, while not impossible, still appears ambiguous. In addition, the ontogenetic fusion of clavicle (epiplastron) and cleithrum would become a turtle synapomorphy, a condition unknown in any other tetrapod. But as noted by Joyce et al. (2006), the presence or absence of a cleithrum in basal turtles is but one character that is unlikely to reverse the signal of large-scale analyses. Whereas a lot still remains to be learned about the origin of turtles, their diapsid status seems to be the best supported hypothesis at this time.

13.4 THE ECOLOGICAL SETTING OF TURTLE ORIGINS

The discussion of the ecological setting of turtle origins focuses on a terrestrial (Joyce & Gauthier, 2004) or, alternatively, an aquatic (Rieppel & Reisz, 1999) environment. The hypothesis of an aquatic origin of turtles was motivated by their possible sister-group relationship with sauropterygians, but because of ambiguous optimization of aquatic versus terrestrial habits on the cladogram, the argument was mostly based on physiological consideration. The development of the turtle shell did not only result in major morphological changes (Rieppel, 2001) but equally important physiological adaptations were required as well. Given that the ribs are embedded in the costal plates of the carapace, turtles had to evolve an entirely new mechanism for lung ventilation (Gaunt & Gans, 1969; Gans & Hughes, 1997). Respiration in turtles depends on volume changes of the thoracico-peritoneal cavity inside the rigid dermal armor, which is achieved by altering the position of the limb flanks. Expansion of the thoracico-peritoneal cavity, supporting inhalation, involves the testocoracoideus muscle (Gaunt & Gans, 1969; Gans & Hughes, 1997), a probable homologue of the serratus ventralis muscle (Walker, 1971), which in other tetrapods suspends the pectoral girdle from the axial skeleton. The evolution of the turtle shell in terrestrial organisms would—in transitional stages—have resulted in conflicting functional roles for the muscle to play in locomotion and respiration. Because lung ventilation in aquatic turtles is not only supported by muscular activity but also, and to a significant degree, by hydrostatic pressure and gravity (Gaunt & Gans, 1969; Gans & Hughes, 1997), Rieppel and Reisz (1999) argued that the mechanisms of turtle respiration might have more plausibly evolved in an aquatic environment than on land. An aquatic origin of turtles might also be indicated by the function of the well-ossified turtle shell as an osmotic barrier (Bentley, 1976; see further discussion in Rieppel & Reisz, 1999). In contrast, Joyce and Gauthier (2004) found terrestrial and aquatic turtles to be characterized by different limb proportions, which indicate that the earliest and most basal fossil turtles—*Proganochelys* and *Palaeochersys* from the Upper Triassic of Germany and Argentina, respectively—were terrestrial. Their conclusion was that turtles had a terrestrial origin but that terrestrial crown turtles are secondary land dwellers because crown turtles had an aquatic origin (Joyce, 2007).

The fossil record as presently known is unlikely to resolve the issue. Considered a terrestrial form by Joyce and Gauthier (2004), *Proganochelys* was considered to have been semi-aquatic (an aquatic bottom-walker) by Gaffney (1990). The earliest known unquestionable turtle carapace fragment (Joyce & Karl, 2006) comes from the Franconian Hauptmuschelkalk (Wild, 1972), which yielded the Bayreuth fauna that has been collected from a section of about 30 to 40 m belonging to the Trochitenkalk Formation and the lower part of the Meißner Formation (*atavus* through *postspinosus* biozone, Anisian, uppermost Illyrian; Hagdorn & Rieppel, 1999). As noted by Wild (1972) and Joyce and Karl (2006), the marine Bayreuth fauna includes allochtonous elements such as lungfish (*Ceratodus*) and labyrinthodont remains. Sedimentary markers (mica and find sand) indicate that these faunal elements may have been transported from the nearby Bohemian Massive not much more than 40 km away (Hagdorn & Rieppel, 1999). The possibility thus remains that *Priscochelys hegnabrunnensis* (Karl, 2005) is an allochtonous element of the marine Bayreuth fauna as well. However, whether this would imply fully terrestrial habits of *Priscochelys* remains open to debate (Hagdorn & Rieppel, 1999; Rieppel & Reisz, 1999). But because the turtle shell is already fully formed in *Priscochelys*, *Proganochelys*, *Palaeochersys*, and all other known fossil turtles, it is the missing intermediate forms that would have to be used in an attempt to further assess the ecological setting of turtle origins on the basis of the fossil record.

13.5 SUMMARY AND CONCLUSIONS

Many of the questions that surround turtle origins remain unanswered at the present time. Whereas evidence for a diapsid status of turtles is accumulating, the precise relationships of turtles within diapsids remains a matter of debate and continued investigation. The ecological setting of turtle origins also remains open for further discussion. The difficulties encountered in research that addresses the origin of turtles no doubt originate from the highly modified body plan that characterizes the group and the absence of intermediate forms, either extant or fossil, that would offer some further clues as to how and where (under which conditions) the turtle body plan evolved. The ontogenetic development of the turtle carapace has become an important field of research for developmental biologists (Burke, 1989; Gilbert et al., 2001; Gilbert et al., 2004; Kuraku et al., 2005), but a link to a gradualistic model of turtle origination and evolution has not yet been successfully established (Rieppel, 2001).

Therefore, there is good reason to expect that ideas about turtle origins and evolution will continue to change. One of the evident directions to go is to pursue more comprehensive combined analyses, adding new data, both molecular and morphological, as they become available. Lee's (2001) is so far the only combined analysis available, and whereas some of the osteological characters included in this analysis remain debatable, it could not include the integumentary characters offered by Hill (2005), nor did it include DNA data other than those offered by Zardoya and Meyer (1998). The other obvious source for further insight into turtle origins is the fossil record. New fossils, such as well-preserved Triassic procolophonids, could hold the potential to yet again turn the tables on turtle relationships—or better yet, fossils of intermediate structure could provide new clues to an understanding of the origin and evolution of the turtle body plan.

REFERENCES

Ax, P., *Das Phylogenetische System*, Stuttgart: Fischer, 1984.

Baur, G., On the morphology of the vertebrate skull, *J. Morphol.*, 3, 467–474, 1889.

Baur, G., Bemerkungen über die Osteologie der Schläfengegend der Höheren Wirbeltiere, *Anat. Anz.*, 10, 315–330, 1895.

Bentley, P.J., Osmoregulation, in *Biology of the Reptilia, Vol. 5*, C. Gans and W.R. Dawson (eds.), London: Academic Press, 1976, 365–412.

Boulenger, G.A., Sur la place des chéloniens dans la classification, *C.R. Acad. Sci. Paris*, 169, 605–607, 1918.

Brochu, C.A., Progress and future directions in archosaur phylogenetics, *J. Paleont.*, 75, 1185–1201, 2001.

Broom, R., On the classification of the reptiles, *Bull. Amer. Mus. Nat. Hist.*, 51, 39–65, 1924.

Burke, A.C., Development of the turtle carapace: implications for the evolution of a novel bauplan, *J. Morph.*, 199, 363–378, 1989.

Burke, A.C., The development and evolution of the turtle body plan: Inferring intrinsic aspects of the evolutionary process from experimental embryology, *Am. Zool.*, 31, 616–627, 1991.

Cao, Y., Sorenson, M.D., Kumazawa, Y., Mindell D.P., and Hasegawa, M., Phylogenetic position of turtles among amniotes: Evidence from mitochondrial and nuclear genes, *Gene*, 259, 139–148, 2000.

Carroll, R.L., Vertebrate Paleontology and Evolution, New York: W.H. Freeman, 1988.

Cope, E.D., On the homologies of the posterior cranial arches in the Reptilia, *Trans. Amer. Phil. Soc., N.S.*, 17, 11–26, 1893.

DeBeer, G.S., *The Development of the Vertebrate Skull*, Oxford: Clarendon Press, 1937.

DeBraga, M., and Rieppel, O., Reptile phylogeny and the interrelationships of turtles, *Zool. J. Linn. Soc.*, 120, 281–354, 1997.

Gaffney, E.S., Phylogenetic relationships of the major groups of amniotes, in *The Terrestrial Environment and the Origin of Land Vertebrates*, A.L. Panchen (ed.), London: Academic Press, 1980, 593.

Gaffney, E.S., The comparative osteology of the Triassic turtle *Proganochelys*, *Bull. Amer. Mus. Nat. Hist.*, 194, 1–263, 1990.

Gaffney, E.S., and McKenna, M.C., A Late Permian captorhinid from Rhodesia, *Am. Mus. Novitates*, 2688, 1–15, 1979.

Gaffney, E.S., and Meylan, P.A.. A phylogeny of the turtles, in *The Phylogeny and Classification of the Tetrapods, Vol. 1: Amphibians, Reptiles, Birds*, M.J. Benton (ed.), Oxford: Clarendon Press, 1988, 157–219.

Gans, C., and Hughes, G.M., The mechanism of lung ventilation in the tortoise *Testudo graeca* Linné, *J. Exp. Biol.*, 47, 1–20, 1997.

Gardiner, B.G., Haematothermia: Warm-blooded amniotes, *Cladistics*, 9, 369–395, 1993.

Gaunt, A.S., and Gans, C., Mechanics of respiration in the snapping turtle, *Chelydra serpentina* (Linné), *J. Morphol.*, 128, 195–228, 1969.

Gauthier, J.A., Kluge, A.G., and Rowe, T., Amniote phylogeny and the importance of fossils, *Cladistics*, 4, 104–209, 1988.

Gilbert, S.F., Loredo, G.A., Brukman, A., and Burke, A.C., Morphogenesis of the turtle shell: The development of a novel structure in tetrapod evolution, *Evol. Dev.*, 3, 47–58, 2001.

Gilbert, S.F., Cebra-Thomas, J.A., and Fraser, T., Working hypothesis for the origin of the turtle shell, *J. Morphol.*, 260, 294, 2004.

Goette, A., Über die Entwicklung des knöchernen Rückenschildes (Carapax) der Schildkröten, *Z. wiss. Zool.*, 66, 407, 1899.

Goodrich, E.S., On the classification of the Reptilia, *Proc. R. Soc. Lond. B*, 89, 261–276, 1916.

Goodrich, E.S., *Studies on the Structure and Development of Vertebrates*, London: Macmillan, 1930.

Gorr, T.A., Mable, B.K., and Kleinschmitt, T., Phylogenetic analysis of reptilian hemoglobins: Trees, rates, and divergences, *J. Mol. Evol.*, 47, 471–485, 1998.

Gregory, W.K., Pareiasaurs versus placodonts as near ancestors to turtles, *Bull. Amer. Mus. Nat. Hist.*, 86, 275–326, 1946.

Hagdorn, H., and Rieppel, O., Stratigraphy of marine reptiles in the Triassic of Central Europe, *Zentralbl. Geol. Paläont. I.*, 1998, 651–678, 1999.

Harris, S.R., Pisani, D., Gower, D,J., and Wilkinson, M., Investigating stagnation in morphologies using consensus data, *Syst. Biol.*, 56, 125–129, 2007.

Hedges, S.B., and Poling, L.L., A molecular phylogeny of reptiles, *Science*, 283, 998–1001, 1999.

Hennig, W., Stammesgeschichte der Chordaten, *Fortschr. Zool. Syst. Evolutforsch.*, 2, 1–208, 1983.

Hill, R.V., Integration of morphological data sets for phylogenetic analysis of Amniota: The importance of integumentary characters and increased taxonomic sampling, *Syst. Biol.*, 54, 530–547, 2005.

Hofsten, N., On the phylogeny of the Reptilia, *Zool. Bidr. Uppsala*, 20, 501–521, 1941.

Iwabe, N., Hara, Y., Kumazawa, Y., Shibamoto, K., Saito, Y., Miyata, T., and Katho,K., Sister group relationship of turtles to the bird-crocodilian clade revealed nuclear DNA-coded proteins, *Mol. Biol. Evol.*, 22, 810–813, 2005.

Jaekel, O., Die Wirbeltierfunde aus dem Keuper von Halberstadt. Serie II. Testudinata. Teil 1. *Stegochelys dux* n. g. n. ap., *Palaeont. Z.*, 2. 88–214, 1915.

Janke, A., Erpenbeck, D., Nilsson, M., and Arnason, U., The mitochondrial genomes of the iguana (*Iguana iguana*) and the caiman (*Caiman crocodylus*): implications for amniote phylogeny, *Proc. R. Soc. Lond. B,* 268, 623-631, 2001.

Joyce, W.G., Phylogenetic relationships of Mesozoic turtles, *Bull. Peabody Mus. Nat. Hist.*, 48, 3–102, 2007.

Joyce, W.G., and Gauthier, J.A., Paleoecology of Triassic stem turtles sheds new light on turtle origins, *Proc. R. Soc. Lond. B*, 271, 1–5, 2004.

Joyce, W.G., and Karl, H.-V., The worlds oldest fossil turtle: fact versus fiction, *Russ. J. Herpetol.*, 13 (Suppl.), 104–111, 2006.

Joyce, W.G., Jenkins, F.A. Jr., and Rowe, T., The presence of cleithra in the basal turtle *Kayentachelys aprix*, *Russ. J. Herpetol.*, 13 (Suppl.), 93–103, 2006.

Kälin, J., Zur Morphogenese des Panzers bei den Schildkröten, *Acta Anat.*, 1, 144–176, 1945.

Karl, H.-V., The homology of the supramarginals in turtles (Reptilia: Chelonii), *Stud. Geol. Salamanticensia*, 41, 63–75, 2005.

Kirsch, J.A.W., and Mayer, G.C., The platypus is not a rodent: DNA hybridization, amniote phylogeny and the palimpsest theory, *Phil. Trans. R. Soc. Lond. B*, 353, 1221–1237, 1998.

Kuhn-Schnyder, E., Observations on temporal openings of reptilian skulls and the classification of reptiles, in *Aspects of Vertebrate History, Essays in Honor of Edwin Harris Colbert*, L.L. Jacobs (ed.), Flagstaff, AZ: Museum of Northern Arizona Press, 1980, 153–175.

Kumazawa, Y., and Nishida, M., Complete mitochondrial DNA sequences of the green turtle and blue-tailed mole skink: Statistical evidence for archosaurian affinity of turtles, *Mol. Biol. Evol.*, 16, 784–792, 1999.

Kuraku, S., Usuda, R., and Kuratani, S., Comprehensive survey of carapacial ridge-specific genes in turtle implies co-option of some regulatory genes in carapace evolution, *Evol. Dev.*, 7, 3–17, 2005.

Lakjer, T., *Studien über die Trigeminus-versorgte Kaumuskulatur der Sauropsiden*, Copenhagen: C.A. Reitzel, 1926.

Laurin, M., and Reisz, R.R., A reevaluation of early amniote phylogeny, *Biol. J. Linn. Soc.*, 113, 165–223, 1995.

Laurin, M., and Reisz, R.R., A new perspective on tetrapod phylogeny, in *Amniote Origins: Completing the Transition to Land*, S.S. Sumida and K.L.M. Martin (eds.), San Diego, CA: Academic Press, 1997, 9–59.

Lee, M.S.Y., The origin of the turtle body plan: Bridging a famous morphological gap, *Science*, 261, 1716–1720, 1993.

Lee, M.S.Y., Historical burden in systematics and the interrelationships of 'parareptiles', *Biol. Rev.*, 70, 459–547, 1995.

Lee, M.S.Y., Correlated progression and the origin of turtles, *Nature*, 379, 811–815, 1996.

Lee, M.S.Y., Pareiasaur phylogeny and the origin of turtles, *Zool. J. Linn. Soc.*, 120, 197–280, 1997a.

Lee, M.S.Y., Reptile relationships turn turtle…and turn back again, *Nature*, 389, 245, 1997b.

Lee, M.S.Y., Molecules, morphology, and the monophyly of diapsid reptiles, *Contr. Zool.*, 70, 1–22, 2001.

Li, Y.-J., and Tsoi, C.-M., Phylogenetic analysis of vertebrate lactate dehydrogenase (LDH) multigene families, *J. Mol. Evol.*, 54, 614–624, 2002.

Lovtrup, S., *The Phylogeny of Vertebrata*, London: John Wiley, 1977.

Lovtrup, S., On the classification of the taxon Tetrapoda, *Syst. Zool.*, 34, 463–470, 1985.

Mannen, H., and Li, S.S.-L., Molecular evidence for a clade of turtles, *Mol. Phyl. Evol.*, 13, 144–148, 1999.

Merck, J.W., A phylogenetic analysis of the euryapsid reptiles, Ph.D. dissertation, University of Texas, Austin, TX, 1997.

Mindell, D.P., Sorenson, M.D., Dimcheff, D.E., Hasegawa, M., Ast, J.C., and Yuri, T., Interordinal relationships of birds and other reptiles based on whole mitochondrial genomes, *Syst. Biol.*, 48, 138–152, 1999.

Nixon, K.C., and Carpenter, J.M., On simultaneous analysis, *Cladistics*, 12, 221–241, 1996.

Olson, E.C., The family Diadectidae and its bearing on the classification of reptiles, *Fieldiana (Geol.)*, 11, 1–53, 1947.

Osborn, H.F., On the primary division of the Reptilia into two subclasses, Synapsida and Diapsida, *Science*, 17, 275–276, 1903.

Page, R.D.M., Extracting species trees from complex gene trees: Reconciled trees and vertebrate phylogeny, *Mol. Phyl. Evol.*, 14, 89–106., 2000.

Platz, J.E., and Conlon, J.M., Reptile relationships turn turtle … and turn back again, *Nature*, 389, 245–246, 1997.

Reisz, R.R., A diapsid reptile from the Pennsylvanian of Kansas, *Spec. Publ. Mus. Nat. Hist. Univ. Ks.*, 7, 1–74, 1981.
Reisz, R.R., and Laurin, M., Owenetta and the origin of turtles, *Nature*, 349, 324–326, 1991.
Remane, A., Die Geschichte der Tiere, in *Die Evolution der Organismen*, Band 1. 2. Aufl., G. Heberer (ed.), Stuttgart: Fischer, 1959, 340–422.
Rest, J.S., Ast, C.J., Austin, C.C., Waddell, P.J., Tibbetts, E.A., Hay, J.M., and Mindell, D.P., Molecular systematics of primary reptilian lineages and the tuatara mitochondrial genome, *Mol. Phyl. Evol.*, 29, 289–297, 2003.
Rieppel, O., The development of the jaw adductor musculature in the turtle *Chelydra serpentina*, *Zool. J. Linn. Soc.*, 98, 27–62, 1990.
Rieppel, O., Osteology of *Simosaurus gaillardoti*, and the phylogenetic interrelationships of stem-group Sauropterygia, *Fieldiana (Geol.)*, n.s. 28, 1–85, 1994.
Rieppel, O., Studies on skeleton formation in reptiles: Implications for turtle relationships, *Zoology*, 98, 298–308, 1995.
Rieppel, O., Testing homology by congruence—the pectoral girdle of turtles, *Proc. R. Soc. Lond. B*, 263, 1395–1398, 1996.
Rieppel, O., Turtles as diapsid reptiles, *Zool. Script.*, 29, 199–212, 2000a.
Rieppel, O., Sauropterygia I: Placodontia, Pachypleurosauria, Nothosauroidea, Pistosauroidea, *Encyclopedia of Paleoherpetology, Vol. 12A*, Munich: Pfeil, 2000b.
Rieppel, O., Turtles as hopeful monsters, *BioEssays*, 23, 987–991, 2001.
Rieppel, O., and DeBraga M., Turtles as diapsid reptiles, *Nature*, 384, 453–455, 1996.
Rieppel, O., and Kearney, M., The poverty of taxonomic characters, *Biol. Philos.*, 22, 95–113, 2006.
Rieppel, O., and Reisz, R.R., The origin and early evolution of turtles, *Ann. Rev. Ecol. Syst.*, 30, 1–22, 1999.
Romer, A.S., *Osteology of the Reptiles*, Chicago: University of Chicago Press, 1956.
Romer, A.S., *Notes and Comments on Vertebrate Paleontology*, Chicago: University of Chicago Press, 1968.
Ruckes, H., Studies in chelonian osteology. Part II. The morphological relationships between girdles, ribs and carapace, *Ann. N.Y. Acad. Sci.*, 31, 81–120, 1929.
Walker, W.F., A structural and functional analysis of walking in the turtle, *Chrysemis picta marginata*, *J. Morphol.*, 134, 195–214, 1971.
Wild, R., Die Wirbeltierfaunen der fränkischen und südalpinen Mitteltrias (ein Vergleich), *Z. Deutsch. Geol. Ges.*, 123, 229–234, 1972.
Williston, S.W., The phylogeny and classification of reptiles, *J. Geol.*, 25, 411–421, 1917.
Vallén, E., Beiträge zur Kenntnis der Ontogenie und der vergleichenden Anatomie des Schildkrötenpanzers, *Acta Zool.*, 23, 1–127, 1942.
Versluys, J., Über die Phylogenie des Panzers der Schildkröten und über die Verwandtschaft der Lederschildkröte (*Dermochelys coriacea*), *Paläont. Z.*, 1, 321–347, 1914.
Völker, H., Über das Stamm-, Gliedmaßen-, und Hautskelett von *Dermochelys coriacea* L., *Zool. Jahrb. Anat.*, 33, 431–552, 1913.
Zangerl, R., The homology of the shell elements in turtles, *J. Morph.*, 65, 383–406, 1939.
Zangerl, R., The turtle shell, in *Biology of the Reptilia, Vol. 1*, C. Gans, T.S. Parsons, and A.d'A. Bellairs (eds.), London: Academic Press, 1969, 311–339.
Zardoya, R., and Meyer, A., Complete mitochondrial genome suggests diapsid affinities of turtles, *Proc. Nat. Acad. Sci.*, 95, 14226–14231, 1998.

Index

Note: Page numbers in **bold** indicate figures and tables.

A

B

D

E

I

J

K

L

M

N

S

T

W